APPLIED
LINEAR
ALGEBRA

SECOND EDITION

APPLIED LINEAR ALGEBRA

Ben Noble □ *University of Wisconsin*

James W. Daniel □ *University of Texas*

PRENTICE-HALL, INC. □ *Englewood Cliffs, N.J. 07632*

Library of Congress Cataloging in Publication Data

NOBLE, BENJAMIN.
Applied linear algebra.

Bibliography: p.
Includes index.
1. Algebras, Linear. I. Daniel, James W., joint author. II. Title.
QA184.N6 1977 512'.5 76-43993
ISBN 0-13-041343-7

10 9 8 7 6 5 4 3 2

Printed in the United States of America

Prentice-Hall International, Inc., *London*
Prentice-Hall of Australia Pty. Limited, *Sydney*
Prentice-Hall of Canada, Ltd., *Toronto*
Prentice-Hall of India Private Limited, *New Delhi*
Prentice-Hall of Japan, Inc., *Tokyo*
Prentice-Hall of Southeast Asia Pte. Ltd., *Singapore*
Whitehall Books Limited, *Wellington, New Zealand*

To Denise, Anna, and John Ben

To Adam and Joshua

CONTENTS

PREFACE

Linear algebra is an essential part of the mathematical toolkit required in the modern study of many areas in the behavioral, natural, physical, and social sciences, in engineering, in business, in computer science, and of course in pure and applied mathematics. The purposes of this book are to develop the fundamental concepts of linear algebra, emphasizing those concepts which are most important in applications, and to illustrate the applicability of these concepts by means of a variety of selected applications. Thus, while we present applications for illustrative and motivative purposes, our main goal is to present the *mathematics that can be applied.*

We have taken great care, however, to present the mathematical theory from a concrete point of view rather than from an abstract one. Because of this approach, we introduce the concrete manipulative matrix and vector algebra (Chapter 1) and elementary row operations (Chapter 3) well before the more abstract notions of vector spaces (Chapter 4) and linear transformations (Chapter 5). Similarly, eigenvalues, eigenvectors, and their associated canonical forms are presented (Chapter 8 and, in more detail, Chapters 9 and 10) as ways to simplify the study of linear transformations which perhaps describe the behavior of very complicated systems; our emphasis on these and other simplifying matrix decompositions (Chapter 9) and their applications (Chapters 9 and 10) is one of the unusual features of our approach.

In addition to presenting a number of applications briefly throughout the book both in the text and in the exercises, we have chosen to collect a variety of applications in Chapter 2 as well; there we assume familiarity with only the basic matrix algebra of Chapter 1. These applications serve to illustrate the ways in which linear algebra arises in applications and to motivate the later study of certain specific aspects of matrices. The efficient solution of systems of linear equations and the basic notions of linear programming can easily be viewed either as applications of linear algebra or as particular aspects of linear algebra which arise in an immense variety of concrete application areas; these important subjects are covered in Chapters 6 and 7.

It is important to note that the presentation throughout the book hinges on the use of elementary row operations; these manipulations as presented in Chapter 3 are used both as tools to develop theoretical notions (Sections 3.5, 3.6, 3.7, 4.5, 5.2, et cetera) and also as the basis of computational methods (Sections 3.2, 3.3, 4.3, 6.2, 7.1, 7.3, et cetera). It is *absolutely essential* that the student master the material on elementary row operations (Sections 3.2 and 3.4).

Although this book is considerably shorter than its predecessor (*Applied Linear Algebra* by Ben Noble; Prentice-Hall, 1969) and contains less peripheral material, it is still possible to teach a variety of courses from this one book; this places on the instructor a special responsibility carefully to select topics to be covered. To facilitate this selection we give some guidelines on the material to be included in various types of courses.

The *classical theoretical linear algebra course* will of course emphasize the mathematical theory; the reason for using our text for such a course is that in addition to the theory it provides applications, a few of which can be used for motivation and illustration. After covering Chapter 1 carefully and one or two topics from Chapter 2 very briefly, the instructor should then concentrate on the mathematical theory in Chapters 3, 4, and 5, covering the theorems and their proofs in detail. In the remaining time we recommend brief presentations of the main results of Chapters 8 and 11, primarily those results indicated in the text and the chapter introductions to be **Key Theorems**. If time permits, we encourage the inclusion of a small amount of applied material selected from Chapters 6, 7, 9, and 10.

The *introductory applied linear algebra course*, emphasizing the *applicability* of the fundamental theoretical results, is of course one of those for which this book was primarily intended. Many variations of such a course are possible, depending, for example, on what types of applications the instructor wishes to emphasize (for engineering? for social sciences? for statistics? et cetera) and on how much theorem-proving the instructor wishes to include (none? only selected **Key Theorems**? et cetera). The syllabus we now describe attempts to provide a good mix of these alternatives. We would cover Chapter 1 carefully and then discuss two applications selected from Chapter 2, say Sections 2.2

and 2.5, but depending on precisely what will be covered later. Next we proceed carefully through Chapter 3, usually "proving by example" rather than devoting too much time to the technical details of proofs, although the **Key Proof** of **Key Theorem 3.6** should be presented because of its importance; most of our time would be spent explaining the theorems indicated in the text to be **Key Theorems**. Chapters 4 and 5 should be covered somewhat more quickly but still carefully with most time spent motivating and explaining the **key** results. At this point the student should have the fundamental concepts and tools of linear algebra, except those involving eigenvalues. To allow time to absorb the preceding material we would very briefly skim Chapter 6, especially Sections 6.2 and 6.5, and quickly cover Sections 7.1, 7.2, and 7.3, perhaps skimming Section 7.5. Next we would complete the necessary conceptual background by covering Chapter 8, as usual concentrating on explanations of **key** results. The first three sections of Chapter 9 and the first four of Chapter 10 develop in detail the ideas outlined in Chapter 8, with Chapter 9 concentrating on what can be accomplished via unitary transformations while Chapter 10 treats the more general similarity transformations; given enough time we would present one of these two groups of sections, probably that in Chapter 9. We also would try to skim some of the advanced applications of these chapters (probably Sections 9.4, 9.5, 9.6, and 10.5), allowing time finally just to introduce the notions of Chapter 11.

It is difficult to settle on a specific format for presenting the material in Chapters 8, 9, and 10; different teachers may well prefer an organization different from that chosen by us. One can view these chapters as containing (1) basic concepts, (2) results on the structure of eigensystems, (3) results on matrix decompositions and canonical forms, and (4) applications. Since we felt that many teachers would concentrate on the simple and important case of normal matrices and unitary transformations, we chose to put (2), (3), and (4) for this case together in one chapter (Chapter 9) with the more general material in another (Chapter 10). A teacher who prefers to present first all the material on canonical forms, then the material on eigensystems, and finally the applications would be forced to jump about among the sections (e.g., 9.2, 9.4, 10.2, 10.3, 9.3, 10.4, 9.5, 9.6, 10.5, 10.6, 10.7) in order to arrange the presentation in that fashion. Our ordering was chosen to reduce such skipping about by the majority of the users of our book.

The above syllabus for an introductory applied linear algebra course includes material from every chapter and therefore quite obviously covers a large number of pages. Instructors must be careful not to become bogged down in technical details but rather to emphasize *concepts* and *techniques* by explaining and illustrating the **key** results and especially emphasizing how they can be used. Only Chapter 3 need be covered with great care, while some care should be exercised with Chapters 4 and 5; the remaining material moves quite quickly.

An *intermediate applied linear algebra course* can easily be built around this text for students who are familiar with the basic concepts of matrix and vector algebra and linear independence. Starting with a quick review and a study of two sections from Chapter 2, we recommend proceeding to a fairly quick coverage of Chapter 3, the last half of Chapter 4, and Chapter 5, depending on just how much linear algebra the students recall. We would cover Chapters 6 and 7 fairly briefly but with more detail than for the introductory applied course, cover Chapter 8 to fill in the holes in the students' knowledge here up to the level described for the introductory applied course, and then treat all of Chapters 9, 10, and 11, as usual, emphasizing ideas and techniques rather than proofs by concentrating on explanations of the **key** results. The obvious difference between this course and the introductory applied course is that the greater assumed background knowledge of the students in the intermediate course allows the material of Chapters 6 through 11 to replace the introductory material.

Several other types of applied courses can be created to fit the interests of the students or instructor. An audience oriented toward statistics and the social and behavioral sciences should probably be sure to include Chapter 7 in its entirety, Sections 8.5, 9.4, 9.5, 9.6, 9.7, and 10.5, and applications selected from Sections 2.2, 2.3, 2.5, and 2.6. An audience oriented more toward engineering might prefer Sections 6.1, 6.2, 6.3, 6.4, 8.4, 9.5, 10.3, 10.5, 10.6, 10.7, 11.5, and 11.6, and applications selected from Sections 2.4 and 2.6.

As indicated, we feel that this book has great flexibility to meet the special desires of knowledgeable instructors. To help those using the book for the first time, we have given, at the start of each chapter, brief statements of what we consider to be **key** results. These are intended to help both the student and the less experienced instructor recognize what are the fundamental conclusions of the chapter; any student understanding the **key** theorems of this text will have a solid foundation in linear algebra, matrix theory, and their applications. **key** material is indicated by the presence of a bold bullet (●) and by the use of the word **key** (as in **Key Theorem 3.6**).

Some remarks on notation are in order. Displayed equations are numbered consecutively within each chapter, so that (6.8) is the eighth numbered equation in Chapter 6. Likewise are corollaries, definitions, figures, lemmas, and theorems consecutively but *independently* numbered within each chapter; thus Definition 8.2 and Theorem 8.4 are respectively the second definition and the fourth theorem of Chapter 8, but Definition 8.2 *need not* precede Theorem 8.4 just because of its lower number. We have included an index to definitions and theorems to facilitate the reader's locating a referenced result. We have similarly numbered consecutively and independently the several hundred exercises indicated by Ex., some of which are presented along with solutions or suggested approaches. Thus Ex. 8.16 is essentially an example in that it is stated as an exercise but its solution is immediately presented in the text, Ex. 8.5 is a problem or exercise but with a solution (or partial solution or hint) that is given not in the text but rather in **Hints and Solutions** near the end of the book, and Ex. 8.6 is a problem or exercise whose solution is not presented or indicated anywhere in this book.

This book has a long history. It is a revision of the first author's earlier (1969) *Applied Linear Algebra*, whose history is detailed in the original's Preface (pp. ix–x); this new edition arose from a desire to create an edition that was more easily teachable and that presented a greater variety of applications at a less technically complex level. In this effort we have been greatly encouraged by Prentice-Hall in the persons of Ed Lugenbeel, John Hunger, Ken Cashman, and Cathy Brenn, whose cooperation and assistance we appreciate.

Madison, Wisconsin
Austin, Texas

BEN NOBLE
JAMES W. DANIEL

COMMON NOTATION

$\mathbf{u}, \mathbf{x}$, boldface small letters refer to vectors (column and row matrices), *1*

$\mathbf{A}, \mathbf{P}$, boldface capital letters refer to matrices other than vectors, *1*

k, α, italic Roman and Greek letters denote scalars, *5*

V, A, italic capital letters denote vector spaces or linear transformations, *108*, *151*

$\mathbf{A} = [a_{ij}]$, matrix, *1*

$\bar{\mathbf{A}} = [\bar{a}_{ij}]$, complex conjugate of **A**, *13*

$\mathbf{A}^T = [a_{ji}]$, transpose of **A**, *11*

$\mathbf{A}^H = [\bar{a}_{ji}]$, hermitian transpose of **A**, *13*

$\mathbf{A}^{-1}$, inverse of **A**, *17*

$\mathbf{A}_e, \mathbf{x}_e, \mathbf{c}_e$, extended vectors in linear programming, *220*

$\mathbf{A}^+$, generalized inverse of **A**, *337*

$\mathbf{A}_{ij}$, cofactor of matrix **A**, *199*

$\mathbf{C}^n$, n-dimensional space of complex column n-vectors, *111*

c(**A**), condition number of matrix **A**, *170*

D, matrix, usually diagonal, *387*

det **A**, determinant of **A**, *198*

$\mathbf{e}_i$, unit column vector, *22*
exp, exponential function: $\exp(x) = e^x$, *400*
$\mathbf{E}_{pq}$, $\mathbf{E}_p(c)$, $\mathbf{E}_{pq}(c)$, elementary matrices, *86*
$\mathbf{I}_m$, $m \times m$ unit matrix, *15*
$\mathbf{J}$, Jordan form, *361*
$\mathbf{J}_i$, Jordan block, *361*
$\mathbf{L}$, $\mathbf{U}$, lower and upper triangular matrices, respectively, *192*
$\mathbf{M}(A)$, matrix representation of linear transformation A, *154*
$\mathbf{P}$, $\mathbf{Q}$, unitary matrices, *282*
$\mathbf{R}$, upper- (or right-) triangular matrix, *317*
$\mathbf{R}_{ij}$, plane rotation matrix, *284*
$\mathbb{R}^n$, n-dimensional space of real column n-vectors, *111*
$\mathbf{T}$, tableau in linear programming, *225*
$\vec{u}$, geometrical vector, *103*
$\vec{u} \cdot \vec{v}$, dot product, *133*
$\mathbf{U}$, $\mathbf{V}$, unitary matrices, *282*
$\mathbf{0}$, $m \times n$ matrix of zeros, *9*
λ, eigenvalue, or parameter in $(\mathbf{A} - \lambda\mathbf{I})\mathbf{x} = \mathbf{b}$, *257, 264*
λ_i, $\mathbf{x}_i$, $i = 1, \ldots, n$, eigenvalues and corresponding eigenvectors of matrix, *264*
$\mathbf{\Lambda}$, diagonal matrix whose ith diagonal element is λ_i, *272*
$\rho(\mathbf{x})$, spectral radius, Rayleigh quotient, *389, 431*
$(\mathbf{u}, \mathbf{v})$, inner product, *134*
$(\mathbf{x}, \mathbf{Ax})$, quadratic form, *419*
$\|\mathbf{x}\|$, $\|\mathbf{x}\|_1$, $\|\mathbf{x}\|_2$, $\|\mathbf{x}\|_\infty$, vector norms, *131*
$\|\mathbf{A}\|$, $\|\mathbf{A}\|_1$, $\|\mathbf{A}\|_2$, $\|\mathbf{A}\|_\infty$, matrix norms, *164, 165*
$\|\mathbf{A}\|_F$, Frobenius norm of a matrix, *328*
$>, \geq, <, \leq$, applied to matrices, *53, 220*
$\mathbf{\Sigma}$, diagonal $m \times n$ matrix of singular values, *325*
σ_i, singular value, *325*

CHAPTER ONE

MATRIX ALGEBRA

All of the material in this first chapter is fundamental: its goal is to introduce matrices and those basic algebraic manipulations which the student must thoroughly understand before proceeding. It is important to practice the addition and multiplication of matrices until these operations become automatic; the examples illustrating Theorem 1.3 will be helpful in this regard. Theorem 1.8 is a **key theorem,** providing the basis for later computational methods; the proofs of Theorems 1.5 and 1.8 have been labeled **key proofs** because they illustrate generally useful arguments.

1.1 INTRODUCTION

From an elementary point of view, matrices provide a convenient tool for systematizing laborious algebraic and numerical calculations. We define a *matrix* to be simply a set of (real or complex) numbers arranged in a rectangular array. Thus

$$\begin{bmatrix} 1 \\ 2 \end{bmatrix}, \qquad \begin{bmatrix} x-a, & 4+b, & 1 \\ -2, & y, & -4 \end{bmatrix}, \qquad [2, b], \qquad [6] \tag{1.1}$$

are matrices. The separate numbers in a given array are called the *elements* of the matrix, and these are, in general, completely independent of each other. The commas separating the elements, as in the above examples, will be omitted if there is no risk of confusion.

The general matrix consists of mn numbers arranged in m rows and n columns, giving the following $m \times n$ (read "m by n") array:

$$\mathbf{A} = [a_{ij}] = \begin{bmatrix} a_{11} & a_{12} & \cdots & a_{1n} \\ a_{21} & a_{22} & \cdots & a_{2n} \\ & \cdots & & \\ a_{m1} & a_{m2} & \cdots & a_{mn} \end{bmatrix}. \tag{1.2}$$

The symbol a_{ij} denotes the number in the ith row and the jth column of the array:

$$\begin{array}{c} \phantom{i\text{th row}} \quad j\text{th column} \\ i\text{th row} \begin{bmatrix} & \cdot & \\ & \cdot & \\ & \cdot & \\ \cdots & a_{ij} & \cdots \\ & \cdot & \\ & \cdot & \\ & \cdot & \end{bmatrix}. \end{array}$$

Thus if $\mathbf{A}$ is the second array in (1.1), then $a_{11} = x - a$, $a_{23} = -4$, etc. We often refer to a_{ij} as the (i, j) element of $\mathbf{A}$. We shall consider matrices whose elements can be real or complex numbers. The notation $[a_{ij}]$ is often convenient since it indicates that the general element of the matrix is a_{ij}. The subscript i runs from 1 to m and j from 1 to n. A comma is used to separate subscripts if there is any risk of confusion—e.g., $a_{p+q,\, r+s}$. We often consider a 1×1 matrix $[p]$ to be identical with the number p, although some care must be exercised in this regard as we shall see in Ex. 1.8.

The use of matrices allows us to consider an array of many numbers as a single object and to denote the array by a single symbol. Relationships between the large sets of numbers arising in applications can then be expressed in a clear and concise way. The more complicated the problem, the more useful matrix symbolism proves to be. In addition, however, as we have seen so often in the history of mathematics, a device that at first sight may appear to be mainly a notational convenience turns out to have extensive ramifications. The systematic application of matrices provides insights that could not be obtained as easily (if at all) by other methods.

1.2 EQUALITY, ADDITION, AND MULTIPLICATION BY A SCALAR

In the previous section, matrices were defined to be rectangular arrays of numbers. In order to work with these arrays, we need to specify rules for comparing and combining matrices. In particular, for matrices, we must develop rules corresponding to those governing the equality, addition, subtraction, multiplication, and division of ordinary numbers. We now state these rules, without attempting to provide any motivation apart from noting that they turn out to be precisely the rules required to deal with arrays of numbers that occur in applications and computational problems. This will be amply illustrated later.

DEFINITION 1.1. The matrices **A** and **B** are said to be *equal* if and only if:

(a) **A** and **B** have the same number of rows and the same number of columns.
(b) All corresponding elements are equal—i.e.,

$$a_{ij} = b_{ij} \qquad (\text{all } i, j).$$

We now consider the addition of matrices.

DEFINITION 1.2. Two matrices can be added if and only if they have the same number of rows and the same number of columns. In this case the *sum* of two $m \times n$ matrices **A** and **B** is the $m \times n$ matrix **C** such that any element of **C** is the sum of the corresponding elements in **A** and **B**, that is, if a_{ij}, b_{ij}, c_{ij} denote the general elements of **A**, **B**, **C**, respectively, then

$$a_{ij} + b_{ij} = c_{ij} \qquad (\text{all } i, j),$$

or, in matrix notation,

$$\mathbf{A} + \mathbf{B} = [a_{ij}] + [b_{ij}] = [a_{ij} + b_{ij}] = [c_{ij}] = \mathbf{C}. \tag{1.3}$$

For example,

$$\begin{bmatrix} x+1 & -1 \\ 2 & y-1 \end{bmatrix} + \begin{bmatrix} -1 & a \\ b & 1 \end{bmatrix} = \begin{bmatrix} x & a-1 \\ 2+b & y \end{bmatrix}.$$

Since the sum of two matrices is formed by simply adding corresponding elements, it is clear that the rules governing the addition of matrices are precisely the same as those governing the addition of ordinary numbers.

Since

$$(a_{ij} + b_{ij}) + c_{ij} = a_{ij} + (b_{ij} + c_{ij}) \qquad (\text{for } 1 \le i \le m \text{ and } 1 \le j \le n),$$
$$a_{ij} + b_{ij} = b_{ij} + a_{ij} \qquad (\text{for } 1 \le i \le m \text{ and } 1 \le j \le n),$$

we have the following theorem.

THEOREM 1.1. Matrix addition is associative and commutative.

(i) $(\mathbf{A} + \mathbf{B}) + \mathbf{C} = \mathbf{A} + (\mathbf{B} + \mathbf{C})$ (*associative law*),
(ii) $\mathbf{A} + \mathbf{B} = \mathbf{B} + \mathbf{A}$ (*commutative law*).

These results state that the order in which matrices are added is not important. We stress this fact because when we consider the multiplication of matrices we shall see that the commutative law is no longer true, although the associative law still holds. The order in which matrices are multiplied is extremely important.

The arguments leading to Theorem 1.1 provide our first contact with the notion of a rigorous proof. If this concept is unfamiliar, our simple proof deserves some study. Note first that Theorem 1.1 is said to hold for *all* matrices, that is, for *all* $m \times n$ matrices for *all* positive integers m and n; our proof therefore must apply to all m, all n, and all $m \times n$ matrices. If we write

$$\begin{bmatrix} 1 & 2 \\ -3 & 4 \end{bmatrix} + \begin{bmatrix} -2 & 6 \\ 1 & 2 \end{bmatrix} = \begin{bmatrix} -1 & 8 \\ -2 & 6 \end{bmatrix} = \begin{bmatrix} -2 & 6 \\ 1 & 2 \end{bmatrix} + \begin{bmatrix} 1 & 2 \\ -3 & 4 \end{bmatrix},$$

then we have only verified Theorem 1.1(ii) for the very special case of $m = n = 2$ and

$$\mathbf{A} = \begin{bmatrix} 1 & 2 \\ -3 & 4 \end{bmatrix}, \qquad \mathbf{B} = \begin{bmatrix} -2 & 6 \\ 1 & 2 \end{bmatrix}.$$

Somewhat more generally if we write

$$\begin{bmatrix} a & b \\ c & d \end{bmatrix} + \begin{bmatrix} e & f \\ g & h \end{bmatrix} = \begin{bmatrix} a+e & b+f \\ c+g & d+h \end{bmatrix} = \begin{bmatrix} e+a & f+b \\ g+c & h+d \end{bmatrix}$$
$$= \begin{bmatrix} e & f \\ g & h \end{bmatrix} + \begin{bmatrix} a & b \\ c & d \end{bmatrix}$$

then we have only verified Theorem 1.1(ii) for all 2×2 matrices, that is, for $m = 2$ and $n = 2$. The only generally valid proof must apply for all m and n, as did our proof immediately preceding the statement of Theorem 1.1. Let us keep clearly in mind this distinction between a general proof and a verification of a very special case.

It is natural to define $-\mathbf{A}$ as the matrix whose elements are $-a_{ij}$, since then

$$\mathbf{A} + (-\mathbf{A}) = [a_{ij}] + [-a_{ij}] = [0] = \mathbf{0},$$

where $\mathbf{0}$ is a *null* matrix—i.e., a matrix whose elements are all zero. The null matrix behaves much like the real number zero (see Ex. 1.2). Having defined $-\mathbf{A}$, we are now able to define what we mean by the *subtraction* of matrices:

$$\mathbf{B} - \mathbf{A} = \mathbf{B} + (-\mathbf{A}) = [b_{ij}] + [-a_{ij}] = [b_{ij} - a_{ij}].$$

From the definition of addition we see that $2\mathbf{A} \equiv \mathbf{A} + \mathbf{A} = [2a_{ij}]$, and in general if n is an integer, then

$$n\mathbf{A} = [na_{ij}].$$

This comment motivates the following definition of what is meant by multiplying a matrix $\mathbf{A}$ by a scalar k. (Real and complex numbers are called *scalars* to distinguish them from arrays of numbers which are *matrices*.)

DEFINITION 1.3. Let $\mathbf{A}$ be a matrix and k a scalar. Then

$$k\mathbf{A} = [ka_{ij}]. \tag{1.4}$$

In sentence form, $k\mathbf{A}$ is the matrix each of whose elements is k times the corresponding element of $\mathbf{A}$. For example,

$$3\begin{bmatrix} 1, & x-1, & 3 \\ -2a, & 4, & -1 \end{bmatrix} = \begin{bmatrix} 3, & 3x-3, & 9 \\ -6a, & 12, & -3 \end{bmatrix}.$$

So far matrices have behaved in much the same way as do symbols representing ordinary numbers. The surprises have been reserved for the next section where we consider the multiplication of two matrices.

EXERCISE 1.1. Prove that the operation of multiplication by a scalar has the following properties:

(a) $(p + q)\mathbf{A} = p\mathbf{A} + q\mathbf{A}$.
(b) $p(\mathbf{A} + \mathbf{B}) = p\mathbf{A} + p\mathbf{B}$.
(c) $p(q\mathbf{A}) = (pq)\mathbf{A}$.
(d) $(-1)\mathbf{A} = -\mathbf{A}$.

EXERCISE 1.2. Prove that

(a) $k\mathbf{0} = \mathbf{0}$.
(b) $0\mathbf{A} = \mathbf{0}$.
(c) $\mathbf{A} + \mathbf{0} = \mathbf{0} + \mathbf{A} = \mathbf{A}$.

1.3 THE MULTIPLICATION OF MATRICES

We first consider the multiplication of two special matrices, namely, the product of a $1 \times m$ *row matrix* (or *row vector*) and an $m \times 1$ *column matrix* (or *column vector*), in that order. This is the basic unit operation that we use later to define the product of rectangular matrices.

DEFINITION 1.4. A row matrix can be multiplied by a column matrix, in that order, if and only if they each have the same number of elements. If

$$\mathbf{u} = [u_1, u_2, \ldots, u_m], \qquad \mathbf{v} = \begin{bmatrix} v_1 \\ v_2 \\ \cdot \\ \cdot \\ \cdot \\ v_m \end{bmatrix},$$

then $\mathbf{uv}$ is defined to be the following 1×1 matrix:

$$\mathbf{uv} = [u_1v_1 + u_2v_2 + \ldots + u_mv_m] = \left[\sum_{j=1}^{m} u_jv_j\right]. \tag{1.5}$$

For example,

$$[4 \quad -1 \quad 3]\begin{bmatrix} 2 \\ 1 \\ -5 \end{bmatrix} = [(4)(2) + (-1)(1) + (3)(-5)] = [-8].$$

We can now state the general rule for the multiplication of two matrices.

DEFINITION 1.5. Two matrices **A** and **B** can be multiplied together in the order **AB** if and only if the number of columns in the first equals the number of rows in the second. The matrices are then said to be *conformable* for the product **AB**. In this case, the (i, k) element of the product **AB** is the element in the 1×1 matrix obtained by multiplying the ith row of **A** by the kth column of **B**.

Pictorially we have, for example, for the (3, 2) element in the product of a 6×5 matrix with a 5×4 matrix:

$$\textit{3rd row}\rightarrow\begin{bmatrix} x & x & x & x & x \\ x & x & x & x & x \\ \cdot & \cdot & \cdot & \cdot & \cdot \\ x & x & x & x & x \\ x & x & x & x & x \\ x & x & x & x & x \end{bmatrix}\begin{bmatrix} x & \cdot & x & x \\ x & \cdot & x & x \\ x & \cdot & x & x \\ x & \cdot & x & x \\ x & \cdot & x & x \end{bmatrix} = \begin{bmatrix} x & x & x & x \\ x & x & x & x \\ x & \cdot & x & x \\ x & x & x & x \\ x & x & x & x \\ x & x & x & x \end{bmatrix}\leftarrow\textit{3rd row.}$$

$$\uparrow \textit{2nd column} \qquad\qquad \uparrow \textit{2nd column}$$

$$\mathbf{A} \quad \times \quad \mathbf{B} \quad = \quad \mathbf{C}$$

By taking the product of each row of **A** with each column of **B** in turn, we see that **AB** must then be 6×4.

Algebraically, if $\mathbf{A} = [a_{ij}]$ is $m \times n$ and $\mathbf{B} = [b_{ij}]$ is $n \times p$, then the two matrices can be multiplied together in the order **AB**, and if the product is denoted by $\mathbf{AB} = \mathbf{C} = [c_{ik}]$, then the matrix **C** is $m \times p$, and

$$c_{ik} = \sum_{j=1}^{n} a_{ij}b_{jk}. \tag{1.6}$$

We give the following examples:

EXERCISE 1.3. $\begin{bmatrix} -1 & 5 \\ 2 & 1 \end{bmatrix}\begin{bmatrix} 4 & 3 \\ 0 & -1 \end{bmatrix} = \begin{bmatrix} -4 & -8 \\ 8 & 5 \end{bmatrix}.$

EXERCISE 1.4. $\begin{bmatrix} a_{11} & a_{12} \\ a_{21} & a_{22} \end{bmatrix}\begin{bmatrix} x_1 \\ x_2 \end{bmatrix} = \begin{bmatrix} a_{11}x_1 + a_{12}x_2 \\ a_{21}x_1 + a_{22}x_2 \end{bmatrix}.$

It is important to practice the row-column procedure for multiplying matrices until it becomes completely automatic. Also, we should be able to pick out immediately the row of **A** and the column of **B** which produce any given element in the product **AB**.

Let us now compare the products **AB** and **BA** for two specific **A**, **B**. If

$$\mathbf{A} = \begin{bmatrix} a_1 \\ a_2 \end{bmatrix}, \qquad \mathbf{B} = [b_1, b_2],$$

then

$$\mathbf{AB} = \begin{bmatrix} a_1b_1 & a_1b_2 \\ a_2b_1 & a_2b_2 \end{bmatrix}, \qquad \mathbf{BA} = [a_1b_1 + a_2b_2].$$

The products **AB** and **BA** are quite different. They do not even have the same numbers of rows and columns. Even if **AB** exists, there is no need for **BA** to exist, for the product **AB** can be formed if **A** is $m \times n$ and **B** is $n \times p$, but in this case the product **BA** cannot be formed unless $m = p$.

To distinguish the order of multiplication of matrices, we say that in the product **AB**, the matrix **A** *premultiplies* **B**, or multiplies **B** on the left; similarly **B** *postmultiplies* **A**, or multiplies **A** on the right. Although $\mathbf{AB} \neq \mathbf{BA}$ in general, it may happen that $\mathbf{AB} = \mathbf{BA}$ for special **A** and **B**. In this case, we say that **A** and **B** *commute*.

Since the commutative law of multiplication is not true in general, when multiplying matrices it is important to retain the order in which matrices appear in the calculation. For example, suppose that we wish to multiply both sides of the equation $\mathbf{X} = \mathbf{A}$ by another matrix **P**. We must multiply both sides either on the left to obtain $\mathbf{PX} = \mathbf{PA}$ (premultiplication by **P**), or on the right to obtain $\mathbf{XP} = \mathbf{AP}$ (postmultiplication by **P**), assuming in each case that it is permissible to form the corresponding products. It is *not* permissible to obtain $\mathbf{PX} = \mathbf{AP}$ or $\mathbf{XP} = \mathbf{PA}$. The truth of these statements is obvious if we think in terms of writing out these equations and operations in longhand. Another way of viewing this is to note that $\mathbf{X} = \mathbf{A}$ implies $\mathbf{X} - \mathbf{A} = \mathbf{0}$. Hence $\mathbf{P}(\mathbf{X} - \mathbf{A}) = \mathbf{0}$, so that $\mathbf{PX} - \mathbf{PA} = \mathbf{0}$ and $\mathbf{PX} = \mathbf{PA}$, where we have assumed the validity of the distributive law (1.7).

Although the commutative law of multiplication is not true for multiplication of matrices, the distributive and associative laws are both valid. We first prove one of the *distributive* laws:

$$\mathbf{A}(\mathbf{B} + \mathbf{C}) = \mathbf{AB} + \mathbf{AC}. \tag{1.7}$$

The left-hand side is obtained by first forming the sum $\mathbf{B} + \mathbf{C}$ and then premultiplying the result by $\mathbf{A}$. The right-hand side is obtained by first forming $\mathbf{AB}$ and $\mathbf{AC}$ and then adding the results. The distributive law states that the two sequences of operations lead to the same final matrix. The proof follows immediately on translating these statements into symbols:

$$\mathbf{A}(\mathbf{B} + \mathbf{C}) = [a_{ij}][b_{jk} + c_{jk}] = \left[\sum_{j=1}^{n} a_{ij}(b_{jk} + c_{jk})\right],$$

$$\mathbf{AB} + \mathbf{AC} = \left[\sum_{j=1}^{n} a_{ij}b_{jk}\right] + \left[\sum_{j=1}^{n} a_{ij}c_{jk}\right] = \left[\sum_{j=1}^{n} a_{ij}b_{jk} + \sum_{j=1}^{n} a_{ij}c_{jk}\right]$$

$$= \left[\sum_{j=1}^{n} a_{ij}(b_{jk} + c_{jk})\right]$$

(The reader should keep in mind the remarks following Theorem 1.1 on the nature of a general proof.)

Similarly, we can prove that

$$(\mathbf{A} + \mathbf{B})\mathbf{C} = \mathbf{AC} + \mathbf{BC}. \tag{1.8}$$

Note that in all cases the order of multiplication must be preserved since the commutative law is not true. As an example of the use of the distributive law, consider

$$(\mathbf{A} + \mathbf{B})(\mathbf{A} - \mathbf{B}) = \mathbf{A}(\mathbf{A} - \mathbf{B}) + \mathbf{B}(\mathbf{A} - \mathbf{B}) = \mathbf{A}^2 - \mathbf{AB} + \mathbf{BA} - \mathbf{B}^2. \tag{1.9}$$

However, we cannot simplify further by cancelling $-\mathbf{AB}$ and $\mathbf{BA}$ in this result. (Why not?)

The *associative* law states that

$$(\mathbf{AB})\mathbf{C} = \mathbf{A}(\mathbf{BC}), \tag{1.10}$$

that is, if we first form $\mathbf{AB}$ and then postmultiply by $\mathbf{C}$, we obtain the same result as if we first form $\mathbf{BC}$ and then premultiply by $\mathbf{A}$. The proof consists of expressing this in symbols:

$$\mathbf{AB} = \left[\sum_{j=1}^{n} a_{ij}b_{jk}\right].$$

$$(\mathbf{AB})\mathbf{C} = \left[\sum_{k=1}^{p}\left\{\sum_{j=1}^{n} a_{ij}b_{jk}\right\}c_{kl}\right] \tag{1.11}$$

$$= \left[\sum_{j=1}^{n}\sum_{k=1}^{p} a_{ij}b_{jk}c_{kl}\right]. \tag{1.12}$$

It is left for the reader to show that precisely the same expression for the general element is obtained if we first form **BC** and then **A(BC)**.

[It is sometimes difficult for beginners to see that the double sums in (1.11) and (1.12) are the same. If we note first of all that c_{kl} in (1.11) is independent of j, we see that this can be taken inside the inner sum so that we have to prove

$$\sum_{k=1}^{p}\left\{\sum_{j=1}^{n} \alpha_{jk}\right\} = \sum_{k=1}^{p}\sum_{j=1}^{n} \alpha_{jk}, \tag{1.13}$$

where $\alpha_{jk} = a_{ij}b_{jk}c_{kl}$. Consider the following array:

$$\begin{bmatrix} \alpha_{11} & \alpha_{12} & \cdots & \alpha_{1p} \\ \alpha_{21} & \alpha_{22} & \cdots & \alpha_{2p} \\ & \cdots & & \\ \alpha_{n1} & \alpha_{n2} & \cdots & \alpha_{np} \end{bmatrix}. \tag{1.14}$$

If we sum the elements in each column and add the results, we obtain the same result as if we simply add together all the elements in (1.14). This is the meaning of (1.13).]

The main point to remember when manipulating matrices is that brackets can be removed and powers can be combined, *as long as the order of multiplication is preserved.* Thus

$$(\mathbf{ABA}^2)\mathbf{A}(\mathbf{AB}^3) = \mathbf{ABA}^4\mathbf{B}^3,$$

but no further simplification is possible. We cannot write this as $\mathbf{A}^5\mathbf{B}^4$. (Why not?)

Consider next the example:

$$\begin{bmatrix} a & 0 \\ b & 0 \end{bmatrix}\begin{bmatrix} 0 & 0 \\ p & q \end{bmatrix} = \begin{bmatrix} 0 & 0 \\ 0 & 0 \end{bmatrix}.$$

This is of the form $\mathbf{AB} = \mathbf{0}$ where **A** and **B** can be nonzero. The importance of this example is that if $\mathbf{AB} = \mathbf{0}$ where $\mathbf{0}$ is a null matrix, then we *cannot* necessarily conclude that either $\mathbf{A} = \mathbf{0}$ or $\mathbf{B} = \mathbf{0}$. As an application, if

$\mathbf{AB} = \mathbf{AC}$ this implies that $\mathbf{A}(\mathbf{B} - \mathbf{C}) = \mathbf{0}$, but we *cannot* conclude that either $\mathbf{A} = \mathbf{0}$ or $\mathbf{B} = \mathbf{C}$. The law of cancellation is not generally true in matrix algebra, although it may be true in special circumstances (see Ex. 1.26).

For emphasis and clarity we restate some of the above results in theorem form.

THEOREM 1.2. The following statements hold for matrix multiplication:

(i) *The commutative law is not, in general, true:*

$$\mathbf{AB} \neq \mathbf{BA}.$$

(ii) *The distributive law is true:*

$$\mathbf{A}(\mathbf{B} + \mathbf{C}) = \mathbf{AB} + \mathbf{AC}.$$
$$(\mathbf{A} + \mathbf{B})\mathbf{C} = \mathbf{AC} + \mathbf{BC}.$$

(iii) *The associative law is true:*

$$\mathbf{A}(\mathbf{BC}) = (\mathbf{AB})\mathbf{C}.$$

(iv) *The cancellation law is not true, in general.* $\mathbf{AB} = \mathbf{0}$ *does not necessarily imply that either* $\mathbf{A} = \mathbf{0}$ *or* $\mathbf{B} = \mathbf{0}$, *and* $\mathbf{AB} = \mathbf{AC}$ *does not necessarily imply that* $\mathbf{B} = \mathbf{C}$.

EXERCISE 1.5. Numerical exercises:

$$[4 \quad 1]\begin{bmatrix} 2 \\ 3 \end{bmatrix} = [11], \qquad \begin{bmatrix} 2 \\ 3 \end{bmatrix}[4 \quad 1] = \begin{bmatrix} 8 & 2 \\ 12 & 3 \end{bmatrix},$$

$$\begin{bmatrix} 3 & -2 \\ 1 & -4 \end{bmatrix}\begin{bmatrix} -1 & 6 \\ 4 & 7 \end{bmatrix} = \begin{bmatrix} -11 & 4 \\ -17 & -22 \end{bmatrix},$$

$$\begin{bmatrix} 2 & 1 \\ 4 & 3 \end{bmatrix}\begin{bmatrix} -1 & 6 \\ 3 & 2 \end{bmatrix}\begin{bmatrix} 7 & 4 \\ -1 & -3 \end{bmatrix} = \begin{bmatrix} -7 & -38 \\ 5 & -70 \end{bmatrix}.$$

[In this last example, check by showing that $(\mathbf{AB})\mathbf{C} = \mathbf{A}(\mathbf{BC})$.]

EXERCISE 1.6. Show that if the third row of $\mathbf{A}$ is four times the first row, then the third row of $\mathbf{AB}$ is also four times its first row.

EXERCISE 1.7. One important case where we can cancel $\mathbf{B}$ from the equation $\mathbf{AB} = \mathbf{CB}$ occurs when this equation is satisfied identically for all $\mathbf{B}$. For example, show that if

$$\begin{bmatrix} a_{11} & a_{12} \\ a_{21} & a_{22} \end{bmatrix}\begin{bmatrix} b_1 \\ b_2 \end{bmatrix} = \begin{bmatrix} c_{11} & c_{12} \\ c_{21} & c_{22} \end{bmatrix}\begin{bmatrix} b_1 \\ b_2 \end{bmatrix}$$

for *all* b_1, b_2, then $\mathbf{A} = \mathbf{C}$.

EXERCISE 1.8. Suppose that $\mathbf{A}$ is $m \times n$ and that p is a scalar quantity. While $p\mathbf{A}$ of course makes sense, show that $[p]\mathbf{A}$ makes sense only when $m = 1$, and that in this case $[p]\mathbf{A} = p\mathbf{A}$. Find the relationship between $p\mathbf{A}$ and $\mathbf{A}[p]$.

In the remainder of this section we consider some special types of matrix. It is convenient to have a special name and symbol for the matrix obtained by interchanging the rows and columns of a given matrix.

DEFINITION 1.6. The *transpose* of the $m \times n$ matrix $\mathbf{A} = [a_{ij}]$ is the following $n \times m$ matrix, denoted by $\mathbf{A}^T$, obtained by interchanging the rows and columns of $\mathbf{A}$:

$$\mathbf{A}^T = \begin{bmatrix} a_{11} & a_{21} & \cdots & a_{m1} \\ a_{12} & a_{22} & \cdots & a_{m2} \\ & \cdots & & \\ a_{1n} & a_{2n} & \cdots & a_{mn} \end{bmatrix}.$$

Note that the (i, j) element of $\mathbf{A}^T$ is a_{ji}.

The transpose matrix possesses the following properties.

THEOREM 1.3

(i) *The transpose of the sum of two matrices is the sum of the transposed matrices:*

$$(\mathbf{A} + \mathbf{B})^T = \mathbf{A}^T + \mathbf{B}^T.$$

(ii) *The transpose of the transpose of a given matrix is identical with the given matrix:*

$$(\mathbf{A}^T)^T = \mathbf{A}.$$

(iii) *The transpose of the product of two matrices is the product of the transposes in the reverse order:*

$$(\mathbf{AB})^T = \mathbf{B}^T\mathbf{A}^T.$$

Proof: The truth of (i) and (ii) is obvious from the definition of transpose, and formal proofs are left to the reader. Part (iii) can be proved in the following way, using formula (1.6) for the general element of a product, and the fact that if the (i, j) element of a matrix is p_{ij}, then the (i, j) element of its transpose is p_{ji}. If $\mathbf{A} = [a_{ij}]$, $\mathbf{B} = [b_{ij}]$, then

$$\text{the } (i, k) \text{ element of } \mathbf{AB} \text{ is } \sum_{j=1}^{n} a_{ij}b_{jk},$$

$$\text{the } (i, k) \text{ element of } (\mathbf{AB})^T \text{ is } \sum_{j=1}^{n} a_{kj}b_{ji}, \tag{1.15}$$

the (i, j) element of $\mathbf{B}^T$ is b_{ji},

the (j, k) element of $\mathbf{A}^T$ is a_{kj},

$$\text{the } (i, k) \text{ element of } \mathbf{B}^T\mathbf{A}^T \text{ is } \sum_{j=1}^{n} b_{ji}a_{kj}. \tag{1.16}$$

Since (1.15) and (1.16) agree, we have the required result.

Let us introduce names for some further special matrices.

DEFINITION 1.7. A matrix for which the number of rows equals the number of columns is known as a *square* matrix. If there are n rows and columns, the matrix is said to be a square matrix of *order n*. The elements a_{ii} $(i = 1, 2, \ldots, n)$ are said to lie on the *principal diagonal* or *main diagonal*. A *symmetric* matrix is a square matrix such that $\mathbf{A}^T = \mathbf{A}$, that is, the elements of the matrix are symmetrically placed about the principal diagonal, $a_{ij} = a_{ji}$. The definition implies that a symmetric matrix is automatically square:

$$\begin{bmatrix} a_{11} & a_{12} & \ldots & a_{1n} \\ a_{12} & a_{22} & \ldots & a_{2n} \\ & \cdots & & \\ a_{1n} & a_{2n} & \ldots & a_{nn} \end{bmatrix}.$$

The product of two symmetric matrices is not in general symmetric, for, if $\mathbf{A}^T = \mathbf{A}$, $\mathbf{B}^T = \mathbf{B}$, we have

$$(\mathbf{AB})^T = \mathbf{B}^T\mathbf{A}^T = \mathbf{BA} \neq \mathbf{AB} \qquad \text{(in general).} \tag{1.17}$$

However, if $\mathbf{A}$ is a symmetric matrix of order n, and $\mathbf{B}$ is a general $n \times m$ matrix, then $\mathbf{B}^T\mathbf{AB}$ is symmetric of order m. For

$$(\mathbf{B}^T\mathbf{AB})^T = \mathbf{B}^T\mathbf{A}^T(\mathbf{B}^T)^T = \mathbf{B}^T\mathbf{AB}, \tag{1.18}$$

which proves the required result. Note that this proof makes use of the general properties already derived for general matrices. We were able to avoid a proof in which the general elements of the matrices were explicitly written out. This illustrates the convenience of using matrix notation, since it is much simpler to argue as above than to compute, for example, the (i, j) element of $\mathbf{B}^T\mathbf{AB}$, namely

$$\sum_{k=1}^{n} \sum_{l=1}^{n} b_{ki}b_{lj}a_{kl}.$$

As we continue to learn about matrices we will take advantage of their notational convenience, whenever possible, by avoiding arguments requiring us to write out explicitly the general matrix elements. The student should follow this general principle as well.

EXERCISE 1.9. If

$$\mathbf{A} = \begin{bmatrix} 1 & -2 \\ -2 & 3 \end{bmatrix}, \quad \mathbf{B} = \begin{bmatrix} -2 & 1 \\ 1 & 1 \end{bmatrix},$$

form $(\mathbf{AB})^T$ and $\mathbf{B}^T\mathbf{A}^T$, and verify that these are the same. Note that although $\mathbf{A}$, $\mathbf{B}$ are symmetric, $\mathbf{AB}$ is not symmetric.

EXERCISE 1.10. If

$$\mathbf{B} = \begin{bmatrix} 1 & 0 & -2 \\ -1 & 3 & 0 \end{bmatrix}, \quad \mathbf{A} = \begin{bmatrix} 1 & -1 \\ -1 & 1 \end{bmatrix},$$

show that

$$\mathbf{B}^T\mathbf{AB} = \begin{bmatrix} 4 & -6 & -4 \\ -6 & 9 & 6 \\ -4 & 6 & 4 \end{bmatrix}.$$

Note that $\mathbf{B}^T\mathbf{AB}$ is symmetric, as it must be [see (1.18)].

EXERCISE 1.11. A *skew* (or *skewsymmetric* or *antisymmetric*) matrix is defined to be a matrix such that $\mathbf{A}^T = -\mathbf{A}$, that is, $a_{ji} = -a_{ij}$. Show that a skew matrix is square, and that the diagonal elements of a skew matrix are zero. Show that if $\mathbf{A}$ is any square matrix, then $\mathbf{A} - \mathbf{A}^T$ is skew. By writing

$$\mathbf{A} = \tfrac{1}{2}(\mathbf{A} + \mathbf{A}^T) + \tfrac{1}{2}(\mathbf{A} - \mathbf{A}^T),$$

show that any square matrix can be decomposed uniquely into the sum of a symmetric matrix and a skew matrix.

DEFINITION 1.8. A *real* matrix is a matrix whose elements are all real. A *complex* matrix has elements that may be complex. An *imaginary* matrix has elements that are all imaginary or zero. The symbol $\bar{\mathbf{A}}$ is used to denote the matrix whose (i, j) element is the complex conjugate $\bar{a}_{ij}$ of the (i, j) element of $\mathbf{A}$. When dealing with complex matrices, it is often useful to employ the *hermitian transpose* $\mathbf{A}^H = \bar{\mathbf{A}}^T$, which is the complex conjugate of the ordinary transpose. A *hermitian* matrix is a matrix such that $\mathbf{A}^H = \mathbf{A}$.

The importance of hermitian matrices will not be apparent until Chapters 8 and 9.

EXERCISE 1.12. If $\mathbf{x}$ is a complex column vector, show that $\mathbf{x}^H\mathbf{x}$ is a real 1×1 matrix.

EXERCISE 1.13. Show that a hermitian matrix is the sum of a real symmetric matrix and an imaginary skewsymmetric matrix [see Ex. 1.11].

EXERCISE 1.14. Show that the results in Theorem 1.3 are true for hermitian transposes as well as for ordinary transposes. Show that $\overline{(\mathbf{AB})} = \bar{\mathbf{A}}\bar{\mathbf{B}}$. Show that $(\bar{\mathbf{A}})^T = (\overline{\mathbf{A}^T})$.

1.4 THE INVERSE MATRIX

Consider the following set of n equations in n unknowns:

$$\begin{aligned} a_{11}x_1 + a_{12}x_2 + \cdots + a_{1n}x_n &= b_1 \\ a_{21}x_1 + a_{22}x_2 + \cdots + a_{2n}x_n &= b_2 \\ &\cdots \\ a_{n1}x_1 + a_{n2}x_2 + \cdots + a_{nn}x_n &= b_n. \end{aligned} \tag{1.19}$$

If we introduce the matrices

$$\mathbf{A} = \begin{bmatrix} a_{11} & a_{12} & \cdots & a_{1n} \\ a_{21} & a_{22} & \cdots & a_{2n} \\ & & \cdots & \\ a_{n1} & a_{n2} & \cdots & a_{nn} \end{bmatrix}, \quad \mathbf{x} = \begin{bmatrix} x_1 \\ x_2 \\ \cdot \\ \cdot \\ \cdot \\ x_n \end{bmatrix}, \quad \mathbf{b} = \begin{bmatrix} b_1 \\ b_2 \\ \cdot \\ \cdot \\ \cdot \\ b_n \end{bmatrix},$$

then, using the definition of matrix multiplication given in Section 1.3, we see that (1.19) can be written in matrix notation as

$$\mathbf{Ax} = \mathbf{b}. \tag{1.20}$$

This can be regarded as the analogue for matrices of the ordinary equation $ax = b$ for real numbers.

To solve $ax = b$ in real numbers, we divide both sides by a, to obtain $x = b/a$. Since $\mathbf{A}$ is an array of numbers, it is not clear how we should proceed to "divide equation (1.20) by $\mathbf{A}$." The clue is to note that the solution of $ax = b$ can be obtained by multiplying both sides of the equation by $1/a$, or a^{-1}, the inverse of a. This is not merely a play on words, since it turns out that the correct method of approach to the solution of (1.20) is to try to find a matrix, say $\mathbf{G}$, such that, when both sides of (1.20) are multiplied by $\mathbf{G}$, to give $\mathbf{GAx} = \mathbf{Gb}$, then the left-hand side of this equation reduces to $\mathbf{x}$, so that we have $\mathbf{x} = \mathbf{Gb}$, and the only possible solution to the equation is found.

DEFINITION 1.9. A *diagonal* matrix is a square matrix in which all the elements off the principal diagonal are zero—i.e., $a_{ij} = 0$ $(i \neq j)$. The *unit* or *identity* matrix, denoted by **I**, is a diagonal matrix whose diagonal elements are all unity:

$$\mathbf{I} = \begin{bmatrix} 1 & 0 & \dots & 0 \\ 0 & 1 & \dots & 0 \\ & & \dots & \\ 0 & 0 & \dots & 1 \end{bmatrix}.$$

If we wish to emphasize that **I** has, say, m rows and columns, we write $\mathbf{I}_m$ in place of **I**.

The important property possessed by the unit matrix (and the reason for its name) is that $\mathbf{IA} = \mathbf{A}$ and $\mathbf{AI} = \mathbf{A}$, for any **A** for which the multiplication is possible. The matrix **I** behaves like the number "1" in ordinary algebra. If **A** is an $m \times n$ matrix, then $\mathbf{I}_m\mathbf{A} = \mathbf{AI}_n = \mathbf{A}$. However, we shall usually write simply $\mathbf{IA} = \mathbf{AI} = \mathbf{A}$, where the orders of the unit matrices are understood to be such that the multiplications are permissible. If **K** is a diagonal matrix whose diagonal elements are all equal to k, we have $\mathbf{KA} = k\mathbf{A}$. For this reason **K** is called a *scalar* matrix (see Ex. 1.8).

We now return to the problem of solving $\mathbf{Ax} = \mathbf{b}$. As suggested earlier, let us find a matrix **G** such that $\mathbf{GAx} = \mathbf{x}$. This will be true if $\mathbf{GA} = \mathbf{I}$, where **I** is the unit matrix of order n. This leads us to the definition below.

DEFINITION 1.10. A matrix **G** such that $\mathbf{GA} = \mathbf{I}$, if such a matrix **G** exists, is called a *left-inverse* of **A**. A matrix **H** such that $\mathbf{AH} = \mathbf{I}$, if it exists, is called a *right-inverse* of **A**.

Inverses (if they exist) can be found by solving sets of simultaneous linear equations. Some simple examples follow.

EXERCISE 1.15. Find a right-inverse for the matrix

$$\mathbf{A} = \begin{bmatrix} 1 & -1 \\ 1 & 2 \end{bmatrix}.$$

SOLUTION: We must find a matrix

$$\mathbf{H} = \begin{bmatrix} x & z \\ y & w \end{bmatrix}, \tag{1.21}$$

such that

$$\begin{bmatrix} 1 & -1 \\ 1 & 2 \end{bmatrix}\begin{bmatrix} x & z \\ y & w \end{bmatrix} = \mathbf{I} = \begin{bmatrix} 1 & 0 \\ 0 & 1 \end{bmatrix},$$

i.e., such that

$$\left.\begin{aligned} x - y &= 1 \\ x + 2y &= 0 \end{aligned}\right\} \quad \text{and} \quad \left.\begin{aligned} z - w &= 0 \\ z + 2w &= 1 \end{aligned}\right\}.$$

These equations are easily solved, and we find

$$\mathbf{H} = \tfrac{1}{3}\begin{bmatrix} 2 & 1 \\ -1 & 1 \end{bmatrix}. \tag{1.22}$$

If a matrix **G** such that **GA** = **I** is determined by a similar procedure, it will be found that **G** is exactly the same matrix as **H** so that **A** has a common left- and right-inverse, namely, (1.22).

EXERCISE 1.16. Does the following matrix have a right-inverse?

$$\mathbf{A} = \begin{bmatrix} 1 & -1 \\ 1 & -1 \end{bmatrix}.$$

SOLUTION: If **H** is again the matrix (1.21), we must solve

$$\left.\begin{aligned} x - y &= 1 \\ x - y &= 0 \end{aligned}\right\} \quad \text{and} \quad \left.\begin{aligned} z - w &= 0 \\ z - w &= 1 \end{aligned}\right\}.$$

These are clearly inconsistent sets of equations that do not possess solutions. Hence the inverse matrix *does not exist*. The important point illustrated here is that *a square matrix need not have an inverse*.

If we wish to find a right-inverse for an $m \times n$ matrix **A**, we must try to find **H** such that $\mathbf{AH} = \mathbf{I}_m$. In order to form the product **AH**, the matrix **H** must have n rows, and the equality of **AH** and $\mathbf{I}_m$ requires that **H** have m columns, so that **H** must be $n \times m$. Similarly, we see that a left-inverse of **A** must also be $n \times m$.

EXERCISE 1.17. Find a right-inverse of

$$\mathbf{A} = \begin{bmatrix} 1 & -1 & 1 \\ 1 & 1 & 2 \end{bmatrix}.$$

SOLUTION: We try to find a matrix such that

$$\begin{bmatrix} 1 & -1 & 1 \\ 1 & 1 & 2 \end{bmatrix}\begin{bmatrix} x & y \\ z & w \\ u & v \end{bmatrix} = \begin{bmatrix} 1 & 0 \\ 0 & 1 \end{bmatrix}.$$

This requires

$$\left.\begin{aligned} x - z + u &= 1 \\ x + z + 2u &= 0 \end{aligned}\right\}, \qquad \left.\begin{aligned} y - w + v &= 0 \\ y + w + 2v &= 1 \end{aligned}\right\}.$$

It is easy to see that u and v can be given any values and then the resulting equations can be solved for x, z, y, w. If we set $u = \alpha$, $v = \beta$, we find that

$$\begin{bmatrix} x & y \\ z & w \\ u & v \end{bmatrix} = \frac{1}{2}\begin{bmatrix} 1-3\alpha & 1-3\beta \\ -1-\alpha & 1-\beta \\ 2\alpha & 2\beta \end{bmatrix}.$$

The matrix on the right is a right-inverse of **A** for any values of α and β. **A** has an infinite number of right-inverses.

EXERCISE 1.18. Show that

$$\begin{bmatrix} 1 & -1 & 1 \\ 1 & -1 & 1 \end{bmatrix}$$

does not have either a left- or a right-inverse.

EXERCISE 1.19. Show that

$$\begin{bmatrix} 1 & -1 & 1 \\ 1 & 1 & 2 \end{bmatrix}$$

does not have a left-inverse.

These exercises on rectangular matrices again illustrate that inverses may or may not exist. Theoretical questions concerning the existence of inverses for a general $m \times n$ matrix are deferred to Section 3.7. We content ourselves here with the following remarks.

THEOREM 1.4. If both a left-inverse and a right-inverse exist, then these are the same, and this common inverse is unique.

Proof: Suppose that **G**, **H** denote the left- and right-inverses of **A**. Then

$$\mathbf{G} = \mathbf{GI} = \mathbf{G(AH)} = \mathbf{(GA)H} = \mathbf{IH} = \mathbf{H}. \tag{1.23}$$

Suppose that there is another left-inverse $\mathbf{G}_1$. The same argument shows that $\mathbf{G}_1 = \mathbf{H}$ so that $\mathbf{G}_1 = \mathbf{G}$.

DEFINITION 1.11. If both a left- and right-inverse exist for a matrix, this common inverse is called *the inverse* of **A** and it is denoted by $\mathbf{A}^{-1}$. (Of course, $\mathbf{A}^{-1}$ is to be interpreted as a single symbol.)

Important results that will become clear later (Section 3.7 and Ex. 3.54) are:

1. A square matrix *either* possesses an inverse *or* it does not possess either a left- or right-inverse.
2. A rectangular (i.e., nonsquare) $m \times n$ matrix *never* possesses an inverse. If $m < n$ it will not possess a left-inverse and it may or may not possess a right-inverse. (If $m > n$, interchange "left" and "right.")

There is no need to memorize these facts since they should become intuitively clear later, but they are included to emphasize that matrices with common left- and right-inverses are the exception rather than the rule. The property of having an inverse is so important that such matrices are distinguished by a special name.

DEFINITION 1.12. A square matrix that possesses an inverse (i.e., a common left- and right-inverse) is said to be *nonsingular*. A square matrix that does not possess an inverse is said to be *singular*.

"Is it singular or nonsingular?" is often the very first question we ask when we are confronted with a square matrix. From the nonsingularity of certain matrices, we can sometimes deduce the nonsingularity of certain other related ones.

THEOREM 1.5. If $\mathbf{A}$ *and* $\mathbf{B}$ *are nonsingular then*

(i) $(\mathbf{A}^{-1})^{-1} = \mathbf{A}$. *(This includes implicitly the result that* $\mathbf{A}^{-1}$ *is nonsingular.)*
(ii) $(\mathbf{AB})^{-1} = \mathbf{B}^{-1}\mathbf{A}^{-1}$. *(This includes implicitly the result that* $\mathbf{AB}$ *is nonsingular.)*

• *Key Proof:* By definition of $\mathbf{A}^{-1}$ we have $\mathbf{A}^{-1}\mathbf{A} = \mathbf{A}\mathbf{A}^{-1} = \mathbf{I}$. Hence $\mathbf{A}^{-1}$ has both left- and right-inverses, namely $\mathbf{A}$, and this proves (i). To prove (ii) we show that $\mathbf{AB}$ has both a left- and right-inverse, namely $\mathbf{B}^{-1}\mathbf{A}^{-1}$. This follows easily since

$$(\mathbf{B}^{-1}\mathbf{A}^{-1})(\mathbf{AB}) = \mathbf{B}^{-1}(\mathbf{A}^{-1}\mathbf{A})\mathbf{B} = \mathbf{B}^{-1}(\mathbf{I})\mathbf{B} = \mathbf{B}^{-1}\mathbf{B} = \mathbf{I},$$

so that

$$(\mathbf{B}^{-1}\mathbf{A}^{-1})(\mathbf{AB}) = \mathbf{I} \tag{1.24}$$

and $\mathbf{B}^{-1}\mathbf{A}^{-1}$ is a *left*-inverse for $\mathbf{AB}$. On the other hand,

$$(\mathbf{AB})(\mathbf{B}^{-1}\mathbf{A}^{-1}) = \mathbf{A}(\mathbf{B}\mathbf{B}^{-1})\mathbf{A}^{-1} = \mathbf{A}(\mathbf{I})\mathbf{A}^{-1} = \mathbf{A}\mathbf{A}^{-1} = \mathbf{I},$$

so that

$$(\mathbf{AB})(\mathbf{B}^{-1}\mathbf{A}^{-1}) = \mathbf{I} \tag{1.25}$$

and $\mathbf{B}^{-1}\mathbf{A}^{-1}$ is a *right*-inverse for $\mathbf{AB}$. Therefore $\mathbf{B}^{-1}\mathbf{A}^{-1}$ is *the* inverse of $\mathbf{AB}$.

The above proof is typical of proofs that a certain matrix $\mathbf{G}$ is the inverse of a certain matrix $\mathbf{H}$; we merely show that $\mathbf{GH} = \mathbf{HG} = \mathbf{I}$. The difficulty is to guess the form of $\mathbf{G}$. In the above theorem this can be done as follows: If we *assume* that $\mathbf{G}(\mathbf{AB}) = \mathbf{I}$ and then try to find the candidates for $\mathbf{G}$, post-multiplication by $\mathbf{B}^{-1}$ and then by $\mathbf{A}^{-1}$ yields $\mathbf{G} = \mathbf{B}^{-1}\mathbf{A}^{-1}$, telling us what theorem to *try* to prove.

The discussion of the inverse was motivated by the suggestion that to solve the equations $\mathbf{Ax} = \mathbf{b}$ we should multiply by $\mathbf{A}^{-1}$ to obtain $\mathbf{x} = \mathbf{A}^{-1}\mathbf{b}$; an example should clarify this.

EXERCISE 1.20. Solve the following set of equations by the formula $\mathbf{x} = \mathbf{A}^{-1}\mathbf{b}$:

$$2x - 3y = -13$$
$$x + 4y = 10.$$

SOLUTION: We have

$$\mathbf{A} = \begin{bmatrix} 2 & -3 \\ 1 & 4 \end{bmatrix}. \qquad \mathbf{A}^{-1} = \tfrac{1}{11}\begin{bmatrix} 4 & 3 \\ -1 & 2 \end{bmatrix}.$$

$$\begin{bmatrix} x \\ y \end{bmatrix} = \tfrac{1}{11}\begin{bmatrix} 4 & 3 \\ -1 & 2 \end{bmatrix}\begin{bmatrix} -13 \\ 10 \end{bmatrix} = \tfrac{1}{11}\begin{bmatrix} -22 \\ 33 \end{bmatrix} = \begin{bmatrix} -2 \\ 3 \end{bmatrix}.$$

Hence $x = -2$, $y = 3$.

From a practical point of view it is easier to solve the set of equations $\mathbf{Ax} = \mathbf{b}$ directly, rather than first to form $\mathbf{A}^{-1}$ and then $\mathbf{A}^{-1}\mathbf{b}$ (see Chapter 6). However, the *idea* of the inverse matrix has many theoretical advantages, such as when considering error analysis. The inverse matrix is also useful in practice when considering equations having coefficient matrices $\mathbf{A}$ of special forms.

EXERCISE 1.21. Show that the general 2×2 matrix $\mathbf{A} = [a_{ij}]$ has an inverse if and only if $\Delta = a_{11}a_{22} - a_{12}a_{21}$ is nonzero. (We shall meet Δ later as the *determinant* of the matrix.) If $\Delta \neq 0$ show that

$$\mathbf{A}^{-1} = \frac{1}{\Delta}\begin{bmatrix} a_{22} & -a_{12} \\ -a_{21} & a_{11} \end{bmatrix}.$$

This result means that the inverse of a 2×2 matrix can be written down immediately by interchanging the diagonal terms a_{11}, a_{22}, changing the signs of the off-diagonal terms a_{12}, a_{21}, and dividing by Δ. Unfortunately, there is no correspondingly simple rule for higher-order matrices.

EXERCISE 1.22. Find $\mathbf{A}^{-1}$ if

$$\mathbf{A} = \begin{bmatrix} -1 & 2 & 1 \\ 0 & 1 & -2 \\ 1 & 4 & -1 \end{bmatrix}.$$

EXERCISE 1.23. Find a left-inverse for the matrix

$$\mathbf{A} = \begin{bmatrix} 1 & -1 \\ 1 & 1 \\ 2 & 3 \end{bmatrix},$$

and show that this left-inverse is not unique. Show that a right-inverse does not exist.

EXERCISE 1.24. Suppose that

$$2x - 3y + u + 4v - 9w = -13,$$
$$x + 4y - 5u + 2v + w = 10,$$

and we wish to express x, y in terms of u, v, w. Write these equations in matrix notation as

$$\mathbf{Ax} = \mathbf{b} + \mathbf{Eu},$$

where

$$\mathbf{x} = \begin{bmatrix} x \\ y \end{bmatrix}, \qquad \mathbf{u} = \begin{bmatrix} u \\ v \\ w \end{bmatrix}.$$

Hence

$$\mathbf{x} = \mathbf{A}^{-1}\mathbf{b} + \mathbf{A}^{-1}\mathbf{Eu}.$$

By writing out **A**, **b**, **E** explicitly and calculating $\mathbf{A}^{-1}\mathbf{b}$ and $\mathbf{A}^{-1}\mathbf{E}$, show that

$$\begin{bmatrix} x \\ y \end{bmatrix} = \begin{bmatrix} -2 \\ 3 \end{bmatrix} + \begin{bmatrix} 1 & -2 & 3 \\ 1 & 0 & -1 \end{bmatrix} \begin{bmatrix} u \\ v \\ w \end{bmatrix}.$$

EXERCISE 1.25. Show that the inverse of a diagonal matrix with nonzero diagonal elements can be written down immediately:

$$\mathbf{K} = \begin{bmatrix} k_1 & 0 & \dots & 0 \\ 0 & k_2 & \dots & 0 \\ & \dots & & \\ 0 & 0 & \dots & k_n \end{bmatrix}, \qquad \mathbf{K}^{-1} = \begin{bmatrix} \frac{1}{k_1} & 0 & \dots & 0 \\ 0 & \frac{1}{k_2} & \dots & 0 \\ & & \dots & \\ 0 & 0 & \dots & \frac{1}{k_n} \end{bmatrix}.$$

EXERCISE 1.26. Show that the cancellation law is true if the appropriate inverse exists. For example, if $\mathbf{AX} = \mathbf{AB}$, and a left-inverse of **A** exists, then $\mathbf{X} = \mathbf{B}$. [Compare Theorem 1.2(iv).]

EXERCISE 1.27. Find all 2×2 matrices **X** such that $\mathbf{X}^2 = \mathbf{I}$, where **I** is the 2×2 unit matrix.

[The answer is that **X** can be any of the following:

$$\pm\mathbf{I}, \qquad \pm\begin{bmatrix} 1 & 0 \\ c & -1 \end{bmatrix}, \qquad \pm\begin{bmatrix} 1 & b \\ 0 & -1 \end{bmatrix}, \qquad \begin{bmatrix} a & b \\ \frac{(1 - a^2)}{b} & -a \end{bmatrix},$$

where a, b, c are arbitrary numbers. Note that a quadratic matrix equation can have an infinite number of solutions, in contrast to the scalar equation $x^2 = 1$ which has only two solutions: $x = \pm 1$. Note also that $\mathbf{X}^2 = \mathbf{I}$ implies $(\mathbf{X} - \mathbf{I})(\mathbf{X} + \mathbf{I}) = \mathbf{0}$, so that if $\mathbf{X} \neq \pm\mathbf{I}$ then both $\mathbf{X} - \mathbf{I}$ and $\mathbf{X} + \mathbf{I}$ must be singular (see Ex. 1.26). The reader should verify that the above matrices satisfy this condition.]

EXERCISE 1.28. Show that if $\mathbf{A}, \mathbf{B}$ and $\mathbf{A} + \mathbf{B}$ possess inverses, then so does $\mathbf{A}^{-1} + \mathbf{B}^{-1}$, and

$$(\mathbf{A}^{-1} + \mathbf{B}^{-1})^{-1} = \mathbf{A}(\mathbf{A} + \mathbf{B})^{-1}\mathbf{B} = \mathbf{B}(\mathbf{A} + \mathbf{B})^{-1}\mathbf{A}.$$

1.5 THE PARTITIONING OF MATRICES

In this section we describe a useful technical device that we employ frequently to facilitate matrix manipulations.

A *submatrix* is a matrix obtained from an original matrix by deleting certain rows and columns. Suppose that we partition a matrix into submatrices by horizontal and vertical lines as illustrated by the following special example:

$$\mathbf{A} = \left[\begin{array}{ccc|c|cc} a_{11} & a_{12} & a_{13} & a_{14} & a_{15} & a_{16} \\ a_{21} & a_{22} & a_{23} & a_{24} & a_{25} & a_{26} \\ \hline a_{31} & a_{32} & a_{33} & a_{34} & a_{35} & a_{36} \end{array}\right]. \tag{1.26}$$

We can write this, using an obvious notation, as

$$\mathbf{A} = \begin{bmatrix} \mathbf{A}_{11} & \mathbf{A}_{12} & \mathbf{A}_{13} \\ \mathbf{A}_{21} & \mathbf{A}_{22} & \mathbf{A}_{23} \end{bmatrix}, \tag{1.27}$$

where

$$\mathbf{A}_{11} = \begin{bmatrix} a_{11} & a_{12} & a_{13} \\ a_{21} & a_{22} & a_{23} \end{bmatrix}, \qquad \mathbf{A}_{22} = [a_{34}], \text{ etc.}$$

Suppose that we partition two $m \times n$ matrices as follows $(a + b + c = m)$:

$$\mathbf{B} = \begin{matrix} \\ a \\ b \\ c \end{matrix} \begin{matrix} n \\ \begin{bmatrix} \mathbf{B}_1 \\ \mathbf{B}_2 \\ \mathbf{B}_3 \end{bmatrix} \end{matrix}, \qquad \mathbf{C} = \begin{matrix} \\ a \\ b \\ c \end{matrix} \begin{matrix} n \\ \begin{bmatrix} \mathbf{C}_1 \\ \mathbf{C}_2 \\ \mathbf{C}_3 \end{bmatrix} \end{matrix}, \tag{1.28}$$

where the numbers of rows and columns in the submatrices are denoted by a, b, c to the left of, and n above, the matrices. By writing the following equations out in detail, we can readily check that

$$\mathbf{B}+\mathbf{C}=\begin{bmatrix}\mathbf{B}_1\\ \mathbf{B}_2\\ \mathbf{B}_3\end{bmatrix}+\begin{bmatrix}\mathbf{C}_1\\ \mathbf{C}_2\\ \mathbf{C}_3\end{bmatrix}=\begin{bmatrix}\mathbf{B}_1+\mathbf{C}_1\\ \mathbf{B}_2+\mathbf{C}_2\\ \mathbf{B}_3+\mathbf{C}_3\end{bmatrix}.$$

Clearly, we find the following general result.

THEOREM 1.6. We can add and subtract partitioned matrices as if the submatrices were ordinary (scalar) elements, provided the matrices are partitioned in the same way, so that it is permissible to form the necessary submatrix additions and subtractions in the final result.

Similarly, suppose that a matrix $\mathbf{A}$ is partitioned as follows:

$$\mathbf{A}=\begin{matrix} \\ p\\ q\end{matrix}\!\!\begin{matrix} a & b & c\\ \left[\begin{matrix}\mathbf{A}_{11}\\ \mathbf{A}_{21}\end{matrix}\right.&\begin{matrix}\mathbf{A}_{12}\\ \mathbf{A}_{22}\end{matrix}&\left.\begin{matrix}\mathbf{A}_{13}\\ \mathbf{A}_{23}\end{matrix}\right]\end{matrix}. \tag{1.29}$$

It is then possible to form the product $\mathbf{AB}$, where $\mathbf{B}$ is assumed to be partitioned as in (1.28), in the following way:

$$\mathbf{AB}=\begin{matrix}p\\ q\end{matrix}\begin{bmatrix}\mathbf{A}_{11}\mathbf{B}_1+\mathbf{A}_{12}\mathbf{B}_2+\mathbf{A}_{13}\mathbf{B}_3\\ \mathbf{A}_{21}\mathbf{B}_1+\mathbf{A}_{22}\mathbf{B}_2+\mathbf{A}_{23}\mathbf{B}_3\end{bmatrix}. \tag{1.30}$$

To check this result we need only visualize writing out, in full detail, all the matrices and operations involved. This brings us to the following general result.

THEOREM 1.7. We can multiply partitioned matrices as if the submatrices were ordinary (scalar) elements, provided that the matrices are partitioned in such a way that the appropriate products can be formed.

Partitioned matrices are useful in several ways. If a physical system can be split into subsystems with interconnections, the behavior of the whole system can often be described by a large matrix partitioned in such a way that the submatrices along the diagonal describe the separate parts of the whole system, and the submatrices off the diagonal describe the interconnections of the subsystems. This can clarify the structure of a complicated system.

An important application of partitioned matrices is given by the next theorem. We require the following definition.

DEFINITION 1.13. The *unit column matrix* $\mathbf{e}_j$ *of order* n is an $n \times 1$ matrix with jth element unity, all other elements zero:

$$\mathbf{e}_1 = \begin{bmatrix} 1 \\ 0 \\ \cdot \\ \cdot \\ \cdot \\ 0 \end{bmatrix}, \quad \mathbf{e}_2 = \begin{bmatrix} 0 \\ 1 \\ \cdot \\ \cdot \\ \cdot \\ 0 \end{bmatrix}, \quad \ldots, \quad \mathbf{e}_n = \begin{bmatrix} 0 \\ 0 \\ \cdot \\ \cdot \\ \cdot \\ 0 \\ 1 \end{bmatrix}. \tag{1.31}$$

In partitioned matrix notation we can write the unit matrix as:

$$\mathbf{I} = [\mathbf{e}_1, \mathbf{e}_2, \ldots, \mathbf{e}_n]. \tag{1.32}$$

The problem of finding an inverse can be reduced to the problem of solving several sets of simultaneous linear equations with the same coefficient matrix. Later, this will turn out to be extremely important from the point of view of developing efficient methods for inverting matrices in practice (see Sections 3.2, 6.4).

● KEY THEOREM 1.8. *If the right-inverse* $\mathbf{H}$ *of an* $m \times n$ *matrix* $\mathbf{A}$ *exists, it is given by the* $n \times m$ *matrix*

$$\mathbf{H} = [\mathbf{x}_1, \mathbf{x}_2, \ldots, \mathbf{x}_m] \tag{1.33}$$

where the $\mathbf{x}_i$ *are the solutions of the equations*

$$\mathbf{A}\mathbf{x}_j = \mathbf{e}_j, \qquad j = 1, 2, \ldots, m, \tag{1.34}$$

where $\mathbf{e}_j$ *denotes the jth unit column matrix of order m.*

● *Key Proof:* Suppose that a right-inverse of $\mathbf{A}$ is denoted by $\mathbf{H} = [\mathbf{x}_1, \ldots, \mathbf{x}_m]$ where $\mathbf{x}_j$ is the jth column of $\mathbf{H}$. Then $\mathbf{AH} = \mathbf{I}$ gives

$$\mathbf{A}[\mathbf{x}_1, \mathbf{x}_2, \ldots, \mathbf{x}_m] = [\mathbf{e}_1, \mathbf{e}_2, \ldots, \mathbf{e}_m].$$

Applying Theorems 1.6, 1.7, we see that the $\mathbf{x}_j$ satisfy (1.34), which proves the theorem.

EXERCISE 1.29. Suppose that the following matrices are partitioned as indicated by the dotted lines:

$$\mathbf{A} = \left[\begin{array}{c:cc:ccc} 2 & 0 & 0 & 4 & -1 & 7 \\ 1 & 0 & 0 & -2 & 0 & 3 \\ \hdashline 1 & 1 & -1 & 0 & 0 & 0 \end{array}\right] \qquad \mathbf{B} = \left[\begin{array}{cc} -1 & 4 \\ \hdashline 2 & 1 \\ -1 & 1 \\ \hdashline 5 & 3 \\ 1 & 2 \\ 0 & -1 \end{array}\right].$$

Write down the matrices $\mathbf{A}_{ij}$, $\mathbf{B}_i$ in the notation of (1.29), (1.28). Show that

$$\mathbf{A}_{11}\mathbf{B}_1 + \mathbf{A}_{12}\mathbf{B}_2 + \mathbf{A}_{13}\mathbf{B}_3 = \begin{bmatrix} 17 & 11 \\ -11 & -5 \end{bmatrix},$$

$$\mathbf{A}_{21}\mathbf{B}_1 + \mathbf{A}_{22}\mathbf{B}_2 + \mathbf{A}_{23}\mathbf{B}_3 = [2 \quad 4].$$

Show that the result given by inserting these in (1.30) is precisely the product obtained by forming **AB** directly.

EXERCISE 1.30. Let **A** be partitioned into its columns, so that

$$\mathbf{A} = [\mathbf{A}_1, \mathbf{A}_2, \ldots, \mathbf{A}_n],$$

where each $\mathbf{A}_i$ is an $m \times 1$ column matrix. Let $\mathbf{x}$ be the $n \times 1$ column matrix

$$\mathbf{x} = \begin{bmatrix} x_1 \\ \cdot \\ \cdot \\ \cdot \\ x_n \end{bmatrix}.$$

Prove that

$$\mathbf{Ax} = \sum_{i=1}^{n} x_i \mathbf{A}_i.$$

MISCELLANEOUS EXERCISES 1

EXERCISE 1.31. If

$$\mathbf{A} = \begin{bmatrix} 1 & 1 & 1 & -1 \\ 1 & -1 & 1 & 1 \\ 1 & 1 & -1 & 1 \\ 1 & -1 & -1 & -1 \end{bmatrix},$$

show that $\mathbf{A}^T\mathbf{A} = \mathbf{A}\mathbf{A}^T = 4\mathbf{I}$.

EXERCISE 1.32. Show by induction that if $k \neq 0$, then

$$\begin{bmatrix} \cos\theta & k\sin\theta \\ -\dfrac{1}{k}\sin\theta & \cos\theta \end{bmatrix}^n = \begin{bmatrix} \cos n\theta & k\sin n\theta \\ -\dfrac{1}{k}\sin n\theta & \cos n\theta \end{bmatrix}.$$

["Show by induction" means (a) verify for $n = 1$; (b) prove that if the result is true for any given n, then it is true for $n + 1$.]

EXERCISE 1.33. Suppose that $\mathbf{A}$ is a matrix in which the third column is equal to twice the first column. Show that the same must be true of any product $\mathbf{BA}$.

EXERCISE 1.34. Construct 2×2 matrices $\mathbf{A}$, $\mathbf{B}$ having no zero entries for which $\mathbf{AB} = \mathbf{0}$.

EXERCISE 1.35. Find 2×2 matrices $\mathbf{X}$, $\mathbf{Y}$, neither of which is null, such that $\mathbf{X}^2 + \mathbf{Y}^2 = \mathbf{0}$.

EXERCISE 1.36. Construct a 2×2 matrix with no zero entries that does not have an inverse.

EXERCISE 1.37. Let $\mathbf{A} = \mathbf{B} + \mathbf{C}$ where $\mathbf{B}$, $\mathbf{C}$ are $n \times n$ matrices such that $\mathbf{C}^2 = 0$ and $\mathbf{BC} = \mathbf{CB}$. Show that, for $p > 0$,

$$\mathbf{A}^{p+1} = \mathbf{B}^p\{\mathbf{B} + (p + 1)\mathbf{C}\}.$$

EXERCISE 1.38. What real matrices satisfy $\mathbf{A}^T\mathbf{A} = \mathbf{0}$? Justify your answer.

EXERCISE 1.39. If $\mathbf{A}$ is a square matrix of order n and $\mathbf{x}$ is an $n \times 1$ column vector, show that $\mathbf{k} = \mathbf{x}^T\mathbf{A}\mathbf{x}$ is a 1×1 matrix. If $\mathbf{x} = \mathbf{P}\mathbf{y}$, show that $\mathbf{k} = \mathbf{y}^T(\mathbf{P}^T\mathbf{A}\mathbf{P})\mathbf{y}$.

EXERCISE 1.40. Show that a product of p nonsingular matrices is nonsingular.

EXERCISE 1.41. Suppose that $\mathbf{A}$ is a 2×2 matrix that commutes with every 2×2 matrix. Show that $\mathbf{A}$ must be a multiple of the unit matrix.

EXERCISE 1.42. Show that if the ith element of a diagonal matrix $\mathbf{D}$ is d_i, then $\mathbf{DA}$ is the matrix obtained by multiplying the ith row of $\mathbf{A}$ by d_i. Similarly, $\mathbf{AD}$ is the matrix obtained by multiplying the jth column of $\mathbf{A}$ by d_j.

EXERCISE 1.43. Detect the flaw in the following argument. "Suppose that $\mathbf{AH} = \mathbf{I}$. Premultiply by $\mathbf{H}$ and postmultiply by $\mathbf{A}$. Then $(\mathbf{HA})^2 = \mathbf{HA}$. Multiply this equation on the left by $(\mathbf{HA})^{-1}$. Then $\mathbf{HA} = \mathbf{I}$, so that $\mathbf{H}$ is a left-inverse of $\mathbf{A}$."

EXERCISE 1.44. Show that

(a) $(\mathbf{A}^T)^{-1} = (\mathbf{A}^{-1})^T$.
(b) If $\mathbf{A}$ is symmetric then $\mathbf{A}^{-1}$ is symmetric.

EXERCISE 1.45. Prove that the inverse of the matrix obtained by interchanging the pth and qth rows of $\mathbf{A}$ is given by interchanging the pth and qth columns of $\mathbf{A}^{-1}$. Prove that the inverse of the matrix obtained by multiplying the pth column of $\mathbf{A}$ by k ($\neq 0$) is given by dividing the pth row of $\mathbf{A}^{-1}$ by k.

EXERCISE 1.46. Let $\mathbf{K}$ be a skewsymmetric matrix ($\mathbf{K}^T = -\mathbf{K}$). If

$$\mathbf{B} = (\mathbf{I} + \mathbf{K})(\mathbf{I} - \mathbf{K})^{-1},$$

show that $\mathbf{B}^T\mathbf{B} = \mathbf{B}\mathbf{B}^T = \mathbf{I}$. (This assumes that $\mathbf{I} - \mathbf{K}$ is nonsingular, and the truth of this will be demonstrated in later exercises. See Exs. 3.40 and 9.22. Prove directly that if $\mathbf{K}$ is 2×2 then $\mathbf{I} - \mathbf{K}$ is nonsingular.)

EXERCISE 1.47. Prove that:

(a) If a matrix has a row or a column of zeros, then it has no inverse.
(b) If $\mathbf{A}$ is any square matrix in which one row is a multiple of another, then this matrix does not have an inverse.

EXERCISE 1.48. If

$$\mathbf{A} = \begin{bmatrix} \alpha & \beta & \delta \\ 0 & \alpha & 0 \\ 0 & \gamma & \epsilon \end{bmatrix},$$

show that, if $\alpha\epsilon \neq 0$,

$$\mathbf{A}^{-1} = \frac{1}{\alpha^2\epsilon}\begin{bmatrix} \alpha\epsilon & \gamma\delta - \beta\epsilon & -\alpha\delta \\ 0 & \alpha\epsilon & 0 \\ 0 & -\alpha\gamma & \alpha^2 \end{bmatrix}.$$

EXERCISE 1.49. Given that

$$\mathbf{a}_1 = \begin{bmatrix} 1 \\ 2 \\ 3 \end{bmatrix}, \quad \mathbf{a}_2 = \begin{bmatrix} 1 \\ 1 \\ -1 \end{bmatrix}, \quad \mathbf{a}_3 = \begin{bmatrix} 5 \\ -4 \\ 1 \end{bmatrix},$$

show that $\mathbf{a}_1^T\,\mathbf{a}_2 = [0]$, $\mathbf{a}_1^T\,\mathbf{a}_3 = [0]$, $\mathbf{a}_2^T\,\mathbf{a}_3 = [0]$. By any method, show that the inverse of $\mathbf{A} = [\mathbf{a}_1\ \mathbf{a}_2\ \mathbf{a}_3]$ is given by

$$\mathbf{A}^{-1} = \begin{bmatrix} \mathbf{b}_1 \\ \mathbf{b}_2 \\ \mathbf{b}_3 \end{bmatrix},$$

where, if $\mathbf{a}_i^T\,\mathbf{a}_i = [k_i]$,

$$\mathbf{b}_i = \frac{1}{k_i}\mathbf{a}_i^T, \qquad i = 1, 2, 3.$$

EXERCISE 1.50. A real matrix $\mathbf{A}$ that satisfies the relations $\mathbf{A}\mathbf{A}^T = \mathbf{A}^T\mathbf{A} = \mathbf{I}$ is said to be *orthogonal*.

(a) Give an example of a 2×2 orthogonal matrix.
(b) Find the general 2×2 orthogonal matrix.

(c) Show that the product of two orthogonal matrices is an orthogonal matrix.
(d) Show that the inverse of an orthogonal matrix is an orthogonal matrix.

EXERCISE 1.51. If $\mathbf{A} = [a_{ij}]$, where the a_{ij} are functions of a variable t, then we define $d\mathbf{A}/dt = [da_{ij}/dt]$. Show that

(a) $\dfrac{d}{dt}(\mathbf{AB}) = \dfrac{d\mathbf{A}}{dt}\mathbf{B} + \mathbf{A}\dfrac{d\mathbf{B}}{dt}$.

(b) By differentiating $\mathbf{Z}^{-1}\mathbf{Z} = \mathbf{I}$,

$$\frac{d\mathbf{Z}^{-1}}{dt} = -\mathbf{Z}^{-1}\frac{d\mathbf{Z}}{dt}\mathbf{Z}^{-1}.$$

EXERCISE 1.52. Suppose that $\mathbf{B} = \mathbf{P}^{-1}\mathbf{AP}$. Show that $\mathbf{B}^m = \mathbf{P}^{-1}\mathbf{A}^m\mathbf{P}$ if m is integral. Deduce that

$$a_n\mathbf{A}^n + a_{n-1}\mathbf{A}^{n-1} + \cdots + a_0\mathbf{I} = \mathbf{0}$$

if and only if

$$a_n\mathbf{B}^n + a_{n-1}\mathbf{B}^{n-1} + \cdots + a_0\mathbf{I} = \mathbf{0}.$$

EXERCISE 1.53. Form $\mathbf{AB}$ where

$$\mathbf{A} = \left[\begin{array}{cc|ccc} 4 & 3 & -2 & 1 & 4 \\ 2 & -5 & 6 & 3 & -1 \end{array}\right], \qquad \mathbf{B} = \left[\begin{array}{cc|c} 0 & -1 & 3 \\ 2 & -1 & 6 \\ \hline 5 & 2 & 1 \\ -3 & 4 & -1 \\ 2 & -1 & 2 \end{array}\right],$$

using

(a) straightforward multiplication,
(b) block multiplication, partitioning as shown by the dotted lines.

EXERCISE 1.54. Show that if $\mathbf{B}$ is the partitioned matrix

$$\mathbf{B} = \begin{bmatrix} \mathbf{A}_{11} & \mathbf{A}_{12} \\ \mathbf{A}_{21} & \mathbf{A}_{22} \end{bmatrix}, \quad \text{then} \quad \mathbf{B}^T = \begin{bmatrix} \mathbf{A}_{11}^T & \mathbf{A}_{21}^T \\ \mathbf{A}_{12}^T & \mathbf{A}_{22}^T \end{bmatrix}.$$

EXERCISE 1.55. If $\mathbf{A}_1, \mathbf{A}_2, \mathbf{A}_3$ are nonsingular matrices, then prove that

$$\begin{bmatrix} \mathbf{A}_1 & \mathbf{0} & \mathbf{0} \\ \mathbf{0} & \mathbf{A}_2 & \mathbf{0} \\ \mathbf{0} & \mathbf{0} & \mathbf{A}_3 \end{bmatrix}^{-1} \quad \text{exists and equals} \quad \begin{bmatrix} \mathbf{A}_1^{-1} & \mathbf{0} & \mathbf{0} \\ \mathbf{0} & \mathbf{A}_2^{-1} & \mathbf{0} \\ \mathbf{0} & \mathbf{0} & \mathbf{A}_3^{-1} \end{bmatrix}.$$

EXERCISE 1.56. Prove that:

(a) $(\mathbf{ABC})^T = \mathbf{C}^T\mathbf{B}^T\mathbf{A}^T$;
(b) If $\mathbf{A}, \mathbf{B}, \mathbf{C}$ are nonsingular, then so is $\mathbf{ABC}$, and $(\mathbf{ABC})^{-1} = \mathbf{C}^{-1}\mathbf{B}^{-1}\mathbf{A}^{-1}$.

EXERCISE 1.57. If

$$\mathbf{X}=\begin{bmatrix}0 & i\\ -i & 0\end{bmatrix},\quad \mathbf{Y}=\begin{bmatrix}0 & 1\\ 1 & 0\end{bmatrix},\quad \mathbf{Z}=\begin{bmatrix}-1 & 0\\ 0 & 1\end{bmatrix},$$

where $i=\sqrt{-1}$, then verify that

$$\mathbf{XY}=-\mathbf{YX}=-i\mathbf{Z},\qquad \mathbf{YZ}=-\mathbf{ZY}=-i\mathbf{X},$$

$$\mathbf{ZX}=-\mathbf{XZ}=-i\mathbf{Y},\qquad \mathbf{X}^2=\mathbf{Y}^2=\mathbf{Z}^2=\mathbf{I}.$$

Show that every 2×2 matrix can be written $\mathbf{A}=a\mathbf{I}+b\mathbf{X}+c\mathbf{Y}+d\mathbf{Z}$.

EXERCISE 1.58. By the *trace* of a square matrix, written tr **A**, we mean the sum of its diagonal elements

$$\operatorname{tr}\mathbf{A}=\sum_{i=1}^{n}a_{ii}.$$

Show that

(a) if k is a scalar, $\operatorname{tr}(k\mathbf{A})=k\operatorname{tr}\mathbf{A}$,
(b) $\operatorname{tr}(\mathbf{A}\pm\mathbf{B})=\operatorname{tr}\mathbf{A}\pm\operatorname{tr}\mathbf{B}$,
(c) $\operatorname{tr}\mathbf{AB}=\operatorname{tr}\mathbf{BA}$,
(d) $\operatorname{tr}(\mathbf{B}^{-1}\mathbf{AB})=\operatorname{tr}\mathbf{A}$,
(e) $\operatorname{tr}(\mathbf{AA}^T)=\sum_{i=1}^{n}\sum_{j=1}^{n}(a_{ij})^2$.

EXERCISE 1.59. Let

$$\begin{bmatrix}\mathbf{A} & \mathbf{u}\\ \mathbf{v}^T & a\end{bmatrix}^{-1}=\begin{bmatrix}\mathbf{B} & \mathbf{p}\\ \mathbf{q}^T & \alpha\end{bmatrix},$$

where **u**, **v**, **p**, **q** are column matrices and **A** has an inverse. Prove that

$$\mathbf{B}=\mathbf{A}^{-1}+\alpha\mathbf{A}^{-1}\,\mathbf{uv}^T\mathbf{A}^{-1},\qquad \alpha=(a-\mathbf{v}^T\mathbf{A}^{-1}\mathbf{u})^{-1},$$

$$\mathbf{p}=-\alpha\mathbf{A}^{-1}\mathbf{u},\qquad \mathbf{q}^T=-\alpha\mathbf{v}^T\mathbf{A}^{-1}.$$

Also $\mathbf{A}^{-1}=\mathbf{B}-(1/\alpha)\mathbf{pq}^T$. (This exercise shows that if the inverse of a given matrix is known, it is easy to compute the inverse of the matrix obtained by either adding a row and column, or omitting a row and column. Matrices of the above type are known as *bordered* matrices.)

EXERCISE 1.60. If, in partitioned form,

$$\mathbf{A}=\begin{bmatrix}\mathbf{P} & \mathbf{Q}\\ \mathbf{R} & \mathbf{S}\end{bmatrix},$$

where $\mathbf{A}$ and $\mathbf{P}$ are nonsingular, prove that

$$\mathbf{A}^{-1} = \begin{bmatrix} \mathbf{X} & -\mathbf{P}^{-1}\mathbf{Q}\mathbf{W} \\ -\mathbf{W}\mathbf{R}\mathbf{P}^{-1} & \mathbf{W} \end{bmatrix},$$

where

$$\mathbf{W} = (\mathbf{S} - \mathbf{R}\mathbf{P}^{-1}\mathbf{Q})^{-1}, \qquad \mathbf{X} = \mathbf{P}^{-1} + \mathbf{P}^{-1}\mathbf{Q}\mathbf{W}\mathbf{R}\mathbf{P}^{-1}. \tag{1.35}$$

Similarly, if $\mathbf{A}$ and $\mathbf{S}$ are nonsingular, prove that

$$\mathbf{A}^{-1} = \begin{bmatrix} \mathbf{X} & -\mathbf{X}\mathbf{Q}\mathbf{S}^{-1} \\ -\mathbf{S}^{-1}\mathbf{R}\mathbf{X} & \mathbf{W} \end{bmatrix},$$

where

$$\mathbf{X} = (\mathbf{P} - \mathbf{Q}\mathbf{S}^{-1}\mathbf{R})^{-1}, \qquad \mathbf{W} = \mathbf{S}^{-1} + \mathbf{S}^{-1}\mathbf{R}\mathbf{X}\mathbf{Q}\mathbf{S}^{-1}. \tag{1.36}$$

If $\mathbf{P}$ and $\mathbf{S}$ are both nonsingular, prove directly that the forms (1.35), (1.36) for $\mathbf{X}$ and $\mathbf{W}$ are equivalent. [The result in Ex. 1.59 is, of course, a special case of (1.35).]

CHAPTER TWO

SOME SIMPLE APPLICATIONS OF MATRICES

This chapter has two goals: (1) to show that complicated sets of relationships can be organized and clarified by the use of matrices, and (2) to demonstrate the need for studying matrix problems and certain properties of matrices.

This chapter contains no theorems (or proofs) and hence no **key** material. Instead the chapter presents a wide variety of matrix applications in diverse fields. These applications will show you why matrices are useful and may motivate you to explore certain kinds of questions about matrices.

The concepts and techniques you will find in this chapter are the ones that you studied in Chapter 1. Thus you should not find the technical material difficult. You may read the entire chapter or you may select sections for study that are appropriate to your background and interests.

2.1 INTRODUCTION

In many applications, the usefulness of matrices arises from their representing an array of many numbers as a single object denoted by a single symbol. Relationships between variables can then be expressed in a clear and concise way. We do nothing that could not be done in terms of the elements of the array, but the introduction of matrices clarifies the important relationships. In problems for which matrices are suitable, the more complicated the problem, the more useful and powerful the idea of the matrix.

In Section 2.2 we consider an application of matrix multiplication to a Markov chain arising as a simple model of a marketing problem in which customers switch from one supplier to another. Similar matrix methods are used in Section 2.3 to study the growth behavior of the total populations of two types of entities under different models of their interactions. Section 2.4 presents a matrix approach to the calculation of strains and stresses in pin-jointed frameworks. The techniques also apply to other network problems, for example, to those of electrical engineering and economics. In Section 2.5 we consider the problem of production planning to maximize profits, resulting in what is called a linear program—a tool of very wide applicability. Finally, Section 2.6 uses matrix notation to synthesize the solution of equations or the approximation of data by the method of least squares.

2.2 A MARKOV-CHAIN EXAMPLE

Let us illustrate the idea of a Markov chain by considering the following simple example. Three dairies X, Y, Z supply all the milk consumed in a certain town. Over a given period, some consumers will switch from one supplier to another for various reasons—e.g., advertising, cost, convenience, dissatisfaction, and so on. We wish to model and analyze the movement of customers from one dairy to another, assuming that constant fractions of consumers switch from any one dairy to any other dairy each month.

Suppose that on, say, December 31, the dairies X, Y, Z have fractions x_0, y_0, z_0 of the total market, and that the corresponding fractions on January 31 are x_1, y_1, z_1. Since X, Y, Z are the only suppliers, we must have

$$\begin{aligned} x_0 + y_0 + z_0 &= 1, \\ x_1 + y_1 + z_1 &= 1. \end{aligned} \tag{2.1}$$

Suppose that in January dairy X retains a fraction a_{11} of its own customers and attracts a fraction a_{12} of Y's customers and a_{13} of Z's. Assuming that the overall number of customers in the town, say N, does not change during the month, the total number that X has on January 31 must equal the number it retains plus the numbers it attracts from Y and Z:

$$(x_1N) = a_{11}(x_0N) + a_{12}(y_0N) + a_{13}(z_0N),$$

or

$$x_1 = a_{11}x_0 + a_{12}y_0 + a_{13}z_0.$$

Similarly,

$$y_1 = a_{21}x_0 + a_{22}y_0 + a_{23}z_0,$$
$$z_1 = a_{31}x_0 + a_{32}y_0 + a_{33}z_0,$$

where, if numbers 1, 2, 3 refer to dairies X, Y, Z, respectively,

$$a_{ii} = \text{fraction of } i\text{'s customers retained by } i,$$
$$a_{ij} = \text{fraction of } j\text{'s customers that switch to } i (i \neq j).$$

In matrix notation these are simply

$$\mathbf{x}_1 = \mathbf{A}\mathbf{x}_0, \tag{2.2}$$

where $\mathbf{A} = [a_{ij}]$ and

$$\mathbf{x}_r = \begin{bmatrix} x_r \\ y_r \\ z_r \end{bmatrix} \qquad (r = 0, 1). \tag{2.3}$$

The elements of **A** have certain properties that follow directly from their definition:

1. Obviously the a_{ij} cannot be negative:

$$a_{ij} \geq 0 \qquad (\text{all } i, j). \tag{2.4}$$

2. Obviously the a_{ij} cannot be greater than unity:

$$a_{ij} \leq 1 \qquad (\text{all } i, j). \tag{2.5}$$

3. Of the customers that X has on December 31, fractions a_{11} stay with X, a_{21} move to Y, and a_{31} move to Z. Since we made the basic assumption that all X's customers are still supplied by X, Y, or Z, this means that the sum of these fractions is unity:

$$a_{11} + a_{21} + a_{31} = 1.$$

Similarly, the sum of the elements in each column of **A** is unity:

$$\sum_{i=1}^{3} a_{ij} = 1 \qquad (j = 1, 2, 3). \tag{2.6}$$

The matrix **A** is known as a *transition* matrix. Note that in much of the literature the transition matrix is defined as the transpose of the matrix **A** introduced above.

We now assume that the fractions of customers who change suppliers in the following months are the same as those given for the first month by the matrix **A**. If the months January, February, March, . . . are denoted by numbers 1, 2, 3 . . . , respectively, and the column vectors giving the fractions of the total number of customers supplied by X, Y, Z at the ends of each month are denoted by $\mathbf{x}_r$, then in addition to (2.2) we have $\mathbf{x}_2 = \mathbf{A}\mathbf{x}_1$, $\mathbf{x}_3 = \mathbf{A}\mathbf{x}_2, \ldots$ and, in general,

$$\mathbf{x}_r = \mathbf{A}\mathbf{x}_{r-1} \qquad (r = 1, 2, 3, \ldots). \tag{2.7}$$

The fractions $\mathbf{x}_r$ at the end of the rth month are easily expressed in terms of those at the end of the first month by substituting for $\mathbf{x}_{r-1}$ in (2.7) in terms of $\mathbf{x}_{r-2}$ and so on:

$$\mathbf{x}_r = \mathbf{A}\mathbf{x}_{r-1} = \mathbf{A}^2\mathbf{x}_{r-2} = \mathbf{A}^3\mathbf{x}_{r-3} = \cdots = \mathbf{A}^r\mathbf{x}_0. \tag{2.8}$$

We now show that if $x_0 + y_0 + z_0 = 1$, then

$$x_r + y_r + z_r = 1 \qquad (r = 1, 2, 3, \ldots), \tag{2.9}$$

where the x_r, y_r, z_r are the components of $\mathbf{x}_r$ as defined in (2.3), and represent fractions of customers supplied by X, Y, Z, respectively, at the end of the rth month. The above equation (2.9) is simply a consistency check. We know that this relation must be true by virtue of the physical interpretation of x_r, y_r, z_r as fractions of a fixed number of customers. On the other hand, the x_r, y_r, z_r are defined by (2.7) so that (2.9) must be deducible from this definition. From (2.7)

$$x_r = a_{11}x_{r-1} + a_{12}y_{r-1} + a_{13}z_{r-1},$$

with two similar equations for y_r and z_r. Adding these three equations we obtain

$$\begin{aligned} x_r + y_r + z_r &= (a_{11} + a_{21} + a_{31})x_{r-1} + (a_{12} + a_{22} + a_{32})y_{r-1} \\ &\quad + (a_{13} + a_{23} + a_{33})z_{r-1} \\ &= x_{r-1} + y_{r-1} + z_{r-1} \\ &= x_0 + y_0 + z_0 = 1, \end{aligned}$$

where we have used (2.6) and (2.1). This proves (2.9).

At this stage we consider a numerical example. Suppose that

$$\mathbf{x}_0 = \begin{bmatrix} 0.2 \\ 0.3 \\ 0.5 \end{bmatrix}, \qquad \mathbf{A} = \begin{bmatrix} 0.8 & 0.2 & 0.1 \\ 0.1 & 0.7 & 0.3 \\ 0.1 & 0.1 & 0.6 \end{bmatrix}. \tag{2.10}$$

On December 31, the dairies X, Y, Z have, respectively, 20%, 30%, 50% of the market. In subsequent months, assuming that the transition matrix does not change, the shares are

$$\mathbf{x}_1 = \mathbf{A}\mathbf{x}_0 = \begin{bmatrix} 0.27 \\ 0.38 \\ 0.35 \end{bmatrix}, \qquad \mathbf{x}_2 = \mathbf{A}\mathbf{x}_1 = \begin{bmatrix} 0.327 \\ 0.398 \\ 0.275 \end{bmatrix}.$$

Similarly, we find, rounding to three decimals at each stage,

$$\mathbf{x}_4 = \begin{bmatrix} 0.397 \\ 0.384 \\ 0.219 \end{bmatrix}, \qquad \mathbf{x}_8 = \begin{bmatrix} 0.442 \\ 0.357 \\ 0.201 \end{bmatrix}, \qquad \mathbf{x}_{16} = \begin{bmatrix} 0.450 \\ 0.350 \\ 0.200 \end{bmatrix}. \tag{2.11}$$

On forming $\mathbf{x}_{17} = \mathbf{A}\mathbf{x}_{16}$, we find that $\mathbf{x}_{17}$ is precisely the same as $\mathbf{x}_{16}$, so that $\mathbf{x}_r$ is the same for all values of r greater than 16. This means that the fractions of the market supplied by X, Y, Z become constant after a sufficiently long time.

To investigate this in the case of a general 3×3 matrix $\mathbf{A} = [a_{ij}]$, where the a_{ij} satisfy (2.6), suppose that for large r the vector $\mathbf{x}_r$ tends to a limiting vector $\mathbf{x}$ where $\mathbf{x}^T = [x, y, z]$. The vector $\mathbf{x}$ must satisfy the equation obtained by setting $\mathbf{x}_{r-1} = \mathbf{x}_r = \mathbf{x}$ in (2.7)—i.e., $\mathbf{x} = \mathbf{A}\mathbf{x}$. Writing these equations out in detail and rearranging slightly, we obtain

$$\begin{aligned} (1 - a_{11})x - a_{12}y - a_{13}z &= 0, \\ -a_{21}x + (1 - a_{22})y - a_{23}z &= 0, \\ -a_{31}x - a_{32}y + (1 - a_{33})z &= 0. \end{aligned} \tag{2.12}$$

In addition, since (2.9) is true for all r, it must also be true for the limiting value as r tends to infinity, so that

$$x + y + z = 1. \tag{2.13}$$

At first sight it seems that (2.12), (2.13) are four equations for three unknowns, but it is apparent from (2.6) that the sum of the three equations in (2.12) is identically zero, so that one of the equations is redundant. The question of how to choose an independent set of equations under these circumstances is discussed in detail later. In the present situation it is clear that we should take (2.13) with two equations from (2.12). For example, with $\mathbf{A}$ defined in (2.10), taking (2.13) with the first two equations in (2.12), we obtain

$$\begin{aligned} x + y + z &= 1 \\ +0.2x - 0.2y - 0.1z &= 0 \\ -0.1x + 0.3y - 0.3z &= 0. \end{aligned} \tag{2.14}$$

The solution of this system is

$$x = 0.45, \qquad y = 0.35, \qquad z = 0.20.$$

This is precisely the result obtained by finding the limit of $\mathbf{A}^r\mathbf{x}_0$ for large r, above [see (2.11)].

One important result of this analysis is that the limiting value $\mathbf{x}$ is completely independent of the initial starting value $\mathbf{x}_0$. This is clear from the way in which $\mathbf{x}$ was found, since Equations (2.12), (2.13) do not depend on $\mathbf{x}_0$.

We found the limiting value of $\mathbf{x}_r$ by computing $\mathbf{A}^r\mathbf{x}_0$ for increasing values of r [see (2.11)]. To understand the final result from a different point of view, we next consider the computed values of $\mathbf{A}^r$ for various r. We find

$$\mathbf{A}^2 = \begin{bmatrix} 0.67 & 0.31 & 0.20 \\ 0.18 & 0.54 & 0.40 \\ 0.15 & 0.15 & 0.40 \end{bmatrix}, \qquad \mathbf{A}^4 = \begin{bmatrix} 0.536 & 0.405 & 0.338 \\ 0.278 & 0.407 & 0.412 \\ 0.188 & 0.188 & 0.250 \end{bmatrix}$$

$$\mathbf{A}^8 = \begin{bmatrix} 0.462 & 0.445 & 0.432 \\ 0.339 & 0.356 & 0.365 \\ 0.199 & 0.199 & 0.203 \end{bmatrix}, \qquad \mathbf{A}^{16} = \begin{bmatrix} 0.450 & 0.450 & 0.450 \\ 0.350 & 0.350 & 0.350 \\ 0.200 & 0.200 & 0.200 \end{bmatrix}.$$

Plainly, as r increases, $\mathbf{A}^r$ tends to a matrix with constant rows; a proof of this for our problem when the matrix $\mathbf{A}$ has strictly positive elements is outlined in Ex. 2.3 below. The behavior of the powers of a matrix $\mathbf{A}$ is very important in a variety of applications; it will arise again, for example, in Section 2.3. In Section 10.5 we will finally have enough tools to allow us to settle this question in general. Returning to the particular problem at hand, we can see that the knowledge that $\mathbf{A}^r$ converges to a matrix with constant rows allows us to show that $\mathbf{x}_r$, which is just $\mathbf{A}^r\mathbf{x}_0$, tends to a limit as r tends to infinity and that this limit is independent of the initial vector $\mathbf{x}_0$. To see this, suppose that

$$\lim_{r\to\infty} \mathbf{A}^r = \begin{bmatrix} \alpha & \alpha & \alpha \\ \beta & \beta & \beta \\ \gamma & \gamma & \gamma \end{bmatrix}.$$

Then for any arbitrary initial vector x_0, we see that

$$\lim_{r\to 0} \mathbf{A}^r\mathbf{x}_0 = \begin{bmatrix} \alpha(x_0 + y_0 + z_0) \\ \beta(x_0 + y_0 + z_0) \\ \gamma(x_0 + y_0 + z_0) \end{bmatrix} = \begin{bmatrix} \alpha \\ \beta \\ \gamma \end{bmatrix}. \tag{2.15}$$

Hence the limiting value of $\mathbf{A}^r\mathbf{x}_0$ as r tends to infinity is independent of the initial vector $\mathbf{x}_0$.

We now return to the specific application considered at the beginning of this section. In order to visualize the results more concretely, suppose that there are 1000 customers altogether on December 31. If the distribution of customers on that date is given by x_0 in (2.10) then X has 200, Y has 300, Z has 500. The analysis following (2.10) then gives the information in Table 2.1 for the distribution of customers at the end of January and February, and also the final equilibrium distribution.

Table 2.1

			Gains			Losses			
Dairy	Customers (Dec. 31)	Customers Retained	From X	From Y	From Z	To X	To Y	To Z	Customers (Jan. 31)
X	200	160	0	60	50	0	20	20	270
Y	300	210	20	0	150	60	0	30	380
Z	500	300	20	30	0	50	150	0	350
	(Jan. 31)								(Feb. 28)
X	270	216	0	76	35	0	27	27	327
Y	380	266	27	0	105	76	0	38	398
Z	350	210	27	38	0	35	105	0	275
				Equilibrium					
X	450	360	0	70	20	0	45	45	450
Y	350	245	45	0	60	70	0	35	350
Z	200	120	45	35	0	20	60	0	200

We see that a fairly complicated interchange of customers between dairies is involved. Z's share of the market is decreasing because the dairy is losing a large proportion of its customers to Y. In practice this would presumably have a significant application in connection with Z's marketing strategy. Dairy Z should try to find out why it is losing customers to Y in order to reduce this loss. The mathematics can indicate the effect of reducing the loss by any given amount. Thus if Z's loss to Y is halved, the transition matrix **A** becomes

$$\mathbf{A} = \begin{bmatrix} 0.8 & 0.2 & 0.1 \\ 0.1 & 0.7 & 0.15 \\ 0.1 & 0.1 & 0.75 \end{bmatrix},$$

and we can readily show that the equilibrium value of Z's share of the market will increase to $\frac{2}{7} = 0.29$ in place of 0.20. Note that although Y's share of the market increases appreciably during January (see Table 2.1), this is misleading since its share eventually drops back to an equilibrium fraction of 0.35.

To sum up, we see that this type of analysis will predict the share of the market a dairy will have at any future time. The precise details of how the dairy will gain and lose customers is predicted. Also predictions can be made concerning the behavior of the market if any changes occur in the fractions of customers that dairies retain or lose. Note, however, that it is necessary to have detailed knowledge of the situation. It is not sufficient to know that dairy X is gaining a net number of 70 customers in January. We need to know that it is retaining 160 out of 200, and gaining 60 from Y, 50 from Z, and so on.

The problem we are considering here is a simple example of a (first-order) Markov chain process. Suppose that a system involves p variables (e.g., the above example involves the $p = 3$ fractions of the market served by dairies X, Y, Z). The state of the system at any instant is described by a *state vector*, say $\mathbf{x}$, which is a $p \times 1$ vector giving the values of the p variables. We are interested in the state vector only at successive instants of time $t_1, t_2, \ldots$. If the corresponding state vectors are denoted by $\mathbf{x}_1, \mathbf{x}_2, \ldots$, the basic assumption is that the state vector $\mathbf{x}_r$ depends only on $\mathbf{x}_{r-1}$ and not on $\mathbf{x}_{r-2}, \mathbf{x}_{r-3}, \ldots$, that is, if $\mathbf{x}_{r-1}$ is known, then $\mathbf{x}_r$ can be found. As illustrated in the example, the vectors $\mathbf{x}_r, \mathbf{x}_{r-1}$ are related by a transition matrix $\mathbf{A}$ by means of which we can write $\mathbf{x}_r = \mathbf{A}\mathbf{x}_{r-1}$. The elements of $\mathbf{A}$ are usually interpreted as transition probabilities, but further discussion lies outside the scope of this section.

EXERCISE 2.1. Three companies A, B, C simultaneously introduce new brands of toothpaste on the market. At the start, the shares of the market are: A, 0.4; B, 0.2; C, 0.4. During the first year, company A retained 85% of its customers, lost 5% to B, 10% to C. Company B retained 75%, and lost 15% to A, 10% to C. Company C retained 90% and lost 5% to A, 5% to B. Assume that buying habits do not change and that the market does not expand or contract. What share of the market will be held by each company after the end of one and two years? What will the final equilibrium shares be?

EXERCISE 2.2. Assume that a person's occupation can be classified as professional, skilled, or unskilled. Assume that 70% of the children of professional people are also professional, 20% are skilled, and 10% are unskilled. Similarly, suppose that 60% of the children of skilled people are also skilled, 20% are professional, and 20% are unskilled. Also 50% of the children of unskilled people are unskilled, 30% are skilled, and 20% are professional. Assume that every person has a child. Set up the transition matrix from one generation to the next. Show that the fractions of the grandchildren of unskilled people who are professional, skilled, and unskilled are 0.30, 0.37, 0.33, respectively. Show that the equilibrium fractions of professional, skilled, and unskilled people are 0.40, 0.37, 0.23, approximately.

EXERCISE 2.3. Let $\mathbf{A} = [a_{ij}]$ be a 3×3 matrix whose elements satisfy (2.4)–(2.6), with the additional restriction that its elements are strictly positive—i.e.,

$$a_{ij} \geq d > 0 \qquad \text{(all } i, j\text{)}.$$

Prove that the following result is true where α, β, γ are constants:

$$\lim_{r\to\infty} \mathbf{A}^r = \begin{bmatrix} \alpha & \alpha & \alpha \\ \beta & \beta & \beta \\ \gamma & \gamma & \gamma \end{bmatrix}. \tag{2.16}$$

SOLUTION: The following proof is elementary but tedious. A straightforward and elegant proof for $n \times n$ matrices will be possible when we have more technical knowledge at our disposal (see Theorem 10.9). Suppose that $\mathbf{A}^r = [p_{ij}]$. Obviously all the p_{ij}'s are positive. The elements of the first row of $(\mathbf{A}^r)\mathbf{A}$ are given by

$$p_{11}a_{1j} + p_{12}a_{2j} + p_{13}a_{3j} \qquad (j = 1, 2, 3).$$

Suppose that M_{r+1}, m_{r+1} are, respectively, the largest and smallest elements in this row. Suppose also that M_r, m_r are the largest and smallest elements in the first row of $\mathbf{A}^r$. We have

$$\begin{aligned} m_{r+1} &= \min_j \{p_{11}a_{1j} + p_{12}a_{2j} + p_{13}a_{3j}\} \\ &= \min_j \{M_r d + (p_{11}a_{1j} + p_{12}a_{2j} + p_{13}p_{3j} - M_r d)\}. \end{aligned} \tag{2.17}$$

The value of M_r is equal to the value of one of the p's, say p_{1k}. We have

$$p_{1k}a_{kj} - M_r d = M_r(a_{kj} - d) \geq m_r(a_{kj} - d).$$

If the two suffixes other than k are denoted by m, n (i.e., m, n, k are 1, 2, 3 in some order) we have

$$\begin{aligned} p_{11}a_{1j} + p_{12}a_{2j} + p_{13}a_{3j} - M_r d &= (p_{1m}a_{mj} + p_{1n}a_{nj}) + (p_{1k}a_{kj} - M_r d) \\ &\geq m_r(a_{mj} + a_{nj}) + m_r(a_{kj} - d) \\ &= m_r(a_{1j} + a_{2j} + a_{3j} - d) \\ &= m_r(1 - d). \end{aligned}$$

Show then that

$$m_{r+1} \geq M_r d + m_r(1 - d). \tag{2.18}$$

By a similar argument, show that

$$M_{r+1} \leq m_r d + M_r(1 - d) \tag{2.19}$$

and then deduce that

$$M_{r+1} - m_{r+1} \leq (1 - 2d)(M_r - m_r). \tag{2.20}$$

Conclude that

$$M_r - m_r \leq (1 - 2d)^r(M_0 - m_0). \tag{2.21}$$

Since $a_{1j} + a_{2j} + a_{3j} = 1$, deduce that the value of d must be less than one-third, leading to the conclusion that

$$\lim_{r \to \infty} (M_r - m_r) = 0. \tag{2.22}$$

On using the notation k, m, n for suffixes introduced above,

$$\begin{aligned} M_{r+1} - M_r &= p_{1m}a_{mj} + p_{1n}a_{nj} + p_{1k}a_{kj} - p_{1k} \\ &= (p_{1m} - p_{1k})a_{mj} + (p_{1n} - p_{1k})a_{nj} \leq 0. \end{aligned}$$

Hence

$$M_0 \geq M_1 \geq M_2 \geq \cdots$$

and, similarly,

$$m_0 \leq m_1 \leq m_2 \leq \cdots.$$

Each sequence is monotone and bounded, since $m_0 \leq m_r \leq M_r \leq M_0$. Hence each of the sequences will have a limit. From (2.22) these limits must be the same. Hence, as indicated in (2.16), we have proved that the elements in the first row of $\mathbf{A}^r$ tend to the same constant value as r tends to infinity. The constancy of the second and third rows in (2.16) is proved similarly.

EXERCISE 2.4. Generalize the theorem in Ex. 2.3 to the case where $\mathbf{A}$ is $n \times n$, assuming that the elements in $\mathbf{A}$ satisfy the same type of conditions as in Ex. 2.3.

2.3 A SIMPLE MODEL OF GROWTH

We want to model and analyze the growth of populations over the passage of time. Imagine that we count the population at certain discrete points in time, such as every month, or every year, or every second, etc., and we let p_i denote the number of individuals in the population at the ith point in time. We are not concerned here with the precise nature of these individuals; they may be people, bacteria, chickens, or whatever. Suppose that somehow, perhaps by observation, we have deduced that the birth rate b and the death rate d are independent of time and that we have somehow found the values of b and d. Thus we assume that the number of individuals born between the ith and $(i + 1)$ points in time is just bp_i, while the number that die is just dp_i.

Algebraically, we can write this assumption as $p_{i+1} - p_i = bp_i - dp_i$, so that

$$p_{i+1} = (1 + b - d)p_i. \tag{2.23}$$

Since (2.23) holds for all i, we of course have

$$p_{i+1} = (1 + b - d)p_i = (1 + b - d)[(1 + b - d)p_{i-1}] = (1 + b - d)^2 p_{i-1},$$

and proceeding in this manner find that

$$p_{i+1} = (1 + b - d)^i p_1, \tag{2.24}$$

where p_1 is the population at the start of our consideration. From (2.24) we easily analyze the behavior of the population over time: if the birth rate exceeds the death rate, then $1 + b - d > 1$ and p_i tends to infinity with i since $(1 + b - d)^i$ does so; if the death rate exceeds the birth rate, however, then $1 + b - d < 1$ and p_i tends to zero.

For example, if (the birth rate) $b = 0.2$ while (the death rate) $d = 0.1$, then $1 + b - d = 1.1$ and starting with $p_1 = 10{,}000$ we find approximately $p_2 = 11{,}000$, $p_3 = 12{,}100$, $p_4 = 13{,}331$, $p_5 = 14{,}641$, $p_{10} = 21{,}435$, $p_{15} = 31{,}384$, $p_{60} = 970{,}137$, $p_{100} = 20{,}483{,}147$, etc.; on the other hand, if $b = 0.1$ while $d = 0.2$, then $1 + b - d = 0.9$ and starting with $p_1 = 10{,}000$ we find approximately $p_2 = 9000$, $p_3 = 8100$, $p_4 = 7290$, $p_5 = 6561$, $p_{10} = 4304$, $p_{15} = 2824$, $p_{60} = 63$, $p_{100} = 2$, etc.

Note that our model is just a model, and we need not get precise quantitative results on actual behavior from it; in particular, if $1 + b - d$ equals, say, 0.9, and $p_1 = 10$, we quickly find that our population contains 8.1 individuals at the third time instant, requiring someone to have a badly split personality. In order to resolve this problem, we only need recall, however, that birth and death rates are merely rough estimates of what actually happens.

Let us now consider a more complex example involving two different populations that interact with one another. We denote the numbers in these populations by F_i and C_i, where we think of F_i as the number of foxes and C_i as the number of chickens. We assume that the chickens, without any foxes to harass them, would have a birth rate exceeding the death rate; for example, we might imagine that $C_{i+1} = 1.2C_i$. Without chickens to feast upon, however, we expect the death rate among foxes to exceed the birth rate; for example, we might have $F_{i+1} = 0.6F_i$. We want to model what happens now when foxes sometimes succeed in killing chickens. Presumably this allows an increase in the fox population proportional to the number of chickens available, so that F_{i+1} might become $0.6F_i + 0.5C_i$, for example. The chickens, on the other hand, would have a depleted population because of the marauding foxes, so that C_{i+1} might become $1.2C_i - kF_i$, where k reflects the kill rate of the chickens by the foxes; we have left k arbitrary to allow us to study the effect on the model of changing the kill rate. If we suppose, specifically, that there are initially 1000 chickens and 100 foxes, we have the model

$$\begin{aligned} F_{i+1} &= 0.6F_i + 0.5C_i \\ C_{i+1} &= -kF_i + 1.2C_i \qquad (\text{for } i \geq 1), \end{aligned} \tag{2.25}$$

with $F_1 = 100$ and $C_1 = 1000$. We want to analyze the behavior of these populations as time passes.

Our first step is to rewrite (2.25) in matrix notation. Let

$$\mathbf{x}_i = \begin{bmatrix} F_i \\ C_i \end{bmatrix},$$

$$\mathbf{A} = \begin{bmatrix} 0.6 & 0.5 \\ -k & 1.2 \end{bmatrix},$$

so that our model becomes

$$\mathbf{x}_{i+1} = \mathbf{A}\mathbf{x}_i \quad (\text{for} \quad i \geq 1), \qquad \mathbf{x}_1 = \begin{bmatrix} 100 \\ 1000 \end{bmatrix}. \tag{2.26}$$

From (2.26) it of course easily follows that

$$\mathbf{x}_{i+1} = \mathbf{A}^i\mathbf{x}_1,$$

so that our study of the behavior of $\mathbf{x}_i$ is equivalent to the study of the behavior of the powers $\mathbf{A}^i$, just as we found in Section 2.2. For the moment we will simply experiment, reserving the precise analysis of this behavior for Section 10.5. It seems intuitively reasonable that if the kill rate is low enough, the chicken population will grow without bound, thus allowing the fox population to explode also. If the kill rate is too large, however, the chickens should be wiped out, and then the foxes will have brought about their own destruction as well.

To demonstrate, we first consider an example with $k = 0.1$, so that

$$\mathbf{A} = \begin{bmatrix} 0.6 & 0.5 \\ -0.1 & 1.2 \end{bmatrix}.$$

The information in Table 2.2 shows roughly the numbers F_i and C_i of foxes and chickens at the ith point in time. Experimentally, we verify what we expected: for a low kill rate, both F_i and C_i tend to infinity. We even find that the populations eventually become equal.

Table 2.2

i	1	2	3	4	5	6	8
F_i	100	560	931	1244	1523	1783	2292
C_i	1000	1190	1372	1553	1739	1934	2367
i	12	16	20	30	100		
F_i	3470	5107	7483	19,409	15,328,199		
C_i	3488	5111	7483	19,409	15,328,199		

For our second example, we consider a larger kill rate, say $k = 0.18$, so that now

$$\mathbf{A} = \begin{bmatrix} 0.6 & 0.5 \\ -0.18 & 1.2 \end{bmatrix}.$$

The information in Table 2.3 shows the development of the populations throughout time. Again the experiment has verified our intuitive expectation: for a higher kill rate, both the chicken and the fox populations die out despite an initial expansion.

Table 2.3

i	1	2	3	4	5	8	12
F_i	100	560	927	1214	1434	1808	1854
C_i	1000	1182	1317	1413	1477	1530	1400
i	16	20	30	40	60	80	100
F_i	1654	1371	713	312	43	3	0
C_i	1177	940	459	193	25	2	0

Obviously, more complex relationships among several populations could be modeled and studied in this same fashion. The key to predicting the population behavior throughout time is to understand the behavior of matrix powers; eventually we will develop this understanding (Chapter 10).

EXERCISE 2.5. The model described by (2.25) is unrealistic when the kill rate is too large; in particular, it sometimes leads to negative values for F_i or C_i, impossible in reality. To illustrate this, let $k = 0.6$ and compute the resulting "population" as shown in Tables 2.2 and 2.3. To illustrate another shortcoming of the model, explain why the equation $F_{i+1} = 0.6\,F_i + 0.5C_i$ in (2.25) should probably be replaced by one like $F_{i+1} = 0.6\,F_i + 5k\,C_i$. What effect would this have on the analysis?

EXERCISE 2.6. Suppose that p_i is the population of a society at the ith point in time. Suppose, as in our development of (2.23), that the number of deaths between the ith and $(i + 1)$ time instants is dp_i, proportional to the size of the population. Suppose, however, that the society is "promiscuous" in the sense that the number of births is proportional to the total number of possible pairings between two members of the society. Show that such a model leads to an equation of the form

$$p_{i+1} = p_i - dp_i + b\frac{p_i(p_i - 1)}{2}.$$

EXERCISE 2.7. As in Ex. 2.6, find an interpretation for the population model

$$p_{i+1} = p_i + bp_i - d\frac{p_i(p_i - 1)}{2}.$$

EXERCISE 2.8. To explain the behavior exhibited in Table 2.2, show from (2.26) that when $k = 0.1$ we have

$$(F_{i+1} - C_{i+1}) = 0.7(F_i - C_i) \quad \text{and} \quad (5C_{i+1} - F_{i+1}) = 1.1(5C_i - F_i).$$

Show then that $F_i - C_i$ tends to zero, that $5C_i - F_i$ tends to infinity, and therefore that F_i and C_i tend to infinity.

EXERCISE 2.9. To explain the behavior exhibited in Table 2.3, show from (2.26) that when $k = 0.18$ we have

$$(0.1C_{i+1} - 0.06F_{i+1}) = 0.9(0.1C_i - 0.06F_i)$$

and

$$(2C_{i+1} - F_{i+1}) = 0.9(2C_i - F_i) + (0.1C_i - 0.06F_i).$$

Show then that $0.1C_i - 0.06F_i$ tends to zero, that $2C_i - F_i$ tends to zero, and therefore that F_i and C_i tend to zero.

2.4 A PLANE PIN-JOINTED FRAMEWORK

In this section we show how the calculation of strains and stresses in a simple type of mechanical structure can be facilitated by using matrices. We solve a simple problem longhand, first of all, by writing out all the equations in detail. We then introduce matrices to express the manipulations in a compact form.

If we wish to skip the technical details of the formulation, we note that the basic equations in longhand form are (2.28), (2.29), (2.30), and the corresponding matrix equations are (2.34), (2.35), (2.36). The final equations that must be solved are (2.32) in longhand, or (2.37) in matrices.

Consider the simple plane framework shown in Figure 2.1. We assume that the framework is *pin-jointed*, so that the members are connected to the wall and connected together at A, loosely, by pins. Thus if the five members were not connected to the wall, they could rotate quite freely around the point A. With this kind of joint there is no tendency to bend the members, and the only forces present are tensile or compressive forces in the members. (In technical language, no bending moments are transmitted.) Let us assume that the weights of the members are negligible and that the lengths of the members are such that, if the external forces F_1 and F_2 are zero, there are no stresses in the members. We wish to determine the motion of the point A when forces are applied.

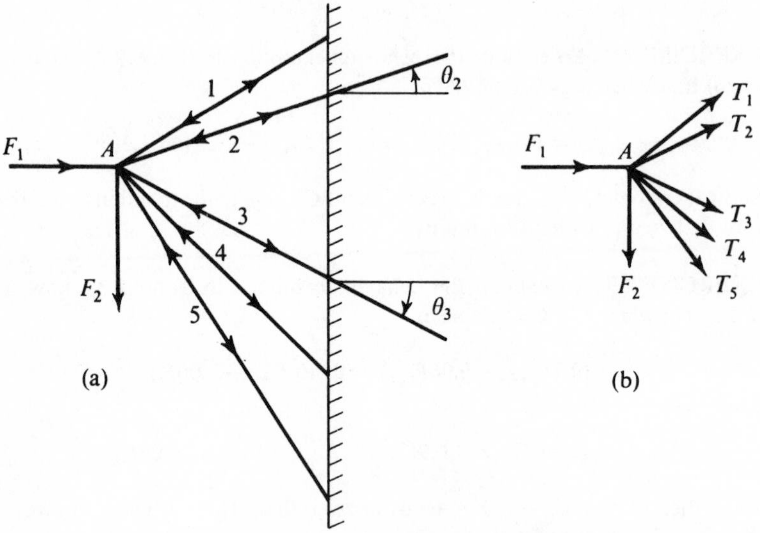

Figure 2.1. A simple plane framework.

The only information required about an individual member is knowledge of its extension when a force is applied along its length. We assume Hooke's law, namely, that the extension is linearly proportional to the force, so that

$$e = kT, \tag{2.27}$$

where T is the force (tension), e is the extension, and k, the *flexibility*, is a factor of proportionality, assumed known.

We number the members 1 to 5 as shown in Figure 2.1, and assume that the angle each member makes with the horizontal is given by θ_i ($i = 1$ to 5). (θ_3, θ_4, θ_5 are negative in the example drawn in Figure 2.1.) The changes in the θ_i produced by the application of forces are considered negligible. Let us denote the flexibilities of the members by k_i, the tensions by T_i, and the extensions by e_i ($i = 1$ to 5). The externally applied forces are denoted by F_1, F_2, in the directions shown.

The equations of force-equilibrium at the joint A are obtained by resolving all forces horizontally and vertically. The arrows in Figure 2.1(a) indicate that the tension in a rod is positive when the rod is being extended, negative when the rod is being compressed. The force diagram at point A is therefore given by Figure 2.1(b), which leads to the following equations of equilibrium:

$$\begin{aligned} -T_1 \cos\theta_1 - T_2 \cos\theta_2 - \cdots - T_5 \cos\theta_5 &= F_1, \\ +T_1 \sin\theta_1 + T_2 \sin\theta_2 + \cdots + T_5 \sin\theta_5 &= F_2. \end{aligned} \tag{2.28}$$

Suppose that, when forces are applied to the framework, point A moves by a distance d_1 horizontally and d_2 vertically, measured in the same directions as the corresponding forces F_1, F_2; our problem is to solve for d_1 and d_2 since these numbers describe the motion of A. The extensions e_i are given in terms of the d_i by the following formulas:

$$e_i = -d_1 \cos\theta_i + d_2 \sin\theta_i \qquad (i = 1, 2, \ldots, 5). \tag{2.29}$$

These equations simply state that the extension of the ith member is given by adding the components obtained by resolving d_1, d_2 along the rod, taking account of sign.

Equations (2.28) give relations between internal and external forces. Equations (2.29) give relations between internal and external displacements. To complete the solution we require relations between internal forces and internal displacements. These are given by writing down Hooke's law (2.27) for each member:

$$e_i = k_i T_i \qquad (i = 1, 2, \ldots, 5). \tag{2.30}$$

We have now derived the basic equations for the problem, namely (2.28) (2.29), (2.30). These are twelve equations in the twelve unknowns T_i ($i = 1$ to 5), e_i ($i = 1$ to 5), and d_i ($i = 1, 2$).

One method of solution is to eliminate the unknown forces T_i and strains e_i, and derive simultaneous linear equations for the unknown displacements d_1, d_2. From (2.29), (2.30).

$$T_i = \frac{-d_1 \cos\theta_i + d_2 \sin\theta_i}{k_i}. \tag{2.31}$$

These values for T_i are then substituted in (2.28) to give two simultaneous equations for d_1, d_2. Equations (2.28) are, using summation signs,

$$-\sum_{i=1}^{5} T_i \cos\theta_i = F_1,$$

$$\sum_{i=1}^{5} T_i \sin\theta_i = F_2.$$

Substitution for T_i from (2.31) gives

$$\begin{aligned} a_{11}d_1 + a_{12}d_2 &= F_1, \\ a_{21}d_1 + a_{22}d_2 &= F_2, \end{aligned} \tag{2.32}$$

where

$$a_{11} = \sum_{i=1}^{5} \frac{\cos^2\theta_i}{k_i}, \qquad a_{22} = \sum_{i=1}^{5} \frac{\sin^2\theta_i}{k_i},$$

$$a_{12} = a_{21} = -\sum_{i=1}^{5} \frac{\cos\theta_i \sin\theta_i}{k_i}.$$

Equations (2.32) are two equations for the two unknowns d_1, d_2. When d_1, d_2 are found from these equations, the tensions in the members can be found from (2.31).

We now express these manipulations in terms of matrices. Define

$$\mathbf{t} = \begin{bmatrix} T_1 \\ T_2 \\ \cdot \\ \cdot \\ \cdot \\ T_5 \end{bmatrix}, \qquad \mathbf{e} = \begin{bmatrix} e_1 \\ e_2 \\ \cdot \\ \cdot \\ \cdot \\ e_5 \end{bmatrix}, \qquad \mathbf{f} = \begin{bmatrix} F_1 \\ F_2 \end{bmatrix}, \qquad \mathbf{d} = \begin{bmatrix} d_1 \\ d_2 \end{bmatrix},$$

$$\mathbf{A} = \begin{bmatrix} -\cos\theta_1, & -\cos\theta_2, & \ldots, & -\cos\theta_5 \\ \sin\theta_1, & \sin\theta_2, & \ldots, & \sin\theta_5 \end{bmatrix}, \tag{2.33}$$

$$\mathbf{K} = \begin{bmatrix} k_1 & 0 & \cdots & 0 \\ 0 & k_2 & \cdots & 0 \\ & & \cdots & \\ 0 & 0 & \cdots & k_5 \end{bmatrix}.$$

Then (2.28), (2.29), (2.30) become, in matrix notation,

$$\mathbf{At} = \mathbf{f}, \tag{2.34}$$

$$\mathbf{e} = \mathbf{A}^T\mathbf{d}, \tag{2.35}$$

$$\mathbf{e} = \mathbf{K}\mathbf{t}. \tag{2.36}$$

The matrix $\mathbf{A}^T$, which occurs in (2.35), is precisely the transpose of the matrix $\mathbf{A}$ in (2.34). This relationship did not spring to our attention when we wrote out the original equations in longhand form (2.28), (2.29); it is a bonus that we receive merely by writing everything in matrix notation. We can use this relationship to check that we have written down (2.28), (2.29) correctly from physical reasoning; this is useful since it is easy to confuse signs. Alternatively, if we are sure that (2.28) is correct, we need not go through a detailed derivation of (2.29) at all.

We wish to set up equations for the desired unknowns $\mathbf{d}$. The $\mathbf{e}$ and $\mathbf{f}$ are also unknowns, so we eliminate these in the following way. From (2.34), (2.36), (2.35), in succession,

$$\mathbf{f} = \mathbf{A}\mathbf{t} = \mathbf{A}\mathbf{K}^{-1}\mathbf{e} = \mathbf{A}\mathbf{K}^{-1}\mathbf{A}^T\mathbf{d},$$

that is,

$$(\mathbf{A}\mathbf{K}^{-1}\mathbf{A}^T)\mathbf{d} = \mathbf{f}. \tag{2.37}$$

These are precisely (2.32). We are of course doing nothing new here; we are merely carrying out in matrix notation the same steps that we previously performed longhand. However, matrix notation makes the structure of the calculation extremely clear.

Although we have confined our attention to a simple specific example, the same procedure can be carried through for more complicated structures. We can generalize the method to deal with three-dimensional structures with many members and many connections (instead of the single connection at A in Figure 2.1). The connections can be rigid as well as pin-jointed, in which case we need to introduce moments and angular displacements. In all cases, we end up with three sets of equations analogous to (2.34)–(2.36). The individual matrices may be very complicated, but the *form* of the matrix equations (2.34)–(2.36) is always the same. It provides a uniform starting point for the analysis of structures, whatever the method used actually to solve the equations.

The above matrix approach to frameworks can be used in many other disciplines where complex systems must be analyzed. The underlying mathematical equations are nearly identical to those for frameworks if, for example, in electrical engineering we attempt to analyze an electrical network or in economics we study the transportation of goods between producers and consumers. In these latter examples, of course, the concepts of electronics or economics replace the corresponding concepts of strain, stress, displacement, etc., in the mechanical structure. Some of these correspondences are indicated in Table 2.4.

Table 2.4

Frameworks Concept	Economics Concept	Electronics Concept
Strain, **e**	Price difference	Voltage difference
Stress, **t**	Flow in branch	Current in branch
Displacement, **d**	Price	Voltage
Force, **f**	Flow at node	Current at node

EXERCISE 2.10. Suppose goods are being produced and consumed in four towns, numbered 1 to 4, connected by six roads, 1 to 6, as illustrated diagrammatically in Figure 2.2. We arbitrarily assign directions along each of the roads as shown, for example, by the arrows in Figure 2.2. We assume that goods flow along branch r at a rate f_r per unit time and that the excess of goods being produced in town s over goods consumed in s, per unit time, is F_s. If no goods are destroyed, F_s must equal the flow of goods leaving along the roads—e.g., for town 2,

$$F_2 = f_2 - f_3 - f_5.$$

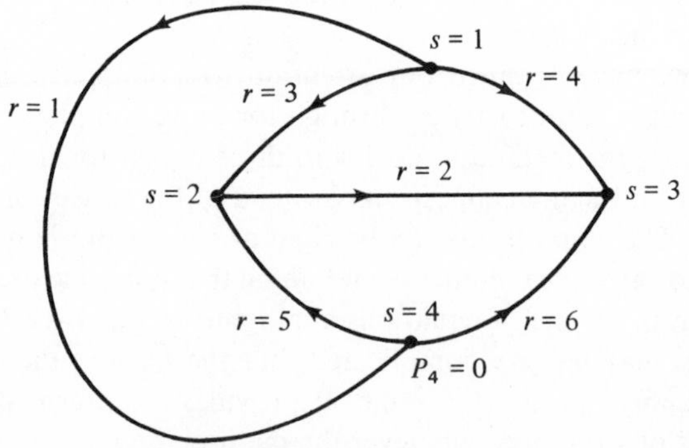

Figure 2.2. A transportation problem.

Suppose that the price of a unit of goods at town s is P_s. Since only price differentials will be important, we can choose the price at one of the towns (namely 4 in Figure 2.2) to be zero and measure all prices relative to this zero level. Denote the price difference between the two towns at the ends of branch r by p_r. Relations between P_s and p_r can be written down directly; thus, for branch 2,

$$P_2 - P_3 = p_2.$$

Show that

$$\mathbf{F} = \mathbf{A}\mathbf{f},$$

$$\mathbf{p} = \mathbf{A}^T\mathbf{P},$$

where

$$\mathbf{F} = \begin{bmatrix} F_1 \\ F_2 \\ F_3 \end{bmatrix}, \quad \mathbf{f} = \begin{bmatrix} f_1 \\ f_2 \\ \cdot \\ \cdot \\ \cdot \\ f_6 \end{bmatrix}, \quad \mathbf{P} = \begin{bmatrix} P_1 \\ P_2 \\ P_3 \end{bmatrix}, \quad \mathbf{p} = \begin{bmatrix} p_1 \\ p_2 \\ \cdot \\ \cdot \\ \cdot \\ p_6 \end{bmatrix},$$

$$\mathbf{A} = \begin{bmatrix} 1 & 0 & 1 & 1 & 0 & 0 \\ 0 & 1 & -1 & 0 & -1 & 0 \\ 0 & -1 & 0 & -1 & 0 & -1 \end{bmatrix}.$$

Complete the analysis by specifying a relationship between $\mathbf{p}$ and $\mathbf{f}$ [compare (2.36)].

The convenience of the matrix analysis of networks stems mainly from (2.37). In practice, of course, for each given network problem the matrix $\mathbf{A}\mathbf{K}^{-1}\mathbf{A}^T$ of (2.37) must be computed and then the system of equations in (2.37) must be solved. Before the advent of modern computers, these computations were performed by hand, and many special tricks were developed to reduce the work involved for special networks; today's computers, however, are so powerful that, instead of special tricks for special problems, what is really needed is a general representation and method for solving general network problems. Our matrix methods are admirably suited for this, since in matrix notation all networks look essentially alike. For the computer to work efficiently, it must apply a method that is systematic, routine, and as general as possible. For network analysis, the computer can be programmed to take a description of the network, for example in terms of such properties of its members as flexibility and force; and from this it can be programmed to generate $\mathbf{K}^{-1}$ automatically, to generate the matrix $\mathbf{A}$ describing the network's operation, to multiply matrices so as to compute the matrix $\mathbf{A}\mathbf{K}^{-1}\mathbf{A}^T$ of (2.37), and finally to solve efficiently and accurately the equations of (2.37). Matrix notation allows quite general networks to be analyzed in this way. Note that crucial steps in this process are the efficient and accurate automatic computation of matrix inverses and of solutions of systems of linear equations. We study these problems in Chapter 6.

EXERCISE 2.11. In the pin-jointed square framework shown in Figure 2.3 the support of A is fixed (though of course the rods are free to rotate), and the support at B can move parallel to the wall. Let

$$\mathbf{f} = [F_i], \qquad \mathbf{t} = [T_i], \qquad \mathbf{d} = [d_i], \qquad \mathbf{e} = [e_i]$$

denote, respectively, external forces, internal tensions in rods, displacements of joints, and extensions—all in the directions shown. Show that

$$\mathbf{f} = \mathbf{CT}, \qquad \mathbf{e} = \mathbf{C}^T\mathbf{d},$$

where

$$C = \begin{bmatrix} 0 & 0 & 0 & 1 & 0 & 1/\sqrt{2} \\ 1 & 0 & 0 & 0 & 0 & 1/\sqrt{2} \\ 0 & 0 & -1 & 0 & -(1/\sqrt{2}) & 0 \\ 0 & 0 & 0 & 1 & 1/\sqrt{2} & 0 \\ 0 & -1 & 0 & 0 & 0 & -(1/\sqrt{2}) \end{bmatrix}.$$

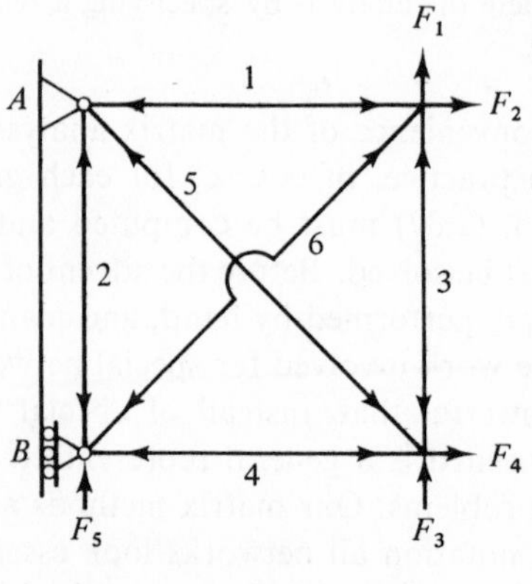

Figure 2.3. A square framework.

EXERCISE 2.12. Any electrical network in which there are two input and two output terminals is called a *two-port* or *fourpole*. Show that, for the series impedance in Figure 2.4(b), we have

$$\begin{bmatrix} v_1 \\ i_1 \end{bmatrix} = \begin{bmatrix} 1 & Z \\ 0 & 1 \end{bmatrix} \begin{bmatrix} v_2 \\ i_2 \end{bmatrix}$$

with Z = impedance and

$$Y = \text{admittance} = \frac{1}{\text{impedance}},$$

and, for the shunt admittance in Figure 2.4(c), we have

$$\begin{bmatrix} v_1 \\ i_1 \end{bmatrix} = \begin{bmatrix} 1 & 0 \\ Y & 1 \end{bmatrix} \begin{bmatrix} v_2 \\ i_2 \end{bmatrix}.$$

If three two-ports are connected in series as in Figure 2.5(a) and the equations relating the input and output voltages and currents are

$$\mathbf{P}_s = \mathbf{A}_s\mathbf{p}_{s+1}, \qquad s = 1, 2, 3, \quad \text{where} \quad \mathbf{p}_s = \begin{bmatrix} v_s \\ i_s \end{bmatrix},$$

show that $\mathbf{p}_1 = \mathbf{A}_1\mathbf{A}_2\mathbf{A}_3\mathbf{p}_4$. Show that the matrix $\mathbf{A}_1\mathbf{A}_2\mathbf{A}_3$ for the "T" and "Π" networks in Figures 2.5(b), (c) are those given on the right of the corresponding diagrams.

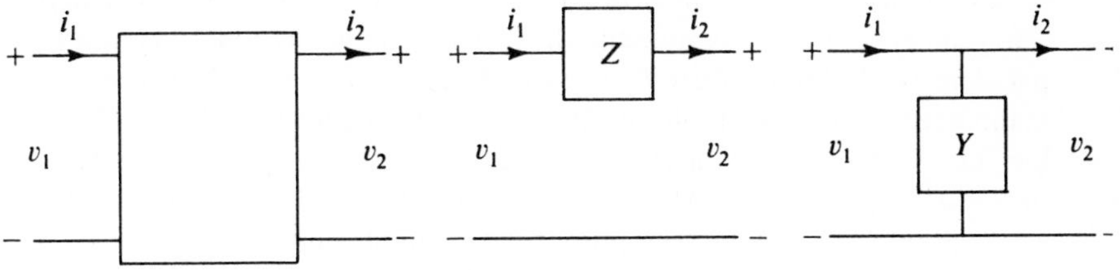

(a) (b) (c)

Figure 2.4. Two-ports.

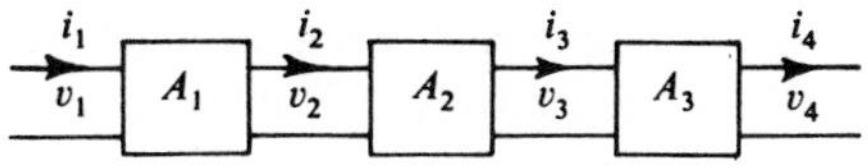

(a)

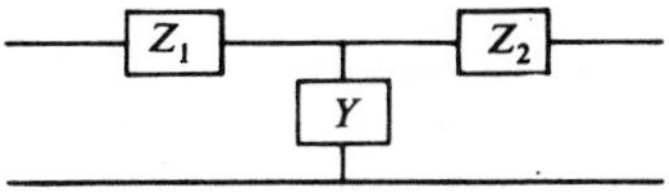

$$\begin{bmatrix} 1 + Z_1Y, & Z_1 + Z_2 + Z_1Z_2Y \\ Y\ , & 1 + Z_2Y \end{bmatrix}$$

(b)

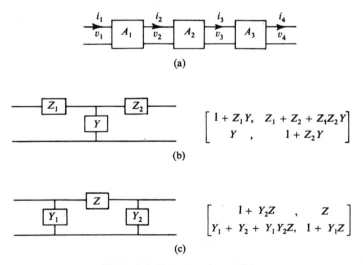

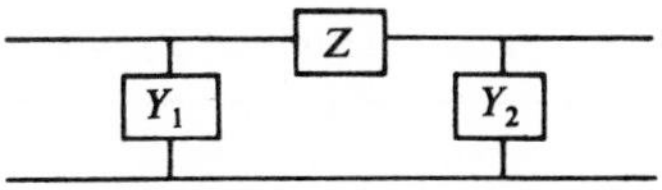

$$\begin{bmatrix} 1 + Y_2Z\ , & Z \\ Y_1 + Y_2 + Y_1Y_2Z, & 1 + Y_1Z \end{bmatrix}$$

(c)

Figure 2.5. Two-ports in a series.

2.5 PRODUCTION PLANNING

In trying to decide how to allocate its manufacturing capability among various products, an industry is usually influenced by many factors, including, of course, the desire to make a reasonable profit. We now consider a very simplified model of such a situation in which we assume that the *sole* desire is to maximize profit.

Suppose that an industrial plant has three types (M_1, M_2, and M_3) of machines, each of which must be used in manufacturing the plant's products, of which there are two types (P_1 and P_2). The problem is to decide how many of each product to produce each week so as to maximize weekly profits. We assume that a fixed profit is made on each unit of each product manufactured, so that the total profit is simply the sum of the profits on each type P_1 and P_2, each of which profit is simply obtained by multiplying the profit per item by the number of items manufactured. Specifically, we assume that the profit per item made for product P_1 is \$40 while that for P_2 is \$60. Clearly our manufacturer should simply produce as much as possible; the word "possible" here is the key, for the manufacturer obviously is limited by the capabilities of the types M_1, M_2, M_3 of machines which must be used. Thus we must assume that we know the amount of time available on each machine and also the amount of time required on each machine to make each type of product. Specifically, suppose that an item of P_1 requires two hours on machines of type M_1 and one hour each on M_2 and M_3, while an item of P_2 requires one hour on each of M_1 and M_2 but three hours on M_3. In addition, suppose that the number of hours available each week on machines of types M_1, M_2, and M_3 are 70, 40, and 90, respectively. All these assumptions are summarized in Table 2.5.

Table 2.5

Machine Type	Hours Needed by One Unit of P_1	Hours Needed by One Unit of P_2	Total Hours Available
M_1	2	1	70
M_1	1	1	40
M_3	1	3	90
Profit per unit of P_1 = \$40		Profit per unit of P_2 = \$60	

Next, let x_1 denote the number of units of P_1 to be produced each week, while x_2 denotes the number of units of P_2. Since each unit of P_1 requires two hours on machines of type M_1 while each unit of P_2 requires one such hour, we will require $2x_1 + x_2$ hours on machines of type M_1; since only 70 hours are available on these machines, we see that we must require

$$2x_1 + x_2 \leq 70.$$

Reasoning similarly from the limited time on machines of type M_2 and M_3 and from the time needed to produce x_1 of P_1 and x_2 of P_2, we see also that we must require

$$x_1 + x_2 \leq 40$$
$$x_1 + 3x_2 \leq 90.$$

Since it is impossible to produce a negative number of units, we also must require

$$x_1 \geq 0, \qquad x_2 \geq 0.$$

To compute the profit resulting from our production plan, recall that the profit on each unit of P_1 is 40 so that x_1 such units yield a profit of $40x_1$; similarly, x_2 units of P_2 yield a profit of $60x_2$, so that our total profit is $40x_1 + 60x_2$. Thus the mathematical version of our planning problem is as follows:

$$\text{maximize} \quad M = 40x_1 + 60x_2 \tag{2.38}$$

where x_1 and x_2 must satisfy the constraints

$$\begin{aligned} 2x_1 + x_2 &\leq 70 \\ x_1 + x_2 &\leq 40 \\ x_1 + 3x_2 &\leq 90 \\ x_1 &\geq 0 \\ x_2 &\geq 0. \end{aligned} \tag{2.39}$$

This is a typical *linear programming problem*, involving the optimization of a certain linear function of some unknowns which are subject to linear constraints restricting the permissible values of the unknowns; we will study such problems closely in Chapter 7, where we will learn how to solve them by matrix methods. For the present we limit ourselves to rewriting the linear program in matrix notation and to using graphical methods to solve the problem.

In order to express the above equations in matrix notation, we require the following concepts. A matrix $\mathbf{P}$ is said to be *greater than* $\mathbf{Q}$, written $\mathbf{P} > \mathbf{Q}$, when $\mathbf{P}$ and $\mathbf{Q}$ have the same numbers of rows and columns, and each element of $\mathbf{P}$ is greater than the corresponding element of $\mathbf{Q}$. Similar definitions hold for $\geq$, $<$, and $\leq$. If $\mathbf{P} > \mathbf{0}$, we say that $\mathbf{P}$ is *positive*. If $\mathbf{P} \geq \mathbf{0}$, we say that $\mathbf{P}$ is *nonnegative*.

If we introduce

$$\mathbf{A} = \begin{bmatrix} 2 & 1 \\ 1 & 1 \\ 1 & 3 \end{bmatrix}, \qquad \mathbf{b} = \begin{bmatrix} 70 \\ 40 \\ 90 \end{bmatrix}, \qquad \mathbf{c} = \begin{bmatrix} 40 \\ 60 \end{bmatrix}, \qquad \mathbf{x} = \begin{bmatrix} x_1 \\ x_2 \end{bmatrix}, \tag{2.40}$$

then our problem described in (2.38), (2.39) can be denoted easily as:

$$\text{maximize} \quad M = \mathbf{c}^T\mathbf{x} \tag{2.41}$$

where $\mathbf{x}$ must satisfy the constraints

$$\mathbf{A}\mathbf{x} \leq \mathbf{b}, \qquad \mathbf{x} \geq \mathbf{0}. \tag{2.42}$$

(Strictly speaking, $\mathbf{c}^T\mathbf{x}$ is a 1×1 matrix, but we adopt the convention that $\mathbf{c}^T\mathbf{x}$ can also be used to denote the element of the matrix.)

The same *form* of equations will hold if we have m different machines producing n products. Then $\mathbf{A} = [a_{ij}]$ is an $m \times n$ matrix, and a_{ij} represents the number of hours on machine i required to produce one unit of product j. The total hours available will be an $m \times 1$ column matrix, and the profit matrix $\mathbf{c}$ and the matrix $\mathbf{x}$ representing the numbers of units produced will be $n \times 1$ column matrices.

EXERCISE 2.13. A balanced diet must contain minimum quantities of nutrients, such as vitamins, minerals, carbohydrates, and so on. Suppose that we wish to determine, from a given number of foods, the lowest-cost diet that satisfies the minimum requirements for a balanced diet. As a specific example, Table 2.6 gives the numbers of units of nutrients 1, 2 contained in three foods 1, 2, 3, together with the costs of the three foods and the minimum number of units of the nutrients required in the balanced diet. If we buy amounts y_i ($i = 1, 2, 3$) of the foods, then we must have

$$2y_1 + y_2 + y_3 \geq 40,$$
$$y_1 + y_2 + 3y_3 \geq 60,$$
$$y_1 \geq 0, \qquad y_2 \geq 0, \qquad y_3 \geq 0,$$

and we wish to *minimize*

$$M = 70y_1 + 40y_2 + 90y_3.$$

Table 2.6

	Food 1	Food 2	Food 3	Minimum Number of Units Required
Nutrient 1	2	1	1	40
Nutrient 2	1	1	3	60
Cost	70	40	90	

In the general case, let a_{ji} be the number of units of nutrient i in one unit of food j. Suppose that y_j units of food j are to be bought, the cost of one unit of food j being b_j. The diet must supply at least c_i units of nutrient i. In an obvious matrix notation, show that we must have

$$\mathbf{A}^T\mathbf{y} \geq \mathbf{c}, \qquad \mathbf{y} \geq 0, \tag{2.43}$$

and that, subject to these conditions, we wish to minimize

$$M = \mathbf{b}^T\mathbf{y}. \tag{2.44}$$

Note that (2.43), (2.44) are analogous to (2.41), (2.42), but one of the inequality signs has been reversed. Also, we minimize (2.44), compared with maximizing (2.41).

(Note that in a practical problem we should take other factors into account. Thus we should provide variety in the diet, and ensure that it is palatable, and so on. Some of these factors can be built into the above formulation of the linear programming method of approach. The size and complexity of realistic models means that it is essential to use a digital computer to solve the resulting equations.)

In Chapter 7 we will learn matrix methods for solving general linear programs of the form in (2.41), (2.42). For the present we describe the specific problem in (2.38), (2.39) geometrically (or graphically) so that perhaps we can visualize matters more easily; actually, we will be able to solve the problem by this approach as well.

To deal with the first inequality in (2.39), we note that $2x_1 + x_2 = 70$ is the equation of a straight line (see Figure 2.6). The inequality $2x_1 + x_2 \leq 70$ means that the point (x_1, x_2) must lie *below* the straight line. Similarly, the other two inequalities in (2.39) define half-planes in which the permissible points (x_1, x_2) must lie. The inequalities $x_1 \geq 0$, $x_2 \geq 0$ mean that (x_1, x_2) must lie in the first quadrant. The net result is that the inequalities confine the point (x_1, x_2) to a polygonal region whose boundary is shaded in Figure 2.6. For any given value of M, the equation (2.38), namely $M = 40x_1 + 60x_2$, defines a straight line. The three dashed lines in Figure 2.6 represent this straight line for $M = 0$, 1200, 2400. The value of M corresponding to any point on a given line is a constant, and the lines are parallel. To maximize M we must go as far as possible in a direction perpendicular to these lines—in the direction of the arrow (near the origin) in Figure 2.6—without leaving the admissible region. In this way, we reach the point P which is the intersection of

$$x_1 + 3x_2 = 90, \qquad x_1 + x_2 = 40,$$

that is, $x_1 = 15$, $x_2 = 25$. The corresponding value of M is 2100. From the

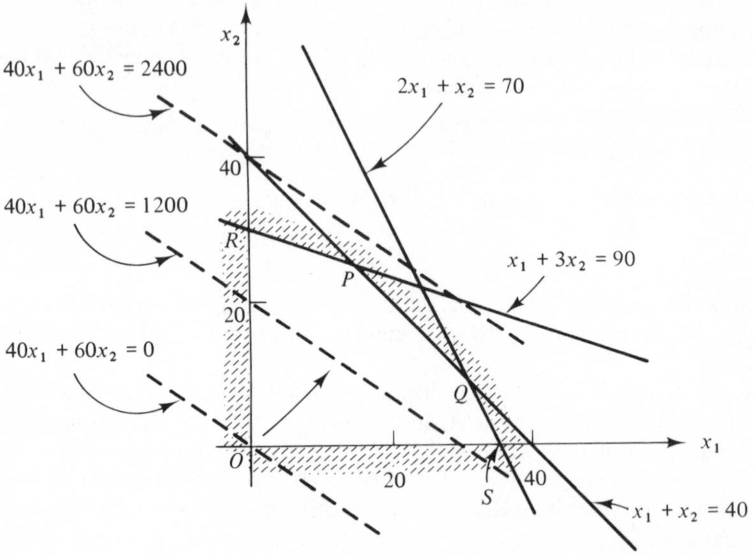

Figure 2.6.

geometrical picture it is obvious that two of the equations in (2.39) are now equalities and one is a strict inequality. In terms of our original production planning problem, we obtain the maximum profit of \$2100 weekly by producing 15 units of P_1 and 25 units of P_2 each week; the machines of type M_2 and M_3 are used to fullest capacity each week, but only 55 hours of time on M_1 machines are used, 15 hours less than capacity. Our manufacturer should probably consider selling some of the M_1 machines since M_1 machine capacity can be reduced with no reduction in profit (see Ex. 2.15).

As we remarked at the close of Section 2.4, the power of matrix representation lies in its ability to denote complex relationships in a simple fashion, thus revealing the common nature of diverse problems. A great variety of practical problems lead to linear programs; matrices allow a common representation for these problems and also allow a common attack on them by computer programs designed to solve the general problem.

EXERCISE 2.14. Suppose that the manufacturer in our main example described in Table 2.5 decides to eliminate unneeded capacity and to reduce to 55 the number of hours per week available on M_1 machines; this of course does not change the maximum profit. Suppose that the money earned by selling these few M_1 machines is used to purchase M_3 machines so as to increase the available M_3 machine time by 10 hours weekly to a total of 100 hours. Find the new optimal production schedule and the associated maximum profit. If instead the money is used to purchase M_2 machines so as to increase available M_2 machine time by 10 hours, then what happens?

EXERCISE 2.15. Suppose that m products are produced by n factories. Let a_{ji} be the number of units of product i produced by factory j in one day, where the production pattern of each factor is fixed. We wish to produce at least c_i units of product i. The cost of keeping factory j in production is b_j per day. We wish to determine the numbers of days y_j that each factory should operate in order to produce at least c_i units of product i, at minimum cost. Show that (2.43), (2.44) express this problem in matrix notation.

2.6 THE METHOD OF LEAST SQUARES

In the applied sciences, we try to express the relationship between variable quantities by means of mathematical expressions. For example, if a body is traveling with constant velocity v, the relation between the time t and the position reached by the body y is given by the linear equation

$$y = \alpha + vt. \tag{2.45}$$

Suppose that we measure y at various times t and obtain the following figures:

t	0	3	5	8	10
y	2	5	6	9	11

These are plotted in Figure 2.7 where it is seen that the points lie approximately on a straight line. There are two reasons why the points may not lie *exactly* on a straight line:

1. Errors of measurement.
2. The velocity may not be constant, so that (2.45) may not be true.

Let us ignore this second possibility—i.e., we will assume that if y and t could be measured exactly, then the relation between y and t would be given by (2.45). Since measurements are subject to experimental error, equation (2.45) is not satisfied exactly by the measured values of y and t, and it is not possible to deduce exact values of α and v in (2.45) from the data. The problem we now wish to consider is how to deduce the "best possible" values of α, v or, in other words, how to fit the "best" straight line to the data plotted in Figure 2.7.

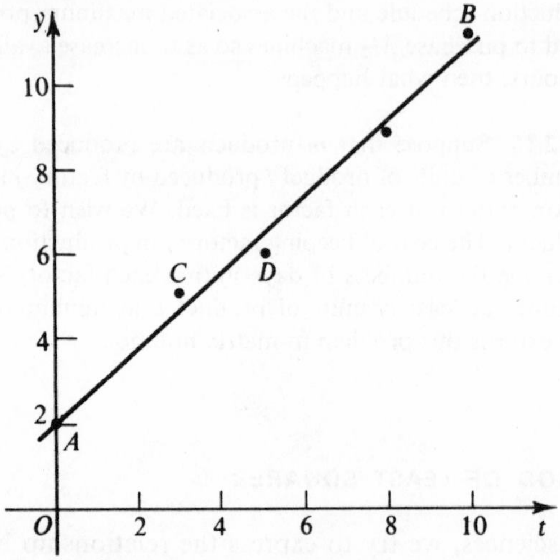

Figure 2.7. Straight-line fitting of data.

If we denote the value of y measured at a time $t = t_i$ by y_i, then the relation between y_i and t_i can be written as [compare (2.45)]

$$y_i = \alpha + vt_i + r_i, \tag{2.46}$$

where r_i is a *residual* resulting from errors in measurement. The criterion we will use to find the unknown parameters α, v is that we minimize the sum of squares of the residuals (hence the term "least squares"):

$$S = \sum_{i=1}^{m} r_i^2 = \sum_{i=1}^{m} (y_i - \alpha - vt_i)^2, \tag{2.47}$$

where m denotes the total number of observations. The residuals r_i are, geometrically, the distances between the points and the line $y = \alpha + vt$, measured parallel to the y-axis in Figure 2.7. It is intuitively reasonable for us to minimize the sum of squares to find the "best" straight line, since this ensures that the line will pass, in some sense, through the middle of the points.

Minimization of (2.47) with respect to α and v gives, as necessary conditions,

$$\begin{aligned} \frac{\partial S}{\partial \alpha} &= -2\sum_{i=1}^{m} (y_i - \alpha - vt_i) = 0, \\ \frac{\partial S}{\partial v} &= -2\sum_{i=1}^{m} t_i(y_i - \alpha - vt_i) = 0. \end{aligned} \tag{2.48}$$

(The reader unfamiliar with calculus should go to the next paragraph.) On rearranging, we obtain two equations in the two unknowns α, v,

$$\begin{aligned} p_{11}\alpha + p_{12}v &= q_1, \\ p_{21}\alpha + p_{22}v &= q_2, \end{aligned} \tag{2.49}$$

where

$$\begin{aligned} &p_{11} = m, \qquad p_{12} = p_{21} = \sum_{i=1}^{m} t_i, \\ &p_{22} = \sum_{i=1}^{m} t_i^2, \qquad q_1 = \sum_{i=1}^{m} y_i, \qquad q_2 = \sum_{i=1}^{m} t_i y_i. \end{aligned} \tag{2.50}$$

Alternatively, we can derive (2.49) without the use of calculus. From (2.47), we have

$$S = \sum_{i=1}^{m} [y_i^2 + \alpha^2 + v^2 t_i^2 - 2\alpha y_i - 2v y_i t_i + 2\alpha v_i t_i] \tag{2.51}$$

which, when we collect terms and use (2.50), becomes

$$S = p_{11}\alpha^2 - 2q_1\alpha + p_{22}v^2 - 2q_2 v + (p_{12} + p_{21})\alpha v + \sum_{i=1}^{m} y_i^2.$$

If we complete the square with respect to α only, this becomes

$$\begin{aligned} S = p_{11}\left\{\alpha + \frac{[-2q_1 + 2p_{12}v]}{2p_{11}}\right\}^2 + \sum_{i=1}^{m} y_i^2 + p_{22}v^2 - 2q_2 v \\ - \frac{[-2q_1 + 2p_{12}v]^2}{4p_{11}}. \end{aligned} \tag{2.52}$$

Since the only term in S in (2.52) involving α is the one enclosed in brackets $\{\ \}$, and since we are to minimize with respect to α, we must choose α so that $\{\ \} = 0$, that is, so that the first equation in (2.49) is satisfied. Similarly, by completing the square with respect to v only, we obtain the second equation in (2.49).

We now express the above analysis in matrix notation. Equation (2.46) can be written

$$\mathbf{y} = \mathbf{A}\mathbf{x} + \mathbf{r}, \tag{2.53}$$

where

$$\mathbf{y} = \begin{bmatrix} y_1 \\ y_2 \\ \cdot \\ \cdot \\ \cdot \\ y_m \end{bmatrix}, \qquad \mathbf{A} = \begin{bmatrix} 1 & t_1 \\ 1 & t_2 \\ \cdot & \cdot \\ \cdot & \cdot \\ \cdot & \cdot \\ 1 & t_m \end{bmatrix}, \qquad \mathbf{x} = \begin{bmatrix} \alpha \\ v \end{bmatrix}, \qquad \mathbf{r} = \begin{bmatrix} r_1 \\ r_2 \\ \cdot \\ \cdot \\ \cdot \\ r_m \end{bmatrix}.$$

In this notation, (2.47) is

$$S = \mathbf{r}^T\mathbf{r} = (\mathbf{y} - \mathbf{A}\mathbf{x})^T(\mathbf{y} - \mathbf{A}\mathbf{x}). \tag{2.54}$$

The definitions (2.50) become

$$\begin{bmatrix} p_{11} & p_{12} \\ p_{21} & p_{22} \end{bmatrix} = \mathbf{A}^T\mathbf{A}, \qquad \begin{bmatrix} q_1 \\ q_2 \end{bmatrix} = \mathbf{A}^T\mathbf{y}. \tag{2.55}$$

Equations (2.49) are then

$$\mathbf{A}^T\mathbf{A}\mathbf{x} = \mathbf{A}^T\mathbf{y}.$$

Summarizing the situation, the original equations (2.53) are usually written as simply

$$\mathbf{A}\mathbf{x} = \mathbf{y} \qquad (y \text{ known}), \tag{2.56}$$

where the error term $\mathbf{r}$ is now omitted. These are a set of m equations in two unknowns, and it will not usually be possible to find $\mathbf{x}$ such that all the equations will be satisfied. They are in general an *inconsistent* set of equations. We conclude from the above analysis that the least squares solution of this set is given by solving

$$\mathbf{A}^T\mathbf{A}\mathbf{x} = \mathbf{A}^T\mathbf{y}, \tag{2.57}$$

which is a set of two equations in two unknowns. *Formally*, (2.57) is obtained by premultiplying (2.56) by $\mathbf{A}^T$.

For the numbers in the example considered at the beginning of this section, we find that (2.57) becomes

$$\begin{bmatrix} 5 & 25 \\ 26 & 198 \end{bmatrix}\begin{bmatrix} \alpha \\ v \end{bmatrix} = \begin{bmatrix} 33 \\ 227 \end{bmatrix}. \tag{2.58}$$

These give $\alpha = 2.01$, $v = 0.88$. The corresponding line is drawn in Figure 2.7.

The above analysis is readily generalized to the case where the dependent variable y depends on n independent variables $t^{(1)}, t^{(2)}, \ldots, t^{(n)}$, where we use an upper suffix to denote different variables. Equation (2.45) becomes

$$y = x_0 + x_1t^{(1)} + x_2t^{(2)} + \cdots + x_nt^{(n)},$$

where, instead of α, v in (2.45) we now have $n + 1$ unknown parameters $x_0, x_1, \ldots, x_n$. Suppose that we have m different measurements of y, each corresponding to a different set of values of the $t^{(j)}$, so that, instead of (2.46) we have

$$y_i = x_0 + x_1 t_i^{(1)} + x_2 t_i^{(2)} + \cdots + x_n t_i^{(n)} + r_i,$$

for $i = 1, 2, \ldots, m$. For simplicity we introduce the notation $t_i^{(j)} = a_{ij}$; then this set of equations can be written in exactly the same form as (2.53), namely

$$\mathbf{y} = \mathbf{Ax} + \mathbf{r},$$

where $\mathbf{r}$ is the same as before, but

$$\mathbf{x} = \begin{bmatrix} x_0 \\ x_1 \\ \cdot \\ \cdot \\ \cdot \\ x_n \end{bmatrix}, \qquad \mathbf{A} = \begin{bmatrix} 1 & a_{11} & \cdots & a_{1n} \\ 1 & a_{21} & \cdots & a_{2n} \\ & & \cdots & \\ 1 & a_{m1} & \cdots & a_{mn} \end{bmatrix}. \tag{2.59}$$

The reader can generalize the previous results, step by step. The sum S of squares of residuals is defined, as before, by

$$S = (\mathbf{y} - \mathbf{Ax})^T(\mathbf{y} - \mathbf{Ax}). \tag{2.60}$$

Instead of (2.48) we have (also, see Ex. 2.16 for "noncalculus" approach):

$$\frac{\partial S}{\partial x_k} = 0 \qquad (k = 0, 1, 2, \ldots, n). \tag{2.61}$$

This gives a set of $n + 1$ equations in $n + 1$ unknowns. In matrix notation the final set of equations is identical with (2.57) in form, namely

$$\mathbf{A}^T\mathbf{Ax} = \mathbf{A}^T\mathbf{y}. \tag{2.62}$$

This again illustrates one of the virtues of matrix notation, namely that the appearance of the final expressions does not depend on the numbers of variables involved—i.e., the complexity of the problem. The notation provides a simple expression for the least squares solution,

$$\mathbf{x} = (\mathbf{A}^T\mathbf{A})^{-1}\mathbf{A}^T\mathbf{y}, \tag{2.63}$$

when the inverse exists; in some cases, however, $\mathbf{A}^T\mathbf{A}$ will not have an inverse. Even when $\mathbf{A}^T\mathbf{A}$ is invertible, computationally one usually obtains large numerical errors from using (2.63) directly. To obtain useful numerical answers, we must use more sophisticated techniques. We return to this matter in Sections 9.5 and 9.6.

EXERCISE 2.16. Derive (2.62) without calculus by "completing the square" with only one variable at a time. More precisely let $\mathbf{x}_1$ minimize

$$(\mathbf{y} - \mathbf{A}\mathbf{x})^T(\mathbf{y} - \mathbf{A}\mathbf{x})$$

with respect to $\mathbf{x}$; then for each unit column vector $\mathbf{e}_j$ (Definition 1.13), if q_1 minimizes $[\mathbf{y} - \mathbf{A}(\mathbf{x}_1 + q\mathbf{e}_j)]^T[\mathbf{y} - \mathbf{A}(\mathbf{x}_1 + q\mathbf{e}_j)]$ with respect to the single scalar variable q, show that we must find that $q_1 = 0$. Expand this quadratic function of q, complete the square with respect to q, conclude that

$$\mathbf{e}_j^T\mathbf{A}^T(\mathbf{A}\mathbf{x}_1 - \mathbf{y}) = 0 \quad \text{for all } j,$$

and finally conclude that $\mathbf{A}^T\mathbf{A}\mathbf{x}_1 = \mathbf{A}^T\mathbf{y}$.

EXERCISE 2.17. Suppose that instead of defining S by (2.47) we set

$$S = \sum_{i=1}^{m} w_i r_i^2, \tag{2.64}$$

where the w_i are given *weighting factors*. (The reason for this terminology will be apparent presently.) Repeat the preceding analysis to show that for α and v we obtain linear equations of the same form as before, namely (2.49), where now

$$p_{11} = \sum_{i=1}^{m} w_i, \qquad p_{12} = p_{21} = \sum_{i=1}^{m} w_i t_i, \ldots$$

with analogous expressions for p_{22}, q_1, and q_2. Hence if, for any value of i, say $i = k$, the factor w_k is large, the contribution of the terms $i = k$ in the sums defining the p_{rs} and q_s will also be large, and if w_k is small, the corresponding contributions will be small. Notice that the kth equation is given more or less weight, depending on whether w_k is large or small, and this is the reason for the term "weighting factor." There are various reasons for giving some equations greater weights than others; for example, we may know that the measurements in some equations are more accurate than in others. Show that the equation analogous to (2.57) for the weighted least-squares solution of $\mathbf{A}\mathbf{x} = \mathbf{y}$ is

$$\mathbf{A}^T\mathbf{W}\mathbf{A}\mathbf{x} = \mathbf{A}^T\mathbf{W}\mathbf{y}, \tag{2.65}$$

where $\mathbf{W}$ is a diagonal matrix of weighting factors:

$$\mathbf{W} = \begin{bmatrix} w_1 & 0 & \cdots & 0 \\ 0 & w_2 & \cdots & 0 \\ & & \cdots & \\ 0 & 0 & \cdots & w_m \end{bmatrix}.$$

EXERCISE 2.18. Clarify the implicit assumptions that have been made in the derivation of (2.49) for the straightline fitting of data y_i measured at times t_i. In particular, discuss the relevance of the following assumptions:

(a) If a measurement were to be made for a given value of t a large number of times, then the average error would be zero.
(b) Measurements made at various values of t have the same accuracy (in statistical language, they have a *common variance*).
(c) Measurements made at various times are independent.
(d) The measurements are made at exactly the times stated—i.e., the error in the measurement of the time t is negligible.

EXERCISE 2.19. Verify the numerical results (2.58).

EXERCISE 2.20. If $\mathbf{x}$ is a $p \times 1$ vector with elements x_i and u is a scalar quantity, we define the following notation:

$$\frac{\partial u}{\partial \mathbf{x}} = \left[\frac{\partial u}{\partial x_i}\right],$$

where this is a $p \times 1$ vector. Show that

(a) If $u = \mathbf{y}^T\mathbf{x}$ where $\mathbf{y}$ is a $p \times 1$ vector, then

$$\frac{\partial u}{\partial \mathbf{x}} = \mathbf{y}.$$

(b) If $\mathbf{A}$ is a $p \times p$ symmetric matrix and $u = \mathbf{x}^T\mathbf{A}\mathbf{x}$, then

$$\frac{\partial u}{\partial \mathbf{x}} = 2\mathbf{A}\mathbf{x}.$$

The condition (2.61) for minimizing S is, in the present notation, $\partial S/\partial \mathbf{x} = \mathbf{0}$. By using (a) and (b), perform this differentiation directly on the expression (2.60) for S, and hence deduce the basic equation (2.62).

MISCELLANEOUS EXERCISES 2

The following essay topics and references to the literature are meant to be suggestive rather than definitive. Projects of this type must take into account the interests of the student and the books available. The essays can be written at various levels of sophistication. The more linear algebra the student knows, the deeper the treatment that can be attempted. The student should not expect to understand everything in the references quoted if he has completed only Chapter 1 of this book. Nevertheless, it is surprising how far matrix algebra alone can take the student.

EXERCISE 2.21. Write an essay on the analysis of mechanical structures by matrix methods. Distinguish between the "equilibrium" and "compatibility" methods of approach to the calculation of structures. Possible references are Argyris [70], Gennaro [80], Pestel and Leckie [104], Robinson [106], and the following:

(a) S. O. Asplund, *Structural Mechanics, Classical and Matrix Methods*, Prentice-Hall (1966).
(b) H. I. Laursen, *Matrix Analysis of Structures*, McGraw-Hill (1966).
(c) R. K. Livesley, *Matrix Methods of Structural Analysis*, Pergamon (1964).
(d) S. J. McMinn, *Matrices for Structural Analysis*, E. & F. Spon, London (1962).
(e) H. C. Martin, *Introduction to Matrix Methods of Structural Analysis*, McGraw-Hill (1966).
(f) P. B. Morice, *Linear Structural Analysis*, Hudson and Thames (1959).
(g) J. Robinson, *Structural Matrix Analysis for the Engineer*, Wiley (1966).
(h) M. F. Rubinstein, *Matrix Computer Analysis of Structures*, Prentice-Hall (1966).
(i) F. Venancio Filho, *Introduction to Matrix Structural Theory and its Applications to Civil and Aircraft Structures*, Ungar (1960).
(j) C. K. Wang, *Matrix Methods of Structural Analysis*, International Textbook Co. (1966).
(k) P. C. Wang, *Numerical Matrix Methods in Structural Mechanics*, Wiley (1966).

EXERCISE 2.22. Write an essay on any aspect of the use of matrices in dynamics. Possible references are Frazer, Duncan, and Collar [38], Goldstein [84], Heading [42], and:

(a) R. M. L. Baker, *Astrodynamics*, Academic (1967).
(b) R. Deutsch, *Orbital Dynamics of Space Vehicles*, Prentice-Hall (1963).
(c) W. C. Nelson and E. E. Loft, *Space Mechanics*, Prentice-Hall (1962).

EXERCISE 2.23. Write an essay on the use of matrix methods in connection with two-port electrical networks, vacuum-tube circuits, and transistor circuits. Possible sources of material are Guillemin [86], Nodelman and Smith [102], v. Weiss [51], and:

(a) A. M. Tropper, *Matrix Theory for Electrical Engineers*, Addison-Wesley and Harrap (1962).
(b) G. Zelinger, *Basic Matrix Algebra and Transistor Circuits*, Pergamon (1966).

EXERCISE 2.24. Write an essay on electrical circuit analysis by matrix methods, with or without special reference to the work of Kron. Possible references are Braae [36], Guillemin [86], Kron [94], [95], [96], LeCorbeiller [98], v. Weiss [51], Zaden and DeSoer [111].

EXERCISE 2.25. Write an essay on the analysis of electric motors using matrices. Possible references are Gibbs [81], and

N. N. Hancock, *Matrix Analysis of Electrical Machinery*, Macmillan (1964).

EXERCISE 2.26. Write a brief essay explaining how the book [73] by Brouwer uses matrices in describing the behavior of optical lenses.

EXERCISE 2.27. Write an essay on the use of matrices to simplify situations involving polarized light based on:

W. A. Shurcliff, *Polarized Light*, Harvard (1962).

EXERCISE 2.28. Write an essay on finite Markov chains. Possible references are [45], [91], [92], [93] by Kemeny et al., and

J. C. Mathews and C. E. Langenhop, *Discrete and Continuous Methods in Applied Mathematics*, Wiley (1966).

EXERCISE 2.29. Write an essay on applications of matrices in the social sciences. Possible references are Johnston, Price, and VanVleck [43], and Kemeny and Snell [91].

EXERCISE 2.30. Write an essay on the use of matrices in economics. A special topic might be Leontief input-output matrix analysis. Possible references are Chenery and Clark [74], Dorfman, Samuelson, and Solow [76], Gale [77], Johnston, Price, and VanVleck [43], Karlin [44], Schwartz [107], and

O. Morganstern, ed., *Economic Analysis Activity*, Wiley (1964).

EXERCISE 2.31. Write an essay on the use of matrices in econometrics. Possible references are Goldberger [83], Johnston [90].

EXERCISE 2.32. Write an essay on applications of matrices in statistics, and in regression. Possible references are Graybill [85], Rao [105], Searle [48] and

(a) N. R. Draper and H. Smith, *Applied Regression Analysis*, Wiley (1962).
(b) O. Kempthorne, *Design and Analysis of Experiments*, Wiley (1952).

EXERCISE 2.33. Write an essay on the uses of matrices in the social sciences based on:

(a) J. Kemeny et al., *Introduction to Finite Mathematics*, 3rd ed., Prentice-Hall (1974).
(b) D. Maki and M. Thompson, *Mathematical Models and Applications*, Prentice-Hall (1973).

CHAPTER THREE

SIMULTANEOUS LINEAR EQUATIONS AND ELEMENTARY OPERATIONS

This chapter is the heart of the entire book. It introduces concepts and techniques that are fundamental to nearly every subsequent chapter. The entire methodology developed in Section 3.2 and generalized in Section 3.4 is the keystone of this structure and therefore must be mastered. Definitions 3.2 and 3.4 are also especially vital. The fundamental results of the chapter are displayed in **Key Theorems 3.4, 3.6, 3.7, 3.8,** and **3.11;** Theorem 3.3 is also designated as **key** since it is heavily used later. The proof of Theorem 3.6 so thoroughly exercises the earlier results and so thoroughly tests our grasp of fundamental concepts that we have labeled it a **key proof.**

3.1 INTRODUCTION

In Chapter 1 we did little more than state certain rules for the manipulation of matrix arrays. In Chapter 2 we illustrated some of the ways in which matrices arise in the applied sciences. In this chapter we study matrices from a more theoretical point of view in order to obtain a deeper understanding of the potentialities and limitations of matrix methods. As illustrated by the examples considered in the last chapter, the utility of matrices in the applied sciences is often connected with the fact that they provide a convenient method for formulating physical problems in terms of a set of simultaneous linear algebraic equations. It is therefore important to understand theoretically the various situations that can arise when solving sets of linear equations. This is the objective of the present chapter.

Each section in this chapter is concerned with ideas that arise naturally from the preceding sections. Sections 3.2–3.3 are concerned mainly with concrete examples; Section 3.4 provides a transition to a more theoretical treatment; Section 3.5 is concerned with basic tools; Section 3.6 applies these tools to determine whether a system of linear equations possesses solutions; and Section 3.7 addresses the related question of the existence of inverses.

3.2 THE METHOD OF GAUSS (OR SUCCESSIVE) ELIMINATION

Consider the following set of equations:

$$\begin{aligned} 2x_1 - 3x_2 + 2x_3 + 5x_4 &= 3 \\ x_1 - x_2 + x_3 + 2x_4 &= 1 \\ 3x_1 + 2x_2 + 2x_3 + x_4 &= 0 \\ x_1 + x_2 - 3x_3 - x_4 &= 0. \end{aligned} \tag{3.1}$$

The *method of Gauss elimination* (or *successive elimination*) consists of reducing this system of four equations in four unknowns to a system of three equations in three unknowns by using one of the equations to eliminate one of the unknowns from the remaining three equations. The resulting set of three equations is reduced to a set of two equations in two unknowns by a similar procedure. The set of two equations is reduced to one equation in one unknown—i.e., one of the unknowns is determined, and the remaining unknowns are found by *back-substitution*, by which we mean that the newly determined unknown is substituted back into the preceding equation, thus allowing the determination of another unknown, and so on; later examples will clarify this entire process.

There is a wide choice as to which unknowns are eliminated and which equations are used to perform the eliminations. We first describe a systematic procedure which uses the first equation to eliminate the first unknown from the remaining three equations, and so on, regardless of whether this is the best procedure in this particular example.

To solve (3.1), we start by using the first equation to eliminate x_1 from the remaining three. Divide the first equation by the coefficient of x_1, obtaining

$$x_1 - 1.5x_2 + x_3 + 2.5x_4 = 1.5. \tag{3.2}$$

Use this equation to eliminate x_1 from the last three equations in (3.1). This may be accomplished in two *apparently* different ways. First, suppose that we solve for x_1 from (3.2) and then substitute this expression for x_1 in the last three equations of (3.1); for example, the second equation of (3.1) becomes

$$(1.5 + 1.5x_2 - x_3 - 2.5x_4) - x_2 + x_3 + 2x_4 = 1,$$

or, after combining terms,

$$0.5x_2 - 0.5x_4 = -0.5.$$

This is one approach for eliminating x_1. Alternatively, we can subtract from each of the later equations an appropriate multiple of the first equation so chosen as to eliminate x_1. For example, if we multiply (in our heads!) the modified first equation of (3.1), namely (3.2), by 0.5 and subtract the result from the second equation, we obtain

$$x_1 - x_2 + x_3 + 2x_4 - 0.5(2x_1 - 3x_2 + 2x_3 + 5x_4) = 1 - 0.5(3),$$

or, after combining terms,

$$0.5x_2 - 0.5x_4 = -0.5,$$

exactly the same result as before. The two procedures *always* give the same result. Since the latter method is easier to organize computationally, not requiring the exchange of terms across the equality sign, we will always use it. Thus, when we say "use equation A to eliminate variable B from equation C", we mean "subtract from equation C that multiple of equation A that will cause the elimination of the explicit appearance of variable B from the resulting new equation." Using this technique on (3.1), we find that (3.1) is replaced by

$$\begin{aligned} x_1 - 1.5x_2 + x_3 + 2.5x_4 &= 1.5 \\ 0.5x_2 \qquad - 0.5x_4 &= -0.5 \\ 6.5x_2 - x_3 - 6.5x_4 &= -4.5 \\ 2.5x_2 - 4x_3 - 3.5x_4 &= -1.5. \end{aligned} \tag{3.3}$$

We replace the second equation by double itself to facilitate the elimination process, obtaining

$$x_2 - x_4 = -1. \tag{3.4}$$

We use this new second equation to eliminate x_2 from the third and fourth equations of (3.3); for example, we replace the third equation by itself minus 6.5 times the newest second equation (3.4). Elimination of x_2 from the final equation results in

$$\begin{aligned} x_1 - 1.5x_2 + x_3 + 2.5x_4 &= 1.5 \\ x_2 \qquad\quad - x_4 &= -1 \\ - x_3 \qquad &= 2 \\ - 4x_3 - x_4 &= 1. \end{aligned} \tag{3.5}$$

Doing this by hand, we of course notice that our third equation in (3.5) luckily does not involve x_4, so that we immediately find $x_3 = -2$. Substituting the value $x_3 = -2$ into the fourth equation of (3.5) gives $x_4 = 7$; substituting the values of x_3 and x_4 into the second equation gives $x_2 = 6$; and finally substituting the values of x_2, x_3, and x_4 into the first equation gives $x_1 = -5$; this process of computing and substituting the values of the variables is called *back-substitution.*

If we were not so observant, however, we would continue our methodical elimination process beyond (3.5). We would multiply the third equation by -1 and then use the result to eliminate x_3 from the last equation. After multiplying the new last equation by -1, we have

$$\begin{aligned} x_1 - 1.5x_2 + x_3 + 2.5x_4 &= 1.5 \\ x_2 \qquad\quad - x_4 &= -1 \\ x_3 \qquad &= -2 \\ x_4 &= 7, \end{aligned} \tag{3.6}$$

from which back-substitution successively generates the (same) values $x_4 = 7$, $x_3 = -2$, $x_2 = 6$, $x_1 = -5$. In general, we cannot hope for the fortuitous elimination of a variable as in (3.5), nor should we expect a computer program to check for such accidents; we therefore prefer the methodical application of elimination until (3.6) is obtained.

In carrying out the elimination procedure described above, there is no need to continue writing down the unknowns x_1, x_2, x_3, x_4, nor the equality signs. The whole procedure can be systematized by operating directly on the following matrix, known as the *augmented matrix*, formed by writing in matrix form the numbers which occur in the original equations (3.1):

$$\begin{bmatrix} 2 & -3 & 2 & 5 & 3 \\ 1 & -1 & 1 & 2 & 1 \\ 3 & 2 & 2 & 1 & 0 \\ 1 & 1 & -3 & -1 & 0 \end{bmatrix}. \tag{3.7}$$

In the notation of partitioned matrices, the augmented matrix for the system $\mathbf{Ax} = \mathbf{b}$ is $[\mathbf{A}, \mathbf{b}]$. Dividing the first row of (3.7) by 2 and subtracting suitable multiples of the result from the other rows, we obtain [compare (3.3)]:

$$\begin{bmatrix} 1 & -1.5 & 1 & 2.5 & 1.5 \\ 0 & 0.5 & 0 & -0.5 & -0.5 \\ 0 & 6.5 & -1 & -6.5 & -4.5 \\ 0 & 2.5 & -4 & -3.5 & -1.5 \end{bmatrix}. \tag{3.8}$$

Similarly, on multiplying the second row by 2 and subtracting suitable multiples of the result from the third and fourth rows, we obtain [compare (3.5)]:

$$\begin{bmatrix} 1 & -1.5 & 1 & 2.5 & 1.5 \\ 0 & 1 & 0 & -1 & -1 \\ 0 & 0 & -1 & 0 & 2 \\ 0 & 0 & -4 & -1 & 1 \end{bmatrix}. \tag{3.9}$$

Finally, multiplying the third row by -1, adding four times the result to the fourth row, and multiplying the result by -1, we obtain [compare (3.6)]:

$$\begin{bmatrix} 1 & -1.5 & 1 & 2.5 & 1.5 \\ 0 & 1 & 0 & -1 & -1 \\ 0 & 0 & 1 & 0 & -2 \\ 0 & 0 & 0 & 1 & 7 \end{bmatrix}. \tag{3.10}$$

Interpreting this matrix in terms of simultaneous linear equations, we find that we have reduced the original equations to the following *triangular set*:

$$\begin{aligned} x_1 - 1.5x_2 + \phantom{0 \cdot{}} x_3 + 2.5x_4 &= 1.5 \\ x_2 + 0 \cdot x_3 - x_4 &= -1 \\ x_3 + 0 \cdot x_4 &= -2 \\ x_4 &= 7, \end{aligned}$$

which is identical with (3.6).

For theoretical reasons it is sometimes convenient to modify the above procedure in the following way. Instead of leaving the back-substitution to the end, we can do it as we go along (in effect) by the following modification, known as the *Gauss-Jordan* method. We explain the procedure in terms of simultaneous linear equations and matrices simultaneously by placing the equations and the corresponding matrices side by side below. After eliminating x_1 from the last three equations in (3.1) we have [compare (3.3), (3.8)]:

$$\begin{aligned} x_1 - 1.5x_2 + x_3 + 2.5x_4 &= 1.5 \\ x_2 \qquad - x_4 &= -1 \\ 6.5x_2 - x_3 - 6.5x_4 &= -4.5 \\ 2.5x_2 - 4x_3 - 3.5x_4 &= -1.5, \end{aligned} \qquad \begin{bmatrix} 1 & -1.5 & 1 & 2.5 & 1.5 \\ 0 & 1 & 0 & -1 & -1 \\ 0 & 6.5 & -1 & -6.5 & -4.5 \\ 0 & 2.5 & -4 & -3.5 & -1.5 \end{bmatrix}. \tag{3.11}$$

We now use the second equation to eliminate x_2 from *all three* of the remaining equations. (In the matrix, we use the second row to reduce the elements in the second column of the remaining rows to zero.)

$$\begin{aligned} x_1 \qquad + x_3 + x_4 &= 0 \\ x_2 \qquad - x_4 &= -1 \\ - x_3 \qquad &= 2 \\ - 4x_3 - x_4 &= 1, \end{aligned} \qquad \begin{bmatrix} 1 & 0 & 1 & 1 & 0 \\ 0 & 1 & 0 & -1 & -1 \\ 0 & 0 & -1 & 0 & 2 \\ 0 & 0 & -4 & -1 & 1 \end{bmatrix}. \tag{3.12}$$

We similarly use the third equation of this new set to eliminate x_3 from *all three* of the remaining equations, and follow the corresponding procedure for the matrix:

$$\begin{aligned} x_1 \qquad + x_4 &= 2 \\ x_2 \qquad - x_4 &= -1 \\ x_3 \qquad &= -2 \\ - x_4 &= -7, \end{aligned} \qquad \begin{bmatrix} 1 & 0 & 0 & 1 & 2 \\ 0 & 1 & 0 & -1 & -1 \\ 0 & 0 & 1 & 0 & -2 \\ 0 & 0 & 0 & -1 & -7 \end{bmatrix}. \tag{3.13}$$

Repetition of the procedure once more gives the final form:

$$\begin{aligned} x_1 \qquad &= -5 \\ x_2 \qquad &= 6 \\ x_3 \qquad &= -2 \\ x_4 &= 7, \end{aligned} \qquad \begin{bmatrix} 1 & 0 & 0 & 0 & -5 \\ 0 & 1 & 0 & 0 & 6 \\ 0 & 0 & 1 & 0 & -2 \\ 0 & 0 & 0 & 1 & 7 \end{bmatrix}. \tag{3.14}$$

This gives the final solution directly.

Summing up, Gauss elimination consists of a reduction to triangular form, followed by back-substitution. The Gauss-Jordan method in essence does the back-substitution as we go along, so that the final solution is obtained directly. If we enumerate the numbers of additions and multiplications involved (see Section 6.2), it is found that Gauss elimination is more efficient than the Gauss-Jordan method in the sense that it requires fewer operations. However, the Gauss-Jordan procedure is often more convenient for *theoretical* discussions, and that is why we introduce it here.

In the above example we work systematically, using the first equation to eliminate the first unknown from the other equations, and so on. At the rth stage we use the rth equation to eliminate the rth unknown from the remaining equations. If we wish at any stage to use a given equation to eliminate one unknown from the other equations, the coefficient of this unknown in that given equation is called the *pivot*. Thus in (3.3) the pivot is the coefficient of x_2 in the second equation, namely 0.5. We emphasize that there is an exact correspondence between the elimination method for the solution of linear equations and row operations on matrices. The pivot that we just defined in connection with the elimination procedure is also, when considering operations on matrices, the element in the matrix that is used to reduce elements in other rows to zero. Thus the pivot used to reduce (3.8) to (3.9) is the (2, 2) element in (3.8), namely 0.5.

There is no need to proceed systematically by always using the *first* equation to eliminate the *first* unknown, as in the above description. In fact a very wide choice of pivots is available. Thus we could use the fourth equation in (3.1) to eliminate x_2 from the first three equations:

$$\begin{aligned} 5x_1 \qquad\quad - 7x_3 + 2x_4 &= 3 \\ 2x_1 \qquad\quad - 2x_3 + x_4 &= 1 \\ x_1 \qquad\quad + 8x_3 + 3x_4 &= 0 \\ x_1 + x_2 - 3x_3 - x_4 &= 0. \end{aligned} \tag{3.15}$$

We can now use the second equation to eliminate x_4 from the first and third:

$$\begin{aligned} x_1 \qquad\quad - 3x_3 \qquad &= 1 \\ 2x_1 \qquad\quad - 2x_3 + x_4 &= 1 \\ -5x_1 \qquad\quad + 14x_3 \qquad &= -3 \\ x_1 + x_2 - 3x_3 - x_4 &= 0. \end{aligned}$$

Using the third equation to eliminate x_1 from the first, we obtain

$$\begin{aligned} - 0.2x_3 \qquad &= 0.4 \\ 2x_1 \qquad\quad - 2x_3 + x_4 &= 1 \\ -5x_1 \qquad\quad + 14x_3 \qquad &= -3 \\ x_1 + x_2 - 3x_3 - x_4 &= 0. \end{aligned}$$

Hence $x_3 = -2$, and back-substitution in the pivotal equations gives $x_1 = -5$, $x_4 = 7$, $x_2 = 6$. This is the same result as before.

From the discussion in the last paragraph it is clear that a wide choice of pivots is available. The only restriction that occurs (at least if we ignore rounding errors; see Section 6.3) can be illustrated by the following example:

$$\begin{aligned} x_2 + x_3 &= 0 \\ x_1 - 5x_2 + 3x_3 &= 0 \\ 2x_1 + x_2 - 4x_3 &= -1. \end{aligned} \tag{3.16}$$

For obvious reasons we cannot use the first equation to eliminate x_1 from the remaining two. From a matrix point of view we wish to reduce the following matrix to a simpler form:

$$\begin{bmatrix} 0 & 1 & 1 & 0 \\ 1 & -5 & 3 & 0 \\ 2 & 1 & -4 & -1 \end{bmatrix}.$$

We cannot use the (1, 1) element as a pivot. We can, however, use the (1, 2) or (1, 3) elements as pivots. We will see later (Section 6.3) that the way in which pivots are chosen is important if we wish to use digital computers to obtain accurate numerical solutions of linear equations.

In the remainder of this section we deal with another point that is of considerable practical importance.

Suppose that we have several sets of equations to solve, say $\mathbf{A}\mathbf{x}_1 = \mathbf{b}_1$, $\mathbf{A}\mathbf{x}_2 = \mathbf{b}_2, \ldots$. (Be careful to distinguish the column vectors $\mathbf{x}_1, \mathbf{x}_2, \ldots$ from the notation $x_1, x_2, \ldots$ for the individual unknowns used previously in this section.) In order to solve several sets of equations $\mathbf{A}\mathbf{x}_i = \mathbf{b}_i$, there is no need to go through the *whole* of the above procedure separately in each case, since the reduction of $\mathbf{A}$ to triangular form in the method of Gauss elimination (or to the unit matrix in the Gauss-Jordan method) is unaffected by the form of $\mathbf{b}$. Thus suppose we wish to solve three sets of equations with the same coefficient matrix as in (3.1), and

$$\mathbf{b}_1 = \begin{bmatrix} 3 \\ 1 \\ 0 \\ 0 \end{bmatrix}, \quad \mathbf{b}_2 = \begin{bmatrix} -5 \\ -2 \\ 0 \\ 5 \end{bmatrix}, \quad \mathbf{b}_3 = \begin{bmatrix} -6 \\ -2 \\ 5 \\ 3 \end{bmatrix}.$$

Then we perform the row operations on the following matrix:

$$\begin{bmatrix} 2 & -3 & 2 & 5 & 3 & -5 & -6 \\ 1 & -1 & 1 & 2 & 1 & -2 & -2 \\ 3 & 2 & 2 & 1 & 0 & 0 & 5 \\ 1 & 1 & -3 & -1 & 0 & 5 & 3 \end{bmatrix}. \tag{3.17}$$

Using the Gauss elimination method, reduction to triangular form gives the matrix [compare (3.10)]:

$$\begin{bmatrix} 1 & -1.5 & 1 & 2.5 & 1.5 & -2.5 & -3 \\ 0 & 1 & 0 & -1 & -1 & 1 & 2 \\ 0 & 0 & 1 & 0 & -2 & -1 & -1 \\ 0 & 0 & 0 & 1 & 7 & -1 & 3 \end{bmatrix}. \tag{3.18}$$

Back-substitution in each set, separately, then gives the solution:

$$\mathbf{x}_1 = \begin{bmatrix} -5 \\ 6 \\ -2 \\ 7 \end{bmatrix}, \quad \mathbf{x}_2 = \begin{bmatrix} 1 \\ 0 \\ -1 \\ -1 \end{bmatrix}, \quad \mathbf{x}_3 = \begin{bmatrix} -2 \\ 5 \\ -1 \\ 3 \end{bmatrix}. \tag{3.19}$$

We will return to this subject in Section 6.4 in more detail.

As an application of the above procedure, suppose next that we wish to find the inverse of an $n \times n$ matrix $\mathbf{A}$. Denote the columns of $\mathbf{A}^{-1}$ by $\mathbf{x}_1, \mathbf{x}_2, \ldots, \mathbf{x}_n$, and the columns of the identity matrix by the unit columns $\mathbf{e}_1, \mathbf{e}_2, \ldots, \mathbf{e}_n$ (see Definition 1.13). Using partitioned matrices we recall that the problem of finding $\mathbf{A}^{-1}$ such that $\mathbf{A}\mathbf{A}^{-1} = \mathbf{I}$ is equivalent to the problem of solving (see **Key Theorem 1.8**)

$$\mathbf{A}[\mathbf{x}_1, \mathbf{x}_2, \ldots, \mathbf{x}_n] = [\mathbf{e}_1, \mathbf{e}_2, \ldots, \mathbf{e}_n],$$

and this is equivalent to solving the n independent sets of equations

$$\mathbf{A}\mathbf{x}_j = \mathbf{e}_j \qquad (j = 1, 2, \ldots, n).$$

These equations all have the same coefficient matrix $\mathbf{A}$ so that they can be solved conveniently by the method described in the last paragraph. We form the augmented matrix

$$[\mathbf{A}, \mathbf{I}],$$

reduce to triangular form, and perform back-substitutions.

As a numerical example, consider the matrix $\mathbf{A}$ consisting of the first four columns of (3.17). This leads to the augmented matrix

$$\begin{bmatrix} 2 & -3 & 2 & 5 & 1 & 0 & 0 & 0 \\ 1 & -1 & 1 & 2 & 0 & 1 & 0 & 0 \\ 3 & 2 & 2 & 1 & 0 & 0 & 1 & 0 \\ 1 & 1 & -3 & -1 & 0 & 0 & 0 & 1 \end{bmatrix}. \tag{3.20}$$

Reduction to triangular form leads to

$$\begin{bmatrix} 1 & -1.5 & 1 & 2.5 & 0.5 & 0 & 0 & 0 \\ 0 & 1 & 0 & -1 & -1 & 2 & 0 & 0 \\ 0 & 0 & 1 & 0 & -5 & 13 & -1 & 0 \\ 0 & 0 & 0 & 1 & -18 & -47 & 4 & -1 \end{bmatrix}. \tag{3.21}$$

The inverse is then obtained by back-substitution,

$$\mathbf{A}^{-1} = [\mathbf{x}_1, \mathbf{x}_2, \mathbf{x}_3, \mathbf{x}_4] = \begin{bmatrix} -14 & 37 & -3 & 1 \\ 17 & -45 & 4 & -1 \\ -5 & 13 & -1 & 0 \\ 18 & -47 & 4 & -1 \end{bmatrix}. \tag{3.22}$$

Various practical aspects of the methods discussed in this section will be considered later. Here we note two results connected with the amount of work involved in solving a set of equations or inverting a matrix. Simple enumeration shows that approximately $\frac{1}{3}n^3$ additions and $\frac{1}{3}n^3$ multiplications are required to solve a set of n equations in n unknowns by the method of successive elimination. This is an important result since it shows that the amount of work involved increases rapidly with the number of equations. Doubling the number of equations multiplies the work by a factor of 8. The other result is that the amount of work involved in inverting an $n \times n$ matrix is approximately *three* times the amount of work involved in solving a single set of n equations in n unknowns (not n times, as might be thought at first sight). We will examine the reason for this later by comparing the work involved in obtaining (3.21) and back-substituting with the amount of work required to solve a single set of equations (see Section 6.2).

EXERCISE 3.1. Find the inverses of the matrices given in Exs. 1.22, 1.48, by carrying out row operations on an augmented matrix, as described in connection with (3.20).

EXERCISE 3.2. Obtain the solutions (3.19) by applying the Gauss-Jordan procedure to (3.17).

EXERCISE 3.3. Obtain the inverse (3.22) by applying the Gauss-Jordan procedure to (3.20).

EXERCISE 3.4. Show that when the Gauss-Jordan procedure is applied successfully in our systematic way to the augmented matrix

$$[\mathbf{A}, \mathbf{I}],$$

we obtain the final result

$$[\mathbf{I}, \mathbf{A}^{-1}]$$

if $\mathbf{A}$ is nonsingular and square.

3.3 THE EXISTENCE OF SOLUTIONS FOR A SET OF EQUATIONS: SOME EXAMPLES

In the last section we considered certain methods for finding the numerical solution of a set of simultaneous linear equations, but we did not pay much attention to the question of whether such a solution existed at all, or whether it was unique. This is the main problem that concerns us for the remainder of this chapter.

Consider first of all the simple equation $ax = b$, where a, x, b are scalars. We tend to say immediately that the solution of this equation is $x = b/a$, but in fact there are *three* possibilities:

1. If $a \neq 0$, then $x = b/a$, and this is the unique solution of the equation, whatever the value of b (which may be zero, in which case the solution is simply $x = 0$).
2. If $a = 0$ there are two possibilities, depending on the value of b:
   (a) If $b \neq 0$, then the equation is $0 \cdot x = b \neq 0$, and no finite solution exists. The solution "x equal to infinity" is not considered to be a permissible solution. We say that "no solution exists," or, alternatively, that "the equation is *inconsistent*" since it implies $0 = b \neq 0$, which is a contradiction.
   (b) If $b = 0$, then *any* number is a solution of the equation, for $0 \cdot x = 0$, whatever value is given to x. Infinity is again excluded, since $0 \cdot \infty$ is meaningless.

It is a striking result that precisely the same possibilities exist in the case of two equations in two unknowns. Thus

$$\begin{aligned} x_1 + x_2 &= 2 \\ x_1 - x_2 &= 0 \end{aligned}$$

have a unique solution,

$$\begin{aligned} x_1 + x_2 &= 2 \\ x_1 + x_2 &= 1 \end{aligned}$$

are inconsistent, and

$$\begin{aligned} x_1 + x_2 &= 2 \\ 2x_1 + 2x_2 &= 4 \end{aligned}$$

have infinitely many solutions, namely $x_1 = k$, $x_2 = 2 - k$, for all k.

An even more striking result is that precisely the same possibilities exist in the general case of n equations in n unknowns, as we shall see. For any specific example it is possible to find out what the situation is by direct solution of the equations or, equivalently, by using the triangular reduction of the augmented matrix described in the last section. Thus consider the equations

$$\begin{aligned} x_1 + 2x_2 - 5x_3 &= 2 \\ 2x_1 - 3x_2 + 4x_3 &= 4 \\ 4x_1 + x_2 - 6x_3 &= 8. \end{aligned} \tag{3.23}$$

The augmented matrix is

$$\begin{bmatrix} 1 & 2 & -5 & 2 \\ 2 & -3 & 4 & 4 \\ 4 & 1 & -6 & 8 \end{bmatrix}.$$

Row operations reduce this to

$$\begin{bmatrix} 1 & 2 & -5 & 2 \\ 0 & 1 & -2 & 0 \\ 0 & 0 & 0 & 0 \end{bmatrix}. \tag{3.24}$$

Note that all the elements in the third row are zero. This means that the third equation has been reduced to $0 \cdot x_3 = 0$, so that a solution of the original equation exists for which $x_3 = k$, where k is any number. Back-substitution then gives $x_2 = 2k$, $x_1 = 2 + k$. This solution can be written in the form

$$\begin{bmatrix} x_1 \\ x_2 \\ x_3 \end{bmatrix} = \begin{bmatrix} 2 \\ 0 \\ 0 \end{bmatrix} + k \begin{bmatrix} 1 \\ 2 \\ 1 \end{bmatrix}. \tag{3.25}$$

If, on the other hand, the third equation in (3.23) had instead been

$$4x_1 + x_2 - 6x_3 = 0,$$

we find, instead of (3.24),

$$\begin{bmatrix} 1 & 2 & -5 & 2 \\ 0 & 1 & -2 & 0 \\ 0 & 0 & 0 & 1 \end{bmatrix}. \tag{3.26}$$

The third equation is now $0 \cdot x_3 = 1$, so that this set of equations is inconsistent.

In the examples considered so far, the number of equations has been exactly equal to the number of unknowns, but the same methods (namely reduction of the augmented matrix to triangular form or, equivalently, direct solution of the equations) can be used to check whether a given set of m equations in n unknowns, for any m, n, possesses no solution, a unique solution, or infinitely many solutions. Thus we can verify the following examples by using the method described above:

EXERCISE 3.5. If the equation

$$x_1 + x_2 + x_3 = 6$$

is added to the set (3.23) as a fourth equation, the resulting system of four equations in three unknowns has the unique solution $x_1 = 3$, $x_2 = 2$, $x_3 = 1$.

EXERCISE 3.6. If the equation

$$3x_1 - x_2 - x_3 = 6$$

is added to the set (3.23), the resulting system of four equations in three unknowns has the same set of solutions as (3.23), namely (3.25).

EXERCISE 3.7. The system of two equations in three unknowns given by the first two equations in (3.23) has the same set of solutions as (3.23), namely (3.25).

The moral of these examples is that it is not, in general, possible to say whether equations have no solution, a unique solution, or infinitely many solutions, merely from knowledge of the *number* of equations and the *number* of unknowns—i.e., m and n. Ten equations in two unknowns can have a unique solution, and two equations in ten unknowns may be inconsistent.

Consider next the following example:

$$\begin{aligned} x_1 + 2x_2 - x_3 &= 2 \\ 2x_1 + 4x_2 + x_3 &= 7 \\ 3x_1 + 6x_2 - 2x_3 &= 7. \end{aligned} \tag{3.27}$$

Using the first equation to eliminate x_1 from the remaining two, we obtain (placing the matrix and equation forms alongside each other):

$$\begin{aligned} x_1 + 2x_2 - x_3 &= 2 \\ 3x_3 &= 3 \\ x_3 &= 1, \end{aligned} \qquad \begin{bmatrix} 1 & 2 & -1 & 2 \\ 0 & 0 & 3 & 3 \\ 0 & 0 & 1 & 1 \end{bmatrix}. \tag{3.28}$$

The unknown x_2 does not appear in the second and third equations. We therefore reduce the coefficient of x_3 in the second equation to unity by dividing this equation by 3. We can then omit the third equation because it simply repeats the second. [In terms of the matrix in (3.28), the (2, 2) and (3, 2) elements are zero. We therefore consider the third column, reduce the (2, 3) elements to unity by multiplying the second row by $\frac{1}{3}$, and subtract the resulting row from the last row, so that the last row is reduced to a row of zeros.]

$$\begin{aligned} x_1 + 2x_2 - x_3 &= 2 \\ x_3 &= 1, \end{aligned} \qquad \begin{bmatrix} 1 & 2 & -1 & 2 \\ 0 & 0 & 1 & 1 \\ 0 & 0 & 0 & 0 \end{bmatrix}. \tag{3.29}$$

The general solution of the original set of equations is therefore given by

$$x_3 = 1, \qquad x_2 = p, \qquad x_1 = 3 - 2p, \tag{3.30}$$

where we have set x_2 equal to p, an arbitrary constant. Instead of setting x_2 equal to an arbitrary constant, we could have set x_1 equal to an arbitrary constant, and then the general solution is

$$x_3 = 1, \qquad x_1 = q, \qquad x_2 = \tfrac{1}{2}(3 - q). \tag{3.31}$$

It is clear that the solutions (3.30) and (3.31) are essentially the same since, if we assign any value to p in (3.30), the same solution will be given by setting $q = 3 - 2p$ in (3.31). However, in more complicated cases the equivalence of different forms of solutions may not be so obvious. It should be noted that in the above example it is not possible to assign an arbitrary value to x_3 since one equation says $x_3 = 1$.

The procedure leading to (3.29) corresponds to what we have previously called "reduction to triangular form," with appropriate modifications for the fact that zero elements in successive matrices mean that we cannot have a strictly triangular form. To see the structure of the solution of (3.29) more clearly, it is convenient to use back-substitution in the equations [in the matrix, use the analog of the Gauss-Jordan procedure to reduce the (1, 3) element to zero]. This gives

$$\begin{aligned} x_1 + 2x_2 \qquad &= 3 \\ x_3 &= 1, \end{aligned} \qquad \begin{bmatrix} 1 & 2 & 0 & 3 \\ 0 & 0 & 1 & 1 \\ 0 & 0 & 0 & 0 \end{bmatrix}.$$

This tells us immediately that either x_1 or x_2 (but not both) can be assigned arbitrary values, but x_3 cannot be given an arbitrary value.

EXERCISE 3.8. Given

$$\begin{aligned}
x - y + z + w &= 1 \\
2x - 2y - 3z - 3w &= 17 \\
-x + y + 2z + 2w &= -10,
\end{aligned}$$

show that it is possible to solve these equations for x, z in terms of y, w and vice versa, but it is not possible to solve for z, w in terms of x, y and vice versa. An alternative, more detailed, statement is that the pairs x, w or x, z or y, w or y, z can be given arbitrary values and the equations can then be solved for the remaining pairs of unknowns. However, the pairs x, y or w, z cannot be given arbitrary values.

EXERCISE 3.9. Find the general solution of the *homogeneous* system $\mathbf{Ax} = \mathbf{0}$, where

$$\mathbf{A} = \begin{bmatrix} 0 & 1 & -2 & 0 & 0 & 2 & 4 \\ 0 & 0 & 0 & 1 & 0 & 0 & 1 \\ 0 & 0 & 0 & 0 & 1 & -1 & 6 \\ 0 & 0 & 0 & 0 & 0 & 0 & 0 \end{bmatrix}.$$

State precisely the situation concerning which combinations of unknowns can be given arbitrary values. (The fact that the right-hand side of the equations is zero does not affect the procedure.)

EXERCISE 3.10. Which of the following sets of equations possess zero, one, or infinitely many solutions?

(a)
$$\begin{aligned}
x - y + z &= 0 \\
2x + y - z &= -3 \\
x + 2y - 2z &= -2.
\end{aligned}$$

(b)
$$\begin{aligned}
x + y + z &= 0 \\
2x + y - z &= -3 \\
-x - 2y - 4z &= -3 \\
2x \qquad - 4z &= -6.
\end{aligned}$$

(c)
$$\begin{aligned}
x - y + z &= 0 \\
2x + y - z &= -3 \\
x - 3y + 4z &= 5.
\end{aligned}$$

3.4 ROW OPERATIONS TO SIMPLIFY MATRICES AND SETS OF EQUATIONS

We have so far considered solving sets of linear equations only for particular examples. We now consider the situation theoretically by generalizing the procedure followed for the specific examples. Suppose that we have a system of m equations in n unknowns:

$$\begin{aligned} a_{11}x_1 + a_{12}x_2 + \cdots + a_{1n}x_n &= b_1 \\ a_{21}x_1 + a_{22}x_2 + \cdots + a_{2n}x_n &= b_2 \\ &\cdots \\ a_{m1}x_1 + a_{m2}x_2 + \cdots + a_{mn}x_n &= b_m. \end{aligned} \tag{3.32}$$

It may happen that all the coefficients a_{i1} in the first "column" are zero. If this is the case we can choose x_1 equal to any arbitrary constant. This will not affect the values of $x_2, \ldots, x_n$, and we can move on to consider x_2. If at least one of the a_{i1} is nonzero, we can choose any one such nonzero coefficient, say a_{p1}, interchange the first and pth equations, and divide the new first equation by a_{p1}. The resulting equation is used to eliminate x_1 from the remaining $m - 1$ equations. The final result is a set of equations of the following form:

$$\begin{aligned} x_1 + \alpha_{12}x_2 + \alpha_{13}x_3 + \cdots + \alpha_{1n}x_n &= \beta_1 \\ a'_{22}x_2 + a'_{23}x_3 + \cdots + a'_{2n}x_n &= b'_2 \\ &\cdots \\ a'_{m2}x_2 + a'_{m3}x_3 + \cdots + a'_{mn}x_n &= b'_m. \end{aligned} \tag{3.33}$$

As we observed throughout the last two sections, we can perform the preceding operations on the augmented matrix rather than on the system of equations, obtaining the same result. More precisely, we write down the augmented matrix corresponding to (3.32), namely

$$\begin{bmatrix} a_{11} & a_{12} & \cdots & a_{1n} & b_1 \\ a_{21} & a_{22} & \cdots & a_{2n} & b_2 \\ & & \cdots & & \\ a_{m1} & a_{m2} & \cdots & a_{mn} & b_m \end{bmatrix}. \tag{3.34}$$

If at least one of the elements in the first column is nonzero, we can choose any one of these elements, say a_{p1}, interchange the first and pth rows, and divide the new first row by a_{p1}. Multiples of the resulting first row are subtracted from the remaining $m - 1$ rows to reduce the last $m - 1$ elements in the first column to zero. The final result is a matrix of the following form [compare (3.33)]:

$$\begin{bmatrix} 1 & \alpha_{12} & \alpha_{13} & \cdots & \alpha_{1n} & \beta_1 \\ 0 & a'_{22} & a'_{23} & \cdots & a'_{2n} & b'_2 \\ & & & \cdots & & \\ 0 & a'_{m2} & a'_{m3} & \cdots & a'_{mn} & b'_m \end{bmatrix}. \tag{3.35}$$

Since these two ways of representing the simplification process are equivalent, while the latter matrix fashion requires less writing, we elect to describe the remainder of the process in the notation for augmented matrices. Therefore we consider the further simplification of (3.34) beyond (3.35). If all the elements in the second column in (3.35) below α_{12} are zero, we proceed to the third column; otherwise we apply the procedure that led to (3.35). More precisely, to simplify (3.35) there are again two possibilities.

1. If at least one of the elements a'_{i2} is nonzero, we can choose any one of these, say a'_{q2}, interchange the second and qth rows, and divide the new second equation by a'_{q2}. Multiples of the resulting second row are subtracted from the remaining $m - 2$ rows to reduce the last $m - 2$ elements in the second column to zero. We then consider the third column.
2. If all the elements a'_{i2} ($i = 2$ to m) are zero, we proceed immediately to consider the third column.

This procedure is then repeated until we have dealt with all of the columns of the augmented matrix.

We now divorce the discussion from consideration of sets of simultaneous linear equations and consider the procedure from the point of view of operations on the rows of a general $m \times n$ matrix $\mathbf{A} = [a_{ij}]$. As illustrated in the above discussion, these operations are of three types.

DEFINTION 3.1. Elementary row operations on matrices (sometimes shortened to *row operations*) are defined as follows.

(a) Interchange of two rows.
(b) Multiplication of any row by a nonzero scalar.
(c) Replacement of the ith row by the sum of the ith row and p times the jth row ($j \neq i$).

The developments given at the beginning of this section were included in order to lead up to the result that any matrix can be reduced to a unique simple "standard" or "normal" form by using elementary row operations. This we now investigate. As we remarked earlier, for theoretical purposes it is more convenient to use Gauss-Jordan elimination than Gauss elimination; we therefore use Gauss-Jordan elimination for the production of a standard simplified form for matrices and systems of equations.

A row or column of a matrix is said to be *nonzero* if at least one element of the row or column is nonzero. Otherwise we talk of a zero row or a zero column, in which all the elements are zero. A zero column cannot be made nonzero by elementary row operations, so that a zero column is already in normal form. We proceed as follows:

1. Suppose that the first nonzero column is the c_1th, where c_1 may be one. By row interchanges, if necessary, we can arrange that the first element of this column is nonzero. Divide the first row by this element,

and subtract multiples of the first row from the other rows (if necessary) so that the 2nd, 3rd, . . . , mth elements of the c_1th column are zero.

2. If, in the matrix obtained in (1), the 2nd, 3rd, . . . , mth rows are all zero, the matrix has been reduced to the required normal form. Otherwise, suppose that the first column in the matrix obtained in (1) which has a nonzero element in the rows below the first row is the c_2th. By definition of c_1 and c_2 we have $c_1 < c_2$. By interchange of the rows below the first, if necessary, we can arrange that the c_2th element of the second row is nonzero. Divide the second row by this element. Subtract suitable multiples of the resulting second row from the remaining $m - 1$ rows (i.e., the first row as well as the 3rd, . . . , mth) so that all the other elements in the c_2th column are reduced to zero. For example, if $m = 4$, $n = 7$, $c_1 = 2$, $c_2 = 4$, we would have, at this stage,

$$\begin{bmatrix} 0 & 1 & X & 0 & X & X & X \\ 0 & 0 & 0 & 1 & X & X & X \\ 0 & 0 & 0 & 0 & X & X & X \\ 0 & 0 & 0 & 0 & X & X & X \end{bmatrix}, \tag{3.36}$$

where the X's represent numbers that may or may not be zero.

3. A repetition of the procedure leads to a state at which, say, the c_kth column, for some known $c_k < n$, has been transformed so that it has a unit in the kth row, for some $k \leq m$, all other elements of the c_kth column being zero. Also, either the kth row is the last (i.e., the mth), or the rows below the kth are zero. In either case, the process terminates. In the above example, for instance, we might obtain

$$\begin{bmatrix} 0 & 1 & X & 0 & 0 & X & X \\ 0 & 0 & 0 & 1 & 0 & X & X \\ 0 & 0 & 0 & 0 & 1 & X & X \\ 0 & 0 & 0 & 0 & 0 & 0 & 0 \end{bmatrix}. \tag{3.37}$$

The X's represent numbers that may be nonzero. We have precisely three nonzero rows so that $k = 3$, and there are $k = 3$ numbers c_1, c_2, c_3, namely $c_1 = 2$, $c_2 = 4$, $c_3 = 5$. Clearly $c_1 < c_2 < c_3$ and in the general case,

$$c_1 < c_2 < \cdots < c_k.$$

These results can be summarized by saying that *any $m \times n$ matrix can be reduced, using elementary row operations, to the standard form given in the following definition:*

DEFINITION 3.2. A matrix is said to be in *row-echelon normal form* (which we sometimes shorten to simply *row-echelon form*) if:

(a) Certain columns numbered $c_1, c_2, \ldots, c_k$ are precisely the unit vectors $\mathbf{e}_1, \mathbf{e}_2, \ldots, \mathbf{e}_k$, where $\mathbf{e}_i$ is defined to be the $m \times 1$ column vector whose ith element is unity, all other elements being zero.
(b) $c_1 < c_2 < \cdots < c_k$.
(c) If a column lies to the left of c_1, then it is a column of zeros. If the cth column lies between the columns numbered c_i and c_{i+1}, then the last $m - i$ elements of the cth column must be zero. If the cth column lies to the right of the column numbered c_k, then the last $(m - k)$ elements of the cth column must be zero.

Note that this definition implies [see (3.37), where $k = 3$]:

1. The last $m - k$ rows of the row-echelon form are zero. The first k rows of the row-echelon form are nonzero.
2. The lower triangle of elements in the (i, j) positions, where $j < i$, are all zero.
3. The first nonzero element in each row is 1. The first $c_i - 1$ elements of the ith row are zero. The c_jth element of the ith row is zero for $i \neq j$.

The point of the developments preceding this definition is that we have given a practical procedure by means of which any $m \times n$ matrix can be reduced, by elementary row operations, to the row-echelon form described in the above definition. The wary or experienced mathematical reader will have noticed that there is considerable freedom in the detailed sequence of operations used to reduce a matrix to a row-echelon form since, for example, we are allowed to interchange rows in any way we like. In the next section we will prove, however, the extremely important result that, whatever sequence of row operations we use, we will always arrive back at *precisely the same* row-echelon form; this unique row-echelon form for a matrix will provide precisely the tool we need to answer a variety of theoretical questions throughout this book. In particular, in Section 3.6 the row-echelon form of the augmented matrix corresponding to a given system of linear equations will allow us easily to analyze, in general, the nature of the sets of solutions, if any, to such equations; it was this question, after all, which motivated the use of the elimination process in the first place. Before reading the general analysis in Section 3.6, however, the reader can obtain some experience in using the row-echelon form of the augmented matrix to study sets of solutions by tackling the following exercise.

EXERCISE 3.11. The following matrices represent the augmented matrix $[\mathbf{A}, \mathbf{b}]$ of a system of linear equations. In each case state whether the equations possess zero, one, or an infinite number of solutions. If infinitely many solutions exist, write down an expression for the general solution.

(a)
$$\begin{bmatrix} 0 & 1 & 0 & 0 & 4 \\ 0 & 0 & 1 & 2 & 5 \\ 0 & 0 & 0 & 0 & 0 \end{bmatrix}$$

(b)
$$\begin{bmatrix} 0 & 1 & 0 & 0 & 4 \\ 0 & 0 & 1 & 0 & 5 \\ 0 & 0 & 0 & 1 & 6 \end{bmatrix}$$

(c)
$$\begin{bmatrix} 1 & 2 & 3 & 4 & 6 \\ 0 & 0 & 0 & 1 & 5 \\ 0 & 0 & 0 & 0 & 1 \end{bmatrix}$$

(d)
$$\begin{bmatrix} 1 & 0 & 0 & 0 & 4 \\ 0 & 1 & 2 & 0 & 5 \\ 0 & 0 & 0 & 1 & 6 \end{bmatrix}$$

(e)
$$\begin{bmatrix} 1 & 3 & 4 & 9 \\ 0 & 2 & 3 & 7 \\ 0 & 0 & 2 & 6 \\ 0 & 0 & 1 & 4 \end{bmatrix}$$

(f)
$$\begin{bmatrix} 1 & 2 & 3 & 4 & 7 \\ 0 & 5 & 7 & 6 & 8 \\ 0 & 5 & 2 & 4 & 7 \\ 0 & 0 & 2 & 1 & 0 \end{bmatrix}.$$

Since we emphasize the power of matrix notation throughout this book, it is interesting to note that the elementary row operations of Definition 3.1 can all be implemented or represented by premultiplication by matrices of special forms.

THEOREM 3.1. The result of performing an elementary row operation on a general $m \times n$ matrix **A** *can also be achieved by forming the product* **EA** *where* **E** *is the matrix obtained by performing the row operation on the $m \times m$ unit matrix* **I**.

Proof: Suppose that **A** is an $m \times n$ matrix and that we interchange the pth and qth rows. The matrix obtained by interchanging the pth and qth rows of the $m \times m$ unit matrix is

$$\mathbf{E} = \begin{bmatrix} 1 & \dots & 0 & \dots & 0 & \dots & 0 \\ & & & \dots & & & \\ 0 & \dots & 0 & \dots & 1 & \dots & 0 \\ & & & \dots & & & \\ 0 & \dots & 1 & \dots & 0 & \dots & 0 \\ & & & \dots & & & \\ 0 & \dots & 0 & \dots & 0 & \dots & 1 \end{bmatrix} \tag{3.38}$$

where we indicate the 1st, pth, qth, and last rows and columns. It is easily seen that the matrix **EA** is in fact precisely the matrix obtained by interchanging the pth and qth rows of **A**. Similarly, the matrix obtained by multiplying the pth row of the unit matrix by the nonzero scalar c is

$$\mathbf{E} = \begin{bmatrix} 1 & \dots & 0 & \dots & 0 \\ & & \dots & & \\ 0 & \dots & c & \dots & 0 \\ & & \dots & & \\ 0 & \dots & 0 & \dots & 1 \end{bmatrix} \tag{3.39}$$

and the matrix **EA** is precisely the matrix obtained by multiplying the pth row of **A** by c. Finally, if we add c times the qth row of the unit matrix to the pth row, we obtain

$$\mathbf{E} = \begin{bmatrix} 1 & \dots & 0 & \dots & 0 & \dots & 0 \\ & & & \dots & & & \\ 0 & \dots & 1 & \dots & c & \dots & 0 \\ & & & \dots & & & \\ 0 & \dots & 0 & \dots & 1 & \dots & 0 \\ & & & \dots & & & \\ 0 & \dots & 0 & \dots & 0 & \dots & 1 \end{bmatrix} \tag{3.40}$$

where we indicate the 1st, pth, qth, and last rows and columns. The matrix **EA** is precisely the matrix obtained by adding c times the qth row of **A** to the pth row. This completes the proof of the theorem.

DEFINITION 3.3. Any matrix **E** obtained by performing a single elementary row operation on the unit matrix **I** is known as an *elementary matrix*. $\mathbf{E}_{pq}$ is the elementary matrix obtained by interchanging the pth and qth rows of **I** [cf. (3.38)]. $\mathbf{E}_p(c)$ is the elementary matrix obtained by multiplying the pth row of **I** by c [cf. (3.39)] with $c \neq 0$. $\mathbf{E}_{pq}(c)$ is the elementary matrix obtained by adding c times the qth row of **I** to the pth [cf. (3.40)].

It is a striking and important fact that each elementary matrix **E** is nonsingular, that is, $\mathbf{H} = \mathbf{E}^{-1}$ exists, and moreover that $\mathbf{E}^{-1}$ is itself an elementary matrix of the same type.

THEOREM 3.2. Each elementary matrix is nonsingular and has an elementary matrix as its inverse. More precisely (see Definition 3.3):

$$\mathbf{E}_{pq}^{-1} = \mathbf{E}_{qp}; \quad \mathbf{E}_p(c)^{-1} = \mathbf{E}_p\left(\frac{1}{c}\right) \qquad (\textit{with } c \neq 0);$$

$$\mathbf{E}_{pq}(c)^{-1} = \mathbf{E}_{pq}(-c).$$

Proof: Recall that in order to show that a particular matrix $\mathbf{H}$ is the inverse of a given matrix $\mathbf{E}$, we only need show that $\mathbf{HE} = \mathbf{EH} = \mathbf{I}$. In our cases, both $\mathbf{H}$ and $\mathbf{E}$ are elementary matrices, so Theorem 3.1 says, for example, that $\mathbf{HE}$ is just the result of performing on $\mathbf{E}$ that row operation described by $\mathbf{H}$. The proofs that $\mathbf{E}_{pq}^{-1} = \mathbf{E}_{qp}$ and $\mathbf{E}_p(c)^{-1} = \mathbf{E}_p(1/c)$ for $c \neq 0$ are then trivial; therefore we consider only whether or not $\mathbf{E}_{pq}(-c)$ is the inverse of $\mathbf{E}_{pq}(c)$. Let $\mathbf{r}_1, \ldots, \mathbf{r}_m$ denote the rows of $\mathbf{I}$ and let $\mathbf{r}'_1, \ldots, \mathbf{r}'_m$ denote the rows of $\mathbf{E}_{pq}(c)$; of course we have

$$\mathbf{r}'_i = \mathbf{r}_i \quad (\text{for } i \neq p), \qquad \mathbf{r}'_p = \mathbf{r}_p + c\mathbf{r}_q. \tag{3.41}$$

If we let the rows of $\mathbf{E}_{pq}(-c)\mathbf{E}_{pq}(c)$ be $\mathbf{r}''_1, \ldots, \mathbf{r}''_m$, since premultiplication by $\mathbf{E}_{pq}(-c)$ corresponds to a row operation, we have that

$$\mathbf{r}''_i = \mathbf{r}'_i \qquad (\text{for } i \neq p), \qquad \mathbf{r}''_p = \mathbf{r}'_p - c\mathbf{r}'_q. \tag{3.42}$$

Combining (3.41) and (3.42) yields

$$\mathbf{r}''_i = \mathbf{r}_i \quad (\text{for } i \neq p), \qquad \mathbf{r}''_p = \mathbf{r}'_p - c\mathbf{r}'_q = (\mathbf{r}_p + c\mathbf{r}_q) - c\mathbf{r}_q = \mathbf{r}_p,$$

so that $\mathbf{E}_{pq}(-c)\mathbf{E}_{pq}(c) = \mathbf{I}$. Replacing c by $-c$, we have $\mathbf{E}_{pq}(c)\mathbf{E}_{pq}(-c) = \mathbf{I}$, so that $\mathbf{E}_{pq}(c)^{-1}$ exists and equals $\mathbf{E}_{pq}(-c)$. Our proof is complete.

We know from Theorem 3.1 that each elementary row operation can be achieved by premultiplication by an elementary matrix, and hence any sequence of row operations can be represented by successive premultiplications by elementary matrices $\mathbf{E}_1, \mathbf{E}_2, \ldots, \mathbf{E}_r$; the overall effect of such successive premultiplications is of course the same as that from one premultiplication by $\mathbf{F} = \mathbf{E}_r\mathbf{E}_{r-1}, \ldots, \mathbf{E}_1$. Since the product of nonsingular matrices is nonsingular (see Theorem 1.5) and each elementary matrix is nonsingular, we have proved the following **key** result.

- *KEY THEOREM 3.3. If the matrix* $\mathbf{B}$ *results from the matrix* $\mathbf{A}$ *by application of a sequence of elementary row operations to* $\mathbf{A}$, *then there is a nonsingular matrix* $\mathbf{F}$ *with* $\mathbf{B} = \mathbf{FA}$.

EXERCISE 3.12. If a matrix $\mathbf{B}$ is derived from a matrix $\mathbf{A}$ by a sequence of elementary row operations, then $\mathbf{B}$ is said to be *row-equivalent* to $\mathbf{A}$, and we write $\mathbf{B} \sim \mathbf{A}$. Prove that $\sim$ is a true *equivalence relation* in the sense that:

(a) $\mathbf{A} \sim \mathbf{A}$ for all $\mathbf{A}$;
(b) if $\mathbf{A} \sim \mathbf{B}$, then $\mathbf{B} \sim \mathbf{A}$;
(c) if $\mathbf{A} \sim \mathbf{B} \sim \mathbf{C}$, then $\mathbf{A} \sim \mathbf{C}$.

EXERCISE 3.13. Consider the matrix

$$\mathbf{A} = \begin{bmatrix} 0 & 1 \\ 2 & 3 \\ 4 & 6 \end{bmatrix}.$$

Use a sequence of elementary row operations to reduce **A** to row-echelon form; for each row operation you use, find the corresponding elementary matrix **E** and verify that premultiplication by **E** in this case gives the same result as the row operation.

EXERCISE 3.14. For each elementary matrix **E** found in Exercise 3.13, find $\mathbf{E}^{-1} = \mathbf{H}$ and verify that $\mathbf{EH} = \mathbf{HE} = \mathbf{I}$.

3.5 ROW-ECHELON FORM AND RANK

Recall from the last section that every $m \times n$ matrix **A** can be reduced to a matrix of row-echelon form by a sequence of elementary row operations. Recall also from Definition 3.2 that a matrix is in row-echelon form when:

1. certain columns numbered $c_1, c_2, \ldots, c_k$ are precisely the unit vectors $\mathbf{e}_1, \mathbf{e}_2, \ldots, \mathbf{e}_k$, where $\mathbf{e}_i$ is defined to be the $m \times 1$ column vector whose ith element is unity, all other elements being zero;
2. $c_1 < c_2 < \cdots < c_k$;
3. if a column lies to the left of c_1, then it is a column of zeros; if the cth column lies between the columns numbered c_i and c_{i+1}, then the last $m - i$ elements of the cth column must be zero; if the cth column lies to the right of the column numbered c_k, then the last $(m - k)$ elements of the cth column must be zero.

   These facts imply that:
4. the last $m - k$ rows of the row-echelon form are zero; the first k rows of the row-echelon form are nonzero;
5. the lower triangle of elements in the (i, j) positions, where $j < i$, is all zero;
6. the first nonzero element in each row is 1; the first $c_i - 1$ elements of the ith row are zero; the c_jth element of the ith row is zero for $i \neq j$.

We are now in a position to prove the **key** result that *precisely the same* row-echelon form results *whatever* sequence of elementary row operations is used to reduce **A** to that form.

● *KEY THEOREM 3.4. The row-echelon form for a matrix is unique. In particular, the number k of nonzero rows in the row-echelon form of a matrix is a constant, regardless of the actual sequence of row operations used to produce the normal form.*

Proof: Suppose that an $m \times n$ matrix **A** is reduced to two row-echelon forms **P** and **P′** by row operations, and suppose that the unit column vectors in the row-echelon forms are in columns numbered $c_1, \ldots, c_r$ for $\mathbf{P}$ and $c'_1, \ldots, c'_s$ for $\mathbf{P}'$; we want to show that $r = s$ and $\mathbf{P} = \mathbf{P}'$. By **Key Theorem 3.3** there are nonsingular matrices $\mathbf{F}$ and $\mathbf{F}'$ such that $\mathbf{P} = \mathbf{FA}$ and $\mathbf{P}' = \mathbf{F}'\mathbf{A}$. Thus $\mathbf{P}' = \mathbf{F}'\mathbf{A} = \mathbf{F}'\mathbf{F}^{-1}\mathbf{FA} = (\mathbf{F}'\mathbf{F}^{-1})\mathbf{P}$, so that

$$\mathbf{P}' = \mathbf{HP}$$

for a nonsingular matrix $\mathbf{H} = \mathbf{F}'\mathbf{F}^{-1}$. Now let $\mathbf{p}_1, \ldots, \mathbf{p}_n$ and $\mathbf{p}'_1, \ldots, \mathbf{p}'_n$ be the columns of $\mathbf{P}$ and $\mathbf{P}'$ respectively, so that

$$\mathbf{p}'_i = \mathbf{H}\mathbf{p}_i \qquad (\text{for } 1 \leq i \leq n). \tag{3.43}$$

By assumption, $\mathbf{p}_i$ is the zero vector for $1 \leq i \leq c_1 - 1$, and hence $\mathbf{p}'_i$ is zero for $1 \leq i \leq c_1 - 1$; this implies that $c_1 \leq c'_1$. Since also

$$\mathbf{p}_i = \mathbf{H}^{-1}\mathbf{p}'_i \qquad (\text{for } 1 \leq i \leq n), \tag{3.44}$$

it follows similarly that $c'_1 \leq c_1$ and therefore $c'_1 = c_1$. Therefore, both $\mathbf{p}_{c_1}$ and $\mathbf{p}_{c'_1}$ equal the unit column vector $\mathbf{e}_1$, so from (3.43) we deduce that

$$\mathbf{e}_1 = \mathbf{H}\mathbf{e}_1.$$

So far we have shown that $\mathbf{p}_i = \mathbf{p}'_i$ for $1 \leq i \leq c_1$; we now treat $c_1 < i \leq c_2$. For $c_1 < i < c_2$, by assumption $\mathbf{p}_i$ is some scalar multiple t_i of $\mathbf{e}_1$; therefore, $\mathbf{p}'_i$, which equals $\mathbf{H}\mathbf{p}_i$, satisfies

$$\mathbf{p}'_i = \mathbf{H}\mathbf{p}_i = \mathbf{H}(t_i\mathbf{e}_1) = t_i(\mathbf{H}\mathbf{e}_1) = t_i\mathbf{e}_1 = \mathbf{p}_i \quad \text{for} \quad c_1 < i < c_2.$$

This in turn implies that $c'_2 \geq c_2$; applying the same argument to (3.44) rather than to (3.43), we also find $c_2 \geq c'_2$ and hence $c_2 = c'_2$. Thus both $\mathbf{p}_{c_2}$ and $\mathbf{p}'_{c'_2}$ equal the unit column vector $\mathbf{e}_2$, so that we now have $\mathbf{p}_i = \mathbf{p}'_i$ for $1 \leq i \leq c_2$, and from (3.43) we have

$$\mathbf{e}_2 = \mathbf{H}\mathbf{e}_2.$$

Continuing in this fashion, we deduce that $r = s$ (whose common value we call k) and that $\mathbf{e}_i = \mathbf{H}\mathbf{e}_i$ for $1 \leq i \leq k$. Since $\mathbf{H}\mathbf{e}_i$ is just the ith column of $\mathbf{H}$, this says that $\mathbf{H}$ has the structure

$$\mathbf{H} = \begin{bmatrix} \mathbf{I}_k & \mathbf{R} \\ \mathbf{0} & \mathbf{S} \end{bmatrix}.$$

By fact (4) on the preceding page, the row-echelon matrices $\mathbf{P}$ and $\mathbf{P}'$ satisfy

$$\mathbf{P} = \begin{bmatrix} \mathbf{Q} \\ \mathbf{0} \end{bmatrix}, \quad \mathbf{P}' = \begin{bmatrix} \mathbf{Q}' \\ \mathbf{0} \end{bmatrix}$$

where **Q** and **Q′** are $k \times n$. From $\mathbf{P}' = \mathbf{HP}$ we get

$$\begin{bmatrix} \mathbf{Q}' \\ \mathbf{0} \end{bmatrix} = \mathbf{P}' = \mathbf{HP} = \begin{bmatrix} \mathbf{I}_k & \mathbf{R} \\ \mathbf{0} & \mathbf{S} \end{bmatrix} \begin{bmatrix} \mathbf{Q} \\ \mathbf{0} \end{bmatrix} = \begin{bmatrix} \mathbf{Q} \\ \mathbf{0} \end{bmatrix} = \mathbf{P}$$

and thus $\mathbf{P}' = \mathbf{P}$, as asserted.

The number of nonzero rows in the (unique) row-echelon form of a matrix (which **Key Theorem 3.4** says is independent of how the row-echelon form is generated) plays a fundamental role in much of our subsequent work in linear algebra. In the next section, for example, it provides a convenient tool to let us describe the nature of the solution sets to systems of linear equations. We therefore give this important number a special name.

DEFINITION 3.4. The number of nonzero rows in the row-echelon form of a matrix is known as its *rank*.

Sometimes we find the concept defined above called "row-rank" rather than "rank." A similar concept, "column-rank," can be introduced by considering *column* operations and *column*-echelon form. Since it turns out that row-rank and column-rank are always equal, we are justified in using the unadorned term "rank." (See **Key Theorem 4.8** and **Key Corollary 4.1.**)

EXERCISE 3.15. Reduce the following matrices to row-echelon form and find the rank of each:

(a)
$$\begin{bmatrix} 1 & 1 & -8 & -14 \\ 3 & -4 & -3 & 0 \\ 2 & -1 & -7 & -10 \end{bmatrix}$$

(b)
$$\begin{bmatrix} 1 & -2 & 1 & 4 \\ 2 & -3 & -1 & 2 \end{bmatrix}$$

(c)
$$\begin{bmatrix} 1 & 2 & 3 & 4 & 7 \\ 0 & 5 & 7 & 6 & 8 \\ 0 & 5 & 2 & 4 & 7 \\ 0 & 0 & 2 & 1 & 0 \end{bmatrix}$$

(d)
$$\begin{bmatrix} 4 & -1 & 2 & 6 \\ -1 & 5 & -1 & -3 \\ 3 & 4 & 1 & 3 \end{bmatrix}.$$

EXERCISE 3.16. Prove that if the rank of a square $m \times m$ matrix **A** is m, then the row-echelon form of **A** is just the unit matrix **I**. Deduce from **Key Theorem 3.3** that **A** is the product of elementary matrices and is nonsingular.

EXERCISE 3.17. Prove that if the square $m \times m$ matrix $\mathbf{A}$ is nonsingular, then its rank is m.

EXERCISE 3.18. If $\mathbf{A}$ is a general $m \times m$ matrix, show that the rank of $[\mathbf{I}, \mathbf{A}]$ is m.

EXERCISE 3.19. If $\mathbf{A}$ is a general $m \times m$ matrix and $\mathbf{B}$ is an $m \times r$ matrix with rank m, show that the rank of $[\mathbf{B}, \mathbf{A}]$ is m.

3.6 SOLUTION SETS FOR EQUATIONS; THE FREDHOLM ALTERNATIVE

Early in this chapter we considered systems of linear equations and, in particular, attempted to discover the nature of the sets of solutions to such equations. Although we found that systems could have zero, exactly one, or infinitely many solutions, we were not able to describe precisely what happens for general systems. We finally have the tools to give this description.

Consider the equation

$$\mathbf{Ax} = \mathbf{b} \tag{3.45}$$

where $\mathbf{A}$ is $m \times n$. In Section 3.4 we attempted to solve (3.45) by simplifying the system of equations or, equivalently, by reducing the augmented matrix $[\mathbf{A}, \mathbf{b}]$ to row-echelon form. We implied earlier that such transformations changed nothing, that is, that the original system and the reduced system had precisely the same solutions. We must prove this.

THEOREM 3.5. Suppose that the system of equations (3.45) *is transformed by row operations into the system of equations*

$$\mathbf{A}'\mathbf{x} = \mathbf{b}'. \tag{3.46}$$

Then the solution sets of (3.45) *and* (3.46) *are identical, that is if* $\mathbf{x}$ *solves* (3.45) *then* $\mathbf{x}$ *solves* (3.46) *and vice versa.*

Proof: We know that $[\mathbf{A}, \mathbf{b}]$ has been transformed into $[\mathbf{A}', \mathbf{b}']$ by elementary row operations, so by **Key Theorem 3.3** there is a nonsingular matrix $\mathbf{F}$ such that

$$[\mathbf{A}', \mathbf{b}'] = \mathbf{F}[\mathbf{A}, \mathbf{b}] = [\mathbf{FA}, \mathbf{Fb}],$$

where the last equality follows from the way in which we multiply partitioned matrices. We therefore know that $\mathbf{A}' = \mathbf{FA}$ and $\mathbf{b}' = \mathbf{Fb}$. Now suppose $\mathbf{x}$ solves (3.45); premultiplication by $\mathbf{F}$ on both sides in (3.45) then immediately tells us that $\mathbf{x}$ solves (3.46). Conversely, if $\mathbf{x}$ solves (3.46) then premultiplication by $\mathbf{F}^{-1}$ tells us that $\mathbf{x}$ solves (3.45), so the two sets of solutions are identical.

This theorem tells us that we can study the set of solutions to (3.45) by studying the set of solutions to the much simpler set of equations obtained by reducing [**A**, **b**] to row-echelon form. As indicated in Section 3.3, we expect only three possible cases:

1. the equations are inconsistent—i.e., no solution exists;
2. the equations are consistent and have a unique solution;
3. the equations are consistent and have infinitely many solutions.

To decide which case holds for a particular set of equations we can proceed as follows. Write down the augmented matrix [**A**, **b**] for the system **Ax** = **b**, and reduce to row-echelon form. We arrive at a matrix of the following form

$$\begin{bmatrix} \mathsf{X} & \mathsf{X} & \dots & \mathsf{X} & \delta_1 \\ 0 & \mathsf{X} & \dots & \mathsf{X} & \delta_2 \\ & & \dots & & \\ 0 & 0 & \dots & 0 & \delta_{k+1} \\ 0 & 0 & \dots & 0 & 0 \\ & & \dots & & \\ 0 & 0 & \dots & 0 & 0 \end{bmatrix} \tag{3.47}$$

where we assume that the row-echelon form of the coefficient matrix **A** has precisely k nonzero rows represented by crosses in the above equation—i.e., k is the *rank* of **A** (recall Definition 3.4). The number δ_{k+1} may be either 1 or 0. If $\delta_{k+1} = 1$, then the rank of the augmented matrix [**A**, **b**] is $k + 1$; if $\delta_{k+1} = 0$, the rank of [**A**, **b**] is k. This leads to the following **key theorem:**

● *KEY THEOREM 3.6. Consider the nonhomogeneous equations* **Ax** = **b** *where* **A** *is* $m \times n$. *Exactly one of the following possibilities must hold:*

(i) *If the rank of the augmented matrix* [**A**, **b**] *is greater than the rank of* **A**, *the system of equations is inconsistent.*
(ii) *If the rank of* [**A**, **b**] *is equal to the rank of* **A**, *this being equal to the number of unknowns, then the equations have a unique solution.*
(iii) *If the rank of* [**A**, **b**] *is equal to the rank of* **A**, *this being less than the number of unknowns, then the equations have infinitely many solutions.*

● *Key Proof:* We already know from Theorem 3.5 that the solution set of **Ax** = **b** is identical with that for the equations corresponding to the augmented matrix in the row-echelon form (3.47). If (i) holds then $\delta_{k+1} = 1$ and the $(k + 1)$ equation corresponding to (3.47) says that $0 = 1$, so the system is inconsistent. If (ii) holds then $k = n$ (= the number of unknowns), the block indicated by the symbols X in (3.47), namely the row-echelon form for **A** alone, forms the unit matrix of order n, and the only solution clearly is

$$x_1 = \delta_1,\ x_2 = \delta_2, \dots, x_n = \delta_n.$$

If (iii) holds, then $k+1 \leq n$, $\delta_{k+1}=0$, and all the $k<n$ unknowns $x_{c_1}, \ldots, x_{c_k}$ are uniquely determined [where $c_1, \ldots, c_k$ are the column numbers of the unit column vectors in the row-echelon form (3.47)] once the remaining $n-k$ unknowns are assigned arbitrary values. This completes the proof.

EXERCISE 3.20. Analyze the solution set of the following system in terms of the parameter α:

$$\begin{aligned} x - 3y &= -2 \\ 2x + y &= 3 \\ 3x - 2y &= \alpha. \end{aligned}$$

SOLUTION: The augmented matrix is

$$\begin{bmatrix} 1 & -3 & -2 \\ 2 & 1 & 3 \\ 3 & -2 & \alpha \end{bmatrix},$$

which is easily reduced to

$$\begin{bmatrix} 1 & 0 & 1 \\ 0 & 1 & 1 \\ 0 & 0 & \alpha - 1 \end{bmatrix}.$$

If $\alpha \neq 1$, we have case (i) of **Key Theorem 3.6** and there are no solutions. If $\alpha = 1$, then $k = n = 2$ and we have case (ii) with the unique solution $x = 1$, $y = 1$.

EXERCISE 3.21. Analyze the solution set of the following system in terms of the parameter α:

$$\begin{aligned} x + y + z &= 4 \\ 2x + y + z &= 5 \\ 3x + 2y + 2z &= \alpha. \end{aligned}$$

SOLUTION: The augmented matrix is

$$\begin{bmatrix} 1 & 1 & 1 & 4 \\ 2 & 1 & 1 & 5 \\ 3 & 2 & 2 & \alpha \end{bmatrix}$$

which is easily reduced to

$$\begin{bmatrix} 1 & 0 & 0 & 1 \\ 0 & 1 & 1 & 3 \\ 0 & 0 & 0 & \alpha - 9 \end{bmatrix}.$$

If $\alpha \neq 9$, then we have case (i) of **Key Theorem 3.6** and there are no solutions. If $\alpha = 9$, then $k = 2 < 3 = n$, $c_1 = 1$, $c_2 = 2$, and under case (iii) we may set z to an arbitrary value p and then find $x = 1, y = 3 - p, z = p$ for all p, giving infinitely many solutions.

It is interesting to consider the case (iii) of infinitely many solutions a little further so as to see how two such solutions differ. If $\mathbf{Az} = \mathbf{b}$ and $\mathbf{Ay} = \mathbf{b}$ then clearly $\mathbf{A}(\mathbf{z} - \mathbf{y}) = \mathbf{Az} - \mathbf{Ay} = \mathbf{b} - \mathbf{b} = \mathbf{0}$, so that the difference $\mathbf{w} = \mathbf{z} - \mathbf{y}$ of two solutions to $\mathbf{Ax} = \mathbf{b}$ must solve the *homogeneous* system

$$\mathbf{Ax} = \mathbf{0}. \tag{3.48}$$

Conversely, if $\mathbf{w}$ is any solution of the homogeneous equation (3.48) and $\mathbf{y}$ is any solution of $\mathbf{Ax} = \mathbf{b}$, then $\mathbf{z} = \mathbf{y} + \mathbf{w}$ also solves $\mathbf{Az} = \mathbf{b}$. We have proved the following **key** result.

- KEY THEOREM 3.7. *If* $\mathbf{y}$ *is any solution of the system* $\mathbf{Ax} = \mathbf{b}$, *then any other vector* $\mathbf{z}$ *also satisfies* $\mathbf{Az} = \mathbf{b}$ *if and only if* $\mathbf{z} = \mathbf{y} + \mathbf{w}$ *for some* $\mathbf{w}$ *solving the homogeneous equation* (3.48). *Equivalently, for any given* $\mathbf{b}$ *for which there is at least one solution to* $\mathbf{Ax} = \mathbf{b}$, *that solution is unique if and only if* $\mathbf{x} = \mathbf{0}$ *is the only solution to* (3.48).

This theorem is striking because it says that for m equations in n unknowns, to study *uniqueness* (but not existence) of solutions to general systems $\mathbf{Ax} = \mathbf{b}$, we need only study uniqueness for the simpler special homogeneous system. A very important **key** theorem says that when $m = n$ even the *existence* of solutions for $\mathbf{Ax} = \mathbf{b}$ for arbitrary $\mathbf{b}$ is equivalent to uniqueness in the simpler special homogeneous problem.

- KEY THEOREM 3.8. (*The Fredholm Alternative*). *Let* $\mathbf{Ax} = \mathbf{b}$ *describe m equations in m unknowns. Then there exists precisely one solution* $\mathbf{x}$ *for every* $\mathbf{b}$ *if and only if* $\mathbf{x} = \mathbf{0}$ *is the unique solution to* $\mathbf{Ax} = \mathbf{0}$. *In other words, either:* $\mathbf{Ax} = \mathbf{b}$ *has precisely one solution for every* $\mathbf{b}$; *or:* $\mathbf{Ax} = \mathbf{0}$ *has a nonzero solution* $\mathbf{x}$; *but not both can hold simultaneously.*

  Proof: The uniqueness aspect has been settled by **Key Theorem 3.7**; we only need prove that if $\mathbf{x} = \mathbf{0}$ is the only solution to $\mathbf{Ax} = \mathbf{0}$, at least one solution to $\mathbf{Ax} = \mathbf{b}$ exists for every $\mathbf{b}$. To do this we first simply apply **Key Theorem 3.6** to the homogeneous equation $\mathbf{Ax} = \mathbf{0}$; since we have assumed this to have the unique solution $\mathbf{x} = \mathbf{0}$, then case (ii) of **Key Theorem 3.6** must hold, so that the rank of $\mathbf{A}$ must equal the number of unknowns, namely m since $\mathbf{A}$ is $m \times m$. But then the rank of $[\mathbf{A}, \mathbf{b}]$ is also m because the same operations that reduce $\mathbf{A}$ to its row-echelon form $\mathbf{I}$ reduce $[\mathbf{A}, \mathbf{b}]$ to the form $[\mathbf{I}, \mathbf{b}']$, clearly making the rank m; thus case (ii) of **Key Theorem 3.6** holds for the system $\mathbf{Ax} = \mathbf{b}$ and a solution must exist, proving our theorem. We have proved incidentally that $\mathbf{A}$ is nonsingular. (Why?)

This theorem is very useful because, for m equations in m unknowns, it allows all questions of existence and uniqueness of solutions to $\mathbf{Ax} = \mathbf{b}$ for arbitrary $\mathbf{b}$ to be reduced to uniqueness for the very special equation $\mathbf{Ax} = \mathbf{0}$. Sometimes this can save much trouble.

EXERCISE 3.22. Let $t_1, \ldots, t_{n+1}$ be $n + 1$ distinct real numbers, and let $b_1, \ldots, b_{n+1}$ be $n + 1$ arbitrary real numbers. Prove that there exists precisely one polynomial p of degree less than or equal to n such that $p(t_i) = b_i$ for $1 \le i \le n + 1$.

SOLUTION: Let the desired polynomial be written as

$$p(t) = x_1 + x_2 t + x_3 t^2 + \cdots + x_{n+1} t^n.$$

Then the equations $p(t_i) = b_i$ for $1 \le i \le n + 1$ can be written as $\mathbf{Ax} = \mathbf{b}$ where

$$\mathbf{x} = \begin{bmatrix} x_1 \\ \cdot \\ \cdot \\ \cdot \\ x_{n+1} \end{bmatrix}, \quad \mathbf{b} = \begin{bmatrix} b_1 \\ \cdot \\ \cdot \\ \cdot \\ b_{n+1} \end{bmatrix},$$

$$\mathbf{A} = \begin{bmatrix} 1 & t_1 & t_1^2 & t_1^3 & \ldots & t_1^n \\ 1 & t_2 & t_2^2 & t_2^3 & \ldots & t_2^n \\ & & & \cdots & & \\ 1 & t_{n+1} & t_{n+1}^2 & t_{n+1}^3 & \ldots & t_{n+1}^n \end{bmatrix}.$$

By the Fredholm Alternative, there is precisely one solution to $\mathbf{Ax} = \mathbf{b}$ if and only if $\mathbf{x} = \mathbf{0}$ is the only solution to $\mathbf{Ax} = \mathbf{0}$, that is, if and only if $p \equiv 0$ is the only polynomial of degree less than or equal to n which vanishes at the $n + 1$ distinct points t_i, $1 \le i \le n + 1$. But we know that any nonzero polynomial of degree n has at most n zeros; therefore $p \equiv 0$ (and $\mathbf{x} = \mathbf{0}$) is the only solution, proving our theorem.

EXERCISE 3.23. Prove that $\mathbf{x} = \mathbf{0}$ is the only solution to $\mathbf{Ax} = \mathbf{0}$ if and only if the rank of $\mathbf{A}$ equals the number of unknowns.

EXERCISE 3.24. Assuming that the equations $\mathbf{Ax} = \mathbf{b}$ have at least one solution, prove that it is unique if and only if the rank of $\mathbf{A}$ equals the number of unknowns.

3.7 INVERSES OF MATRICES

As we discovered in Chapter 1, the question of whether a matrix $\mathbf{A}$ has an inverse (left-, right-, or two-sided) can be resolved by considering systems of equations; for example, **Key Theorem 1.8** says that $\mathbf{A}$ has a right inverse $\mathbf{R}$

if and only if we can solve the equations $\mathbf{Ax}_j = \mathbf{e}_j$ for $1 \leq j \leq m$. Since Section 3.6 gave us the tools for analyzing the existence of solutions to linear systems, we are now prepared to consider, in general, the existence of inverses. First we describe one aspect of the significance of left- and right-inverses.

Suppose that the $m \times n$ matrix $\mathbf{A}$ has a left-inverse $\mathbf{L}$, necessarily $n \times m$, so that $\mathbf{LA} = \mathbf{I}$. Given any system of equations $\mathbf{Ax} = \mathbf{b}$ which has solutions $\mathbf{x}_1$ and $\mathbf{x}_2$, premultiplication of $\mathbf{A}(\mathbf{x}_1 - \mathbf{x}_2) = \mathbf{b} - \mathbf{b} = \mathbf{0}$ by $\mathbf{L}$ gives

$$\mathbf{x}_1 - \mathbf{x}_2 = \mathbf{LA}(\mathbf{x}_1 - \mathbf{x}_2) = \mathbf{L0} = \mathbf{0},$$

so that $\mathbf{x}_1 = \mathbf{x}_2$ and the solution is unique. Thus, whenever a left-inverse exists, systems $\mathbf{Ax} = \mathbf{b}$ can have *at most* one solution. By **Key Theorem 3.6**(ii), we can then relate existence of a left-inverse to statements about rank.

Suppose, instead, that the $m \times n$ matrix $\mathbf{A}$ has a right-inverse $\mathbf{R}$, necessarily $n \times m$, so that $\mathbf{AR} = \mathbf{I}$. Then, given any $\mathbf{b}$, the vector $\mathbf{x}_0 = \mathbf{Rb}$ satisfies

$$\mathbf{Ax}_0 = \mathbf{ARb} = \mathbf{b},$$

and hence every system $\mathbf{Ax} = \mathbf{b}$ has *at least* one solution. By **Key Theorem 3.6**(ii, iii), we can then relate existence of a right-inverse to statements about rank.

We now relate rank and inverses precisely; for simplicity we state separately the theorems for $m \leq n$, $m \geq n$, and the very important square case $m = n$.

THEOREM 3.9. Let $\mathbf{A}$ *be* $m \times n$ *and have rank* k. $\mathbf{A}$ *has a right-inverse* $\mathbf{R}$ *if and only if* $k = m \leq n$.

Proof: Suppose that $k = m \leq n$; to see if a right-inverse exists, we simply try to solve the equations $\mathbf{AR} = \mathbf{I}$, whose augmented matrix is $[\mathbf{A}, \mathbf{I}]$. Since $k = m$, when we perform the row operations on $[\mathbf{A}, \mathbf{I}]$ which reduce $\mathbf{A}$ to row-echelon form, the last row of the reduced $\mathbf{A}$ is not zero; but then we know, as in the proof of **Key Theorem 3.6,** that solutions of the systems of equations exist, so $\mathbf{R}$ exists. Conversely, suppose that a right-inverse $\mathbf{R}$ exists, and let $\mathbf{F}$ be that nonsingular matrix guaranteed by **Key Theorem 3.3** to reduce $\mathbf{A}$ to its row-echelon form $\mathbf{H}$, that is,

$$\mathbf{FA} = \mathbf{H} \text{ in row-echelon form.}$$

Since $\mathbf{AR} = \mathbf{I}$, premultiplication by $\mathbf{F}$ gives

$$\mathbf{HR} = \mathbf{F}. \tag{3.49}$$

The last row of $\mathbf{F}$ is nonzero since $\mathbf{F}$ is nonsingular (see Ex. 3.55). Therefore the last (mth) row of the row-echelon matrix $\mathbf{H}$ is not zero because of (3.49). Since $\mathbf{H}$ is a row-echelon matrix this says $k = m$; since always $k \leq n$, we have $k = m \leq n$ as asserted.

THEOREM 3.10. Let **A** *be* $m \times n$ *and have rank* k. **A** *has a left-inverse* **L** *if and only if* $k = n \leq m$.

Proof: Suppose that $k = n \leq m$. Since $k = n$ and $n \leq m$, the row-echelon form of **A** is just

$$\mathbf{H} = \begin{bmatrix} \mathbf{I}_n \\ \mathbf{0} \end{bmatrix}.$$

Let **F** be the matrix of **Key Theorem 3.3** such that **FA** = **H**, and let **L** be the matrix consisting of the first n rows of **F**, so that

$$\mathbf{F} = \begin{bmatrix} \mathbf{L} \\ \mathbf{P} \end{bmatrix}.$$

From **FA** = **H** and the rules for multiplying partitioned matrices we find **LA** = **I** as required. Conversely, suppose that there exists an **L** such that **LA** = **I**; then if $\mathbf{Ax} = \mathbf{0}$ it follows that $\mathbf{x} = \mathbf{LAx} = \mathbf{0}$ and the equation $\mathbf{Ax} = \mathbf{0}$ has the unique solution $\mathbf{x} = \mathbf{0}$. By **Key Theorem 3.6**, this happens precisely when $k = n$, the number of unknowns; since always $k \leq m$, we have $k = n \leq m$ as asserted.

● *KEY THEOREM 3.11. Let* **A** *be a square* $m \times m$ *matrix.* **A** *is nonsingular if and only if the rank of* **A** *is* m.

Proof: This follows immediately from Theorems 3.9 and 3.10, and Theorem 1.4.

EXERCISE 3.25. Prove that each nonsingular matrix can be written as a product of elementary matrices.

EXERCISE 3.26. If **A** and **B** are $m \times m$ and **AB** is nonsingular, prove that both **A** and **B** are nonsingular. If **A** is nonsingular, prove that **AB** is nonsingular if and only if **B** is nonsingular.

EXERCISE 3.27. If **A** is $m \times m$ and nonsingular and **B** is $m \times p$, prove that the ranks of **B** and of **AB** are equal.

MISCELLANEOUS EXERCISES 3

EXERCISE 3.28. Find the value of α for which the following equations possess solutions:

$$\begin{aligned} x - 3y + 2z &= 4 \\ 2x + y - z &= 1 \\ 3x - 2y + z &= \alpha. \end{aligned}$$

EXERCISE 3.29. Show that the following equations for x, y, z possess a nonzero solution if $\alpha + \beta + \gamma = 0$:

$$\begin{aligned} x \qquad + y\cos\gamma + z\cos\beta &= 0 \\ x\cos\gamma + y \qquad + z\cos\alpha &= 0 \\ x\cos\beta + y\cos\alpha + z \qquad &= 0. \end{aligned}$$

EXERCISE 3.30. Show that if a, b, c are distinct nonzero numbers, the equations

$$\begin{aligned} ax + by + cz &= 1 \\ a^2x + b^2y + c^2z &= 1 \\ a^3x + b^3y + c^3z &= 1 \end{aligned}$$

possess the solution

$$x = \frac{(b-1)(1-c)}{a(c-a)(a-b)}, \qquad y = \frac{(1-c)(a-1)}{b(a-b)(b-c)}, \qquad z = \frac{(a-1)(1-b)}{c(b-c)(c-a)}.$$

EXERCISE 3.31. Show that if **A** is a square matrix of order m and rank m, it possesses a unique right inverse. Show that if the rank of **A** is less than m, it does not possess a right-inverse.

EXERCISE 3.32. Show that if **A** is a general $m \times n$ matrix, then:

(a) If $m > n$, **A** has no right-inverse.
(b) If $m < n$, there are exactly two possibilities:
  (i) If the rank of **A** is m, there exists an infinite number of right-inverses.
  (ii) If the rank of **A** is less than m, there exist no right-inverses.

EXERCISE 3.33. Let **B** and **C** denote 3×4 matrices and let

$$\mathbf{A} = \begin{bmatrix} 1 & -2 & 3 \\ -2 & 5 & -6 \\ 2 & -3 & 6 \end{bmatrix}.$$

What conclusions can we draw from the equation $\mathbf{AB} = \mathbf{AC}$?

EXERCISE 3.34. Find all vectors **b** for which the equation $\mathbf{Ax} = \mathbf{b}$ can be solved, and find the corresponding general solution, for

$$\mathbf{A} = \begin{bmatrix} 4 & -1 & 2 & 6 \\ -1 & 5 & -1 & -3 \\ 3 & 4 & 1 & 3 \end{bmatrix}.$$

EXERCISE 3.35. Reduce the following matrix to row-echelon form:

$$\mathbf{A} = \begin{bmatrix} 1 & -2 & 3 & 1 \\ 2 & k & 6 & 6 \\ -1 & 3 & k-3 & 0 \end{bmatrix}.$$

Suppose that this is the augmented matrix $[\mathbf{A}, \mathbf{b}]$ for a system $\mathbf{Ax} = \mathbf{b}$ of three equations in three unknowns. Deduce from the row-echelon form that for $k = 0$ the system has an infinite number of solutions, for another value of k the system is contradictory, and for all other values of k the system has a unique solution. For the case $k = 0$, what is the general form of the solution?

EXERCISE 3.36. Let $t_1, \ldots, t_p$ be distinct real numbers, and let $b_1, \ldots, b_p$, $b'_1, \ldots, b'_p$ be arbitrary real numbers. Prove that there exists precisely one polynomial q of degree less than or equal to $n = 2p - 1$ such that $q(t_i) = b_i$ and $q'(t_i) = b'_i$ for $1 \le i \le p$.

EXERCISE 3.37. Prove that if $\mathbf{B}$ is a submatrix of $\mathbf{A}$, then the rank of $\mathbf{B}$ is not greater than the rank of $\mathbf{A}$.

EXERCISE 3.38. If $\mathbf{B}$, $\mathbf{C}$ are of ranks r, s, respectively, and

$$\mathbf{A} = \begin{bmatrix} \mathbf{B} & \mathbf{0} \\ \mathbf{0} & \mathbf{C} \end{bmatrix}.$$

show that $\mathbf{A}$ is of rank $r + s$.

EXERCISE 3.39. If $\mathbf{A}$, $\mathbf{B}$ are $m \times n$ matrices with the same rank, show that there exist nonsingular matrices $\mathbf{E}$, $\mathbf{F}$ such that

$$\mathbf{EAF} = \mathbf{B}.$$

EXERCISE 3.40. If $\mathbf{K}$ is skewsymmetric, prove that $\mathbf{x}^T\mathbf{Kx} = [0]$ for all $\mathbf{x}$ and thereby deduce that $\mathbf{I} + \mathbf{K}$ is nonsingular.

EXERCISE 3.41. Prove that the condition for the n straight lines

$$a_i x + b_i y = c_i \qquad (i = 1, \ldots, n)$$

to pass through one point is that the matrices $[\mathbf{a}, \mathbf{b}]$ and $[\mathbf{a}, \mathbf{b}, \mathbf{c}]$ have the same rank, where $\mathbf{a} = [a_i]$, $\mathbf{b} = [b_i]$, $\mathbf{c} = [c_i]$ are $n \times 1$ column vectors.

EXERCISE 3.42. Prove that three points (x_i, y_i), $i = 1, 2, 3$, in the plane are collinear if and only if the rank of the following matrix is less than three:

$$\begin{bmatrix} x_1 & y_1 & 1 \\ x_2 & y_2 & 1 \\ x_3 & y_3 & 1 \end{bmatrix}.$$

EXERCISE 3.43. A matrix $\mathbf{A}$ is said to be *divisor of zero* if $\mathbf{A} \neq \mathbf{0}$ and if there exists a matrix $\mathbf{B} \neq \mathbf{0}$ such that $\mathbf{AB} = \mathbf{0}$, or there exists a matrix $\mathbf{C} \neq \mathbf{0}$ such that $\mathbf{CA} = \mathbf{0}$. Show that an arbitrary matrix $\mathbf{A}$ is a divisor of zero if and only if the rank of $\mathbf{A}$ is less than $\max(m, n)$. (*Hint:* Consider $\mathbf{Ax} = \mathbf{0}$, and $\mathbf{y}^T\mathbf{A} = \mathbf{0}$, where $\mathbf{x}$, $\mathbf{y}$ are column vectors.) Deduce that every rectangular matrix is a divisor of zero, and a square matrix is a divisor of zero only if it is singular.

EXERCISE 3.44. Prove that every matrix of rank r is a sum of r matrices of rank 1.

EXERCISE 3.45. If $\mathbf{A}$ is a square matrix of rank 1, prove that $\mathbf{A}^2 = c\mathbf{A}$ for some constant c.

EXERCISE 3.46. Show that $\mathbf{AB}$ is singular if $\mathbf{A}$ is $m \times n$ and $\mathbf{B}$ is $n \times m$ with $n < m$.

EXERCISE 3.47. Suppose that $\mathbf{AB} = \mathbf{0}$. If $\mathbf{A}$ is $n \times n$ and $\mathbf{B}$ is $n \times p$, show that either $\mathbf{B} = \mathbf{0}$ or $\mathbf{A}$ is singular. If $\mathbf{A}$ is $m \times n$ and $\mathbf{B}$ is $n \times n$, show that either $\mathbf{A} = \mathbf{0}$ or $\mathbf{B}$ is singular. If $\mathbf{A}$ and $\mathbf{B}$ are both $n \times n$, show that $\mathbf{A} = \mathbf{0}$ or $\mathbf{B} = \mathbf{0}$ or both $\mathbf{A}$ and $\mathbf{B}$ are singular.

EXERCISE 3.48. A *block diagonal* matrix is a square partitioned matrix in which the diagonal elements are square matrices and all other elements are null matrices. Show that a block diagonal matrix is nonsingular if and only if all the diagonal matrix elements are nonsingular.

EXERCISE 3.49. Prove that the following statements are equivalent ($\mathbf{A}$ is a square matrix):

(a) $\mathbf{A}$ has a left-inverse.
(b) $\mathbf{A}$ has a right-inverse.
(c) $\mathbf{A}$ has an inverse.
(d) $\mathbf{A}$ is row-equivalent to the unit matrix.
(e) $\mathbf{A}$ can be expressed as a product of elementary matrices.

EXERCISE 3.50. Show that the only square matrices of order n that commute with all other square matrices are of the form $k\mathbf{I}$, where k is a scalar.

EXERCISE 3.51. Show that interchanging two rows (columns) of $\mathbf{A}$ interchanges the corresponding columns (rows) of $\mathbf{A}^{-1}$. Show that if the sth column of $\mathbf{A}$ is the unit vector $\mathbf{e}_i$, then the ith column of $\mathbf{A}^{-1}$ is $\mathbf{e}_s$.

EXERCISE 3.52. One of the difficulties in analyzing mechanical structures by computer is to provide an automatic procedure for deciding which members of a framework are redundant, if any. A method for doing this by systematic use of the concept of the rank of a matrix has been described by Robinson [106]. Write a short essay on the method.

EXERCISE 3.53. Suppose that **A** is a square matrix. Use Theorem 3.10 and **Key Theorem 3.11** to prove that if **L** is a left inverse of **A** then **A** is nonsingular and $\mathbf{A}^{-1} = \mathbf{L}$; prove the analogous statement for **A** having a right inverse **R**.

EXERCISE 3.54. Deduce the following results from the theorems in Section 3.7, and make the statements more precise by stating them in terms of rank:

(a) A square matrix *either* posseses a (two-sided) inverse *or* it has neither a left-inverse nor a right-inverse.

(b) A nonsquare $m \times n$ matrix (i.e., $m \neq n$) *never* possesses a (two-sided) inverse. If $m < n$ then it cannot have a left-inverse; it may or may not have a right-inverse. If $m > n$, then the preceding sentence holds with "left" and "right" interchanged.

EXERCISE 3.55. Use Ex. 3.47 to show that a square matrix having a zero row or column must be singular.

CHAPTER FOUR

VECTORS AND VECTOR SPACES

This chapter continues the development of fundamental theoretical concepts and tools needed to understand the applications, only we introduce here a more geometrical flavor. As with the row-reduction of Chapter 3, a single fundamental notion dominates this chapter, namely, the notion of linear dependence (or linear independence). The most important concepts are presented in **Key Theorem 4.4** and in **Key Theorem 4.8** and its **Key Corollary 4.1,** while crucial techniques appear in **Key Theorems 4.7** and **4.12** and in **Key Proof 4.12.**

4.1 GEOMETRICAL VECTORS IN THREE-DIMENSIONAL SPACE

So far, the theoretical development has been concerned with the manipulation of arrays of numbers. Our present situation can be compared with that of a student who has learned the rules for the manipulation of the symbols and equations representing points, lines, and planes in three-dimensional cartesian geometry, without knowing anything about the underlying geometrical picture represented by the symbols. In order to obtain a deeper appreciation of the significance of matrices, it is necessary to understand the idea of a *vector space*. We introduce this concept by discussing geometrical vectors in three-dimensional "physical" space.

Later, in Section 4.5, we will return to matrices and show how matrix manipulations are useful for solving some problems involving vectors; we will also describe some matrix results using the language of vectors. Chapter 5 presents another approach, that of describing vector-space concepts in terms of matrices. For the moment, however, we nearly ignore matrices as tools and use them only to provide some examples.

To discuss vectors geometrically, we first choose a system of three perpendicular axes, which are labeled x, y, z (Figure 4.1). A vector is represented by a line OP starting from the origin O of coordinates and pointing in the direction of the vector, the length OP being equal to the magnitude of the vector. Instead of specifying the direction and length of OP we could equally have specified the coordinates (x, y, z) of the point P relative to the given coordinate system. We introduce the notation $\vec{u}$ (i.e., a symbol with an arrow above it) to represent a geometrical vector. We shall write $\vec{u} = (x_1, y_1, z_1)$ to denote that the components of $\vec{u}$ are given by x_1, y_1, z_1.

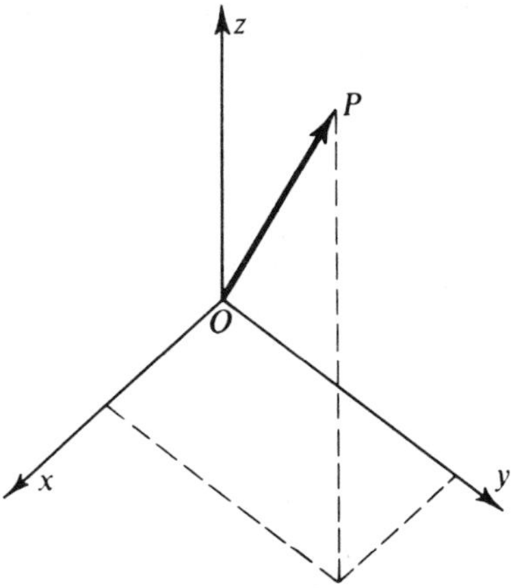

Figure 4.1

Two standard operations on geometrical vectors are:

1. *Multiplication of a geometrical vector by a scalar, say k.* This leaves the direction of the vector unchanged, but multiplies its length by $|k|$, the absolute value of k. In terms of coordinates, the vector (x_1, y_1, z_1) is changed to (kx_1, ky_1, kz_1):

$$k\vec{u} = (kx_1, ky_1, kz_1).$$

2. *Addition of two geometrical vectors.* This is performed by the parallelogram law. In terms of coordinates, if we add vectors $\vec{u} = (x_1, y_1, z_1)$

and $\vec{v} = (x_2, y_2, z_2)$, we obtain a vector whose coordinates are the sums of those of $\vec{u}$ and $\vec{v}$:

$$\vec{u} + \vec{v} = (x_1 + x_2, y_1 + y_2, z_1 + z_2).$$

Geometrical vectors obey the commutative and associative laws of addition:

$$\vec{u} + \vec{v} = \vec{v} + \vec{u},$$
$$(\vec{u} + \vec{v}) + \vec{w} = \vec{u} + (\vec{v} + \vec{w}).$$

Instead of thinking in terms of coordinates (x, y, z) we can think in terms of column matrices. Suppose that we define

$$\mathbf{u} = \begin{bmatrix} x_1 \\ y_1 \\ z_1 \end{bmatrix}, \qquad \mathbf{v} = \begin{bmatrix} x_2 \\ y_2 \\ z_2 \end{bmatrix}.$$

Then the definitions of multiplication of a matrix by a scalar, and of addition of matrices, give

(1a) $$k\mathbf{u} = \begin{bmatrix} kx_1 \\ ky_1 \\ kz_1 \end{bmatrix}.$$

(2a) $$\mathbf{u} + \mathbf{v} = \begin{bmatrix} x_1 + x_2 \\ y_1 + y_2 \\ z_1 + z_2 \end{bmatrix}.$$

Similarly, we have

$$\mathbf{u} + \mathbf{v} = \mathbf{v} + \mathbf{u},$$
$$(\mathbf{u} + \mathbf{v}) + \mathbf{w} = \mathbf{u} + (\mathbf{v} + \mathbf{w}).$$

There is clearly a correspondence or parallelism between the quantities $\vec{u}$, $\vec{v}$ and $\mathbf{u}$, $\mathbf{v}$. It can be shown, by establishing a strict correspondence between all the properties of geometrical vectors and of 3×1 column matrices, that any result which is true when expressed in terms of geometrical vectors is also true when expressed in terms of column matrices, and vice versa. This means that we can give a geometrical interpretation of results involving column matrices.

The reason for developing this correspondence between geometrical vectors and 3×1 column matrices is that we can now give an intuitive background for the more abstract developments discussed later, assuming that the reader has a minimum background in vector analysis and analytic geometry.

The position of a point in a plane can be specified by choosing two perpendicular axes and determining the coordinates (x, y) of the point with reference to the axes. Two numbers are required to specify the point, and we say that the plane forms a *two-dimensional space*. Similarly, the position of a point on a line can be specified by giving the distance of the point from a fixed point or origin on the line, and we say that the line forms a *one-dimensional space*. A definition of precisely what is meant by a k-dimensional space will be given later (Section 4.4, Definition 4.8), but the ordinary geometrical interpretation for $k = 1, 2, 3$, as just described, will suffice for our present purposes.

If we consider all possible multiples of a vector $\vec{u}$, that is, if we consider the set of all vectors of the form $\alpha\vec{u}$ where α is a scalar, we specify a line (or one-dimensional space) in three-dimensional space. Similarly, vectors of the form $\alpha\vec{u} + \beta\vec{v}$, which are sums of multiples of two vectors $\vec{u}$ and $\vec{v}$ that are not collinear, define a plane (or two-dimensional space) in three-dimensional space. By taking suitable multiples of $\vec{u}$ and $\vec{v}$ we can reach any point in this plane, but we cannot move outside the plane. In order to reach a point outside the plane, we must introduce a third vector $\vec{w}$ which does not lie in the plane defined by $\vec{u}$ and $\vec{v}$. By taking sums of multiples of vectors of the form $\alpha\vec{u} + \beta\vec{v} + \gamma\vec{w}$ we can reach any point in three-dimensional space.

It is convenient to introduce terminology that will be defined precisely in later sections of this chapter. If we are, for instance, dealing with a plane or two-dimensional space, in a three-dimensional space, we say that the plane is a *subspace* of the three-dimensional space. More generally, an m-dimensional space which is contained in an n-dimensional space $(m \leq n)$ is said to be an *m-dimensional subspace*. (We allow the possibility $m = n$ for convenience.) As a further example, a line lying in a plane is a one-dimensional subspace of the plane.

Consider the geometrical set defined by all sums of multiples of nonzero vectors $\vec{u}$ and $\vec{v}$,

$$\vec{x} = \alpha\vec{u} + \beta\vec{v}.$$

Any such vector $\vec{x}$ in the set is said to be a *linear combination* of $\vec{u}$ and $\vec{v}$. If $\vec{u}$ and $\vec{v}$ are not collinear, then the set forms a plane. If $\vec{u}$ and $\vec{v}$ are collinear, we have $\vec{u} = k\vec{v}$ for some nonzero constant k, the set forms a line, and we say that $\vec{u}$ and $\vec{v}$ are linearly dependent on one another.

So far we have implicitly assumed that the vectors involved have nonzero lengths. Although it is convenient to make this assumption in the geometrical discussion, it is a nuisance later, and from now on we allow the possibility that any vector in any discussion may be the zero vector $\vec{0}$ represented by the point at the origin, with zero length and no direction. A detailed discussion of linear dependence and independence when zero vectors may be present is postponed until later (Ex. 4.9). We now say quite generally that a set of two vectors $\vec{u}$, $\vec{v}$ is *linearly dependent* if constants α, β, not both zero, can be found such that $\alpha\vec{u} + \beta\vec{v} = \vec{0}$. If $\{\vec{u}, \vec{v}\}$ is linearly dependent and neither vector is zero, both the constants α, β must be nonzero, and $\vec{v} = -(\alpha/\beta)\vec{u}$, that is, $\vec{u}$, $\vec{v}$ have the same (or directly opposite) direction. If nonzero vectors $\vec{u}$, $\vec{v}$ do not point in the same (or directly opposite) direction, then we cannot find nonzero constants α, β such that $\alpha\vec{u} + \beta\vec{v} = \vec{0}$ and we say that $\{\vec{u}, \vec{v}\}$ is a *linearly independent* set. Similar definitions hold for sets of three or more vectors. Thus $\{\vec{u}, \vec{v}, \vec{w}\}$ is said to be linearly dependent if and only if three constants α, β, γ, at least one of which is nonzero, exist such that $\alpha\vec{u} + \beta\vec{v} + \gamma\vec{w} = \vec{0}$.

If $\vec{u}$, $\vec{v}$ are nonzero vectors that do not have the same (or directly opposite) direction, the collection of vectors $\alpha\vec{u} + \beta\vec{v}$ forms a plane through $\vec{0}$ and is a two-dimensional subspace of three-dimensional space, as we mentioned above. We say that the subspace is *generated* or *spanned* by $\vec{u}$, $\vec{v}$. Also the vectors $\vec{u}$, $\vec{v}$ are said to form a *basis* (or set of base vectors) for the subspace. Any set of two vectors lying in the plane can be chosen as a basis, provided that the set is linearly independent. The set must be linearly independent because if it is linearly dependent there are two possibilities: at least one of $\vec{u}$, $\vec{v}$ is zero, or both $\vec{u}$, $\vec{v}$ are nonzero in which case they must have the same (or directly opposite) direction. In neither case can $\vec{u}$, $\vec{v}$ define a plane. It is intuitively clear that any three coplanar vectors $\vec{p}$, $\vec{q}$, $\vec{r}$ must be related by an equation of the form

$$\alpha\vec{p} + \beta\vec{q} + \gamma\vec{r} = \vec{0},$$

with at least one of α, β, γ nonzero—i.e., any set of three vectors in a two-dimensional space is linearly dependent.

Similar results hold for three-dimensional space. We summarize the results. The definitions of "span," "basis," etc., should be clear from the discussion of the two-dimensional case in the last paragraph.

1. Three vectors in an independent set in three-dimensional space are noncoplanar.
2. Any linearly independent set of three vectors spans a three-dimensional space and therefore can be chosen as a basis for the space.
3. Any basis for a three-dimensional space must be a linearly independent set of three vectors.
4. Any set of four or more vectors in three-dimensional space is linearly dependent.

4.2 GENERAL VECTOR SPACES

So far, we have described vectors and vector spaces as geometric entities, although we have seen that they correspond simply to the nongeometric column matrices with one, two, or three entries. In this section we want to discuss very briefly what is really at the heart of the concept of vector space. To do this we define an abstract model of what it means to be a vector space, and we indicate how this abstract notion can be used to define properties or concepts that are valid in all concrete vector spaces. We will quickly return, however, to simple extensions of the geometric notions of vectors discussed in the last section. We do this partly because you probably will find it simpler to think in these more concrete terms; partly because these special cases are sufficient to handle many applications; and partly because much of what we develop for the special vector spaces is true in general.

The basic property of vectors, from our present point of view, is that we can form linear combinations of them. There are various other mathematical entities, or objects, such that we wish to consider linear combinations of these entities. We already observed geometrical vectors in two- and three-dimensional space, where vectors are defined as directed line-segments. While at first sight it might seem that geometrical vectors in three-dimensional space and 3×1 column matrices are mathematical objects of completely different kinds, we indicated in Section 4.1 that this is not the case, since the operations obey the same laws for geometrical vectors and column matrices.

Let us assume that we are dealing with a collection or set of mathematical entities in connection with which certain operations called "addition" and "multiplication by a scalar" are defined. We will not define the precise nature of these entities, so that we will not specify how these operations are actually carried out. However, we will specify axioms or rules that the operations must obey. In this way, we can isolate the fundamental assumptions without considering the concrete details of particular cases. The advantage of this procedure is that we can work out the consequences of the assumptions once and for all. If a particular set is known to obey the fundamental axioms, then we can immediately say that all the theorems in the general theory apply. Our point is to simplify and clarify the theory by discarding unnecessary, irrelevant properties of concrete mathematical objects that confuse the situation.

We proceed to list the assumptions used to define an *abstract vector space*. We will call the entities in an abstract vector space *vectors*, regardless of the fact that they may be quite different from the geometric vectors introduced earlier. Similarly we will use the word *scalars* for the entities that can multiply vectors, although they may be different from the numbers we have used earlier. No attempt is made to set down a minimum number of assumptions. It is easier to see the implications of the assumptions if we allow a certain amount of redundancy at the expense of mathematical elegance.

We now consider specific detail. We assume that we have a set of elements called *scalars*, denoted by Greek letters $\alpha, \beta, \gamma \ldots,$, that form a *field* F, by which we mean that they satisfy the following conditions:

To every pair of scalars α, β there corresponds a scalar $\alpha + \beta$ in F called the *sum* of α and β such that:

1. Addition is commutative and associative,

$$\alpha + \beta = \beta + \alpha, \qquad \alpha + (\beta + \gamma) = (\alpha + \beta) + \gamma.$$

2. There exists a unique element 0 in F called *zero* such that $\alpha + 0 = \alpha$ for every α in F.
3. For every α in F there corresponds a unique scalar $(-\alpha)$ such that $\alpha + (-\alpha) = 0$.

To every pair of scalars α, β there corresponds a scalar $\alpha\beta$ called the *product* of α and β such that:

1. Multiplication is commutative and associative,

$$\alpha\beta = \beta\alpha, \qquad \alpha(\beta\gamma) = (\alpha\beta)\gamma.$$

2. There exists a unique nonzero scalar 1 (called *one* or *unity* or the *identity* element) such that $\alpha 1 = \alpha$ for every α in F.
3. To each nonzero α in F there corresponds a unique scalar α^{-1} called the *inverse* of α such that $\alpha\alpha^{-1} = 1$.
4. Multiplication is distributive with respect to addition.

$$\alpha(\beta + \gamma) = \alpha\beta + \alpha\gamma, \qquad (\beta + \gamma)\alpha = \beta\alpha + \gamma\alpha.$$

In previous sections, we dealt sometimes with the field of real numbers and sometimes with the field of complex numbers. Another example of a field is the set of all rational numbers. An example of a set that is not a field is the set of all positive and negative integers, since the multiplicative inverse of an integer is not in general an integer.

We can now define what we mean by an abstract vector space.

DEFINITION 4.1. An *abstract vector space* (or *linear space*) consists of:

(a) A field F of *scalars*.
(b) A set V of entities called *vectors*.
(c) An operation called *vector addition* that associates a *sum* $\mathbf{x} + \mathbf{y}$ in V with each pair of vectors $\mathbf{x}, \mathbf{y}$ in such a way that:
    (1) Addition is commutative and associative:

$$\mathbf{x} + \mathbf{y} = \mathbf{y} + \mathbf{x}, \qquad \mathbf{x} + (\mathbf{y} + \mathbf{z}) = (\mathbf{x} + \mathbf{y}) + \mathbf{z}.$$

(2) There exists in V a unique vector $\mathbf{0}$ called the zero vector such that $\mathbf{x} + \mathbf{0} = \mathbf{x}$ for all $\mathbf{x}$ in V.
(3) For each vector $\mathbf{x}$ in V there is a unique vector $(-\mathbf{x})$ such that $\mathbf{x} + (-\mathbf{x}) = \mathbf{0}$.

(d) An operation called *multiplication by a scalar* that associates with each scalar α in F and vector $\mathbf{x}$ in V, a vector $\alpha\mathbf{x}$ in V is called the *product* of α and $\mathbf{x}$ such that:
(1) Multiplication by a scalar is associative and is distributive with respect to scalar addition:

$$\alpha(\beta\mathbf{x}) = (\alpha\beta)\mathbf{x}, \qquad (\alpha + \beta)\mathbf{x} = \alpha\mathbf{x} + \beta\mathbf{x}.$$

(2) Multiplication by a scalar is distributive with respect to vector addition:

$$\alpha(\mathbf{x} + \mathbf{y}) = \alpha\mathbf{x} + \alpha\mathbf{y}.$$

(3) If 1 is the identity element in F then $1\mathbf{x} = \mathbf{x}$.

Unless stated otherwise, we will usually assume that the field associated with a vector space is the field $\mathbb{R}$ of real numbers or the field $\mathbb{C}$ of complex numbers.

We now give examples of linear spaces:

1. Free physical vectors in three-dimensional space. The vectors in the space are specified by giving a direction and a length. By saying that vectors are "free," we mean that the line-segment defining the vector may start at any point in space. The rule for addition is the usual parallelogram law. The rule for multiplication by a scalar is that the direction of the vector remains unchanged and its length is multiplied by the scalar.
2. Fixed physical vectors in three-dimensional space, where by "fixed" we mean that the vectors all start from the origin of coordinates. The rules for vector addition and multiplication by a scalar are as in (1) with the additional proviso that the results of these operations are vectors that also start from the origin.
3. $m \times n$ matrices (in particular, column vectors and row vectors) with the usual rules for addition and multiplication by a scalar. The field can be either the field of real numbers or the field of complex numbers.
4. The set of vectors with an infinite number of components $[x_1, x_2, \ldots]$, with the obvious definitions for addition and multiplication by a real or complex scalar. (This is not a particularly useful vector space unless restrictions are placed on the behavior of x_i as $i \to \infty$.)
5. The space of real-valued functions $f(x)$, $a \leq x \leq b$, that are continuous in this range. This is denoted by $C[a, b]$, with the usual rules for addition and multiplication by a real scalar. This is a vector space since the sum of two continuous functions is continuous, and a multiple of a continuous function is continuous.

6. The space of functions that are k times differentiable, and whose kth derivatives are continuous, in $a \leq x \leq b$, denoted $C^{(k)}[a, b]$.
7. The space of polynomials $P_n(x) = a_0 + a_1 x + \cdots + a_n x^n$, for any n (so that the space contains polynomials of arbitrarily high degree).
8. The space of polynomials $P_n(x)$, $n \leq N$, where N is fixed.
9. The set of all solutions of a given linear homogeneous differential equation

$$a_0(x)\frac{d^n y}{dx^n} + a_1(x)\frac{d^{n-1}y}{dx^{n-1}} + \cdots + a_n(x)y = 0.$$

10. The set of all solutions y of an integral equation of the form

$$\int_0^a K(x, t)y(t)\,dt + \lambda y(x) = 0, \qquad 0 \leq x \leq a,$$

where $K(x, t)$ and λ are given, and $K(x, t)$ is continuous in x, t.

11. If V and W are any two vector spaces over the same field, with vectors denoted by $\mathbf{v}$, $\mathbf{w}$, respectively, the *product space* of V and W over the same field is the space of vectors $\mathbf{u} = (\mathbf{v}, \mathbf{w})$, where, if a second vector in the product space is $(\mathbf{v}', \mathbf{w}')$, the laws of addition and scalar multiplication are given by

$$(\mathbf{v}, \mathbf{w}) + (\mathbf{v}', \mathbf{w}') = (\mathbf{v} + \mathbf{v}', \mathbf{w} + \mathbf{w}'), \quad \lambda(\mathbf{v}, \mathbf{w}) = (\lambda\mathbf{v}, \lambda\mathbf{w}).$$

EXERCISE 4.1. Are the following sets abstract vector spaces, under suitable laws of addition and scalar multiplication (to be stated if the answer is yes)? If not, why not?

(a) Fixed vectors in three-dimensional space with end points in the first quadrant.
(b) Free vectors in a plane except for vectors perpendicular to a given line.
(c) Ratios of polynomials $P_m(x)/Q_n(x)$ for all (finite) m, n.
(d) Ratios of polynomials $P_m(x)/Q_n(x)$ for $m \leq M$, $n \leq N$.
(e) The set of all solutions of the differential equation

$$\frac{dy}{dx} + \alpha y = 1.$$

(f) The set of vectors with an infinite number of components $[x_1, x_2, \ldots]$ with convergent sum of absolute values $|x_1| + |x_2| + \cdots$.

EXERCISE 4.2. In the examples of abstract vector spaces (1)–(11), define the rules for addition and multiplication by a scalar where these have not been specified in detail. Also, prove that if $\mathbf{x}$, $\mathbf{y}$ are vectors in the space, then $\mathbf{x} + \mathbf{y}$ and $\lambda\mathbf{y}$ are also in the space, where this has not been done in the examples.

In examples (1)–(11) above, it is obvious that many of the vector spaces are related. For example, every polynomial of degree $n \leq N$ [see (8)] is also a polynomial of arbitrary degree [see (7)], which is also a k-times differentiable function [see (6)], which in turn is also continuous [see (5)]: This situation is very common and the relationship is given a special name.

DEFINITION 4.2. If V and W are both vector spaces over the same field and if every element of V is also an element of W, then and only then do we say that V is a *subspace* of W.

In this terminology, the vector space of (8) is a subspace of that of (7), etc.

EXERCISE 4.3. Prove that the set of all 3×1 matrices whose (1, 1) element is zero is a subspace of the set of all 3×1 matrices.

The special case of $n = 1$ (or of $m = 1$ instead) in (3) above is so common and important that we denote it by a special symbol.

DEFINITION 4.3. The vector space of all $m \times 1$ real column matrices, considered over the field of real numbers, is denoted by $\mathbb{R}^m$. If we allow *complex* $m \times 1$ matrices, over the complex field, the space is denoted by $\mathbb{C}^m$.

Throughout the rest of this book we will primarily concern ourselves only with the vector spaces $\mathbb{R}^m$ and $\mathbb{C}^m$. Many of the results we will obtain are valid in the generality of abstract vector spaces, however.

EXERCISE 4.4. Let $\mathbf{A}$ be a given $m \times n$ real matrix. Prove that the set of all solutions $\mathbf{x}$ to $\mathbf{Ax} = \mathbf{0}$ is a subspace of $\mathbb{R}^n$.

EXERCISE 4.5. Let $\mathbf{A}$ be a given $m \times n$ real matrix and let $\mathbf{b}$ be a nonzero vector in $\mathbb{R}^m$. Prove that the set of all solutions $\mathbf{x}$ to $\mathbf{Ax} = \mathbf{b}$ is *not* a subspace of $\mathbb{R}^n$.

EXERCISE 4.6. Let $\{\mathbf{u}_1, \ldots, \mathbf{u}_s\}$ be a subset of a vector space. Prove that the set of all linear combinations $\alpha_1\mathbf{u}_1 + \cdots + \alpha_s\mathbf{u}_s$ forms a subspace as the $\{\alpha_i\}$ range over the scalar field.

SOLUTION: Let

$$\mathbf{u} = \alpha_1\mathbf{u}_1 + \alpha_2\mathbf{u}_2 + \cdots + \alpha_s\mathbf{u}_s,$$
$$\mathbf{v} = \beta_1\mathbf{u}_1 + \beta_2\mathbf{u}_2 + \cdots + \beta_s\mathbf{u}_s.$$

Then

$$\mathbf{u} + \mathbf{v} = (\alpha_1 + \beta_1)\mathbf{u}_1 + (\alpha_2 + \beta_2)\mathbf{u}_2 + \cdots + (\alpha_s + \beta_s)\mathbf{u}_s,$$
$$\alpha\mathbf{u} = (\alpha\alpha_1)\mathbf{u}_1 + (\alpha\alpha_2)\mathbf{u}_2 + \cdots + (\alpha\alpha_s)\mathbf{u}_s.$$

Each of these is a linear combination of $\mathbf{u}_1, \ldots, \mathbf{u}_s$, so that the set of all linear combinations is a vector space.

4.3 LINEAR DEPENDENCE AND INDEPENDENCE

The idea of linear independence has already been introduced by means of geometrical considerations. We now approach this concept from a purely algebraic point of view.

If we wish to deal with a collection of column vectors with, say, three elements apiece (that is, with $\mathbb{R}^3$ or $\mathbb{C}^3$), such as

$$\begin{bmatrix} 1 \\ -1 \\ 0 \end{bmatrix}, \quad \begin{bmatrix} 1 \\ 4 \\ -2 \end{bmatrix}, \quad \begin{bmatrix} 0 \\ 0 \\ 5 \end{bmatrix}, \quad \begin{bmatrix} 2 \\ 3 \\ -4 \end{bmatrix}, \quad \ldots, \tag{4.1}$$

it is often convenient to think of them as consisting of sums of multiples of certain basic column vectors, so that computations with "general" vectors can be reduced to hopefully simpler computations involving only the basic vectors. Thus if we define the "unit vectors"

$$\mathbf{e}_1 = \begin{bmatrix} 1 \\ 0 \\ 0 \end{bmatrix}, \quad \mathbf{e}_2 = \begin{bmatrix} 0 \\ 1 \\ 0 \end{bmatrix}, \quad \mathbf{e}_3 = \begin{bmatrix} 0 \\ 0 \\ 1 \end{bmatrix} \tag{4.2}$$

then the first vector in (4.1) is $\mathbf{e}_1 - \mathbf{e}_2$, the second is $\mathbf{e}_1 + 4\mathbf{e}_2 - 2\mathbf{e}_3$, and similarly any 3×1 column vector can be expressed as sums of multiples of the three basic vectors $\mathbf{e}_1, \mathbf{e}_2, \mathbf{e}_3$. It is not necessary to choose the $\mathbf{e}_i$ defined above; we could just as easily choose

$$\mathbf{e}'_1 = \begin{bmatrix} 1 \\ 2 \\ 1 \end{bmatrix}, \quad \mathbf{e}'_2 = \begin{bmatrix} 1 \\ 0 \\ -1 \end{bmatrix}, \quad \mathbf{e}'_3 = \begin{bmatrix} 1 \\ -2 \\ 1 \end{bmatrix}. \tag{4.3}$$

In order to express the first vector in (4.1) in terms of these vectors, we have to find numbers $\alpha_1, \alpha_2, \alpha_3$ such that

$$\begin{bmatrix} 1 \\ -1 \\ 0 \end{bmatrix} = \alpha_1 \begin{bmatrix} 1 \\ 2 \\ 1 \end{bmatrix} + \alpha_2 \begin{bmatrix} 1 \\ 0 \\ -1 \end{bmatrix} + \alpha_3 \begin{bmatrix} 1 \\ -2 \\ 1 \end{bmatrix}.$$

This gives a set of three equations in three unknowns, the solution of which is found to be $\alpha_1 = 0, \alpha_2 = \alpha_3 = \frac{1}{2}$. Hence, the first vector in (4.1) is $\frac{1}{2}(\mathbf{e}'_2 + \mathbf{e}'_3)$.

However, the vectors in (4.1) *cannot* be expressed in terms of the set

$$\mathbf{e}_1'' = \begin{bmatrix} 1 \\ 1 \\ 1 \end{bmatrix}, \quad \mathbf{e}_2'' = \begin{bmatrix} 1 \\ 0 \\ -1 \end{bmatrix}, \quad \mathbf{e}_3'' = \begin{bmatrix} 1 \\ -1 \\ -3 \end{bmatrix}. \tag{4.4}$$

For instance, suppose that we try to express the first vector in (4.1) in terms of this set, by writing

$$\begin{bmatrix} 1 \\ -1 \\ 0 \end{bmatrix} = \alpha_1 \begin{bmatrix} 1 \\ 1 \\ 1 \end{bmatrix} + \alpha_2 \begin{bmatrix} 1 \\ 0 \\ -1 \end{bmatrix} + \alpha_3 \begin{bmatrix} 1 \\ -1 \\ -3 \end{bmatrix}.$$

This gives a set of three equations in the three unknowns α_1, α_2, α_3. The equation obtained by eliminating α_2 between the first and third of these equations, together with the second equation, gives

$$\alpha_1 - \alpha_3 = \frac{1}{2}$$

$$\alpha_1 - \alpha_3 = -1.$$

These are contradictory equations, so that it is *not* possible to express the first vector in (4.1) in terms of the three vectors (4.4). From a more fundamental point of view, the reason for the difficulty can be seen directly from (4.4), because $\mathbf{e}_3'' = 2\mathbf{e}_2'' - \mathbf{e}_1''$. Since we are expressing vectors as sums of multiples of the basic vectors $\mathbf{e}_1''$, $\mathbf{e}_2''$, $\mathbf{e}_3''$, but $\mathbf{e}_3''$ can be expressed in terms of $\mathbf{e}_1''$ and $\mathbf{e}_2''$, there is no point in including $\mathbf{e}_3''$ as one of the basic vectors. We are therefore trying to express the vectors in (4.1) as sums of multiples of only two vectors $\mathbf{e}_1''$ and $\mathbf{e}_2''$. This will not in general be possible, since it will involve finding a solution for a set of three equations in two unknowns.

So far we have discussed only specific examples. The ideas that have been introduced can be formulated in general terms by means of the following definition. Although, at this stage, the concept that is defined may seem to be straightforward—and even superficial—it is extremely powerful and fundamental.

DEFINITION 4.4. A set of vectors $\mathbf{v}_1, \mathbf{v}_2, \ldots, \mathbf{v}_s$ is said to be *linearly dependent* if there exist numbers $\alpha_1, \alpha_2, \ldots, \alpha_s$, not all zero, such that

$$\alpha_1\mathbf{v}_1 + \alpha_2\mathbf{v}_2 + \cdots + \alpha_s\mathbf{v}_s = \mathbf{0}. \tag{4.5}$$

If the set is not linearly dependent, it is said to be *linearly independent.*

For example, the three vectors in (4.4) form a linearly dependent set, since we showed that $\mathbf{e}_1'' - 2\mathbf{e}_2'' + \mathbf{e}_3'' = \mathbf{0}$. On the other hand, if we wish to find out whether the three vectors in (4.2) form a linearly dependent set, we try to find constants α_i such that

$$\alpha_1 \begin{bmatrix} 1 \\ 0 \\ 0 \end{bmatrix} + \alpha_2 \begin{bmatrix} 0 \\ 1 \\ 0 \end{bmatrix} + \alpha_3 \begin{bmatrix} 0 \\ 0 \\ 1 \end{bmatrix} = \begin{bmatrix} \alpha_1 \\ \alpha_2 \\ \alpha_3 \end{bmatrix} = \begin{bmatrix} 0 \\ 0 \\ 0 \end{bmatrix}.$$

This automatically implies that $\alpha_1 = \alpha_2 = \alpha_3 = 0$ so that the set is linearly independent. Similarly, to find out whether the three vectors in (4.3) form a linearly independent set, we set

$$\alpha_1 \begin{bmatrix} 1 \\ 2 \\ 1 \end{bmatrix} + \alpha_2 \begin{bmatrix} 1 \\ 0 \\ -1 \end{bmatrix} + \alpha_3 \begin{bmatrix} 1 \\ -2 \\ 1 \end{bmatrix} = \begin{bmatrix} 0 \\ 0 \\ 0 \end{bmatrix}.$$

This implies

$$\begin{aligned} \alpha_1 + \alpha_2 + \alpha_3 &= 0 \\ 2\alpha_1 \qquad - 2\alpha_3 &= 0 \\ \alpha_1 - \alpha_2 + \alpha_3 &= 0; \end{aligned}$$

it is readily verified that the only solution is $\alpha_1 = \alpha_2 = \alpha_3 = 0$, so that the set of three vectors in (4.3) is linearly independent.

It is tempting to speak of "sets of linearly independent vectors" or even just of "linearly independent vectors"; this is incorrect. Inspection of Definition 4.4 reveals that it is the *collection* or *set* which is linearly dependent or independent, not the vectors themselves. While it may seem pedantic to worry about this point, it is important. If we have two sets of green apples, then the union of the two sets is also a set of green apples; analogously if we could say that

$$\left\{ \begin{bmatrix} 1 \\ 0 \end{bmatrix}, \begin{bmatrix} 0 \\ 1 \end{bmatrix} \right\}$$

is a set of "linearly independent vectors" and

$$\left\{ \begin{bmatrix} 1 \\ 0 \end{bmatrix}, \begin{bmatrix} 1 \\ 1 \end{bmatrix} \right\}$$

is a set of "linearly independent vectors," then it would follow as before that the union

$$\left\{ \begin{bmatrix} 1 \\ 0 \end{bmatrix}, \begin{bmatrix} 0 \\ 1 \end{bmatrix}, \begin{bmatrix} 1 \\ 1 \end{bmatrix} \right\}$$

is a set of "linearly independent vectors"; unfortunately, in fact,

$$\left\{\begin{bmatrix}1\\0\end{bmatrix}, \begin{bmatrix}0\\1\end{bmatrix}, \begin{bmatrix}1\\1\end{bmatrix}\right\}$$

is a linearly dependent *set* while

$$\left\{\begin{bmatrix}1\\0\end{bmatrix}, \begin{bmatrix}0\\1\end{bmatrix}\right\}$$

and

$$\left\{\begin{bmatrix}1\\0\end{bmatrix}, \begin{bmatrix}1\\1\end{bmatrix}\right\}$$

are both linearly independent *sets.*

We will, however, associate the words "dependent" and "independent" with individual vectors in the following way.

DEFINITION 4.5. If $\mathbf{v}_1, \ldots, \mathbf{v}_n$ are vectors and $\alpha_1, \ldots, \alpha_n$ are scalars, then $\alpha_1\mathbf{v}_1 + \cdots + \alpha_n\mathbf{v}_n$ is called a *linear combination* of the vectors $\mathbf{v}_1, \ldots, \mathbf{v}_n$. A vector $\mathbf{v}$ is said to be *linearly dependent on the vectors* $\mathbf{v}_1, \ldots, \mathbf{v}_n$ if and only if $\mathbf{v}$ can be written as a linear combination of the vectors $\mathbf{v}_1, \ldots, \mathbf{v}_n$; otherwise $\mathbf{v}$ is said to be *linearly independent of the vectors.*

EXERCISE 4.7. According to our calculations earlier in this section,

$$\begin{bmatrix}1\\-1\\0\end{bmatrix} = \tfrac{1}{2}(\mathbf{e}_2' + \mathbf{e}_3')$$

so that the first vector in (4.1) is linearly dependent on the vectors $\mathbf{e}_2', \mathbf{e}_3'$ in (4.3); on the other hand, show that the first vector is linearly independent of $\mathbf{e}_1'', \mathbf{e}_2'', \mathbf{e}_3''$ defined in (4.4).

We conclude this section by stating some direct consequences of the above definitions.

THEOREM 4.1.

(i) *If among the q vectors $\mathbf{u}_1, \ldots, \mathbf{u}_q$ there is a subset of $p < q$ vectors that is linearly dependent, then the entire set is linearly dependent.*

(ii) *If the set of q vectors $\mathbf{u}_1, \ldots, \mathbf{u}_q$ is linearly independent, then any subset of $p < q$ vectors is linearly independent.*

(iii) *A set of nonzero vectors $\mathbf{u}_1, \ldots, \mathbf{u}_p$ is linearly dependent if and only if one of the $\mathbf{u}_k$, for some k, is linearly dependent on the remaining vectors $\mathbf{u}_j, j \neq k$. In particular, the set is linearly dependent if and only if one of the $\mathbf{u}_k$, for some k, is a linear combination of the preceding $\mathbf{u}_1, \ldots, \mathbf{u}_{k-1}$.*

(iv) *If any vector is linearly dependent on a set of vectors* $\mathbf{u}_1, \ldots, \mathbf{u}_q$ *and if* $\mathbf{u}_m$ *is linearly dependent on the remaining vectors* $\mathbf{u}_i$, *then the given vector is linearly dependent on the set with* $\mathbf{u}_m$ *omitted.*

Proof: To prove (i), by reordering the vectors if necessary we can assume, without loss of generality, that the set of the first p vectors is linearly dependent, so that constants α_i exist, not all zero, such that

$$\alpha_1\mathbf{u}_1 + \alpha_2\mathbf{u}_2 + \cdots + \alpha_p\mathbf{u}_p = \mathbf{0}.$$

If we take $\alpha_{p+1} = \cdots = \alpha_q = 0$, we see that this implies that constants α_i exist, not all zero, such that

$$\alpha_1\mathbf{u}_1 + \alpha_2\mathbf{u}_2 + \cdots + \alpha_q\mathbf{u}_q = \mathbf{0}.$$

This proves part (i). Part (ii) is left as an exercise. To prove (iii), suppose that the set is linearly dependent. Then an equation of the form

$$\alpha_1\mathbf{u}_1 + \alpha_2\mathbf{u}_\alpha + \cdots + \alpha_k\mathbf{u}_k = \mathbf{0}$$

holds, where some of the α_i are nonzero, and, without loss of generality, we can assume that $\alpha_k \neq 0$. Hence

$$\mathbf{u}_k = -\frac{\alpha_1}{\alpha_k}\mathbf{u}_1 - \cdots - \frac{\alpha_{k-1}}{\alpha_k}\mathbf{u}_{k-1}$$

and $\mathbf{u}_k$ has been expressed as a linear combination of the preceding vectors. Conversely, if

$$\mathbf{u}_k = \gamma_1\mathbf{u}_1 + \cdots + \gamma_{k-1}\mathbf{u}_{k-1},$$

then

$$\gamma_1\mathbf{u}_1 + \cdots + \gamma_{k-1}\mathbf{u}_{k-1} - \mathbf{u}_k = \mathbf{0},$$

and the set is linearly dependent. To prove (iv), if $\mathbf{u}_m$ is linearly dependent on the remaining vectors $\mathbf{u}_i$, then we can write

$$\mathbf{u}_m = \sum_{r=1}^{q}{}' \gamma_r\mathbf{u}_r,$$

where the prime denotes that the term $r = m$ in the sum is omitted. Hence, if

$$\mathbf{u} = \sum_{r=1}^{q} \alpha_r\mathbf{u}_r,$$

we can substitute the above expression for $\mathbf{u}_m$ to obtain

$$\mathbf{u} = \sum_{r=1}^{q}{}' (\alpha_r + \alpha_m\gamma_r)\mathbf{u}_r,$$

and $\mathbf{u}$ has been expressed as a linear combination of the set with $\mathbf{u}_m$ omitted.

Although it certainly is not apparent from the development so far, the concepts of linear dependence and independence which we just discussed are perhaps the most fundamental in linear algebra. We will see that these concepts provide new ways of viewing rank of matrices, existence and uniqueness of solutions of systems of linear equations, and other concepts to follow. They provide the key to understanding means for representing general members of vector spaces in a simple way, and they even help us to understand numerical difficulties that arise in the practical solution of linear equations, linear programs, and other applied problems. Thus it is absolutely crucial to grasp these fundamental tools; several of the next sections provide ample opportunity for us to become accustomed to using these tools.

EXERCISE 4.8. Prove part (ii) of Theorem 4.1.

EXERCISE 4.9. Note that Definition 4.4 of linear dependence does not specify that the vectors are nonzero. Prove that

(a) If one of the vectors $\mathbf{v}_1, \ldots, \mathbf{v}_p$ is zero, the set is linearly dependent.
(b) If a set of vectors is linearly independent, this implies that all the vectors are nonzero.
(c) A single vector $\mathbf{v}$ forms a linearly dependent set if $\mathbf{v} = \mathbf{0}$, and a linearly independent set if $\mathbf{v} \neq \mathbf{0}$.

EXERCISE 4.10. If $\mathbf{u}_1$ and $\mathbf{u}_2$ are linearly independent of each other and $\mathbf{w}_1 = a\mathbf{u}_1 + b\mathbf{u}_2$, $\mathbf{w}_2 = c\mathbf{u}_1 + d\mathbf{u}_2$, show that $\mathbf{w}_1$ and $\mathbf{w}_2$ are linearly independent of each other if and only if $ad \neq bc$.

EXERCISE 4.11. Show that if the set $\{\mathbf{u}, \mathbf{v}, \mathbf{w}\}$ is linearly independent, then so is the set $\{\mathbf{u} + \mathbf{v}, \mathbf{v} + \mathbf{w}, \mathbf{w} + \mathbf{u}\}$.

EXERCISE 4.12. If

$$\begin{aligned} \mathbf{v}_1 &= \mathbf{u}_1 + \mathbf{u}_2 + \mathbf{u}_3 \\ \mathbf{v}_2 &= \mathbf{u}_1 + \alpha\mathbf{u}_2 \\ \mathbf{v}_3 &= \mathbf{u}_2 + \beta\mathbf{u}_3, \end{aligned}$$

where $\mathbf{u}_1, \mathbf{u}_2, \mathbf{u}_3$ are given and form a linearly independent set, find the condition that must be satisfied by α, β in order to ensure that $\{\mathbf{v}_1, \mathbf{v}_2, \mathbf{v}_3\}$ is linearly independent.

EXERCISE 4.13. Given an upper triangular matrix $\mathbf{A}$ with nonzero diagonal elements (i.e., $a_{ij} = 0$ if $i > j$, and $a_{ii} \neq 0$), show that the set of rows of $\mathbf{A}$ is linearly independent. If any of the diagonal elements is zero, show that the set is dependent.

EXERCISE 4.14. If a vector $\mathbf{u}$ can be expressed in the form

$$\mathbf{u} = \alpha_1\mathbf{u}_1 + \cdots + \alpha_p\mathbf{u}_p,$$

where $\{\mathbf{u}_1, \ldots, \mathbf{u}_p\}$ is linearly independent, show that the coefficients α_i are uniquely determined. Conversely, if the expression for $\mathbf{u}$ in this form is unique, show that the vectors $\mathbf{u}_i$ form a linearly independent set. (See Theorem 4.3.)

EXERCISE 4.15. Express a general vector $[x, y, z]$ as a linear combination of $[1, 2, 1]$, $[1, 0, -1]$, $[1, -2, 1]$.

EXERCISE 4.16. Which one of the following matrices is linearly independent of the others:

$$\begin{bmatrix} 1 & -1 \\ -1 & 2 \end{bmatrix}, \quad \begin{bmatrix} -1 & 2 \\ 3 & 1 \end{bmatrix}, \quad \begin{bmatrix} 2 & -3 \\ -3 & 2 \end{bmatrix}, \quad \begin{bmatrix} 1 & 1 \\ 1 & 6 \end{bmatrix}?$$

4.4 BASIS AND DIMENSION

At the close of the previous section we mentioned that the concepts of linear dependence and independence can help us understand the problems involved in finding simple ways of representing or dealing with vector spaces. In many applications we find that we are not directly interested in, say, $\mathbb{R}^n$, the space of *all* real $n \times 1$ matrices, but rather in some subspace of $\mathbb{R}^n$ defined as the set of all linear combinations of some given set of vectors (see Ex. 4.6). Often it will be more convenient for us to write elements in this subspace as linear combinations of vectors from a set containing as *few* vectors as possible, and often we will even want these vectors to be of simple forms. In this section and the next we address ourselves to this task.

Recall from Ex. 4.6 that the set of all linear combinations of a set of vectors is a linear subspace. This and the notions in the preceding paragraph lead naturally to the following definition.

DEFINITION 4.6. If a subspace W of a vector space V consists of the set of all linear combinations of a finite set of vectors $\mathbf{u}_1, \ldots, \mathbf{u}_s$ from V, then the set is said to *span* the space W. We also say that the set *generates* the space W.

For example, the set of vectors

$$\begin{bmatrix} 1 \\ 0 \\ 0 \end{bmatrix}, \quad \begin{bmatrix} 0 \\ 1 \\ 0 \end{bmatrix}, \quad \begin{bmatrix} 0 \\ 0 \\ 1 \end{bmatrix}, \tag{4.6}$$

spans the vector space $\mathbb{R}^3$. This space is also generated by the set

$$\begin{bmatrix} 1 \\ 0 \\ 0 \end{bmatrix}, \quad \begin{bmatrix} 0 \\ 1 \\ 0 \end{bmatrix}, \quad \begin{bmatrix} 0 \\ 0 \\ 1 \end{bmatrix}, \quad \begin{bmatrix} 1 \\ -1 \\ 1 \end{bmatrix}, \quad \begin{bmatrix} -2 \\ 0 \\ 3 \end{bmatrix}. \tag{4.7}$$

Note that the fact that the last two vectors of (4.7) are linearly dependent on the first three is immaterial since the definition did not say that the set which generates a space must be independent. However, we will be interested in spanning sets containing as few vectors as possible, and we will see that the question of whether the set is linearly independent is then important. The following theorem is useful in this connection.

THEOREM 4.2.

(i) *If* $\mathbf{u}_1, \ldots, \mathbf{u}_s$ *span a vector space and one of these vectors, say* $\mathbf{u}_m$, *is linearly dependent on the others, then the vector space is spanned by the set obtained by omitting* $\mathbf{u}_m$ *from the original set.*

(ii) *If* $\mathbf{u}_1, \ldots, \mathbf{u}_s$ *are not all zero and span a vector space, we can always select from these a linearly independent set that spans the same space.*

Proof: If $\mathbf{u}$ is any vector in the space, it can be expressed as a linear combination of $\mathbf{u}_1, \ldots, \mathbf{u}_s$. From Theorem 4.1(iv) it can therefore be expressed as a linear combination of the set with $\mathbf{u}_m$ omitted. Hence, the set with $\mathbf{u}_m$ omitted spans the space, proving (i). For (ii), if the original set of vectors is linearly independent then we are done; if not, by Theorem 4.1(iii) and our own (i) we can reduce the set to one containing $s - 1$ vectors while still spanning the space. Continuing to cast out vectors in this fashion, either we obtain a linearly independent subset of at least two vectors spanning the space (thus proving the theorem) or we are reduced to a set of one vector $\mathbf{u}_i$ spanning the space. Since not all the original vectors were zero, the spanned space is not just the zero vector; therefore $\mathbf{u}_i$ is not zero, $\{\mathbf{u}_i\}$ is linearly independent, and the proof is complete.

Linearly independent spanning sets will turn out to be those smallest possible spanning sets that we desire for efficient representation of vector spaces. We therefore give them a special name.

DEFINITION 4.7. A *basis* for a vector space is a linearly independent set of vectors that spans the space.

For example, the set of vectors (4.6) forms a basis for the vector space $\mathbb{R}^3$ since these vectors are linearly independent, and any 3×1 column vector can be expressed as a sum of multiples of the three vectors in the set. However, the set (4.7) does not form a basis, since it is linearly dependent.

One of the important features of the representation of the vectors in a vector space in terms of the vectors in a basis is the following easy consequence of the linear independence of a basis.

THEOREM 4.3. The expression for any vector in a vector space in terms of the vectors in a basis for the space is unique.

Proof: If $\mathbf{u}_1, \ldots, \mathbf{u}_s$ forms a basis and

$$\mathbf{u} = a_1\mathbf{u}_1 + \cdots + a_s\mathbf{u}_s = b_1\mathbf{u}_1 + \cdots + b_s\mathbf{u}_s,$$

then

$$\mathbf{0} = \mathbf{u} - \mathbf{u} = (a_1 - b_1)\mathbf{u}_1 + \cdots + (a_s - b_s)\mathbf{u}_s;$$

since a basis is a linearly independent set, we must have $a_1 = b_1, \ldots, a_s = b_s$ and the theorem is proved.

It should be clear that the *set* of vectors constituting a basis is not unique. Thus, in the case of $\mathbb{R}^3$, we could take the set of vectors

$$\begin{bmatrix} 1 \\ 2 \\ 1 \end{bmatrix}, \quad \begin{bmatrix} 1 \\ 0 \\ -1 \end{bmatrix}, \quad \begin{bmatrix} 1 \\ -2 \\ 1 \end{bmatrix}$$

as a basis just as easily as the unit column vectors $\mathbf{e}_1, \mathbf{e}_2, \mathbf{e}_3$. However it turns out that the *number* of vectors in a basis is unique, as we now prove.

● KEY THEOREM 4.4. *Suppose that a vector space has a basis containing n vectors* $\mathbf{u}_1, \ldots, \mathbf{u}_n$.

(i) *If a set of vectors* $\mathbf{v}_1, \ldots, \mathbf{v}_m$ *in that space is linearly independent, then* $m \leq n$.
(ii) *Every basis for the space contains exactly n vectors.*

Proof: We first prove (i). Consider the set of vectors obtained by adding $\mathbf{v}_1$ in front of the $\mathbf{u}$'s:

$$\mathbf{v}_1, \mathbf{u}_1, \mathbf{u}_2, \ldots, \mathbf{u}_n.$$

Since the $\mathbf{u}$'s span the vector space, $\mathbf{v}_1$ can be expressed as a linear combination of the $\mathbf{u}$'s. Hence the above set is linearly dependent, and, from Theorem 4.1(iii), one of the vectors can be expressed as a linear combination of preceding vectors. This means that one of the $\mathbf{u}$'s can be expressed as a linear combination of the preceding vectors. By relabeling the $\mathbf{u}$'s, if necessary, we can arrange that this is $\mathbf{u}_n$. From Theorem 4.2, the vector space is spanned by the set obtained by omitting $\mathbf{u}_n$, that is, by the set

$$\mathbf{v}_1, \mathbf{u}_1, \ldots, \mathbf{u}_{n-1}. \tag{4.8}$$

We next add the vector $\mathbf{v}_2$ in front of the new set,

$$\mathbf{v}_2, \mathbf{v}_1, \mathbf{u}_1, \ldots, \mathbf{u}_{n-1}.$$

Since the set (4.8) spans the space, $\mathbf{v}_2$ can be expressed as a linear combination of the remaining vectors. As before, this means that one of the vectors can be expressed as a linear combination of the preceding. Since $\mathbf{v}_1$ and $\mathbf{v}_2$ are independent, one of the $\mathbf{u}$'s can be expressed as a linear combination of the preceding vectors. By relabeling, if necessary, we can arrange that this is $\mathbf{u}_{n-1}$, and the vector space is spanned by the set obtained by omitting this vector:

$$\mathbf{v}_2, \mathbf{v}_1, \mathbf{u}_1, \ldots, \mathbf{u}_{n-2}. \tag{4.9}$$

Suppose that $m > n$. By repeating the above procedure we shall arrive at the set

$$\mathbf{v}_n, \mathbf{v}_{n-1}, \ldots, \mathbf{v}_1, \tag{4.10}$$

which spans the space, and we still have the vectors $\mathbf{v}_{n+1}, \ldots, \mathbf{v}_m$ left over. Since the set (4.10) spans the space, the vector $\mathbf{v}_{n+1}$ can be expressed as a linear combination of $\mathbf{v}_1, \mathbf{v}_2, \ldots, \mathbf{v}_n$. This contradicts the original assumption that the v's are linearly independent. Hence, we cannot have $m > n$; we must have $m \leq n$ as stated in the theorem. For (ii), if another basis contains m vectors, then since a basis must be linearly independent we deduce from (i) that $m \leq n$; by reversing the roles of the two bases, we also find $n \leq m$ so that $n = m$ as stated.

The unique number of vectors in a basis is so important that we give it a special name:

DEFINITION 4.8. The number of vectors in a basis for a vector space is known as the *dimension* of the space. We say that we are dealing with an *m-dimensional space* if and only if the space has a basis of m vectors. When a space has a basis consisting of some finite number of vectors we say that the space is *finite-dimensional.*

The point of introducing the word finite-dimensional is that in more general contexts vector spaces occur that have bases possessing an infinite number of elements (Ex. 4.26). We restrict our attention to finite-dimensional spaces.

If a vector space consists only of the null vector $\mathbf{x} = \mathbf{0}$, there are no linearly independent sets composed of vectors in the space, and the dimension of the space is 0.

EXERCISE 4.17. Prove that $\mathbb{R}^n$ and $\mathbb{C}^n$ are n-dimensional.

SOLUTION: We only need show that the set of n unit column vectors

$$\mathbf{e}_1, \ldots, \mathbf{e}_n$$

is a basis for $\mathbb{R}^n$ (and $\mathbb{C}^n$). Clearly this set is linearly independent. To see that it also spans $\mathbb{R}^n$ (and $\mathbb{C}^n$) we only need note that every vector

$$\mathbf{x} = \begin{bmatrix} x_1 \\ \cdot \\ \cdot \\ \cdot \\ x_n \end{bmatrix}$$

in $\mathbb{R}^n$ (and $\mathbb{C}^n$) can be represented as $\mathbf{x} = x_1\mathbf{e}_1 + \cdots + x_n\mathbf{e}_n$ where the x_i are in $\mathbb{R}$ (or $\mathbb{C}$), completing the proof.

EXERCISE 4.18. Use **Key Theorem 4.4** to prove the following, essentially a restatement of that theorem: in an n-dimensional vector space, every set with more than n vectors is linearly dependent, and no set with fewer than n vectors spans the space.

SOLUTION: Suppose that a linearly independent set contains m vectors; then by **Key Theorem 4.4**(i), $m \leq n$, so if a set has m vectors with $m > n$ then the set is linearly dependent. For the second part, there is nothing to prove if $n = 0$; for $n > 0$, if a set of $m < n$ vectors spans the space then $m > 0$ and not all the vectors in the set can be zero, so by Theorem 4.2(ii) this set has a subset of $p \leq m < n$ vectors which is independent and still spans the space, that is, the subset is a basis and by **Key Theorem 4.4**(ii) $p = n$, a contradiction.

EXERCISE 4.19. Find a basis for the subspace of $\mathbb{R}^4$ consisting of all vectors $[x_i]$ such that $x_1 - x_2 = 0$, $x_2 - x_3 = 0$, and $x_1 - 2x_2 + x_3 = 0$. Find the dimension of this subspace.

So far, the only way we have for knowing that a vector space is finite-dimensional, say of dimension n, is explicitly to exhibit a basis containing n vectors. Our next result gives an independent characterization of dimension.

THEOREM 4.5. A vector space has finite dimension k if and only if k is the maximum number of vectors in a linearly independent subset of the space.

Proof: If the space has dimension k, then the desired conclusion follows immediately from Definition 4.8 and **Key Theorem 4.4** (see Ex. 4.18). It remains to prove the theorem in the opposite direction. Suppose that the set $\mathbf{u}_1, \ldots, \mathbf{u}_k$ is linearly independent, and that k is the maximum number of vectors in such a set. Then any other vector $\mathbf{u}$, together with these k vectors, forms a linearly dependent set, and $\mathbf{u}$ can be expressed as a linear combination of $\mathbf{u}_1, \ldots, \mathbf{u}_k$, that is, the $\mathbf{u}_1, \ldots, \mathbf{u}_k$ span the space. Therefore, the set of k vectors span the space. Since it is linearly independent it constitutes a basis for the space and the dimension of the space is k.

We have one remaining theoretical result concerning dimension and basis before we turn to applying our theory to matrices and to developing techniques for explicitly generating bases; even this result addresses itself to finding bases, however.

THEOREM 4.6. Let $\mathbf{u}_1, \ldots, \mathbf{u}_k$ form an independent set in an n-dimensional space. Then there exist vectors $\mathbf{u}_{k+1}, \ldots, \mathbf{u}_n$ such that the set of vectors $\mathbf{u}_1, \ldots, \mathbf{u}_n$ is a basis for the space.

Proof: We know of course that $k \leq n$. If the set does not span the space then there is a vector $\mathbf{u}_{k+1}$ linearly independent of the set so that now the set of vectors $\mathbf{u}_1, \ldots, \mathbf{u}_{k+1}$ is linearly independent. We can continue in this way adding vectors while maintaining linear independence; since n is the maximum possible number of such vectors by Theorem 4.5, this process must stop, and this can only happen when the set spans the space and necessarily contains n vectors as asserted.

EXERCISE 4.20. Prove Theorem 4.6 in detail.

EXERCISE 4.21. Use Theorem 4.5 to prove that the dimension of any subspace of an n-dimensional space cannot exceed n.

EXERCISE 4.22. Consider the subspace of $\mathbb{R}^4$ consisting of the vectors $[x_i]$ for which $x_1 + x_2 = x_3 + x_4$. Show that the set of two vectors

$$\begin{bmatrix} 1 \\ 0 \\ 1 \\ 0 \end{bmatrix}, \quad \begin{bmatrix} 0 \\ 1 \\ 0 \\ 1 \end{bmatrix}$$

is a linearly independent set in this subspace. Extend the set to form a basis for the subspace as guaranteed by Theorem 4.6. (After attempting this, see Ex. 4.28.)

EXERCISE 4.23. If $\mathbf{u}_1, \ldots, \mathbf{u}_p$ span a vector space V, but V cannot be spanned by the set of vectors obtained by omitting any vector from this set, show that the set forms a basis for V.

EXERCISE 4.24. Show that a set of vectors is a basis for a vector space V if and only if each vector in V can be expressed uniquely as a linear combination of vectors in the set.

EXERCISE 4.25. Find the dimension of the vector space of all real $m \times n$ matrices $[a_{ij}]$ for which $a_{ij} = 0$ whenever $i < j - 1$ or $j < i$.

EXERCISE 4.26. Consider the vector space of all polynomials of arbitrary degree with real coefficients. By considering the infinite set of polynomials $1, x, x^2, x^3, \ldots$, use Theorem 4.5 to prove that this space is not finite-dimensional.

4.5 MATRICES, RANK, BASIS, AND DIMENSION

The previous section on basis and dimension is theoretical and raises many questions as to precisely how we will accomplish some of the computations guaranteed by theorems. Given a spanning set of vectors, *how* can we find a linearly independent subset spanning the same space (Theorem 4.2)? *How* can we find a basis for a given space and hence its dimension? *How* can we extend a linearly independent set to form a basis (Theorem 4.6)? In this section we develop the techniques we need by using vector space language to express some facts about matrices and the rank of matrices; as a bonus we obtain still more equivalent ways of defining rank. The fundamental key to our development lies in using the linear algebra concepts of linear dependence and independence to describe the effects of elementary row operations.

- *KEY THEOREM 4.7. Let a sequence of elementary row operations transform an $m \times n$ matrix* **A** *into a matrix* **B**. *Then:*

  (i) *a given collection of columns of* **A** *is linearly dependent (independent) if and only if the set of corresponding columns of* **B** *is linearly dependent (independent);*

  (ii) *a $1 \times n$ row-matrix is a linear combination of the rows of* **A** *if and only if it is a linear combination of the rows of* **B** *and hence the sets of rows of* **A** *and of* **B** *span the some space.*

Proof: Let **F** be a nonsingular $m \times m$ matrix with $\mathbf{FA} = \mathbf{B}$ (see **Key Theorem 3.3**). (i): If $\mathbf{c}_1, \ldots, \mathbf{c}_n$ denotes the $m \times 1$ columns of an $m \times n$ matrix **C** and $\mathbf{x} = [x_i]$ is an $n \times 1$ column vector, then

$$\mathbf{Cx} = \sum_{i=1}^{n} x_i \mathbf{c}_i.$$

Therefore, a set of columns numbered $p_1, \ldots, p_r$ from **A** is linearly dependent if and only if there is a nonzero $n \times 1$ vector $\mathbf{x} = [x_i]$ with $x_i = 0$ whenever i is not one of the numbers $p_1, \ldots, p_r$ for which $\mathbf{Ax} = \mathbf{0}$. Since **F** is nonsingular, $\mathbf{Ax} = \mathbf{0}$ if and only if $\mathbf{FAx} = \mathbf{0}$, which in turn is equivalent to the dependence of the set of columns numbered $p_1, \ldots, p_r$ from $\mathbf{FA} = \mathbf{B}$ as asserted. (ii): a $1 \times n$ row-vector **x** is a linear combination of the rows of **A** if and only if there is a $1 \times m$ row-vector **y** with $\mathbf{x} = \mathbf{yA}$. Since $\mathbf{B} = \mathbf{FA}$ and **F** is nonsingular, $\mathbf{x} = \mathbf{yA} = \mathbf{yF}^{-1}\mathbf{B} = \mathbf{y}'\mathbf{B}$ (with $\mathbf{y}' = \mathbf{yF}^{-1}$) if and only if **x** is also a linear combination of the rows of **B**, completing our proof.

This theorem is a useful computational tool because the row-echelon form of a matrix is so simple as to allow us easily to answer questions of linear dependence and independence.

EXERCISE 4.27. Find the dimension of the subspace of $\mathbb{R}^4$ spanned by the vectors

$$\begin{bmatrix} 1 \\ -1 \\ -1 \\ 2 \end{bmatrix}, \begin{bmatrix} -1 \\ 2 \\ 3 \\ 1 \end{bmatrix}, \begin{bmatrix} 2 \\ -3 \\ -3 \\ 2 \end{bmatrix}, \begin{bmatrix} 1 \\ 1 \\ 1 \\ 6 \end{bmatrix}$$

and find all possible subsets of these vectors that can be used as a basis for the subspace.

SOLUTION: The 4×4 matrix with these as columns is

$$\begin{bmatrix} 1 & -1 & 2 & 1 \\ -1 & 2 & -3 & 1 \\ -1 & 3 & -3 & 1 \\ 2 & 1 & 2 & 6 \end{bmatrix}$$

and can be reduced by row operations to the row-echelon form

$$\begin{bmatrix} 1 & 0 & 0 & 5 \\ 0 & 1 & 0 & 0 \\ 0 & 0 & 1 & -2 \\ 0 & 0 & 0 & 0 \end{bmatrix}.$$

Clearly the first three columns form a linearly independent set while the fourth column is linearly dependent on that set; the dimension is therefore three since this set of three vectors is a basis. The subset of the original vectors numbered 1, 2, 3 forms a basis; so do the sets of columns numbered 1, 2, 4 and 2, 3, 4, but *not* 1, 3, 4.

EXERCISE 4.28. We consider the problem of extending the set of two vectors in Ex. 4.22 to form a basis for the subspace of that problem. The augmented matrix of the equations defining the subspace there is

$$[1 \quad 1 \quad -1 \quad -1 \quad 0]$$

which is of course in row-echelon form, so that we may assign x_2, x_3, and x_4 arbitrary values a, b, c and find that $x_1 = -a + b + c$. Thus the general element of the space is

$$\begin{bmatrix} -a + b + c \\ a \\ b \\ c \end{bmatrix}$$

and we must adjoin vectors of this form to the set given in Ex. 4.22. Adjoining this one vector and then forming a 4×3 matrix from the resulting set yields

$$\begin{bmatrix} 1 & 0 & -a + b + c \\ 0 & 1 & a \\ 1 & 0 & b \\ 0 & 1 & c \end{bmatrix}$$

that row operations reduce to

$$\begin{bmatrix} 1 & 0 & -a + b + c \\ 0 & 1 & a \\ 0 & 0 & a - c \\ 0 & 0 & c - a \end{bmatrix}.$$

To make the columns form an independent set we must take $a - c \neq 0$; choosing $a = 1$, $b = 1$, $c = 0$, we have a linearly independent set of the three vectors

$$\begin{bmatrix}1\\0\\1\\0\end{bmatrix}, \begin{bmatrix}0\\1\\0\\1\end{bmatrix}, \begin{bmatrix}0\\1\\1\\0\end{bmatrix}$$

in the subspace. To verify that these vectors form a basis for the subspace, we show that any vector of the subspace is linearly dependent on these three; to do this, we form the 4 × 4 matrix of the above three vectors and a vector of the correct general form:

$$\mathbf{A} = \begin{bmatrix}1 & 0 & 0 & -\alpha+\beta+\gamma\\ 0 & 1 & 1 & \alpha\\ 1 & 0 & 1 & \beta\\ 0 & 1 & 0 & \gamma\end{bmatrix}.$$

Row operations reduce this to the row-echelon form:

$$\mathbf{B} = \begin{bmatrix}1 & 0 & 0 & -\alpha+\beta+\gamma\\ 0 & 1 & 0 & \gamma\\ 0 & 0 & 1 & \alpha-\gamma\\ 0 & 0 & 0 & 0\end{bmatrix}.$$

Since the fourth column of **B** is clearly linearly dependent on the linearly independent set of the first three columns of **B**, by **Key Theorem 4.7** the same is true for the columns of **A**, that is, the fourth column of **A** is linearly dependent on the set of the first three columns of **A**. Therefore the linearly independent set of three vectors we have found spans the subspace and therefore is a basis.

In **Key Theorem 4.7**(ii) we spoke of linear combinations of the rows of a matrix; the following definition naturally arises.

DEFINITION 4.9. The *row space* of an $m \times n$ matrix **A** is that subspace of $\mathbb{R}^m$ (or $\mathbb{C}^m$) spanned by the set of rows of **A**. Similarly, the *column space* of **A** is the space spanned by the set of columns of **A**.

In terms of these concepts **Key Theorem 4.7** provides us quite readily with a new meaning for the rank of a matrix **A**.

● *KEY THEOREM 4.8. Let* **A** *be an $m \times n$ matrix. Then:*

(i) *the rank k of* **A** *equals the dimension of the column space of* **A**, *a basis for which is provided by the set of columns from* **A** *corresponding to the unit columns in the row-echelon form of* **A**; *and*

(ii) *the rank k of* **A** *equals the dimension of the row space of* **A**, *a basis for which is provided by the non-zero rows (that is, the first k rows) in the row-echelon form of* **A**.

Proof: Let the unit column vectors $\mathbf{e}_1, \ldots, \mathbf{e}_k$ in the row-echelon form of $\mathbf{A}$ fall in columns numbered $c_1, \ldots, c_k$ (see Definition 3.2).

(i) The set of k unit column vectors is clearly linearly independent so by **Key Theorem 4.7** the corresponding set of columns of $\mathbf{A}$ is linearly independent. If we adjoin any other column of the row-echelon form to the set of unit columns we clearly get a linearly dependent set, so by **Key Theorem 4.7** the analogous statement holds for the columns of $\mathbf{A}$, which guarantees that the columns numbered $c_1, \ldots, c_k$ of $\mathbf{A}$ span the column space. Hence these columns form a basis and the column space has dimension k as asserted.

(ii) Since k of the columns in the row-echelon form are the unit columns $\mathbf{e}_1, \ldots, \mathbf{e}_k$, the only linear combination of the first k rows of that form that can be zero is the trivial "zero" linear combination, so those rows form a linearly independent set and hence clearly a basis for the row space of the row-echelon form; since by **Key Theorem 4.7** this row space and that of $\mathbf{A}$ are identical, we are done.

This theorem tells us that the dimensions of the row space and column space are equal, both being the rank. Since the row space of $\mathbf{A}$ is essentially the column space of $\mathbf{A}^T$, we have the following result.

- *KEY COROLLARY 4.1. For any matrix* $\mathbf{A}$, *the rank of* $\mathbf{A}$ *equals the rank of* $\mathbf{A}^T$.

Since the rows in the row-echelon form of a matrix are often fairly simple (contain many zeros), **Key Theorem 4.8**(ii) provides a simple device for finding a *convenient* basis for a space.

EXERCISE 4.29. Find a simple basis for the space generated by the vectors

$$\begin{bmatrix} 1 \\ 3 \\ -7 \end{bmatrix}, \quad \begin{bmatrix} 2 \\ -1 \\ 0 \end{bmatrix}, \quad \begin{bmatrix} 3 \\ -1 \\ -1 \end{bmatrix}, \quad \begin{bmatrix} 4 \\ -3 \\ 2 \end{bmatrix}.$$

SOLUTION: We write these vectors as the *rows* of a matrix

$$\begin{bmatrix} 1 & 3 & -7 \\ 2 & -1 & 0 \\ 3 & -1 & -1 \\ 4 & -3 & 2 \end{bmatrix}$$

and reduce it to row-echelon form, obtaining

$$\begin{bmatrix} 1 & 0 & -1 \\ 0 & 1 & -2 \\ 0 & 0 & 0 \\ 0 & 0 & 0 \end{bmatrix}.$$

According to **Key Theorem 4.8**(ii) the first two rows of this provide our simple basis; rewriting as columns, we obtain the simple basis

$$\begin{bmatrix} 1 \\ 0 \\ -1 \end{bmatrix}, \quad \begin{bmatrix} 0 \\ 1 \\ -2 \end{bmatrix}.$$

EXERCISE 4.30. Find whether the set of vectors

$$\begin{bmatrix} 1 \\ -1 \\ 1 \end{bmatrix}, \quad \begin{bmatrix} 1 \\ 1 \\ -3 \end{bmatrix}, \quad \begin{bmatrix} 1 \\ 2 \\ -5 \end{bmatrix}$$

generates the same subspace as in Ex. 4.29.

SOLUTION: Use the same technique as in Ex. 4.29 to obtain the row-echelon form

$$\begin{bmatrix} 1 & 0 & -1 \\ 0 & 1 & -2 \\ 0 & 0 & 0 \end{bmatrix}$$

and hence precisely the same basis as before. Therefore the same subspace is generated.

EXERCISE 4.31. Find the dimension of the vector space spanned by the following vectors, and find a basis for the space (see Ex. 4.27).

$$\begin{bmatrix} 1 \\ -1 \\ -1 \\ 2 \end{bmatrix}, \quad \begin{bmatrix} -1 \\ 2 \\ 3 \\ 1 \end{bmatrix}, \quad \begin{bmatrix} 2 \\ -3 \\ -3 \\ 2 \end{bmatrix}, \quad \begin{bmatrix} 1 \\ 1 \\ 1 \\ 7 \end{bmatrix}.$$

EXERCISE 4.32. Does either of the following sets provide a basis for the set of 4×1 column vectors $[x_i]$ such that $x_1 + x_2 - x_3 - x_4 = 0$?

(a) $\begin{bmatrix} 1 \\ -1 \\ -1 \\ 1 \end{bmatrix}, \quad \begin{bmatrix} 1 \\ 0 \\ 1 \\ 0 \end{bmatrix}, \quad \begin{bmatrix} 0 \\ 1 \\ 0 \\ 1 \end{bmatrix}.$ (b) $\begin{bmatrix} 1 \\ -1 \\ 1 \\ -1 \end{bmatrix}, \quad \begin{bmatrix} 1 \\ 0 \\ 1 \\ 0 \end{bmatrix}, \quad \begin{bmatrix} 0 \\ 1 \\ 0 \\ 1 \end{bmatrix}.$

EXERCISE 4.33. For what values of k will the vectors $[3 - k, -1, 0]$, $[-1, 2 - k, -1]$, $[0, -1, 3 - k]$ span a *two*-dimensional space?

EXERCISE 4.34. Show that the sets

$$\{[1, 0, -1], [1, 1, 0], [0, 1, 1]\} \quad \text{and} \quad \{[2, 1, -1], [1, 2, 1]\}$$

span the same vector space. Show that the set $\{[2, 1, -1], [1, -1, 0]\}$ does *not* span this space.

EXERCISE 4.35. Show that a square matrix is nonsingular if and only if its rows (columns) form an independent set.

The results of this section allow for yet another characterization of rank.

EXERCISE 4.36. Show that the rank k of a matrix $\mathbf{A}$ is the order of the largest nonsingular square submatrix of $\mathbf{A}$ by completing the details of the following outline of the proof: Suppose $\mathbf{A}$ has rank k and consider a $p \times p$ submatrix with $p > k$. Show that the p rows of the submatrix form a dependent set since the p rows of $\mathbf{A}$ do so (since $p > k$ = rank of $\mathbf{A}$); deduce that the submatrix is singular. Choose any k rows of $\mathbf{A}$ forming an independent subset; then the rank of the matrix of these k rows is k so there are some k columns from this matrix forming an independent set. Deduce that the resulting $k \times k$ submatrix is nonsingular. Conversely, suppose that k is the maximum order of nonsingular submatrices and choose a nonsingular $k \times k$ submatrix. Since the rows of the submatrix form a linearly independent set, show that the corresponding k rows of $\mathbf{A}$ form a linearly independent set and deduce then that the rank of $\mathbf{A}$ is not less than k. Next suppose that some subset of p rows of $\mathbf{A}$ is linearly independent; deduce as in the first part of the proof of this theorem that some $p \times p$ submatrix is nonsingular, so that $p \leq k$. Since k is thus the maximum number of rows of $\mathbf{A}$ in any linearly independent set, deduce that k equals the rank of $\mathbf{A}$.

EXERCISE 4.37. Prove that the rank of $\mathbf{AB}$ is less than or equal to the rank of both $\mathbf{A}$ and $\mathbf{B}$ by showing its rank to be less than or equal to that of $\mathbf{B}$ and then using **Key Corollary 4.1.**

One awkward feature of the rank of matrices is that it is not a continuous function: if a matrix changes slightly then its rank, being an integer, either does not change at all or changes by at least one. Similarly a vector may be linearly independent of a set while a slight perturbation of the vector is linearly dependent on the set. This difficulty is important in practice since vectors that are created by measurements contain measurement errors or computational rounding errors which cause us to think of our vectors as being subject to small changes.

EXERCISE 4.38. The vector

$$\begin{bmatrix} 1 \\ 1 \\ 1.0001 \end{bmatrix}$$

clearly is linearly independent of the set of vectors

$$\begin{bmatrix} 1 \\ 0 \\ 0.5 \end{bmatrix}, \quad \begin{bmatrix} 0 \\ 1 \\ 0.5 \end{bmatrix},$$

while the slightly perturbed vector

$$\begin{bmatrix} 1 \\ 1 \\ 1 \end{bmatrix}$$

is linearly dependent on the set. If the vectors were all subject to measurement errors and if it were very important to determine whether we had linear independence, this would be very difficult to determine.

In the above discussion we used the intuitive concept of "small change in a vector"; the next section allows us to make this notion precise.

4.6 VECTOR NORMS

In numerous instances it is important to have some notion of the "size" of a vector, where we are referring to the *size* of its components rather than to the *number* of components. At the end of the preceding section, for example, we spoke rather loosely of a perturbation vector

$$\begin{bmatrix} 0 \\ 0 \\ 0.0001 \end{bmatrix}$$

as being "small." In one dimension, we measure the size of a vector (real or complex number) x most commonly by $|x|$. In two dimensions, we measure the usual geometric length of a vector $[x_i]$ via the Pythagorean Theorem as $(x_1^2 + x_2^2)^{1/2}$; in three dimensions the analogous formula is $(x_1^2 + x_2^2 + x_3^2)^{1/2}$. Generally, we would be inclined to measure the size of an $n \times 1$ vector $[x_i]$ as $(x_1^2 + x_2^2 + \cdots + x_n^2)^{1/2}$, but there are other possible ways which may be more natural for measuring size in a given situation. If n is large, for example, this formula could make a vector each of whose components was "small" have a "large" size, perhaps contradicting the interpretation that is appropriate in the particular situation. Therefore we need to be able to consider more general concepts of size. In Section 2.3, for example, where the two components of a vector $\mathbf{x}$ in $\mathbb{R}^2$ represented the numbers of chickens and of foxes in an environment, either $\|x\|_1 = |x_1| + |x_2|$ representing the total number of individuals in the society or $\|x\|_\infty = \max\{|x_1|, |x_2|\}$ representing the maximum single population might be more meaningful than the "length" $\|x\|_2 = (x_1^2 + x_2^2)^{1/2}$. We will use the word *norm* to describe such a general concept so long as it satisfies certain natural properties that most of us tend to associate with length. Intuitively, a small norm for a vector should indicate that *all* of its components are small, while a large norm should indicate that *at least one* of its components is large.

DEFINITION 4.10. A *vector norm* of $\mathbf{x}$ is a nonnegative number denoted $\|\mathbf{x}\|$, associated with $\mathbf{x}$, satisfying:

(a) $\|\mathbf{x}\| > 0$ for $\mathbf{x} \neq \mathbf{0}$, and $\|\mathbf{x}\| = 0$ precisely when $\mathbf{x} = \mathbf{0}$.
(b) $\|k\mathbf{x}\| = |k|\|\mathbf{x}\|$ for any scalar k.
(c) $\|\mathbf{x} + \mathbf{y}\| \leq \|\mathbf{x}\| + \|\mathbf{y}\|$ (the triangle inequality).

The third condition is called the triangle inequality because it is a generalization of the fact that the length of any side of a triangle is less than or equal to the sum of the lengths of the other two sides.

We state that each of the following quantities defines a vector norm.

$$\|\mathbf{x}\|_1 = |x_1| + |x_2| + \cdots + |x_n|, \tag{4.11}$$

$$\|\mathbf{x}\|_2 = (|x_1|^2 + |x_2|^2 + \cdots + |x_n|^2)^{1/2} \tag{4.12}$$

$$\|\mathbf{x}\|_\infty = \max_i |x_i|. \tag{4.13}$$

The only difficult point in proving that these are actually norms lies in proving that $\|\cdot\|_2$ satisfies the triangle inequality. To do this we use the hermitian transpose $\mathbf{x}^H$ of a vector, introduced in Definition 1.8; this arises naturally since $\|\mathbf{x}\|_2 = (\mathbf{x}^H\mathbf{x})^{1/2}$.

EXERCISE 4.39. Prove the validity of the *Schwarz inequality*

$$|\mathbf{x}^H\mathbf{y}| \leq \|\mathbf{x}\|_2\|\mathbf{y}\|_2 \tag{4.14}$$

SOLUTION: We have for any complex α

$$\begin{aligned} 0 \leq \|\mathbf{x} + \alpha\mathbf{y}\|_2^2 &= \mathbf{x}^H\mathbf{x} + \alpha\mathbf{x}^H\mathbf{y} + \bar{\alpha}\mathbf{y}^H\mathbf{x} + \alpha\bar{\alpha}\mathbf{y}^H\mathbf{y} \\ &= \|\mathbf{x}\|_2^2 + \alpha\mathbf{x}^H\mathbf{y} + \bar{\alpha}\mathbf{y}^H\mathbf{x} + |\alpha|^2\|\mathbf{y}\|_2^2. \end{aligned} \tag{4.15}$$

The desired inequality is clearly true if $\mathbf{x}^H\mathbf{y} = 0$. When $\mathbf{x}^H\mathbf{y} \neq 0$, we choose $\alpha = -\|\mathbf{x}\|_2^2/\mathbf{x}^H\mathbf{y}$ and then (4.15) becomes

$$-\|\mathbf{x}\|_2^2 + \frac{\|\mathbf{x}\|_2^4\,\|\mathbf{y}\|_2^2}{|\mathbf{x}^H\mathbf{y}|^2} \geq 0$$

from which the Schwarz inequality is immediate.

EXERCISE 4.40. Prove the triangle inequality for $\|\cdot\|_2$ by applying the Schwarz inequality in the expansion of $\|\mathbf{x}+\mathbf{y}\|_2^2 = (\mathbf{x}^H + \mathbf{y}^H)(\mathbf{x}+\mathbf{y})$.

We note that

$$\|\mathbf{x}\|_\infty \le \|\mathbf{x}\|_1 \le n\|\mathbf{x}\|_\infty$$
$$\|\mathbf{x}\|_\infty \le \|\mathbf{x}\|_2 \le \sqrt{n}\,\|\mathbf{x}\|_\infty.$$

From the Schwarz inequality applied to vectors with elements $|x_i|$ and 1 respectively, we see that $\|\mathbf{x}\|_1 \le \sqrt{n}\,\|\mathbf{x}\|_2$. Also, by inspection, $\|\mathbf{x}\|_2^2 \le \|\mathbf{x}\|_1^2$. Hence

$$\frac{1}{\sqrt{n}}\|\mathbf{x}\|_2 \le \|\mathbf{x}\|_\infty \le \|\mathbf{x}\|_2,$$
$$\|\mathbf{x}\|_2 \le \|\mathbf{x}\|_1 \le \sqrt{n}\,\|\mathbf{x}\|_2,$$
$$\frac{1}{n}\|\mathbf{x}\|_1 \le \|\mathbf{x}\|_\infty \le \|\mathbf{x}\|_1 \tag{4.16}$$

In a sense, therefore, the 1, 2, and ∞ norms are equivalent. In fact it can be shown that *any* two norms $\|\cdot\|$ and $|||\cdot|||$ on a finite-dimensional space are equivalent in the sense that positive constants c_1 and c_2 exist for which $c_1\|\mathbf{x}\| \le |||\mathbf{x}||| \le c_2\|\mathbf{x}\|$ for all $\mathbf{x}$.

EXERCISE 4.41. Prove that the expressions in (4.11), (4.12), (4.13) are indeed vector norms.

EXERCISE 4.42. For each of the following vectors $\mathbf{x}$, compute $\|\mathbf{x}\|_1$, $\|\mathbf{x}\|_2$, and $\|\mathbf{x}\|_\infty$:

$$\begin{bmatrix} 1 \\ 2 \end{bmatrix}; \quad \begin{bmatrix} -1 \\ 2 \end{bmatrix}; \quad \begin{bmatrix} 1 \\ 2 \\ 3 \end{bmatrix}; \quad \sqrt{-1}\begin{bmatrix} -1 \\ 2 \\ -3 \end{bmatrix}; \quad \begin{bmatrix} 0 \\ 2 \\ 0 \\ 0 \end{bmatrix}.$$

EXERCISE 4.43. Let S be the set of points (a line) in $\mathbb{R}^2$ of the form

$$\begin{bmatrix} 1 \\ t \end{bmatrix}$$

as t ranges over all real numbers. For each of the norms $\|\cdot\|_1$, $\|\cdot\|_2$, $\|\cdot\|_\infty$, find the points in S that are closest to $\mathbf{0}$.

EXERCISE 4.44. Suppose that $\|\cdot\|$ is a vector norm on $\mathbb{R}^m$ and that $\mathbf{A}$ is an $m \times n$ matrix having rank n. Prove that $|||\mathbf{x}|||$ defined by

$$|||\mathbf{x}||| = \|\mathbf{A}\mathbf{x}\|$$

is a vector norm on $\mathbb{R}^n$.

EXERCISE 4.45. For any vector norm $||\cdot||$, prove that

$$\big|||\mathbf{x}|| - ||\mathbf{y}||\big| \leq ||\mathbf{x} - \mathbf{y}||.$$

4.7 INNER PRODUCTS AND ORTHOGONALITY

Although the notion of an abstract vector is quite general, in the last section we were able to bring this notion closer to our physical intuition by speaking of the length of a vector, more formally known as its norm. In this section we bring geometry further into the picture by seeing how to introduce the concept of *angle* into certain vector spaces. First we consider ordinary physical three-dimensional space with which we are familair (see Section 4.1).

As in Figure 4.2, let U, V be two points with coordinates (u_1, u_2, u_3), (v_1, v_2, v_3), the vectors from O to U and V being represented by $\vec{u}$, $\vec{v}$ respectively. The *scalar product* or *dot product* of $\vec{u}$ and $\vec{v}$ is defined to be

$$\vec{u}\cdot\vec{v} = u_1v_1 + u_2v_2 + u_3v_3.$$

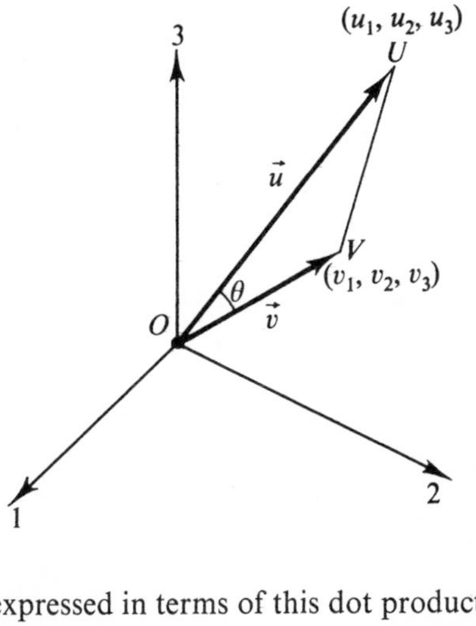

Figure 4.2. Vectors in three-dimensional space.

The physical length of a vector can be expressed in terms of this dot product. If $|\vec{u}|$ represents the length of $\vec{u}$ we have

$$|\vec{u}| = (u_1^2 + u_2^2 + u_3^2)^{1/2} = (\vec{u}\cdot\vec{u})^{1/2}.$$

The distance between U and V can also be expressed in terms of the dot product:

$$\begin{aligned} UV^2 &= (u_1 - v_1)^2 + (u_2 - v_2)^2 + (u_3 - v_3)^2 \\ &= |\vec{u} - \vec{v}|^2 = |\vec{u}|^2 + |\vec{v}|^2 - 2\vec{u}\cdot\vec{v}. \end{aligned} \tag{4.17}$$

The well-known formula for expressing a side of a triangle in terms of the opposite angle, say θ, and the sides adjacent to that angle, gives

$$UV^2 = |\vec{u}|^2 + |\vec{v}|^2 - 2|\vec{u}||\vec{v}|\cos\theta. \tag{4.18}$$

Comparing (4.17) and (4.18) we see that

$$\cos\theta = \frac{\vec{u}\cdot\vec{v}}{|\vec{u}||\vec{v}|}. \tag{4.19}$$

Using these results as a guide, we are in a position to introduce the concept of angle into vector spaces. The basic idea is to generalize the notion of the scalar product. We first suppose that our vectors lie in $\mathbb{R}^n$ and hence have real elements. (We give the following definition for clarity, although it will be superseded almost immediately by the more general Definition 4.12.)

DEFINITION 4.11. The *inner product* (or *scalar product*) of two real vectors **u**, **v** in $\mathbb{R}^n$ is the scalar quantity (**u**, **v**) defined by

$$(\mathbf{u}, \mathbf{v}) = u_1v_1 + u_2v_2 + \cdots + u_nv_n. \tag{4.20}$$

We recall that the *length* or *norm* of **u**, denoted by $||\mathbf{u}||_2$, is defined to be

$$||\mathbf{u}||_2 = (\mathbf{u}, \mathbf{u})^{1/2} = +(u_1^2 + u_2^2 + \cdots + u_n^2)^{1/2} \tag{4.21}$$

The *angle* θ *between* **u** *and* **v** is defined by

$$\cos\theta = \frac{(\mathbf{u}, \mathbf{v})}{||\mathbf{u}||_2||\mathbf{v}||_2}$$

Next consider what happens when the elements of the vectors are complex. If we use Definition 4.11, Equation (4.21), to find the length of a vector, then with $i = \sqrt{-1}$ and

$$\mathbf{u} = \begin{bmatrix} 1 \\ i \end{bmatrix},$$

we find

$$||\mathbf{u}||_2 = (1 - 1)^{1/2} = 0,$$

that is, a nonzero vector has zero length, and we have lost property (a) in Definition 4.10. It will turn out to be convenient for the norm $||\cdot||_2$ to be defined in terms of the inner product nonetheless; to allow this we slightly modify our definition of the inner product.

DEFINITION 4.12. The *inner product* (or *scalar product*) of two $n \times 1$ column vectors **u**, **v** with real or complex elements is the scalar quantity (**u**, **v**) defined by

$$(u, v) = \bar{u}_1v_1 + \bar{u}_2v_2 + \cdots + \bar{u}_nv_n,$$

where $\bar{u}_i$ is the complex conjugate of u_i. The *norm* or *length* of $\mathbf{u}$, denoted by $\|\mathbf{u}\|_2$, can then be represented by

$$\|\mathbf{u}\|_2 = (\mathbf{u}, \mathbf{u})^{1/2} = (\bar{u}_1 u_1 + \bar{u}_2 u_2 + \cdots + \bar{u}_n u_n)^{1/2}$$

as in (4.12).

Note:

(a) We could equally have defined $(\mathbf{u}, \mathbf{v})$ as the sum of the products $u_i \bar{v}_i$, with the complex conjugates on the second term. It does not matter which definition we use as long as we use the same convention consistently. We use the above convention because we shall often form the inner product of both sides of an equation, say $\mathbf{u} = \mathbf{v}$, with a third vector $\mathbf{w}$, to give $(\mathbf{w}, \mathbf{u}) = (\mathbf{w}, \mathbf{v})$ and it seems slightly more natural to take the complex conjugate of $\mathbf{w}$ rather than that of the equation $\mathbf{u} = \mathbf{v}$, if we are writing the result out explicitly. However, this is a matter of taste.

(b) Definition 4.12 reduces to the previous Definition 4.11 in the real case.

It is convenient to summarize certain important properties of the inner product. The results are very simple but they are useful because they avoid the necessity of going back to the definition every time we use inner products.

THEOREM 4.9. The inner product has the following properties:

(i) *Hermitian linearity:*

$$(\alpha\mathbf{u} + \beta\mathbf{w}, \mathbf{v}) = \bar{\alpha}(\mathbf{u}, \mathbf{v}) + \bar{\beta}(\mathbf{w}, \mathbf{v}),$$
$$(\mathbf{u}, \alpha\mathbf{v} + \beta\mathbf{w}) = \alpha(\mathbf{u}, \mathbf{v}) + \beta(\mathbf{u}, \mathbf{w}).$$

(ii) *Hermitian symmetry:*

$$(\mathbf{u}, \mathbf{v}) = \overline{(\mathbf{v}, \mathbf{u})}.$$

(iii) *Positive-definiteness:*

$$(\mathbf{u}, \mathbf{u}) > 0 \quad \text{if} \quad \mathbf{u} \neq \mathbf{0}, \qquad (\mathbf{u}, \mathbf{u}) = 0 \quad \textit{implies} \quad \mathbf{u} = \mathbf{0}.$$

(iv) *If α and β are real and $\mathbf{u}$ and $\mathbf{v}$ are in $\mathbb{R}^n$, then the complex conjugation symbols $^-$ may be removed in (i) and (ii).*

Proof: The proof follows directly from the definition.

Often it is important to consider the inner product between a vector $\mathbf{u}$ and special vectors of the form $\mathbf{Av}$, where $\mathbf{A}$ is an $n \times n$ matrix. The following simple facts often help in this regard. We recall from Definition 1.8 that $\mathbf{A}^H$ denotes the hermitian transpose of $\mathbf{A}$, namely, that matrix whose (i, j) element is the complex conjugate of the (j, i) element of $\mathbf{A}$; for convenience we restate the properties of the hermitian transpose also. Note that $(\mathbf{u}, \mathbf{v}) = \mathbf{u}^H\mathbf{v}$, disregarding the fact that technically $\mathbf{u}^H\mathbf{v}$ is a 1×1 matrix rather than a scalar.

THEOREM 4.10.

(i) $(\mathbf{A}^H)^H = \mathbf{A}$.
(ii) $(\mathbf{AB})^H = \mathbf{B}^H\mathbf{A}^H$
(iii) $(\mathbf{u}, \mathbf{Av}) = (\mathbf{A}^H\mathbf{u}, \mathbf{v})$
(iv) *If* $\mathbf{A}$ *is hermitian, so that* $\mathbf{A}^H = \mathbf{A}$, *then* $(\mathbf{u}, \mathbf{Av}) = (\mathbf{Au}, \mathbf{v})$.

Proof: (See the proof of Theorem 1.3.) The only nontrivial statement is (iii). For this,

$$(\mathbf{u}, \mathbf{Av}) = \mathbf{u}^H(\mathbf{Av}) = (\mathbf{u}^H\mathbf{A})\mathbf{v} = (\mathbf{A}^H\mathbf{u})^H\mathbf{v} = (\mathbf{A}^H\mathbf{u}, \mathbf{v})$$

as asserted.

So far we have only considered the notion of an inner product on specific concrete vector spaces, but more generality is possible.

DEFINITION 4.13. If V is an abstract vector space over the real or complex scalar field, then an *inner product* $(\cdot, \cdot)$ on V is a function which assigns to each ordered pair of vectors $\mathbf{u}, \mathbf{v}$ an element $(\mathbf{u}, \mathbf{v})$ of the scalar field in such a way that (i), (ii), and (iii) of Theorem 4.9 hold. In terms of the inner product $(\cdot, \cdot)$, we define the induced norm $\|\cdot\|$ via $\|\mathbf{v}\| = (\mathbf{v}, \mathbf{v})^{1/2}$.

EXERCISE 4.46. By duplicating the proof in Ex. 4.39, prove the Schwarz inequality $|(\mathbf{u}, \mathbf{v})| \leq \|\mathbf{u}\|\|\mathbf{v}\|$ if $\|\cdot\|$ is induced by the inner product $(\cdot, \cdot)$ on the vector space V.

EXERCISE 4.47. Let V be the vector space of continuous real-valued functions defined on [0, 1] [see (5) of Section 4.2], and let

$$(f, g) = \int_0^1 f(t)g(t)\,dt.$$

Prove that $(\cdot, \cdot)$ defines an inner product on V. Write out the Schwarz inequality.

In elementary geometry an especially important role is played by *perpendicular* (or *orthogonal*) lines and right angles. If θ is a right angle, then its cosine is zero; so if two vectors in physical space meet at right angles we know that $\vec{u} \cdot \vec{v} = 0$ by (4.19). This naturally motivates the following.

DEFINITION 4.14. Let V be a vector space with an inner product $(\cdot, \cdot)$. Two vectors $\mathbf{u}$ and $\mathbf{v}$ are said to be *orthogonal* when $(\mathbf{u}, \mathbf{v}) = 0$. A set S of vectors is said to be *orthogonal* when every pair of vectors from the set is orthogonal. If in addition each vector $\mathbf{v}$ in the set S satisfies $\|\mathbf{v}\| = 1$, then S is said to be *orthonormal.* If a nonzero vector $\mathbf{v}$ in a set is replaced by $\mathbf{v}' = \mathbf{v}/\|\mathbf{v}\|$ so that $\|\mathbf{v}'\| = 1$, then $\mathbf{v}$ is said to have been *normalized* and $\mathbf{v}'$ is a *normalized vector*.

In physical three-dimensional space it is intuitively clear that the set of all vectors orthogonal to a given vector $\mathbf{p}$ is just a plane through the origin.

More generally, the subset V_0 of all vectors $\mathbf{v}$ in a vector space V which satisfy $(\mathbf{p}, \mathbf{v}) = 1$ for a fixed $\mathbf{p}$ is usually called a *hyperplane*; see Ex. 4.94.

In physical three-dimensional space two vectors pointing in the same (or directly opposite) direction form a linearly dependent set, while intuitively it would appear that perpendicular vectors form linearly independent sets (in fact, "as independent as possible"). Two of the reasons for the importance of orthogonal sets, more generally, are that they are linearly independent and that other vectors are easily represented in terms of them.

THEOREM 4.11. Let $S = \{\mathbf{v}_1, \ldots, \mathbf{v}_n\}$ be an orthogonal set in a vector space V with inner product $(\cdot, \cdot)$.

(i) *If $\mathbf{v} = \alpha_1\mathbf{v}_1 + \cdots + \alpha_n\mathbf{v}_n$, then $\alpha_i(\mathbf{v}_i, \mathbf{v}_i) = (\mathbf{v}_i, \mathbf{v})$ for $i = 1, 2, \ldots, n$.*
(ii) *If the vectors of S are all nonzero, so in particular if S is orthonormal, then S is linearly independent.*

Proof: For (i),

$$(\mathbf{v}_i, \mathbf{v}) = (\mathbf{v}_i, \alpha_1\mathbf{v}_1 + \cdots + \alpha_n\mathbf{v}_n) = \alpha_1(\mathbf{v}_i, \mathbf{v}_1) + \cdots + \alpha_n(\mathbf{v}_i, \mathbf{v}_n) = \alpha_i(\mathbf{v}_i, \mathbf{v}_i)$$

as desired. For (ii), if $\mathbf{0} = \alpha_1\mathbf{v}_1 + \cdots + \alpha_n\mathbf{v}_n$ then by (i) we have $\alpha_i(\mathbf{v}_i, \mathbf{v}_i) = (\mathbf{0}, \mathbf{v}_i) = 0$ so that $\alpha_i = 0$ since $\|\mathbf{v}_i\| \neq 0$, completing the proof.

EXERCISE 4.48. Show that the set of vectors $\begin{bmatrix} 1 \\ -2 \end{bmatrix}, \begin{bmatrix} 4 \\ 2 \end{bmatrix}$ is linearly independent in $\mathbb{R}^2$.

SOLUTION: The vectors are nonzero and orthogonal since $1 \cdot 4 + (-2) \cdot 2 = 0$; by Theorem 4.11(ii) the set is linearly independent.

EXERCISE 4.49. Represent $\mathbf{x}$ in $\mathbb{R}^2$ as a linear combination of the vectors $\mathbf{v}_1 = \begin{bmatrix} 1 \\ -2 \end{bmatrix}, \mathbf{v}_2 = \begin{bmatrix} 4 \\ 2 \end{bmatrix}$.

SOLUTION: If

$$\mathbf{x} = \begin{bmatrix} x_1 \\ x_2 \end{bmatrix} = \alpha_1\mathbf{v}_1 + \alpha_2\mathbf{v}_2,$$

then by Theorem 4.11(i) we have $\alpha_1(1 + 4) = (\mathbf{v}_1, \mathbf{x}) = x_1 - 2x_2$ and $\alpha_2(16 + 4) = (\mathbf{v}_2, x) = 4x_1 + 2x_2$ so that

$$\alpha_1 = \frac{x_1 - 2x_2}{5} \text{ and } \alpha_2 = \frac{2x_1 + x_2}{10}.$$

The convenience with which we can find the numbers α_i describing a vector $\mathbf{x}$ as $\mathbf{x} = \alpha_1\mathbf{v}_1 + \cdots + \alpha_n\mathbf{v}_n$ for orthonormal sets of vectors makes it attractive to find an orthonormal basis for a space. In fact, we can go farther than this.

- *KEY THEOREM 4.12.* (*Gram-Schmidt Orthogonalization*). *Let vectors* $\mathbf{u}_1, \ldots, \mathbf{u}_s$ *form a linearly independent set in a vector space* V *with inner product* $(\cdot, \cdot)$. *Then we can construct an orthonormal set of vectors* $\mathbf{x}_1, \ldots, \mathbf{x}_s$ *such that for each* i *with* $1 \leq i \leq s$ *the set of vectors* $\mathbf{x}_1, \ldots, \mathbf{x}_i$ *spans precisely the same subspace as does the set of vectors* $\mathbf{u}_1, \ldots, \mathbf{u}_i$.
- *Key Proof:* We shall first form an orthogonal set of nonzero vectors $\mathbf{v}_1, \ldots, \mathbf{v}_s$, and then normalize them to yield

$$\mathbf{x}_i = \frac{\mathbf{v}_i}{\|\mathbf{v}_i\|}.$$

To start, clearly we must have $\mathbf{v}_1$ a multiple of $\mathbf{u}_1$ in order that they span the same space, so we take

$$\mathbf{v}_1 = \mathbf{u}_1, \; \mathbf{x}_1 = \frac{\mathbf{v}_1}{\|\mathbf{v}_1\|}.$$

We next choose the second vector $\mathbf{u}_2$ from the original set, and subtract from it a multiple of $\mathbf{x}_1$,

$$\mathbf{v}_2 = \mathbf{u}_2 - \alpha_1 \mathbf{x}_1,$$

where α_1 is chosen in such a way that $\mathbf{x}_1$ and $\mathbf{v}_2$ are orthogonal as required by the theorem:

$$0 = (\mathbf{x}_1, \mathbf{v}_2) = (\mathbf{x}_1, \mathbf{u}_2) - \alpha_1(\mathbf{x}_1, \mathbf{x}_1).$$

Since $(\mathbf{x}_1, \mathbf{x}_1) = 1$, this gives $\alpha_1 = (\mathbf{x}_1, \mathbf{u}_2)$ and

$$\mathbf{v}_2 = \mathbf{u}_2 - (\mathbf{x}_1, \mathbf{u}_2)\mathbf{x}_1, \qquad \mathbf{x}_2 = \frac{\mathbf{v}_2}{\|\mathbf{v}_2\|}.$$

Since $\mathbf{u}_1$ and $\mathbf{u}_2$ are linear combinations of $\mathbf{x}_1$ and $\mathbf{x}_2$, and vice versa, the same spaces are spanned as asserted. The only way in which this construction would fail would be if $\mathbf{v}_2$ were identically zero. This, however, would imply the linear dependence of $\mathbf{u}_1, \mathbf{u}_2$, whereas it has been assumed in the theorem that the set of $\mathbf{u}_1, \mathbf{u}_2$ is independent. We next form $\mathbf{v}_3$ by subtracting multiples of $\mathbf{x}_1$ and $\mathbf{x}_2$ from the third given vector $\mathbf{u}_3$,

$$\mathbf{v}_3 = \mathbf{u}_3 - \alpha_2 \mathbf{x}_2 - \beta_1 \mathbf{x}_1,$$

where α_2, β_1 are determined in such a way that $\mathbf{v}_3$ is orthogonal to $\mathbf{x}_1, \mathbf{x}_2$. This gives

$$(\mathbf{x}_1, \mathbf{v}_3) = (\mathbf{x}_1, \mathbf{u}_3) - \beta_1 = 0, \qquad (\mathbf{x}_2, \mathbf{v}_3) = (\mathbf{x}_2, \mathbf{u}_3) - \alpha_2 = 0.$$

Hence,

$$\mathbf{v}_3 = \mathbf{u}_3 - (\mathbf{x}_2, \mathbf{u}_3)\mathbf{x}_2 - (\mathbf{x}_1, \mathbf{u}_3)\mathbf{x}_1, \qquad \mathbf{x}_3 = \frac{\mathbf{v}_3}{\|\mathbf{v}_3\|}.$$

Again, $\{\mathbf{u}_1, \mathbf{u}_2, \mathbf{u}_3\}$ and $\{\mathbf{x}_1, \mathbf{x}_2, \mathbf{x}_3\}$ span the same space. Clearly the formula for the general $\mathbf{v}_r$ is

$$\mathbf{v}_r = \mathbf{u}_r - (\mathbf{x}_{r-1}, \mathbf{u}_r)\mathbf{x}_{r-1} - (\mathbf{x}_{r-2}, \mathbf{u}_r)\mathbf{x}_{r-2} - \cdots - (\mathbf{x}_1, \mathbf{u}_r)\mathbf{x}_1.$$

The only way that this procedure could break down would be if $\mathbf{v}_r$ were identically zero for some r. From the method of formation of $\mathbf{v}_r$ it is clear that $\mathbf{v}_r$ is a linear combination of $\mathbf{u}_1, \ldots, \mathbf{u}_r$ with a nonzero coefficient of $\mathbf{u}_r$. If $\mathbf{v}_r$ is identically zero, this means that $\{\mathbf{u}_1, \ldots, \mathbf{u}_r\}$ is linearly dependent, contradicting the assumption in the theorem.

When solving simple numerical examples by hand, two points should be noted in connection with the implementation of this theorem:

1. It is convenient to work throughout in terms of the $\mathbf{v}_r$, leaving the normalizations to the end, as illustrated in the following example. This means that the formulas we use are not quite the same as those developed in the proof.
2. In any case, one should not try to memorize formulas for the method, since the principle behind the procedure tells us what to do at each stage.

EXERCISE 4.50. Form an orthonormal set from

$$\mathbf{u}_1 = \begin{bmatrix} 1 \\ 1 \\ 1 \\ -1 \end{bmatrix}, \quad \mathbf{u}_2 = \begin{bmatrix} 2 \\ -1 \\ -1 \\ 1 \end{bmatrix}, \quad \mathbf{u}_3 = \begin{bmatrix} -1 \\ 2 \\ 2 \\ 1 \end{bmatrix}.$$

SOLUTION: We set $\mathbf{v}_1 = \mathbf{u}_1$ and then choose α so that

$$\mathbf{v}_2' = \mathbf{u}_2 - \alpha \mathbf{v}_1$$

is orthogonal to $\mathbf{v}_1$. (The reason for the prime will appear in a moment.) This gives $\alpha = (\mathbf{v}_1, \mathbf{u}_2)/(\mathbf{v}_1, \mathbf{v}_1)$, and we find $\alpha = -\frac{1}{4}$. For ease in computation, we compute $4\mathbf{v}_2'$ since then no fractions are involved.

$$4\mathbf{v}_2' = 4\mathbf{u}_2 + \mathbf{v}_1 = \begin{bmatrix} 9 \\ -3 \\ -3 \\ 3 \end{bmatrix} \quad \text{and we choose} \quad \mathbf{v}_2 = \begin{bmatrix} 3 \\ -1 \\ -1 \\ 1 \end{bmatrix}.$$

The point here is that we are leaving normalizations to the end, so that we can choose, as $\mathbf{v}_2$, any convenient multiple of $\mathbf{v}_2'$. In a similar way we choose β, γ so that

$$\mathbf{v}_3' = \mathbf{u}_3 - \beta \mathbf{v}_1 - \gamma \mathbf{v}_2$$

is orthogonal to $\mathbf{v}_1$ and $\mathbf{v}_2$. This gives $\beta = \frac{1}{2}$, $\gamma = -\frac{1}{2}$, so we compute $2\mathbf{v}_3'$, again to avoid fractions:

$$2\mathbf{v}_3' = 2\mathbf{u}_3 - \mathbf{v}_1 + \mathbf{v}_2 = \begin{bmatrix} 0 \\ 2 \\ 2 \\ 4 \end{bmatrix} \quad \text{and we choose} \quad \mathbf{v}_3 = \begin{bmatrix} 0 \\ 1 \\ 1 \\ 2 \end{bmatrix}.$$

Normalization of $\mathbf{v}_1, \mathbf{v}_2, \mathbf{v}_3$ gives, finally,

$$\mathbf{x}_1 = \frac{1}{2}\begin{bmatrix} 1 \\ 1 \\ 1 \\ -1 \end{bmatrix}, \quad \mathbf{x}_2 = \frac{1}{2\sqrt{3}}\begin{bmatrix} 3 \\ -1 \\ -1 \\ 1 \end{bmatrix}, \quad \mathbf{x}_3 = \frac{1}{\sqrt{6}}\begin{bmatrix} 0 \\ 1 \\ 1 \\ 2 \end{bmatrix}.$$

It is instructive to interpret the Gram-Schmidt procedure geometrically when $\mathbf{u}_1, \mathbf{u}_2, \mathbf{u}_3$ are 3×1 column vectors in three-dimensional space. The usual geometrical interpretation of scalar products, already given, leads to the picture in Figure 4.3. The vectors $\overrightarrow{OP}$, $\overrightarrow{OQ}$ represent $\mathbf{u}_1$, $\mathbf{u}_2$, respectively,

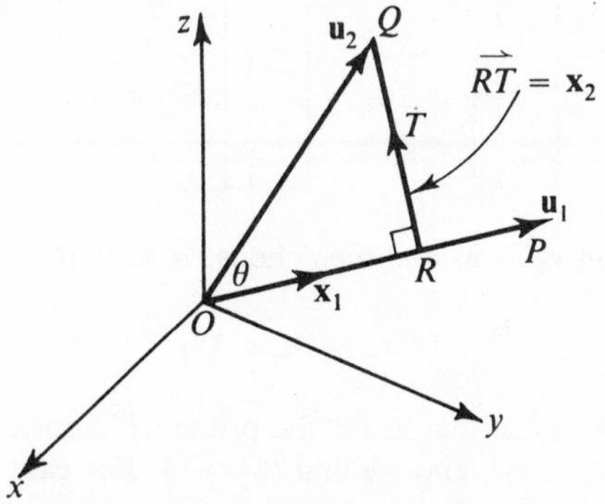

Figure 4.3. Gram-Schmidt orthogonalization.

and $\mathbf{x}_1$ is a unit vector along OP. The line QR is perpendicular to OP so that

$$OR = OQ \cos\theta = (\mathbf{x}_1, \mathbf{u}_2),$$

where θ is the angle between $\mathbf{u}_1$ and $\mathbf{u}_2$. Hence, $(\mathbf{x}_1, \mathbf{u}_2)\mathbf{x}_1$ is a vector in the direction of $\mathbf{x}_1$ of length OR, that is, it is the vector $\overrightarrow{OR}$ in Figure 4.3. The vector $\overrightarrow{OR}$ is known as the *projection* of $\overrightarrow{OQ}$ on $\overrightarrow{OP}$. we see that

$$\mathbf{v}_2 = \mathbf{u}_2 - (\mathbf{x}_1, \mathbf{u}_2)\mathbf{x}_1$$

is the vector $\overrightarrow{RQ}$ which, geometrically, is perpendicular to $\overrightarrow{OP}$, as we should expect. The vector $\mathbf{x}_2$ is a unit vector in the direction of $\overrightarrow{RQ}$. Summing up, the vector $\mathbf{x}_1$ is a unit vector in the direction of $\mathbf{u}_1$. The vector $\mathbf{x}_2$ is a unit vector in the plane of $\mathbf{u}_1$ and $\mathbf{u}_2$ perpendicular to $\mathbf{u}_1$. Similarly $\mathbf{x}_3$ is a unit vector perpendicular to the plane of $\mathbf{u}_1$ and $\mathbf{u}_2$. We are constructing a set of three perpendicular unit vectors in three-space, starting from $\mathbf{u}_1$, $\mathbf{u}_2$, $\mathbf{u}_3$. The condition that $\mathbf{u}_1$, $\mathbf{u}_2$, $\mathbf{u}_3$ are linearly independent means geometrically that $\mathbf{u}_1$, $\mathbf{u}_2$, $\mathbf{u}_3$ are not coplanar, and this is a necessary condition for the construction to succeed.

Since **Key Theorem 4.12** was motivated by our desire to construct a basis that formed an orthonormal set, we state the existence of such an orthonormal set as a corollary.

COROLLARY 4.2. Every finite dimensional vector space V with an inner product has an orthonormal basis.

Proof: By virtue of being finite dimensional V has a basis $\{\mathbf{u}_1, \ldots, \mathbf{u}_s\}$. **Key Theorem 4.12** then gives us the orthonormal basis $\{\mathbf{x}_1, \ldots, \mathbf{x}_s\}$.

EXERCISE 4.51. In Ex. 4.50, verify that $\{\mathbf{u}_1\}$ and $\{\mathbf{x}_1\}$ span the same space, that $\{\mathbf{u}_1, \mathbf{u}_2\}$ and $\{\mathbf{x}_1, \mathbf{x}_2\}$ span the same space, and that $\{\mathbf{u}_1, \mathbf{u}_2, \mathbf{u}_3\}$ and $\{\mathbf{x}_1, \mathbf{x}_2, \mathbf{x}_3\}$ span the same space.

EXERCISE 4.52. Apply the Gram-Schmidt procedure to the vectors

$$\mathbf{u}_1 = \frac{1}{\sqrt{10}}\begin{bmatrix} 1 \\ 2 \\ 2 \\ 1 \end{bmatrix}, \qquad \mathbf{u}_2 = \frac{1}{2}\begin{bmatrix} 1 \\ 1 \\ -1 \\ -1 \end{bmatrix}, \qquad \mathbf{u}_3 = \frac{1}{\sqrt{10}}\begin{bmatrix} 2 \\ -1 \\ -1 \\ 2 \end{bmatrix}.$$

Make a statement about the analogous general situation.

EXERCISE 4.53. Find an orthonormal basis for $\mathbb{R}^n$.

In numerical work it was discovered that the Gram-Schmidt process, described in the proof of **Key Theorem 4.12**, often gives very inaccurate results in that the computed vectors may be far from orthogonal (see Ex. 4.91). An improved process called *modified Gram-Schmidt*, which is mathematically equivalent but computationally superior (see Ex. 4.92), will now be sketched. Given the vectors $\mathbf{u}_1, \ldots, \mathbf{u}_s$, the standard Gram-Schmidt process first modifies $\mathbf{u}_1$ to compute $\mathbf{x}_1$, leaving $\mathbf{u}_2, \ldots, \mathbf{u}_s$ alone. As the second step it modifies $\mathbf{u}_2$ to compute $\mathbf{x}_2$, but leaves $\mathbf{u}_3, \ldots, \mathbf{u}_s$ alone; at the general ith step it modifies $\mathbf{u}_i$ to compute $\mathbf{x}_i$ but leaves $\mathbf{u}_{i+1}, \ldots, \mathbf{u}_s$ alone. The *modified* Gram-Schmidt procedure, on the other hand, changes each of the remaining vectors at each step. Thus in the first step $\mathbf{x}_1$ is computed from $\mathbf{u}_1$, and also

$(\mathbf{x}_1, \mathbf{u}_i)\mathbf{x}_1$ is subtracted from $\mathbf{u}_i$ to produce $\mathbf{u}_i^{(1)}$ for $i = 2, \ldots, s$ to make $\mathbf{u}_i^{(1)}$ orthogonal to $\mathbf{x}_1$. At the second step $\mathbf{u}_2^{(1)}$ is normalized to produce $\mathbf{x}_2$, and $(\mathbf{x}_2, \mathbf{u}_i^{(1)})\mathbf{x}_2$ is subtracted from $\mathbf{u}_i^{(1)}$ to produce $\mathbf{u}_i^{(2)}$ for $i = 3, \ldots, s$ to make $\mathbf{u}_i^{(2)}$ orthogonal to $\mathbf{x}_2$ as well as to $\mathbf{x}_1$. At the general jth step, $\mathbf{u}_j^{(j-1)}$ is normalized to produce $\mathbf{x}_j$, and $(\mathbf{x}_j, \mathbf{u}_i^{(j-1)})\mathbf{x}_j$ is subtracted from $\mathbf{u}_i^{(j-1)}$ to produce $\mathbf{u}_i^{(j)}$ for $i = j + 1, \ldots, s$ to make $\mathbf{u}_i^{(j)}$ orthogonal to $\mathbf{x}_j$ as well as to $\mathbf{x}_{j-1}, \mathbf{x}_{j-2}, \ldots, \mathbf{x}_1$. The process ceases when $\mathbf{x}_s$ is computed.

To see the difference involved, we apply the modified process to the vectors in Ex. 4.50. We first normalize $\mathbf{u}_1$ to obtain

$$\mathbf{x}_1 = \frac{1}{2}\begin{bmatrix} 1 \\ 1 \\ 1 \\ -1 \end{bmatrix},$$

just as in Ex. 4.50. We then modify $\mathbf{u}_2$ and $\mathbf{u}_3$ to obtain $\mathbf{u}_2^{(1)}$ and $\mathbf{u}_3^{(1)}$. For example,

$$\mathbf{u}_2^{(1)} = \mathbf{u}_2 - (\mathbf{x}_1, \mathbf{u}_2)\mathbf{x}_1 = \begin{bmatrix} 2 \\ -1 \\ -1 \\ 1 \end{bmatrix} - \left(-\frac{1}{2}\right)\frac{1}{2}\begin{bmatrix} 1 \\ 1 \\ 1 \\ -1 \end{bmatrix} = \frac{1}{4}\begin{bmatrix} 9 \\ -3 \\ -3 \\ 3 \end{bmatrix}.$$

Similarly,

$$\mathbf{u}_3^{(1)} = \mathbf{u}_3 - (\mathbf{x}_1, \mathbf{u}_3)\mathbf{x}_1 = \frac{3}{2}\begin{bmatrix} -1 \\ 1 \\ 1 \\ 1 \end{bmatrix}.$$

Next we normalize $\mathbf{u}_2^{(1)}$ to obtain

$$\mathbf{x}_2 = \frac{1}{2\sqrt{3}}\begin{bmatrix} 3 \\ -1 \\ -1 \\ 1 \end{bmatrix},$$

just as in Ex. 4.50. We then modify $\mathbf{u}_3^{(1)}$ to obtain

$$\mathbf{u}_3^{(2)} = \mathbf{u}_3^{(1)} - (\mathbf{x}_2, \mathbf{u}_3^{(1)})\mathbf{x}_2 = \frac{3}{2}\begin{bmatrix} -1 \\ 1 \\ 1 \\ 1 \end{bmatrix} - (-\sqrt{3})\frac{1}{2\sqrt{3}}\begin{bmatrix} 3 \\ -1 \\ -1 \\ 1 \end{bmatrix} = \begin{bmatrix} 0 \\ 1 \\ 1 \\ 2 \end{bmatrix}.$$

Finally we finish by normalizing $\mathbf{u}_3^{(2)}$ to obtain $\mathbf{x}_3$ as

$$\mathbf{x}_3 = \frac{1}{\sqrt{6}}\begin{bmatrix} 0 \\ 1 \\ 1 \\ 2 \end{bmatrix},$$

just as in Ex. 4.50. In summary: precisely the same results were found as in Ex. 4.50. In many cases, however, the modified Gram-Schmidt procedure will be more accurate than the standard one, as illustrated in Exs. 4.91 and 4.92.

MISCELLANEOUS EXERCISES 4

EXERCISE 4.54. Let $\vec{u} = (-1, 2, 3)$ and $\vec{v} = (3, -2, 14)$ be two vectors in three-dimensional physical space. Find $\vec{u} + \vec{v}$, $3\vec{u}$, $-\vec{v}$, $-2\vec{u} + 3\vec{v}$, and $-\vec{u} - \vec{v}$.

EXERCISE 4.55. If $\vec{u}$ and $\vec{v}$ are two general three-dimensional physical vectors, show geometrically that $\vec{u} + \vec{v} = \vec{v} + \vec{u}$.

EXERCISE 4.56. Let $\vec{u} = (-8, -9)$ and $\vec{v} = (16, 18)$ be two vectors in the two-dimensional plane space. Find $-3\vec{u}$, $2\vec{u} + \vec{v}$, $-\vec{v}$, and $\vec{u} + \vec{v}$.

EXERCISE 4.57. If $\vec{u}$, $\vec{v}$, and $\vec{w}$ are three general two-dimensional plane vectors, show geometrically that $(\vec{u} + \vec{v}) + \vec{w} = \vec{u} + (\vec{v} + \vec{w})$.

EXERCISE 4.58. Find the straight line formed by all multiples of the physical vector $\vec{u} = (1, 2, -4)$.

EXERCISE 4.59. Find the equation $y = mx + b$ satisfied by all two-dimensional vectors (x, y) making up the line of multiples of $(-3, 2)$.

EXERCISE 4.60. Find all physical vectors (x, y, z) of the form $\alpha\vec{u} + \beta\vec{v}$, where α and β range over the real numbers while $\vec{u} = (4, -2, -1)$ and $\vec{v} = (-4, 0, 3)$.

EXERCISE 4.61. Suppose that $\vec{u}$ and $\vec{v}$ are collinear physical vectors; show geometrically that as α and β vary over the real numbers, the set of all vectors of the form $\alpha\vec{u} + \beta\vec{v}$ forms not a *plane* but a *line*.

EXERCISE 4.62. Let $\vec{0}$ be the zero three-dimensional physical vector and $\vec{u}$ any other physical vector. Show that $\vec{0} + \vec{u} = \vec{u} + \vec{0} = \vec{u}$.

EXERCISE 4.63. In three-dimensional physical space, show that each of $\{(1, 1, 0), (0, 1, 1)\}$ and $\{(1, 0, -1), (2, 3, 1)\}$ is a basis for the space spanned by $\{(1, 1, 0), (0, 1, 1)\}$. Find another basis.

EXERCISE 4.64. Show that the set Q of rational numbers forms a field.

EXERCISE 4.65. Show that as r_1 and r_2 vary over the rational numbers the set of all numbers of the form $r_1 + r_2\sqrt{-1}$ forms a field.

EXERCISE 4.66. Show that the set $\{0, 1, 2\}$ forms a field if operations are defined via

$$\begin{array}{lll} 0+0=0, & 0+1=1+0=1, & 0+2=2+0=2, \\ 1+2=2+1=0, & 1+1=2, & 2+2=1, \\ 0\cdot 0=0, & 0\cdot 1=1\cdot 0=0, & 0\cdot 2=2\cdot 0=0, \\ 1\cdot 1=1, & 1\cdot 2=2\cdot 1=2, & 2\cdot 2=1. \end{array}$$

Find -2 and -1.

EXERCISE 4.67. Let V be a vector space over a field F. Prove that $0\mathbf{x} = \mathbf{0}$ for all $\mathbf{x}$ in V by writing

$$\begin{aligned} \mathbf{0} &= 0\mathbf{x} + (-(0\mathbf{x})) = (0+0)\mathbf{x} + (-(0\mathbf{x})) = (0\mathbf{x} + 0\mathbf{x}) + (-(0\mathbf{x})) \\ &= 0\mathbf{x} + (0\mathbf{x} + (-(0\mathbf{x}))) = 0\mathbf{x} + \mathbf{0} = 0\mathbf{x}. \end{aligned}$$

EXERCISE 4.68. Let V be a vector space over a field F. Prove that $(-1)\mathbf{x} = -\mathbf{x}$ for all $\mathbf{x}$ in V.

EXERCISE 4.69. Show that the set $\mathbb{C}$ of complex numbers is a vector space over the field $\mathbb{R}$ of reals.

EXERCISE 4.70. Show that the field $\mathbb{C}$ is a vector space over the field $\mathbb{C}$.

EXERCISE 4.71. Describe the vector $\mathbf{0}$ in each of the sample vector spaces (1)–(11) in Section 4.2.

EXERCISE 4.72. Prove that the vector space $\mathbb{R}$ of real numbers over the field $\mathbb{R}$ is a subspace of the vector space $\mathbb{C}$ of complex numbers over the field $\mathbb{R}$.

EXERCISE 4.73. Verify the statement in Section 4.3 that

$$\left\{ \begin{bmatrix} 1 \\ 2 \\ 1 \end{bmatrix}, \begin{bmatrix} 1 \\ 0 \\ -1 \end{bmatrix}, \begin{bmatrix} 1 \\ -2 \\ 1 \end{bmatrix} \right\}$$

is a linearly independent set. For what values of x is

$$\left\{\begin{bmatrix}1\\2\\1\end{bmatrix}, \begin{bmatrix}1\\x\\-1\end{bmatrix}, \begin{bmatrix}1\\-2\\1\end{bmatrix}\right\}$$

linearly dependent?

EXERCISE 4.74. Let p vectors of order n be arranged as the columns of an $n \times p$ matrix $\mathbf{A}$. Prove that a necessary and sufficient condition for these vectors to form a linearly dependent set is that the rank of $\mathbf{A}$ is less than p.

EXERCISE 4.75. Find a basis for the subspace of $\mathbb{R}^4$ generated by the vectors

$$\begin{bmatrix}1\\2\\-1\\0\end{bmatrix}, \begin{bmatrix}4\\8\\-4\\-3\end{bmatrix}, \begin{bmatrix}0\\1\\3\\4\end{bmatrix}, \begin{bmatrix}2\\5\\1\\4\end{bmatrix}.$$

EXERCISE 4.76. Extend the set

$$\begin{bmatrix}1\\0\\-1\\0\end{bmatrix}, \begin{bmatrix}1\\1\\0\\0\end{bmatrix}, \begin{bmatrix}1\\2\\-1\\4\end{bmatrix}$$

to form a basis in $\mathbb{R}^4$.

EXERCISE 4.77. Consider the subspace of $\mathbb{R}^4$ consisting of all 4×1 column vectors $\mathbf{x}$ with $x_1 + x_2 + x_3 = 0$. Extend the following set to form a basis for the space:

$$\begin{bmatrix}1\\-2\\1\\0\end{bmatrix}, \begin{bmatrix}1\\-1\\0\\0\end{bmatrix}.$$

EXERCISE 4.78. Show that

$$\begin{bmatrix}2\\-1\\1\end{bmatrix}, \begin{bmatrix}1\\2\\-3\end{bmatrix} \text{ and } \begin{bmatrix}3\\1\\-2\end{bmatrix}, \begin{bmatrix}-1\\3\\4\end{bmatrix}$$

generate the same subspace of $\mathbb{R}^3$.

EXERCISE 4.79. Find the dimension and a basis for the column space of

$$\mathbf{A} = \begin{bmatrix} 1 & 1 & 2 & -3 & 3 \\ 3 & 4 & -1 & 2 & 15 \\ 4 & 5 & 1 & 1 & 18 \\ -2 & -3 & 3 & -5 & -12 \end{bmatrix}.$$

EXERCISE 4.80. Given a set of basis vectors $\mathbf{u}_1, \ldots, \mathbf{u}_m$ and any other vector $\mathbf{v}$ such that

$$\mathbf{v} = \alpha_1\mathbf{u}_1 + \cdots + \alpha_m\mathbf{u}_m,$$

show that a new basis can be obtained by replacing any vector $\mathbf{u}_r$ for which $\alpha_r \neq 0$ by $\mathbf{v}$. If

$$\mathbf{u}_1 = \begin{bmatrix} 1 \\ 2 \\ 1 \end{bmatrix}, \quad \mathbf{u}_2 = \begin{bmatrix} 1 \\ 0 \\ -1 \end{bmatrix}, \quad \mathbf{u}_3 = \begin{bmatrix} 1 \\ -2 \\ 1 \end{bmatrix}, \quad \mathbf{v} = \begin{bmatrix} 1 \\ 2 \\ -3 \end{bmatrix},$$

which of the following sets form a basis for $\mathbb{R}^3$?

(a) $\mathbf{v}, \mathbf{u}_2, \mathbf{u}_3$.
(b) $\mathbf{u}_1, \mathbf{v}, \mathbf{u}_3$.
(c) $\mathbf{u}_1, \mathbf{u}_2, \mathbf{v}$.

EXERCISE 4.81. If $\mathbf{A}$ is $m \times n$ and $\mathbf{B}$ is $n \times p$, show that the columns of $\mathbf{AB}$ are in the column space of $\mathbf{A}$.

EXERCISE 4.82. Let V be a subspace of $\mathbb{R}^n$. Show that the set W of vectors $\mathbf{u}$ in $\mathbb{R}^n$ for which $\mathbf{u}^T\mathbf{x} = 0$ for all $\mathbf{x}$ in V is also a subspace of $\mathbb{R}^n$. Find a basis for this subspace W when $n = 3$ and V is the subspace of $\mathbb{R}^3$ spanned by $\mathbf{x}_1 = [2, 1, -3]^T$, $\mathbf{x}_2 = [-1, 0, -2]^T$.

EXERCISE 4.83. Construct a system of three homogeneous equations in five unknowns with solution space generated by the vectors

$$\mathbf{u}_1 = [1 \quad 1 \quad 1 \quad 1 \quad 1]^T, \qquad \mathbf{u}_2 = [1 \quad -1 \quad 1 \quad -1 \quad 1]^T.$$

Describe a general procedure for finding a set of $n - p$ homogeneous equations in n unknowns whose solution space is a given p-dimensional subspace of $\mathbb{R}^n$.

EXERCISE 4.84. Are the vectors $[-3, \ -1, \ 15, \ 6]$, $[1, \ 0, \ -1, \ 0]$, $[1, \ 1, \ 1, \ 0]$ in the subspace of $\mathbb{R}^4$ spanned by the vectors $[-1, \ 3, \ 5, \ 2]$, $[2, \ -1, \ 0, \ 1]$, $[1, \ -8, \ 5, \ 3]$? Explain.

EXERCISE 4.85. Let V denote the subspace of $\mathbb{R}^4$ spanned by

$$\mathbf{u}_1 = \begin{bmatrix} 1 \\ 2 \\ 2 \\ 1 \end{bmatrix}, \quad \mathbf{u}_2 = \begin{bmatrix} 1 \\ 0 \\ 2 \\ 0 \end{bmatrix}, \quad \mathbf{u}_3 = \begin{bmatrix} 2 \\ 0 \\ 4 \\ -3 \end{bmatrix}.$$

(a) Show that these vectors form a basis for V.
(b) If $\mathbf{x} = [x_1, x_2, x_3, x_4]^T$ is an arbitrary vector in V, find $\boldsymbol{\alpha} = [\alpha_i]$ such that $\mathbf{x} = \alpha_1\mathbf{u}_1 + \alpha_2\mathbf{u}_2 + \alpha_3\mathbf{u}_3$.
(c) Show that the following vectors span the same space:

$$\mathbf{v}_1 = \begin{bmatrix} 0 \\ 2 \\ 0 \\ 1 \end{bmatrix}, \quad \mathbf{v}_2 = \begin{bmatrix} 2 \\ 1 \\ 4 \\ -1 \end{bmatrix}, \quad \mathbf{v}_3 = \begin{bmatrix} 1 \\ 2 \\ 2 \\ 4 \end{bmatrix}.$$

(d) If $\mathbf{x}$ in (b) is written in the form $\mathbf{x} = \beta_1\mathbf{v}_1 + \beta_2\mathbf{v}_2 + \beta_3\mathbf{v}_3$, find a matrix $\mathbf{K}$ such that $\boldsymbol{\alpha} = \mathbf{K}\boldsymbol{\beta}$ where $\boldsymbol{\alpha} = [\alpha_1, \alpha_2, \alpha_3]^T$ and $\boldsymbol{\beta} = [\beta_1, \beta_2, \beta_3]^T$.
(e) Why does general theory tell us that the matrix $\mathbf{K}$ in (d) must be nonsingular?

EXERCISE 4.86. Prove that the following expressions for $\mathbf{x}$ are equivalent (the α_i are arbitrary constants).

$$\mathbf{x} = \begin{bmatrix} 1 \\ 2 \\ -1 \\ 3 \end{bmatrix} + \alpha_1 \begin{bmatrix} 1 \\ -3 \\ 4 \\ 1 \end{bmatrix} + \alpha_2 \begin{bmatrix} 2 \\ 1 \\ 1 \\ -5 \end{bmatrix} + \alpha_3 \begin{bmatrix} 0 \\ 4 \\ -3 \\ -2 \end{bmatrix},$$

$$\mathbf{x} = \begin{bmatrix} 4 \\ 4 \\ 1 \\ -3 \end{bmatrix} + \alpha_4 \begin{bmatrix} 1 \\ 1 \\ 1 \\ -1 \end{bmatrix} + \alpha_5 \begin{bmatrix} -1 \\ 2 \\ -1 \\ 4 \end{bmatrix} + \alpha_6 \begin{bmatrix} 3 \\ -1 \\ 2 \\ -9 \end{bmatrix}.$$

EXERCISE 4.87. Prove that in a vector space V with inner product $(\cdot, \cdot)$ we have $\|\mathbf{x} + \mathbf{y}\|^2 = \|\mathbf{x}\|^2 + \|\mathbf{y}\|^2$ if and only if $\mathbf{x}$ and $\mathbf{y}$ are orthogonal.

EXERCISE 4.88. Let $\mathbf{v}_1, \ldots, \mathbf{v}_n$ form an orthonormal set, and let $\mathbf{v}$ be a vector. Find a vector $\mathbf{v}_0$ of the form $\mathbf{v}_0 = \mathbf{v} - \alpha_1\mathbf{v}_1 - \cdots - \alpha_n\mathbf{v}_n$ that is orthogonal to each $\mathbf{v}_i$ for $i = 1, \ldots, n$.

EXERCISE 4.89. Consider the vector space and inner product of Ex. 4.47, and let the vectors $\mathbf{u}_1$, $\mathbf{u}_2$, $\mathbf{u}_3$ be given by: $\mathbf{u}_1(t) = 1$, $\mathbf{u}_2(t) = t$, $\mathbf{u}_3(t) = t^2$. Apply the Gram-Schmidt procedure to produce an orthonormal set.

EXERCISE 4.90. Provide the details to prove in **Key Theorem 4.12** that $\mathbf{x}_1, \ldots, \mathbf{x}_i$ spans the same space as does $\mathbf{u}_1, \ldots, \mathbf{u}_i$ for $i = 1, \ldots, s$.

EXERCISE 4.91. Let ϵ be a very small number such that $1 + \epsilon$ is represented exactly on our computer but such that $1 + \epsilon + \epsilon^2$ is rounded to $1 + \epsilon$; on a ten-figure decimal machine, for example, $\epsilon = 10^{-5}$ will do nicely. Show that if the standard Gram-Schmidt process is implemented on this computer to orthonormalize the vectors

$$\mathbf{u}_1 = \begin{bmatrix} 1 \\ 1+\epsilon \\ 1 \\ 1 \end{bmatrix}, \quad \mathbf{u}_2 = \begin{bmatrix} 1 \\ 1 \\ 1+\epsilon \\ 1 \end{bmatrix}, \quad \mathbf{u}_3 = \begin{bmatrix} 1 \\ 1 \\ 1 \\ 1+\epsilon \end{bmatrix},$$

then the results will be very inaccurate in that $\mathbf{x}_2$ and $\mathbf{x}_3$ will be far from orthogonal.

Hint: Since $(\mathbf{u}_1, \mathbf{u}_1) = 1 + (1+\epsilon)^2 + 1 + 1 = 1 + (1 + 2\epsilon + \epsilon^2) + 1 + 1$ will be computed to equal $4 + 2\epsilon$, $\mathbf{x}_1$ will be computed as $\mathbf{u}_1/\sqrt{4 + 2\epsilon}$. One can then deduce that $\mathbf{x}_2$ will be the normalized version of $[0, -\epsilon, \epsilon, 0]^T$, namely

$$\mathbf{x}_2 = \frac{1}{\sqrt{2}} \begin{bmatrix} 0 \\ -1 \\ 1 \\ 0 \end{bmatrix}.$$

The procedure then leads to $\mathbf{x}_3$ as the normalization of $[0, -\epsilon, 0, \epsilon]^T$, namely

$$\mathbf{x}_3 = \frac{1}{\sqrt{2}} \begin{bmatrix} 0 \\ -1 \\ 0 \\ 1 \end{bmatrix}.$$

One finds $(\mathbf{x}_1, \mathbf{x}_2) = (\mathbf{x}_1, \mathbf{x}_3) = -\epsilon/\sqrt{8 + 4\epsilon}$, but $(\mathbf{x}_2, \mathbf{x}_3) = \frac{1}{2}$ so that $\mathbf{x}_2$ and $\mathbf{x}_3$ are far from orthogonal.

EXERCISE 4.92. Provide the details for using the modified Gram-Schmidt procedure in Ex. 4.91 so as to obtain

$$\mathbf{x}_1 = \frac{1}{\sqrt{4+2\epsilon}}\begin{bmatrix} 1 \\ 1+\epsilon \\ 1 \\ 1 \end{bmatrix}, \qquad \mathbf{u}_2^{(1)} = \begin{bmatrix} 0 \\ -\epsilon \\ \epsilon \\ 0 \end{bmatrix}, \qquad \mathbf{u}_3^{(1)} = \begin{bmatrix} 0 \\ -\epsilon \\ 0 \\ \epsilon \end{bmatrix},$$

$$\mathbf{x}_2 = \frac{1}{\sqrt{2}}\begin{bmatrix} 0 \\ -1 \\ 1 \\ 0 \end{bmatrix}, \qquad \mathbf{u}_3^{(2)} = \frac{1}{2}\begin{bmatrix} 0 \\ -\epsilon \\ -\epsilon \\ 2\epsilon \end{bmatrix},$$

$$\mathbf{x}_3 = \frac{1}{\sqrt{6}}\begin{bmatrix} 0 \\ -1 \\ -1 \\ 2 \end{bmatrix}.$$

One thus has $(\mathbf{x}_1, \mathbf{x}_2) = -\epsilon/\sqrt{8+4\epsilon}$, $(\mathbf{x}_1, \mathbf{x}_3) = -\epsilon/\sqrt{24+12\epsilon}$, and $(\mathbf{x}_2, \mathbf{x}_3) = 0$, so that the vectors are very nearly orthonormal, as opposed to what we found in Ex. 4.91.

EXERCISE 4.93. Prove that, in the absence of rounding error so that computations are performed exactly, the modified Gram-Schmidt procedure produces precisely the same vectors $\mathbf{x}_1, \ldots, \mathbf{x}_s$ as does the standard procedure of **Key Theorem 4.12.**

EXERCISE 4.94. Let V be a vector space of dimension n having the inner product $(\cdot, \cdot)$. For a fixed vector $\mathbf{p} \neq \mathbf{0}$, let V_0 be the set of all vectors $\mathbf{v}$ in V which satisfy $(\mathbf{p}, \mathbf{v}) = 0$. Prove that V_0 is a linear subspace of V and that V_0 has dimension $n - 1$. If V_1 is the set of all vectors $\mathbf{v}$ in V for which $(\mathbf{v}, \mathbf{p}) = 1$, show that V_1 is just the subspace V_0 translated by adding $\mathbf{p}/\|\mathbf{p}\|$ to every vector in V_0; V_1 is called a *hyperplane*.

CHAPTER FIVE

MATRICES AND LINEAR TRANSFORMATIONS

This chapter recasts many of the facts we have learned into a more general framework so as to emphasize one particulary important use of matrices; **Key Theorem 5.1** starts this process, while the results of Section 5.2 carry these ideas further. The later sections are of more practical computational importance and can be studied independently of Section 5.2, especially **Key Theorems 5.6, 5.7, 5.8** (and its preceding **Key Lemma 5.1**), and **5.9.**

5.1 LINEAR TRANSFORMATIONS

Many problems in applied mathematics involve the study of *transformations*, that is, the way in which certain input data is transformed into output data. In this fashion we can view, for example, the complex factors that produce certain profits for a company from a certain pricing and manufacturing structure as a transformation of inputs (price and production data) into outputs (profit structure). In many mathematical models of such complex situations, the transformations involved turn out to be *linear* in the sense that the sum of two inputs is transformed into the sum of their individual outputs, and a multiple of an input is transformed into that multiple of the original output. The linear transformation of (2.7), for example, describes the transformation of one market pattern into another after one month, while (2.26) describes the transformation of one population distribution into another during one unit of time. The applied mathematician attempts to deduce properties of the relevant transformations so as to learn about the properties of the underlying real structure being modeled. In this chapter we develop some of the basic language and some of the basic facts relevant to linear transformations.

DEFINITION 5.1. A *linear transformation* or *linear operator* A from an (abstract) vector space V, called the domain, to an (abstract) vector space W, called the range, is a correspondence that assigns to every (abstract) vector $\mathbf{x}$ in V an (abstract) vector $A(\mathbf{x})$ in W, such that

$$A(\alpha\mathbf{x} + \beta\mathbf{z}) = \alpha A(\mathbf{x}) + \beta A(\mathbf{z}) \tag{5.1}$$

for any scalars α, β and any vectors $\mathbf{x}$, $\mathbf{z}$ in V.

Note that the symbol A for a linear transformation or operator is not the same as the symbol $\mathbf{A}$ for a matrix; the abstract concept of a linear transformation does not require matrices.

We shall use the words "*transformation*" and "*operator*" interchangeably. The word transformation is closer to the geometrical picture in which the equation $\mathbf{y} = A(\mathbf{x})$ is interpreted in terms of "*transforming*" a vector $\mathbf{x}$ into a vector $\mathbf{y}$. (The word *mapping* is also used in this connection.) On the other hand, we can equally think of A as "*operating*" on $\mathbf{x}$ to produce $\mathbf{y}$, as in differentiation or integration, for example.

Let us look at some examples. The idea of a linear operator on an abstract vector space covers a wide variety of situations.

1. If V is the vector space $\mathbb{R}^n$ of $n \times 1$ real column vectors, and W is $\mathbb{R}^m$, then $A(\mathbf{x}) = \mathbf{A}\mathbf{x}$, where $\mathbf{A}$ is any $m \times n$ real matrix, is a linear transformation.
2. If $V = W$ is the space of continuous functions of a variable t in $a \le t \le b$, so that $\mathbf{x} = \mathbf{x}(t)$, a scalar function of t, then $A(\mathbf{x}) = f(t)\mathbf{x}(t)$, where $f(t)$ is a fixed continuous function independent of $\mathbf{x}$, is a linear transformation.
3. If $V = W$ is the same space as in (2), and K is continuous, then

$$A(\mathbf{x}) = \mathbf{x}(t) + \int_a^b K(t, s)\mathbf{x}(s)\, ds$$

is a linear transformation.

4. If V is the space of differentiable functions of t, and W the space of continuous functions, then differentiation is a linear transformation:

$$A(\mathbf{x}) = \frac{d\mathbf{x}(t)}{dt}.$$

Similarly, if V is the space of twice-differentiable functions of t, then

$$A(\mathbf{x}) = \frac{d}{dt}p(t)\frac{d\mathbf{x}(t)}{dt} - q(t)\mathbf{x}(t)$$

is a linear transformation if p is continuously differentiable and q is continuous.

If A is a linear transformation from V to W and B is a linear transformation from W to Z, it is easy to see that BA is a linear transformation from V to Z, where BA is defined by $(BA)(\mathbf{x}) = B(A(\mathbf{x}))$. Similarly, we can easily define scalar multiplication so that αA is defined by $(\alpha A)(\mathbf{x}) = \alpha(A(\mathbf{x}))$, and if A and B both are linear transformations from V to W, we can define $A + B$ via $(A + B)(\mathbf{x}) = A(\mathbf{x}) + B(\mathbf{x})$. Thus general linear transformations can be added and multiplied by scalars essentially as though they were vectors; in fact, the set of all linear transformations from V to W is itself a vector space (see Ex. 5.49).

Our example (1) above shows that an $m \times n$ real (or complex) matrix $\mathbf{A}$ defines, in a natural way, a linear transformation A from $\mathbb{R}^n$ to $\mathbb{R}^m$ (or $\mathbb{C}^n$ to $\mathbb{C}^m$). Perhaps surprisingly, the converse is also true.

● *KEY THEOREM 5.1.*

(i) *Let* $\mathbf{A}$ *be an* $m \times n$ *matrix. Then* $\mathbf{A}$ *defines a linear transformation* A *from* $\mathbb{R}^n$ *to* $\mathbb{R}^m$ *(or* $\mathbb{C}^n$ *to* $\mathbb{C}^m$*) by* $A(\mathbf{x}) = \mathbf{Ax}$.

(ii) *Conversely, let* A *be a linear transformation from a finite-dimensional space* V *to a finite-dimensional space* W*, both having* $\mathbb{R}$ *(or* $\mathbb{C}$*) as the field of scalars; then* A *can be represented by a matrix* $\mathbf{A}$ *with respect to fixed bases for* V *and* W *(the vectors in each basis being written in a fixed order).*

Proof: (i) is covered in (1) above and is straightforward. For (ii), let the vectors $\mathbf{v}_1, \ldots, \mathbf{v}_n$ form a basis for V and the vectors $\mathbf{w}_1, \ldots, \mathbf{w}_m$ form a basis for W (always writing the basis vectors in precisely the same order). Since every $\mathbf{v}$ in V and $\mathbf{w}$ in W can be uniquely represented as

$$\mathbf{v} = \sum_{i=1}^{n} v_i\mathbf{v}_i, \qquad \mathbf{w} = \sum_{j=1}^{m} w_j\mathbf{w}_j,$$

where $v_1, \ldots, v_n$ and $w_1, \ldots, w_m$ are scalars, if we define the $m \times n$ matrix $\mathbf{A} = [a_{ij}]$ by representing $A(\mathbf{v}_j)$ in terms of $\mathbf{w}_1, \ldots, \mathbf{w}_m$ as

$$A(\mathbf{v}_j) = \sum_{i=1}^{m} a_{ij}\mathbf{w}_i, \tag{5.2}$$

then

$$A(\mathbf{v}) = A\Big(\sum_{j=1}^{n} v_j\mathbf{v}_j\Big) = \sum_{j=1}^{n} v_j(A(\mathbf{v}_j))$$

$$= \sum_{j=1}^{n} v_j\Big[\sum_{i=1}^{m} a_{ij}\mathbf{w}_i\Big] = \sum_{i=1}^{m}\Big(\sum_{j=1}^{n} a_{ij}v_j\Big)\mathbf{w}_i$$

$$= \sum_{i=1}^{m} w_i\mathbf{w}_i$$

where

$$w_i = \sum_{j=1}^{m} a_{ij}v_j.$$

Thus if $\hat{\mathbf{w}}$ is the element of $\mathbb{R}^m$ (or $\mathbb{C}^m$) defined as $\hat{\mathbf{w}} = [w_i]$ and $\hat{\mathbf{v}}$ is the element of $\mathbb{R}^n$ (or $\mathbb{C}^n$) defined as $\hat{\mathbf{v}} = [v_i]$, then the relationships

$$A(\mathbf{v}) = \mathbf{w}$$

and

$$\mathbf{A}\hat{\mathbf{v}} = \hat{\mathbf{w}}$$

are equivalent, proving (ii) of the theorem.

The above theorem indicates that the abstract notion of a linear transformation is really very concrete: for finite-dimensional vector spaces over the field $\mathbb{R}$ (or $\mathbb{C}$), we can think of all linear transformations as being defined by $m \times n$ matrices and as operating from $\mathbb{R}^n$ to $\mathbb{R}^m$ (or from $\mathbb{C}^n$ to $\mathbb{C}^m$).

EXERCISE 5.1. Let V be the space of real polynomials of degree at most three and W be the space of real polynomials of degree at most two [see example (8) of Section 4.2.]; let A be the linear transformation defined by $A(\mathbf{v}) = d/dt[\mathbf{v}(t)]$. Choose bases for V and W and find the matrix representation of A.

SOLUTION: Let us take the sets

$$\{\mathbf{v}_1, \mathbf{v}_2, \mathbf{v}_3, \mathbf{v}_4\} = \{1, t, t^2, t^3\}$$

and

$$\{\mathbf{w}_1, \mathbf{w}_2, \mathbf{w}_3\} = \{1, t, t^2\}$$

as our respective bases. Thus any $\mathbf{v}$ in V is represented as

$$\mathbf{v} = v_1 + v_2t + v_3t^2 + v_4t^3$$

while any $\mathbf{w}$ in W is

$$\mathbf{w} = w_1 + w_2t + w_3t^2.$$

We have

$$A(\mathbf{v}_1) = \frac{d}{dt}(1) = 0 = 0 + 0t + 0t^2 = 0\mathbf{w}_1 + 0\mathbf{w}_2 + 0\mathbf{w}_3,$$

so that $a_{11} = a_{21} = a_{31} = 0$ by (5.2). Similarly,

$$A(\mathbf{v}_2) = \frac{d}{dt}(t) = 1 = \mathbf{w}_1 = 1\mathbf{w}_1 + 0\mathbf{w}_2 + 0\mathbf{w}_3,$$

so that $a_{12} = 1$, $a_{22} = a_{32} = 0$,

$$A(\mathbf{v}_3) = \frac{d}{dt}(t^2) = 2t = 2\mathbf{w}_2 = 0\mathbf{w}_1 + 2\mathbf{w}_2 + 0\mathbf{w}_3,$$

so that $a_{13} = 0$, $a_{23} = 2$, $a_{33} = 0$, and

$$A(\mathbf{v}_4) = \frac{d}{dt}(t^3) = 3t^2 = 3\mathbf{w}_3 = 0\mathbf{w}_1 + 0\mathbf{w}_2 + 3\mathbf{w}_3,$$

so that $a_{14} = a_{24} = 0$, $a_{34} = 3$. Thus with respect to these bases, A is represented by the 3×4 matrix

$$\mathbf{A} = \begin{bmatrix} 0 & 1 & 0 & 0 \\ 0 & 0 & 2 & 0 \\ 0 & 0 & 0 & 3 \end{bmatrix}.$$

EXERCISE 5.2. Let V be as in Ex. 5.1, but let $W = V$ and let A be defined by $A(\mathbf{v}) = d/dt\,[t\mathbf{v}(t)]$. Using the basis $\{1, t, t^2, t^3\}$ for both V and W, find the matrix $\mathbf{A}$ representing A.

EXERCISE 5.3. Let V and W be fixed finite-dimensional vector spaces over the same field $\mathbb{R}$ or $\mathbb{C}$ with fixed given bases. For any linear transformation A from V to W let $\mathbf{M}(A)$ denote the matrix representing A with respect to these bases via **Key Theorem 5.1.** Prove that $\mathbf{M}(A + B) = \mathbf{M}(A) + \mathbf{M}(B)$, $\mathbf{M}(\alpha A) = \alpha\mathbf{M}(A)$.

EXERCISE 5.4. Let $\mathbf{A}$ be an $m \times n$ real matrix and define the transformation A from $\mathbb{R}^n$ to $\mathbb{R}^m$ via $A(\mathbf{x}) = \mathbf{A}\mathbf{x}$. Choose the bases of unit column vectors $\mathbf{e}_1, \ldots, \mathbf{e}_n$ and $\mathbf{e}_1, \ldots, \mathbf{e}_m$ (always in the same order) for $\mathbb{R}^n$ and $\mathbb{R}^m$. Show that the matrix $\mathbf{M}(A)$ representing A with respect to these bases is just $\mathbf{A}$. What happens if the basis of vectors $\mathbf{e}_1, \ldots, \mathbf{e}_n$ is replaced by $\mathbf{e}_n, \ldots, \mathbf{e}_1$, in that order?

5.2 EQUATIONS AND INVERSES

Let A be a linear transformation from a vector space V to a vector space W; in applications we very often need to study the solvability of such equations as $A(\mathbf{v}) = \mathbf{w}_0$, given $\mathbf{w}_0$. If A, V, and W are as in example (1) of Section 5.1, for instance, such an equation merely represents a system of m linear equations in n unknowns; the importance of this problem has been amply demonstrated in preceding chapters. If A, V, and W are as in example (4) of Section 5.1, then the equation represents a *differential equation* for an unknown function

$\mathbf{v}$; this problem arises in an immense variety of application areas. The general notion of an abstract linear transformation allows us to study the solvability of such diverse equations in one unified framework; the student may be surprised to learn that this is a fairly simple task because *most of the results of Sections 3.6 and 3.7 on m equations in n unknowns carry over without essential change to the more general situation.* Our task really is just to describe those earlier concrete results in the language of vector spaces and linear transformations.

We begin to study the question of the solvability of an equation

$$A(\mathbf{v}) = \mathbf{w}_0 \qquad \text{(given } \mathbf{w}_0\text{)}. \tag{5.3}$$

In Section 3.6 we considered m linear equations in n unknowns as described by

$$\mathbf{Ax} = \mathbf{b}, \tag{5.4}$$

and we discovered that there were exactly three possibilities:

1. (5.4) has *no* solutions,
2. (5.4) has precisely *one* solution,
3. (5.4) has *infinitely many* solutions;

moreover, in **Key Theorem 3.6** we could describe precisely when a certain one of the three possibilities would hold. For the moment we combine (2) and (3) above and ask: *when does at least one solution exist*? We see from (ii) and (iii) of **Key Theorem 3.6** that *at least one solution exists for* (5.4) *if and only if the rank of the augmented matrix* $[\mathbf{A}, \mathbf{b}]$ *equals the rank of* $\mathbf{A}$; we want to find an analogous condition for (5.3). The reason that this characterization holds for the solvability of $\mathbf{Ax} = \mathbf{b}$ is of course simple: the equation $\mathbf{Ax} = \mathbf{b}$ is equivalent to writing

$$x_1\mathbf{a}_1 + \cdots + x_n\mathbf{a}_n = \mathbf{b},$$

where

$$\mathbf{A} = [\mathbf{a}_1, \ldots, \mathbf{a}_n], \qquad \mathbf{x} = [x_i]$$

which says that $\mathbf{b}$ is linearly dependent on the vectors $\mathbf{a}_1, \ldots, \mathbf{a}_n$ (the columns of $\mathbf{A}$) and hence adjoining $\mathbf{b}$ to the matrix $\mathbf{A}$ cannot increase the rank. Recall from Definition 4.9 that the set of *all* vectors of the form $x_1\mathbf{a}_1 + \cdots + x_n\mathbf{a}_n$ ($= \mathbf{Ax}$) as $\mathbf{x}$ varies is called the *column space* of $\mathbf{A}$, and that its dimension, according to **Key Theorem 4.8**, is just the rank of $\mathbf{A}$. We can therefore now use vector space terminology to rephrase (ii) and (iii) of **Key Theorem 3.6** as to when $\mathbf{Ax} = \mathbf{b}$ has at least one solution: (5.4) *has at least one solution if and only if the dimension of the column space of* $\mathbf{A}$ *equals the dimension of the space generated by adjoining* $\mathbf{b}$ *to the column space of* $\mathbf{A}$, that is, the space of all linear combinations of **b** and vectors in the column space. Having phrased **Key Theorem 3.6** in this language, it is clear that the column space of all vectors of the form **Ax** is the key to analyzing the existence of solutions to (5.4); to pave the way for the analogous analysis for the abstract equation (5.3), we give a definition.

DEFINITION 5.2. Let A be a linear transformation from a vector space V to a vector space W. Then the *exact range space* (sometimes simply called the *range space*) of A is that linear subspace W_0 of W of all vectors of the form $A(\mathbf{v})$ as $\mathbf{v}$ varies over V. If $W_0 = W$ we say that A maps V *onto* W.

It is stated in the definition that the exact range space W_0 actually is a linear subspace of W (see Ex. 5.47). For example, if $\mathbf{w}_0$ is in W_0 [so that $\mathbf{w}_0 = A(\mathbf{v}_0)$ for some $\mathbf{v}_0$ in V], then for any scalar α we have

$$\alpha\mathbf{w}_0 = \alpha A(\mathbf{v}_0) = A(\alpha\mathbf{v}_0)$$

which is then in W_0 because $\alpha\mathbf{v}_0$ is in V.

The following straightforward exercise observes that the exact range space does in fact generalize the notion of column space.

EXERCISE 5.5. Let the $m \times n$ matrix **A** of rank k define the linear transformation A in the usual way of **Key Theorem 5.1**(i). Show that the exact range space of A is the same as the column space of **A** and hence has dimension k.

Just as the column space of **A** was the key to the solvability of (5.4), the exact range space of A is the key to the solvability of (5.3).

THEOREM 5.2. Let A be a linear transformation from a vector space V to a vector space W, and let $\mathbf{w}_0$ be given in W.

(i) *The equation* (5.3), *namely*

$$A(\mathbf{v}) = \mathbf{w}_0,$$

has at least one solution if and only if $\mathbf{w}_0$ is in the exact range space W_0 of A, and (5.3) *is solvable for every $\mathbf{w}_0$ in W if and only if the exact range space W_0 equals W (so that A maps V onto W).*

(ii) *If the exact range space W_0 is finite dimensional, then $A(\mathbf{v}) = \mathbf{w}_0$ is solvable if and only if the dimension of W_0 is equal to the dimension of the space W_1 of all linear combinations of $\mathbf{w}_0$ with vectors in W_0.*

Proof: The first statement (i) is a trivial consequence of the definition of W_0. For (ii), let $\mathbf{w}_1, \ldots, \mathbf{w}_n$ form a basis for the exact range space W_0. If (5.3) is solvable, then $\mathbf{w}_0$ is in W_0, $\mathbf{w}_0$ is therefore expressible as a linear combination of $\mathbf{w}_1, \ldots, \mathbf{w}_n$, and hence any linear combination of $\mathbf{w}_0$ with vectors in W_0 would still be in W_0 because it would still merely be a linear combination of $\mathbf{w}_1, \ldots, \mathbf{w}_n$; thus any vector in W_1 is also in W_0 (and vice versa by definition of W_1), so

$W_0 = W_1$ and certainly their dimensions are equal. Conversely, suppose the dimensions of W_0 and W_1 were equal. If $\{\mathbf{w}_1, \ldots, \mathbf{w}_n, \mathbf{w}_0\}$ were a linearly independent set in W_1, then by Theorem 4.5 the dimension of W_1 would be at least $n + 1$, which exceeds the dimension n of W_0 in contradiction of our assumption. Thus $\{\mathbf{w}_1, \ldots, \mathbf{w}_n, \mathbf{w}_0\}$ must be linearly dependent, and hence by Theorem 4.1(iii) $\mathbf{w}_0$ is linearly dependent on $\{\mathbf{w}_1, \ldots, \mathbf{w}_n\}$ since this latter set is linearly independent. Since $\mathbf{w}_0$ is linearly dependent on the basis $\{\mathbf{w}_1, \ldots, \mathbf{w}_n\}$ for W_0, $\mathbf{w}_0$ is in the exact range space W_0 and hence (5.3) has a solution, completing the proof.

From the remarks preceding the theorem, it should be clear that Theorem 5.2(ii) directly extends **Key Theorem 3.6**(ii), (iii) so far as existence is concerned:

EXERCISE 5.6. Use Theorem 5.2 to prove that $\mathbf{Ax} = \mathbf{b}$ is solvable if and only if the rank of $\mathbf{A}$ equals the rank of $[\mathbf{A}, \mathbf{b}]$.

So far we have treated only the question of the existence of *at least one* solution to $A(\mathbf{v}) = \mathbf{w}_0$, generalizing the analysis for $\mathbf{Ax} = \mathbf{b}$. We now follow the same program in order to treat the *uniqueness* of solutions to $A(\mathbf{v}) = \mathbf{w}_0$.

The question of the uniqueness of solutions to $\mathbf{Ax} = \mathbf{b}$ was treated in **Key Theorem 3.6**(ii) and in **Key Theorem 3.7**: *if the equation* $\mathbf{Ax} = \mathbf{b}$ *in* (5.4) *is solvable, then it has at most one solution for an* $m \times n$ *matrix* $\mathbf{A}$ *if and only if the rank of* $\mathbf{A}$ *equals n which in turn holds if and only if the only solution to* $\mathbf{Ax} = \mathbf{0}$ *is* $\mathbf{x} = \mathbf{0}$. We saw in our above analysis of the existence of at least one solution that the dimension of the exact range space of a linear transformation A plays the role of the rank of a matrix $\mathbf{A}$, so it seems clear how to generalize the first part of the preceding uniqueness result for $\mathbf{Ax} = \mathbf{b}$; to generalize the second part, we extend in the obvious way the study of the homogeneous equation $\mathbf{Ax} = \mathbf{0}$.

DEFINITION 5.3. Let A be a linear transformation from a vector space V to a vector space W. The *null space* (or *kernel*) of A is that linear subspace V_0 of V consisting of all solutions to the equation $A(\mathbf{v}) = \mathbf{0}$.

We remark that the null space V_0 actually is a linear subspace of V as stated in the definition (see Ex. 5.48); for example, if $\mathbf{v}_0$ is in V_0 [so that $A(\mathbf{v}_0) = \mathbf{0}$], then for any scalar α we have

$$A(\alpha\mathbf{v}_0) = \alpha A(\mathbf{v}_0) = \alpha\mathbf{0} = \mathbf{0},$$

and hence $\alpha\mathbf{v}_0$ is also in V_0.

When the linear transformation A is defined in the usual way from a matrix $\mathbf{A}$, it is clear that the null space of A is just the set of solutions to $\mathbf{Ax} = \mathbf{0}$; just as this set is part of the key to analyzing the uniqueness of solutions to $\mathbf{Ax} = \mathbf{b}$, the null space is part of the key for analyzing the uniqueness of solutions to $A(\mathbf{v}) = \mathbf{w}_0$.

THEOREM 5.3. Let A be a linear transformation from a vector space V to a vector space W and let W_0 denote the exact range space.

(i) *The equation* (5.3), *namely,*

$$A(\mathbf{v}) = \mathbf{w}_0,$$

has at most one solution for a given $\mathbf{w}_0$ in the exact range space W_0 if and only if it has at most one solution for every $\mathbf{w}_0$ in W_0, and in particular if and only if the null space V_0 of A equals $\{\mathbf{0}\}$.

(ii) *Moreover, if V has finite dimension n, then $A(\mathbf{v}) = \mathbf{w}_0$ has at most one solution for $\mathbf{w}_0$ in W_0 if and only if the dimension of the range space W_0 also equals n.*

Proof: If $\mathbf{v}_1 \neq \mathbf{v}_2$ in V both solve $A(\mathbf{v}) = \mathbf{w}_0$, then

$$\mathbf{0} = \mathbf{w}_0 - \mathbf{w}_0 = A(\mathbf{v}_1) - A(\mathbf{v}_2) = A(\mathbf{v}_1 - \mathbf{v}_2)$$

and the null space V_0 contains the nonzero vector $\mathbf{v}_1 - \mathbf{v}_2$. Conversely, if $\mathbf{v}_0 \neq \mathbf{0}$ is in V_0 and $\mathbf{v}_1$ solves $A(\mathbf{v}) = \mathbf{w}_0$ for $\mathbf{w}_0$ in W_0, then

$$\mathbf{w}_0 = \mathbf{w}_0 + \mathbf{0} = A(\mathbf{v}_1) + A(\mathbf{v}_0) = A(\mathbf{v}_1 + \mathbf{v}_0)$$

and $\mathbf{v}_1 + \mathbf{v}_0 \neq \mathbf{v}_1$ also solves $A(\mathbf{v}) = \mathbf{w}_0$, so there is more than one solution. From this (i) follows easily. For (ii), let $\mathbf{v}_1, \ldots, \mathbf{v}_n$ form a basis for V; since every $\mathbf{v}$ in V is a linear combination of $\mathbf{v}_1, \ldots, \mathbf{v}_n$, it is clear that every $\mathbf{w} = A(\mathbf{v})$ in the range space W_0 is a linear combination of the n vectors $A(\mathbf{v}_1), \ldots, A(\mathbf{v}_n)$. In other words, $\{A(\mathbf{v}_1), \ldots, A(\mathbf{v}_n)\}$ spans W_0, so that it follows from **Key Theorem 4.4** as in Ex. 4.18 that the dimension of W_0 is at most n; in fact, the dimension of W_0 equals n if and only if $\{A(\mathbf{v}_1), \ldots, A(\mathbf{v}_n)\}$ is linearly independent. This of course is equivalent to saying that

$$a_1A(\mathbf{v}_1) + \cdots + a_nA(\mathbf{v}_n) = \mathbf{0}$$

only when $a_1 = \cdots = a_n = 0$, that is,

$$A(\mathbf{v}) = \mathbf{0}$$

with $\mathbf{v} = a_1\mathbf{v}_1 + \cdots + a_n\mathbf{v}_n$ only when $\mathbf{v} = \mathbf{0}$. Thus the dimension of W_0 equals n if and only if V_0 equals $\{\mathbf{0}\}$. From (i), the validity of (ii) then follows, and the proof is completed.

From the remarks preceding the theorem, it should be clear that Theorem 5.3 directly extends **Key Theorem 3.7** and the uniqueness part of **Key Theorem 3.6(ii)**.

EXERCISE 5.7. Use Theorem 5.3 to prove **Key Theorem 3.7.**

The proof of Theorem 5.3 seems to indicate that as the dimension of the range space decreases the dimension of the null space increases; in fact their sum is a constant, as we now demonstrate in a special case.

THEOREM 5.4. Let the $m \times n$ matrix $\mathbf{A}$ *define the linear transformation A from $\mathbb{R}^n$ to $\mathbb{R}^m$ (or $\mathbb{C}^n$ to $\mathbb{C}^m$) via $A(\mathbf{x}) = \mathbf{Ax}$, and let k be the rank of* $\mathbf{A}$. *Then the null space of A has dimension $n - k$. If $k = n$ then the null space is just $\{\mathbf{0}\}$; if $k < n$, a basis for the null space of A can be constructed from the row-echelon form of* $\mathbf{A}$ *as detailed in the proof.*

Proof: The row-echelon form of $\mathbf{A}$ has k unit column vectors. For simplicity in notation (and without loss of generality) we can assume that the unknowns have been renumbered if necessary so that the unit columns in the row-echelon form are the first k columns. (See Exs. 5.9 and 5.38.) Suppose then that the row-echelon form of $\mathbf{A}$ is

$$\begin{bmatrix} \mathbf{I}_k & \mathbf{B} \\ \mathbf{0} & \mathbf{0} \end{bmatrix}.$$

The following set of equations is equivalent to $\mathbf{Ax} = \mathbf{0}$:

$$\begin{bmatrix} \mathbf{I}_k & \mathbf{B} \\ \mathbf{0} & \mathbf{0} \end{bmatrix} \begin{bmatrix} \mathbf{x}_1 \\ \mathbf{x}_2 \end{bmatrix} = \begin{bmatrix} \mathbf{0} \\ \mathbf{0} \end{bmatrix},$$

where $\mathbf{B}$ is $k \times (n - k)$, $\mathbf{x}_1$ is $k \times 1$ and $\mathbf{x}_2$ is $(n - k) \times 1$. This gives

$$\mathbf{x}_1 = -\mathbf{B}\mathbf{x}_2, \tag{5.5}$$

so that we can choose $\mathbf{x}_2$ arbitrarily, and express $\mathbf{x}_1$ in terms of the arbitrary $\mathbf{x}_2$. More formally, suppose that

$$\mathbf{x}_2 = \begin{bmatrix} \alpha_1 \\ \cdot \\ \cdot \\ \cdot \\ \alpha_{n-k} \end{bmatrix}.$$

If we denote the columns of $\mathbf{B}$ by $\mathbf{b}_1, \ldots, \mathbf{b}_{n-1}$, (5.5) gives

$$\mathbf{x}_1 = -\alpha_1\mathbf{b}_1 - \alpha_2\mathbf{b}_2 - \cdots - \alpha_{n-k}\mathbf{b}_{n-k}.$$

Also

$$\mathbf{x}_2 = \alpha_1\mathbf{e}_1 + \alpha_2\mathbf{e}_2 + \cdots + \alpha_{n-k}\mathbf{e}_{n-k},$$

where $\mathbf{e}_j$ is the $(n - k) \times 1$ unit vector with jth element equal to unity and other elements zero. Hence every solution of $\mathbf{Ax} = \mathbf{0}$ can be written in the form

$$\mathbf{x} = \alpha_1\mathbf{u}_1 + \alpha_2\mathbf{u}_2 + \cdots + \alpha_{n-k}\mathbf{u}_{n-k}$$

where

$$\mathbf{u}_j = \begin{bmatrix} -\mathbf{b}_j \\ \mathbf{e}_j \end{bmatrix}.$$

The set of vectors $\mathbf{u}_j$ is clearly linearly independent and spans the null space. Hence it forms a basis for the space. If $k = n$, then $\mathbf{B}$ is nonexistent, and the only solution is $\mathbf{x} = \mathbf{0}$. This completes the proof.

EXERCISE 5.8. Find a basis for the null space of the homogeneous equations

$$\begin{aligned} x_1 + 2x_2 + x_3 + 2x_4 - 3x_5 &= 0 \\ 3x_1 + 6x_2 + 4x_3 - x_4 + 2x_5 &= 0 \\ 4x_1 + 8x_2 + 5x_3 + x_4 - x_5 &= 0 \\ -2x_1 - 4x_2 - 3x_3 + 3x_4 - 5x_5 &= 0. \end{aligned}$$

SOLUTION: The row-echelon form of the coefficient matrix is

$$\begin{bmatrix} 1 & 2 & 0 & 9 & -14 \\ 0 & 0 & 1 & -7 & 11 \\ 0 & 0 & 0 & 0 & 0 \\ 0 & 0 & 0 & 0 & 0 \end{bmatrix}.$$

The rank is 2, so that the dimension of the null space is 3. To find a basis for the null space, we note that the row-echelon form tells us that the original system is equivalent to

$$\begin{aligned} x_1 + 2x_2 \qquad\quad + 9x_4 - 14x_5 &= 0 \\ x_3 - 7x_4 + 11x_5 &= 0, \end{aligned}$$

so that x_1 and x_3 will be determined once we assign arbitrary values to x_2, x_4, x_5. If we let $x_2 = \alpha_1$, $x_4 = \alpha_2$, $x_5 = \alpha_3$, where $\alpha_1, \alpha_2, \alpha_3$ are arbitrary, we find $x_1 = -2\alpha_1 - 9\alpha_2 + 14\alpha_3$ and $x_3 = 7\alpha_2 - 11\alpha_3$ which, in vector notation, describes the null space as all vectors $\mathbf{x}$ of the form

$$\mathbf{x} = \alpha_1 \begin{bmatrix} -2 \\ 1 \\ 0 \\ 0 \\ 0 \end{bmatrix} + \alpha_2 \begin{bmatrix} -9 \\ 0 \\ 7 \\ 1 \\ 0 \end{bmatrix} + \alpha_3 \begin{bmatrix} 14 \\ 0 \\ -11 \\ 0 \\ 1 \end{bmatrix}.$$

Therefore the desired basis is the set of vectors

$$\begin{bmatrix} -2 \\ 1 \\ 0 \\ 0 \\ 0 \end{bmatrix}, \quad \begin{bmatrix} -9 \\ 0 \\ 7 \\ 1 \\ 0 \end{bmatrix}, \quad \begin{bmatrix} 14 \\ 0 \\ -11 \\ 0 \\ 1 \end{bmatrix}.$$

EXERCISE 5.9. Use the more formal method in the proof of Theorem 5.4 to derive the basis in Ex. 5.8. That proof assumes that the unit columns are the *first k* columns of the row-echelon form, and this is *not* the case here. To satisfy that assumption, you must temporarily interchange the second and third columns of the row-echelon form, which amounts to interchanging temporarily the variables x_2 and x_3. After you find the basis for the null space after the interchanges, remember to interchange the values for the x_2 and x_3 variables again to return to the original problem.

Another way to state Theorem 5.4 is to say that the dimension of the null space of A plus the dimension of the exact range space of A equals n, the dimension of the space upon which A is defined; in the exercises at the end of the chapter we show how to extend this slightly to the following. (See Exs. 5.36, 5.37, 5.39, and 5.40.)

COROLLARY 5.1. Let A be a linear transformation from a finite-dimensional vector space V to a vector space W. Then the dimension of the null space of A plus the dimension of the exact range space of A equals the dimension of V.

In Section 3.7 we used results on the solvability of $\mathbf{Ax} = \mathbf{b}$ to deduce results on the existence of inverses for $\mathbf{A}$; we can do the same sort of thing for linear transformations by means of our theorems on the solvability of equations. For convenience, we study only the question of the existence of a two-sided inverse.

In the concrete case of matrices, we know that only square matrices are invertible, and that the invertibility of $\mathbf{A}$ is equivalent to the existence of precisely one solution for every equation $\mathbf{Ax} = \mathbf{b}$. Since we know how to characterize both existence and uniqueness for the more general $A(\mathbf{v}) = \mathbf{w}_0$ by Theorem 5.2 and Theorem 5.3, we hope to characterize the existence of an inverse for a linear transformation; of course, we need to define what we mean by an inverse.

DEFINITION 5.4. An *inverse* B of a linear transformation A from a vector space V to a vector space W is a linear transformation from W to V for which AB is the identity transformation on W and BA is the identity transformation on V, that is, $(AB)(\mathbf{w}) = \mathbf{w}$ for all $\mathbf{w}$ in W and $(BA)(\mathbf{v}) = \mathbf{v}$ for all $\mathbf{v}$ in V. We write $B = A^{-1}$.

If the linear transformation A from V to W is such that (5.3) always has exactly one solution, then it is clear how to define $B = A^{-1}$; given any $\mathbf{w}_0$ in W, define $B(\mathbf{w}_0)$ to be that unique element $\mathbf{v}_0$ in V which solves $A(\mathbf{v}) = \mathbf{w}_0$. By the definition of $\mathbf{v}_0$ then,

$$(AB)(\mathbf{w}_0) = A[B(\mathbf{w}_0)] = A(\mathbf{v}_0) = \mathbf{w}_0$$

as required by Definition 5.4. To check that also $BA(\mathbf{v}_0) = \mathbf{v}_0$ for all $\mathbf{v}_0$, recall that $B(\mathbf{w}_0)$ is the *unique* solution to $A(\mathbf{v}) = \mathbf{w}_0$; since $\mathbf{v}_0$ certainly solves $A(\mathbf{v}) = \mathbf{w}_0$ when $\mathbf{w}_0$ is chosen to be $A(\mathbf{v}_0)$, we have

$$\mathbf{v}_0 = B(\mathbf{w}_0) = B[A(\mathbf{v}_0)] = (BA)(\mathbf{v}_0)$$

as required. In this way we see that we can define a B that acts like an inverse, but the technical question remains of whether or not B is a linear transformation. If we can handle this technical point, then we can characterize invertible linear transformations just as we characterized invertible matrices.

THEOREM 5.5. Let A be a linear transformation from a vector space V to a vector space W.

(i) *A has an inverse linear transformation $B = A^{-1}$ if and only if the null space V_0 of A equals $\{\mathbf{0}\}$ and the exact range space W_0 of A equals W.*
(ii) *If V has finite dimension n, then A^{-1} exists if and only if the null space V_0 of A equals $\{\mathbf{0}\}$ and W has dimension n.*

Proof: If A^{-1} exists then $A(\mathbf{v}) = \mathbf{0}$ implies

$$\mathbf{0} = A^{-1}(\mathbf{0}) = A^{-1}[A(\mathbf{v})] = (A^{-1}A)(\mathbf{v}) = \mathbf{v},$$

and $A[A^{-1}(\mathbf{w}_0)] = \mathbf{w}_0$ for all $\mathbf{w}_0$ in W; therefore every equation $A(\mathbf{v}) = \mathbf{w}_0$ has a unique solution and it follows from Theorem 5.2 and Theorem 5.3 that $V_0 = \{\mathbf{0}\}$ and $W_0 = W$. If $\mathbf{v}_1, \ldots, \mathbf{v}_n$ is a basis for V it follows easily that $A(\mathbf{v}_1), \ldots, A(\mathbf{v}_n)$ is a basis for W, and hence W has dimension n; thus if A^{-1} exists then the appropriate conclusions of (i) and (ii) are valid. Conversely, if $V_0 = \{\mathbf{0}\}$ and $W_0 = W$ then by Theorem 5.2 and Theorem 5.3 every equation $A(\mathbf{v}) = \mathbf{w}_0$ is uniquely solvable; if V has dimension n, then again by Theorem 5.2 and Theorem 5.3 every equation $A(\mathbf{v}) = \mathbf{w}_0$ is uniquely solvable. We can then define B as in the paragraph preceding this theorem, and we only need show that B is a linear transformation as defined by Definition 5.1. Given any $\mathbf{w}_0$ in W, by definition $B(\mathbf{w}_0)$ solves $A(\mathbf{v}) = \mathbf{w}_0$ and hence certainly $\alpha B(\mathbf{w}_0)$ solves $A(\mathbf{v}) = \alpha\mathbf{w}_0$ since

$$A[\alpha B(\mathbf{w}_0)] = \alpha A[B(\mathbf{w}_0)] = \alpha\mathbf{w}_0$$

for every scalar α. But by definition $B(\alpha\mathbf{w}_0)$ is the unique solution to $A(\mathbf{v}) = \alpha\mathbf{w}_0$; since $\alpha B(\mathbf{w}_0)$ solves this, we must have $\alpha B(\mathbf{w}_0) = B(\alpha\mathbf{w}_0)$ as required in Definition 5.1. Similarly we can show $B(\mathbf{w}_1 + \mathbf{w}_2) = B(\mathbf{w}_1) + B(\mathbf{w}_2)$, completing the proof.

EXERCISE 5.10. Use Theorem 5.5 to prove **Key Theorem 3.11.**

EXERCISE 5.11. Complete the proof of Theorem 5.5 by showing that $B(\mathbf{w}_1 + \mathbf{w}_2) = B(\mathbf{w}_1) + B(\mathbf{w}_2)$ for all $\mathbf{w}_1$, $\mathbf{w}_2$ in W.

EXERCISE 5.12. In the notation of Ex. 5.3, show that $\mathbf{M}(A^{-1}) = [\mathbf{M}(A)]^{-1}$, where V and W have the same dimension.

5.3 NORMS OF MATRICES AND LINEAR TRANSFORMATIONS

As we indicated earlier, the applied mathematician often studies the properties of linear transformations in an attempt to understand their effects on their inputs. One common problem is to understand the "size" of the transformation in the sense of how the "size" of the output compares with the "size" of the input, that is, to measure the effects of the transformation on "size." For example, in Section 2.3 we might want to know by how much the total population of chickens and foxes can be amplified in a single time period. We have already, in Section 4.6, developed precise ways to speak of the size of a vector, and it turns out that it is rather easy in the context of our previous ideas to describe the size of the effect of a linear transformation; since matrices can define linear transformations, this will provide a way of measuring the size of the effects of a matrix (as a linear transformation).

Let A be a linear transformation from a vector space V with a norm $\|\cdot\|_V$ to a vector space W with a norm $\|\cdot\|_W$. For each nonzero $\mathbf{v}$ in V, the quotient $\|A(\mathbf{v})\|_W/\|\mathbf{v}\|_V$ measures the magnification caused by the transformation A; an upper bound on this quotient would therefore measure the overall size of A.

EXERCISE 5.13. Let the $m \times n$ matrix $\mathbf{A}$ define the linear transformation A from $\mathbb{R}^n$ to $\mathbb{R}^m$ (or $\mathbb{C}^n$ to $\mathbb{C}^m$) by $A(\mathbf{v}) = \mathbf{A}\mathbf{v}$ as usual, and let both $\mathbb{R}^n$ and $\mathbb{R}^m$ be given the norm $\|\cdot\|_\infty$ of (4.13) in Section 4.6. Define, for $\mathbf{A} = [a_{ij}]$,

$$\alpha = \max_i \sum_{j=1}^n |a_{ij}|.$$

Show that $\|A(\mathbf{v})\|_\infty/\|\mathbf{v}\|_\infty \leq \alpha$ for all $\mathbf{v} \neq 0$, and that for some special $\mathbf{v}$ we have $\|A(\mathbf{v})\|_\infty/\|\mathbf{v}\|_\infty = \alpha$.

SOLUTION: Suppose that the maximum defining α is attained at $i = i_0$, so that

$$\alpha = \sum_{j=1}^n |a_{i_0 j}| = \max_i \sum_{j=1}^n |a_{ij}|.$$

For any $\mathbf{v} = [v_j]$ in $\mathbb{R}^n$ (or $\mathbb{C}^n$), we have

$$\begin{aligned}\|A(\mathbf{v})\|_\infty &= \|\mathbf{A}\mathbf{v}\|_\infty = \max_i |(\mathbf{A}\mathbf{v})_i| = \max_i \left| \sum_{j=1}^n a_{ij}v_j \right| \\ &\leq \max_i \sum_{j=1}^n (|a_{ij}|\,|v_j|) \leq \max_i \sum_{j=1}^n (|a_{ij}| \max_p |v_p|) \\ &\leq \alpha \|\mathbf{v}\|_\infty,\end{aligned}$$

and hence $\|A(\mathbf{v})\|_\infty/\|\mathbf{v}\|_\infty \le \alpha$ for all nonzero $\mathbf{v}$, as asserted. If we can now exhibit a special $\mathbf{v}$ for which $\|A(\mathbf{v})\|_\infty/\|\mathbf{v}\|_\infty = \alpha$, then the proof will be finished. Define v_j to be that complex number having absolute value unity and for which $a_{i_0j}v_j = |a_{i_0j}|$, and define $\mathbf{v} = [v_j]$; if $\mathbf{A}$ is real then so is $\mathbf{v}$. Clearly $\|\mathbf{v}\|_\infty = 1$ so that $\|A(\mathbf{v})\|_\infty \le \alpha\|\mathbf{v}\|_\infty \le \alpha$, but also

$$|(A(\mathbf{v}))_{i_0}| = \left|\sum_{j=1}^{n} a_{i_0j}v_j\right| = \left|\sum_{j=1}^{n} |a_{i_0j}|\right| = \alpha$$

so that in fact $\|A(\mathbf{v})\|_\infty = \alpha$ and the result is proved.

It turns out that in every finite-dimensional space V the quotient $\|A(\mathbf{v})\|_W/\|\mathbf{v}\|_V$ cannot become arbitrarily large as $\mathbf{v}$ varies, and in fact the quotient has a maximum value achieved for some special vector $\mathbf{v}$. In *some* infinite-dimensional spaces the quotient cannot become arbitrarily large, although it may not actually have a greatest value generated by some special $\mathbf{v}$; in this case there is a number, called the *supremum*, that is the least of all the numbers which exceed the quotients for all $\mathbf{v}$. Thus in the finite-dimensional case and in some infinite-dimensional cases, there is a smallest number which we can use to bound the quotient $\|A(\mathbf{v})\|_W/\|\mathbf{v}\|_V$, and this number serves to measure the size of the effects of the linear transformation A; if this number is small, then *every* vector $\mathbf{v}$ is reduced in size by A, while if the number is large, then *some* vectors are greatly increased in size by A. By analogy with the vector norms of Section 4.6 which measure the size of a vector, we call this number measuring the size of A (as an operator) an *operator norm* of A.

DEFINITION 5.5. Let A be a linear transformation from a vector space V to a vector space W equipped with norms $\|\cdot\|_V$, $\|\cdot\|_W$. The *operator norm* $\|\cdot\|_{V,W}$ is defined via

$$\|A\|_{V,W} = \operatorname*{supremum}_{\substack{\mathbf{v}\neq 0\\ \mathbf{v} \text{ in } V}} \{\|A(\mathbf{v})\|_W/\|\mathbf{v}\|_V\}$$

EXERCISE 5.14. Consider $\mathbf{A}$, A, V, and W as in Ex. 5.13. Show that

$$\|A\|_{V,W} = \max_i \sum_{j=1}^{n} |a_{ij}|.$$

The operator norm of a linear transformation shares many properties with vector norms: for example the triangle inequality and the other properties of Definition 4.10 hold; in addition, some special properties hold which follow from the fact that we are talking about *transformations*.

● *KEY THEOREM 5.6. Let V, W, and Z be vector spaces and let A, B, and C be linear transformations from V to W, V to W, and from W to Z, respectively. Then:*

(i) $||A(\mathbf{v})||_W \leq ||A||_{V,W}\, ||\mathbf{v}||_V$

(ii) $||A||_{V,W} \geq 0$, $||A||_{V,W} = 0$ *if and only if* $A(\mathbf{v}) = \mathbf{0}$ *for all* $\mathbf{v}$;

(iii) $||kA||_{V,W} = |k|\, ||A||_{V,W}$ *for all real or complex scalars* k;

(iv) $||A + B||_{V,W} \leq ||A||_{V,W} + ||B||_{V,W}$

(v) $||CA||_{V,Z} \leq ||C||_{W,Z}\, ||A||_{V,W}$

(vi) $||I||_{V,V} = 1$, *where* $I(\mathbf{v}) = \mathbf{v}$ *for all* $\mathbf{v}$ *in* V.

Proof: (i) follows immediately from the definition of operator norm if we divide each side of (i) by $||\mathbf{v}||_V$; (ii) also follows trivially from Definition 5.5. For (iii), we need merely observe that k can be factored out of the supremum in Definition 5.5. For (iv), we write

$$||(A + B)\mathbf{v}||_W = ||A(\mathbf{v}) + B(\mathbf{v})||_W \leq ||A(\mathbf{v})||_W + ||B(\mathbf{v})||_W$$

by the triangle inequality for vector norms. Using (i), we then deduce

$$||(A + B)(\mathbf{v})||_W \leq (||A||_{V,W} + ||B||_{V,W})||\mathbf{v}||_V$$

so the quotient defining the norm of $A + B$ does not exceed the upper bound $||A||_{V,W} + ||B||_{V,W}$; since $||A + B||_{V,W}$ is the supremum, that is the *least of all upper bounds* for the quotient, (iv) is proved. A similar argument yields (v); (vi) is straightforward.

In this book we are primarily concerned with the vector spaces $\mathbb{R}^n$ and $\mathbb{C}^n$ and with linear transformations A defined by matrix multiplication: $A(\mathbf{x}) = \mathbf{A}\mathbf{x}$. In this case we most commonly use the vector norms $||\cdot||_1$, $||\cdot||_2$, $||\cdot||_\infty$ defined in (4.11), (4.12), (4.13) in Section 4.6. It is natural and useful to consider the operator norms induced by these vector norms on linear transformations defined by matrices.

DEFINITION 5.6. Let $\mathbf{A}$ be an $m \times n$ matrix, and let A be the linear transformation $A(\mathbf{x}) = \mathbf{A}\mathbf{x}$ defined from $\mathbb{C}^n$ to $\mathbb{C}^m$ by $\mathbf{A}$. By the norms $||\mathbf{A}||_1$, $||\mathbf{A}||_2$, $||\mathbf{A}||_\infty$ we mean the corresponding norms of A induced by using the appropriate vector norm in both the domain $\mathbb{C}^n$ and the range $\mathbb{C}^m$. That is

$$||\mathbf{A}||_1 = \max_{x \neq 0} \left\{\frac{||\mathbf{A}\mathbf{x}||_1}{||\mathbf{x}||_1}\right\},$$

$$||\mathbf{A}||_2 = \max_{x \neq 0} \left\{\frac{||\mathbf{A}\mathbf{x}||_2}{||\mathbf{x}||_2}\right\},$$

$$||\mathbf{A}||_\infty = \max_{x \neq 0} \left\{\frac{||\mathbf{A}\mathbf{x}||_\infty}{||\mathbf{x}||_\infty}\right\}.$$

We have already found in Ex. 5.13 and Ex. 5.14 that $||\mathbf{A}||_\infty$ can easily be computed directly from the elements a_{ij} of the matrix $\mathbf{A}$; a very similar argument verifies a similar formula for $||\mathbf{A}||_1$. The computation of $||\mathbf{A}||_2$ involves a concept we have not encountered yet; we will not validate the formula for

$\|\mathbf{A}\|_2$ until Ex. 11.35 in Section 11.6. For these reasons we state the following important results without proof.

- *KEY THEOREM 5.7. Let* $\mathbf{A}$ *be an* $m \times n$ *matrix. Then:*

(i) $\|\mathbf{A}\|_1 = \max_j \sum_{i=1}^{m} |a_{ij}|$ *(maximum absolute column sum)*;

(ii) $\|\mathbf{A}\|_\infty = \max_i \sum_{j=1}^{n} |a_{ij}|$ *(maximum absolute row sum)*;

(iii) $\|\mathbf{A}\|_2 = [$*maximum eigenvalue of* $\mathbf{A}^H\mathbf{A}]^{1/2}$
$=$ *maximum singular value of* $\mathbf{A}$.

EXERCISE 5.15. Consider the linear transformation in Section 2.3 describing the transformation of the populations of foxes (x_1) and chickens (x_2) over a period of time; in particular, let

$$\mathbf{A} = \begin{bmatrix} 0.6 & 0.5 \\ -0.1 & 1.2 \end{bmatrix}.$$

If we are interested in the *total* number of individuals in the environment, then $\|\mathbf{x}\|_1 = |x_1| + |x_2|$ seems natural, and we find $\|\mathbf{A}\|_1 = 1.7$ so that the total population may be multiplied by 1.7 in one time period. If we are interested in the *maximum* number of animals of any one kind, then $\|\mathbf{x}\|_\infty = \max\{|x_1|, |x_2|\}$ seems natural, and we find that $\|\mathbf{A}\|_\infty = 1.3$ so that the maximum single population can be multiplied by 1.3 in any one time period. For the other similar example of that section, where

$$\mathbf{A} = \begin{bmatrix} 0.6 & 0.5 \\ -0.18 & 1.2 \end{bmatrix},$$

find $\|\mathbf{A}\|_1$ and $\|\mathbf{A}\|_\infty$.

Since we can compare vector norms [see (4.16)] we can easily deduce comparisons for operator norms. For example, if $\mathbf{A}$ is $m \times n$, using (4.16) we find that

$$\|\mathbf{Ax}\|_1 \le m\|\mathbf{Ax}\|_\infty \le m\|\mathbf{A}\|_\infty\|\mathbf{x}\|_\infty \le m\|\mathbf{A}\|_\infty\|\mathbf{x}\|_1$$

so that $\|\mathbf{A}\|_1 \le m\|\mathbf{A}\|_\infty$. By similar arguments we obtain

$$\frac{1}{\sqrt{m}}\|\mathbf{A}\|_2 \le \|\mathbf{A}\|_\infty \le \sqrt{n}\,\|\mathbf{A}\|_2,$$

$$\frac{1}{\sqrt{n}}\|\mathbf{A}\|_2 \le \|\mathbf{A}\|_1 \le m^{1/2}\|\mathbf{A}\|_2, \qquad (5.6)$$

$$\frac{1}{n}\|\mathbf{A}\|_\infty \le \|\mathbf{A}\|_1 \le m\|\mathbf{A}\|_\infty.$$

Because of the relations (5.6), if in any *one* of our three norms the norms of an infinite sequence of matrices $\mathbf{A}_i$ tend to zero then they do so in *all* norms. Also since $\|\mathbf{A}\|_\infty \leq n \max_{i,j} |a_{ij}|$ and $\max_{i,j} |a_{ij}| \leq \|\mathbf{A}\|_\infty$, the norms of a sequence $\mathbf{A}_i$ tend to zero if and only if every element of the matrices $\mathbf{A}_i$ is tending to zero. Thus the norms provide us with a simple device for discussing the *convergence* of a sequence of matrices. (See Definition 10.4 in Section 10.5.)

EXERCISE 5.16. In Section 2.3 we found it important to understand the behavior of the sequence of powers $\mathbf{A}^i$ of a matrix $\mathbf{A}$. From **Key Theorem 5.6**(v) we know that for any operator norm $\|\mathbf{A}^2\| \leq \|\mathbf{A}\|^2$ and more generally $\|\mathbf{A}^i\| \leq \|\mathbf{A}\|^i$. In Section 2.3 for

$$\mathbf{A} = \begin{bmatrix} 0.6 & 0.5 \\ -0.1 & 1.2 \end{bmatrix}$$

we have $\|\mathbf{A}\|_1 = 1.7$ and $\|\mathbf{A}\|_\infty = 1.3$, so it might appear reasonable to guess that $\mathbf{A}^i$ tends to infinity; Table 2.2 bears this out. When

$$\mathbf{A} = \begin{bmatrix} 0.6 & 0.5 \\ -0.18 & 1.2 \end{bmatrix}$$

we have $\|\mathbf{A}\|_1 = 1.7$ and $\|\mathbf{A}\|_\infty = 1.38$ and we might again guess that $\mathbf{A}^i$ tends to infinity, but in this case Table 2.3 indicates that actually $\mathbf{A}^i$ tends to zero! Thus, while the condition $\|\mathbf{A}\| < 1$ certainly guarantees that $\mathbf{A}^i$ tends to zero, the condition $\|\mathbf{A}\| \geq 1$ tells us nothing about $\mathbf{A}^i$; we will have to wait until Section 10.5 when we have more powerful tools before we can understand the behavior of $\mathbf{A}^i$.

EXERCISE 5.17. Prove **Key Theorem 5.7**(i).

EXERCISE 5.18. For each of the following matrices $\mathbf{A}$, compute $\|\mathbf{A}\|_1$ and $\|\mathbf{A}\|_\infty$:

(a) $\begin{bmatrix} -3 & 2 & 1 \end{bmatrix}$

(b) $\begin{bmatrix} -3 \\ 2 \\ 1 \end{bmatrix}$

(c) $\begin{bmatrix} 4 & -7 \\ -6 & 1 \end{bmatrix}$

(d) $\begin{bmatrix} \sqrt{-1} & 3 & 2 \\ 1 & -\sqrt{-1} & 2 \end{bmatrix}.$

EXERCISE 5.19. Prove (5.6).

EXERCISE 5.20. Prove that $\mathbf{A}^i$ tends to zero if $\|\mathbf{A}\|_1 < 1$.

EXERCISE 5.21. Show that $\|\mathbf{I}\|_1 = \|\mathbf{I}\|_\infty = 1$ by using the formulas in **Key Theorem 5.7.**

5.4 INVERSES OF PERTURBED MATRICES

As we have mentioned before, in practical situations it is important to realize that inaccuracies in measuring data cause us to be dealing with vectors, matrices, etc., which are slightly perturbed away from their "true" values for exact measurements. One problem this might cause is that a "true" matrix $\mathbf{A}$ is nonsingular while the measured perturbed matrix $\mathbf{A}_0$ is singular. Now that we have operator norms to allow us to speak precisely about the size of these perturbations, we can rule out this possibility if the perturbations are sufficiently small. We prove first a special case.

- *KEY LEMMA 5.1 (Banach Lemma). Let* $\mathbf{P}$ *be an* $n \times n$ *matrix and let* $\|\cdot\|$ *denote any of the operator norms* $\|\cdot\|_1, \|\cdot\|_2, \|\cdot\|_\infty$ *and its corresponding vector norm. (Actually,* **any** *operator norm will do.) If* $\|\mathbf{P}\| < 1$ *then* $\mathbf{I} + \mathbf{P}$ *is nonsingular and*

$$\frac{1}{1 + \|\mathbf{P}\|} \leq \|(\mathbf{I}+\mathbf{P})^{-1}\| \leq \frac{1}{1 - \|\mathbf{P}\|}.$$

Proof: By **Key Theorem 3.8** or **Key Theorem 3.11** we know that $\mathbf{I} + \mathbf{P}$ is nonsingular if and only if the only solution to $(\mathbf{I} + \mathbf{P})\mathbf{x} = \mathbf{0}$ is $\mathbf{x} = \mathbf{0}$. Suppose that $(\mathbf{I} + \mathbf{P})\mathbf{x} = \mathbf{0}$ so that $\mathbf{x} = -\mathbf{P}\mathbf{x}$. Then $\|\mathbf{x}\| = \|-\mathbf{P}\mathbf{x}\| = \|\mathbf{P}\mathbf{x}\| \leq \|\mathbf{P}\|\|\mathbf{x}\|$, and since $\|\mathbf{P}\| < 1$ this is a contradiction unless $\mathbf{x} = \mathbf{0}$ as we sought to prove. Therefore $\mathbf{B} \equiv (\mathbf{I} + \mathbf{P})^{-1}$ exists. From

$$\mathbf{I} = \mathbf{B}(\mathbf{I} + \mathbf{P}) = \mathbf{B} + \mathbf{B}\mathbf{P}$$

we find

$$1 = \|\mathbf{I}\| = \|\mathbf{B}(\mathbf{I} + \mathbf{P})\| \leq \|\mathbf{B}\|\|\mathbf{I} + \mathbf{P}\| \leq \|\mathbf{B}\|(1 + \|\mathbf{P}\|)$$

so that

$$\frac{1}{1 + \|\mathbf{P}\|} \leq \|\mathbf{B}\|$$

as desired. Also

$$\mathbf{B} = \mathbf{I} - \mathbf{B}\mathbf{P}$$

so that

$$\|\mathbf{B}\| = \|\mathbf{I} - \mathbf{B}\mathbf{P}\| \leq 1 + \|\mathbf{B}\mathbf{P}\| \leq 1 + \|\mathbf{B}\|\|\mathbf{P}\|$$

from which

$$\|\mathbf{B}\| \leq \frac{1}{1 - \|\mathbf{P}\|}$$

follows, completing the proof.

As a simple example of **Key Lemma 5.1,** we note that

$$\mathbf{A} = \begin{bmatrix} 1.1 & -0.6 \\ 0.8 & 0.9 \end{bmatrix}$$

is nonsingular since $\mathbf{A} = \mathbf{I} + \mathbf{P}$ with

$$\mathbf{P} = \begin{bmatrix} 0.1 & -0.6 \\ 0.8 & -0.1 \end{bmatrix}$$

and, for example, $||\mathbf{P}||_\infty = 0.9 < 1$. Moreover,

$$\frac{1}{1.9} \leq ||\mathbf{A}^{-1}||_\infty \leq \frac{1}{0.1}.$$

We can now easily prove the **Key** general result.

● KEY THEOREM 5.8. *Let* $\mathbf{A}$ *and* $\mathbf{R}$ *be* $n \times m$ *matrices with* $\mathbf{A}$ *being nonsingular, and let* $||\cdot||$ *denote any of the operator norms* $||\cdot||_1, ||\cdot||_2, ||\cdot||_\infty$. *If* $\alpha = ||\mathbf{A}^{-1}\mathbf{R}|| < 1$ (or $\alpha = ||\mathbf{R}\mathbf{A}^{-1}|| < 1$) (*so in particular if* $||\mathbf{R}|| < ||\mathbf{A}^{-1}||^{-1}$), *then* $\mathbf{A} + \mathbf{R}$ *is nonsingular and*

$$||(\mathbf{A} + \mathbf{R})^{-1}|| \leq \frac{||\mathbf{A}^{-1}||}{1 - \alpha}.$$

Proof: We take the case of $\alpha = ||\mathbf{A}^{-1}\mathbf{R}|| < 1$; the other case is similar. Since $\mathbf{A}$ is nonsingular, we can write

$$\mathbf{A} + \mathbf{R} = \mathbf{A}(\mathbf{I} + \mathbf{P}), \qquad \mathbf{P} = \mathbf{A}^{-1}\mathbf{R}.$$

By **Key Lemma 5.1,** $\mathbf{I} + \mathbf{P}$ is nonsingular and $||(\mathbf{I} + \mathbf{P})^{-1}|| < (1 - \alpha)^{-1}$, and the result follows easily since

$$[\mathbf{A}(\mathbf{I} + \mathbf{P})]^{-1} = (\mathbf{I} + \mathbf{P})^{-1}\mathbf{A}^{-1}.$$

EXERCISE 5.22. Suppose $\mathbf{P}$ is an $n \times n$ matrix $[p_{ij}]$ such that $\sum_{j=1}^{n} |p_{ij}| < 1$ for all i. Prove that $\mathbf{I} + \mathbf{P}$ is nonsingular.

EXERCISE 5.23. Complete the proof of **Key Theorem 5.8.**

EXERCISE 5.24. Suppose that the $n \times n$ matrix $\mathbf{A} = [a_{ij}]$ is *strictly row-diagonally-dominant*, that is,

$$|a_{ii}| > \sum_{j=1}^{n}{}' |a_{ij}| \qquad \text{(for all } i\text{)}$$

where the prime indicates that the term for $j = i$ is omitted in the sum. Prove that $\mathbf{A}$ is nonsingular (see Ex. 5.22).

EXERCISE 5.25 Show that **Key Theorem 5.8** is true when $\|\cdot\|$ is *any* operator norm.

5.5 CONDITION OF LINEAR EQUATIONS

We have seen that one problem in linear algebra which arises in many applications is the solution of sets of linear equations $\mathbf{Ax} = \mathbf{b}$. Taking the viewpoint again that in practice our data such as $\mathbf{A}$ and $\mathbf{b}$ may be subject to measurement errors, we have to ask what effect this has on the solution to the system of equations. More precisely, if $\mathbf{A}$ is changed to $\mathbf{A} + \boldsymbol{\delta}\mathbf{A}$ and $\mathbf{b}$ is changed to $\mathbf{b} + \boldsymbol{\delta}\mathbf{b}$, can we say anything about the change from $\mathbf{x}$ solving $\mathbf{Ax} = \mathbf{b}$ to $\mathbf{x} + \boldsymbol{\delta}\mathbf{x}$ solving $(\mathbf{A} + \boldsymbol{\delta}\mathbf{A})(\mathbf{x} + \boldsymbol{\delta}\mathbf{x}) = \mathbf{b} + \boldsymbol{\delta}\mathbf{b}$? In numerical analysis, whenever "small" changes in the data can lead to "large" changes in the solution, a problem is said to be *ill-conditioned*; otherwise it is said to be *well-conditioned.* Exactly how changes are measured and what "small" and "large" mean will vary with the problem. Since we can use norms to measure the size of the perturbations $\boldsymbol{\delta}\mathbf{b}$ and $\boldsymbol{\delta}\mathbf{A}$ to our data, we can hope to understand something about the *condition* of the solution of a system of linear equations. We will not discuss more complicated ways of measuring the size of perturbations, such as considering their magnitude *relative* to unperturbed values.

- *KEY THEOREM 5.9. Let* $\mathbf{A}$ *be an* $n \times n$ *matrix and let* $\|\cdot\|$ *denote any of the operator norms* $\|\cdot\|_1$, $\|\cdot\|_2$, $\|\cdot\|_\infty$ *and its corresponding vector norm. Let* $\mathbf{A}$ *be nonsingular and let* $\mathbf{x}$ *solve* $\mathbf{Ax} = \mathbf{b}$. *Suppose that changes from* $\mathbf{A}$ *to* $\mathbf{A} + \boldsymbol{\delta}\mathbf{A}$ *and from* $\mathbf{b}$ *to* $\mathbf{b} + \boldsymbol{\delta}\mathbf{b}$, *where* $\|(\boldsymbol{\delta}\mathbf{A})\mathbf{A}^{-1}\| < 1$, *produce a change in the solution from* $\mathbf{x}$ *to* $\mathbf{x} + \boldsymbol{\delta}\mathbf{x}$. *Then*

$$\frac{\|\boldsymbol{\delta}\mathbf{x}\|}{\|\mathbf{x}\|} \leq M \cdot c(\mathbf{A}) \cdot \left\{\frac{\|\boldsymbol{\delta}\mathbf{b}\|}{\|\mathbf{b}\|} + \frac{\|\boldsymbol{\delta}\mathbf{A}\|}{\|\mathbf{A}\|}\right\} \tag{5.7}$$

where

$$M = [1 - \|(\boldsymbol{\delta}\mathbf{A})\mathbf{A}^{-1}\|]^{-1}$$

and

$$c(\mathbf{A}) = \|\mathbf{A}\|\,\|\mathbf{A}^{-1}\|$$

is the condition number of $\mathbf{A}$.

Proof: We remark first that since $\|(\boldsymbol{\delta}\mathbf{A})\mathbf{A}^{-1}\| < 1$, by **Key Theorem 5.8** $\mathbf{A} + \boldsymbol{\delta}\mathbf{A}$ is nonsingular so that the solution $\mathbf{x} + \boldsymbol{\delta}\mathbf{x}$ to the perturbed problem exists. In fact, the change $\boldsymbol{\delta}\mathbf{x}$ solves

$$(\mathbf{A} + \boldsymbol{\delta}\mathbf{A})\boldsymbol{\delta}\mathbf{x} = \mathbf{b} + \boldsymbol{\delta}\mathbf{b} - \mathbf{Ax} - \boldsymbol{\delta}\mathbf{Ax} = \boldsymbol{\delta}\mathbf{b} - \boldsymbol{\delta}\mathbf{Ax}$$

so that

$$\boldsymbol{\delta}\mathbf{x} = (\mathbf{A} + \boldsymbol{\delta}\mathbf{A})^{-1}(\boldsymbol{\delta}\mathbf{b} - \boldsymbol{\delta}\mathbf{Ax}).$$

Setting $\mathbf{R} = \boldsymbol{\delta}\mathbf{A}$ in **Key Theorem 5.8** and identifying M in the present theorem with $1/(1 - \alpha)$ in the bound on $\|(\mathbf{A} + \mathbf{R})^{-1}\|$ in **Key Theorem 5.8,** we obtain

$$\begin{aligned}\|\boldsymbol{\delta}\mathbf{x}\| &\leq M\cdot\|\mathbf{A}^{-1}\|\cdot\|\boldsymbol{\delta}\mathbf{b} - \boldsymbol{\delta}\mathbf{A}\mathbf{x}\| \\ &\leq M\cdot\|\mathbf{A}^{-1}\|\,[\|\boldsymbol{\delta}\mathbf{b}\| + \|\boldsymbol{\delta}\mathbf{A}\|\,\|\mathbf{x}\|].\end{aligned}$$

Therefore

$$\begin{aligned}\frac{\|\boldsymbol{\delta}\mathbf{x}\|}{\|\mathbf{x}\|} &\leq M\cdot\|\mathbf{A}^{-1}\|\left[\frac{\|\boldsymbol{\delta}\mathbf{b}\|}{\|\mathbf{x}\|} + \|\boldsymbol{\delta}\mathbf{A}\|\right] \\ &\leq M\cdot\|\mathbf{A}^{-1}\|\cdot\left[\frac{\|\boldsymbol{\delta}\mathbf{b}\|}{\|\mathbf{b}\|/\|\mathbf{A}\|} + \|\boldsymbol{\delta}\mathbf{A}\|\right]\end{aligned}$$

since $\mathbf{b} = \mathbf{A}\mathbf{x}$ implies $\|\mathbf{b}\| \leq \|\mathbf{A}\|\,\|\mathbf{x}\|$. Simplifying, we obtain

$$\frac{\|\boldsymbol{\delta}\mathbf{x}\|}{\|\mathbf{x}\|} \leq M\cdot c(\mathbf{A})\cdot\left[\frac{\|\mathbf{b}\|}{\|\boldsymbol{\delta}\mathbf{b}\|} + \frac{\|\boldsymbol{\delta}\mathbf{A}\|}{\|\mathbf{A}\|}\right].$$

as required, completing the proof.

It is easy to interpret **Key Theorem 5.9** in terms of ill-conditioning. If we decide to measure the change from one quantity $\mathbf{y}$ to another quantity $\mathbf{y} + \boldsymbol{\delta}\mathbf{y}$ in terms of $\|\boldsymbol{\delta}\mathbf{y}\|/\|\mathbf{y}\|$, then **Key Theorem 5.9** compares the change (measured in this sense) in the solution $\mathbf{x}$ of $\mathbf{A}\mathbf{x} = \mathbf{b}$ to the changes in the data $\mathbf{A}$ and $\mathbf{b}$; other measures of the size of a change may be more appropriate in certain circumstances, of course. If the perturbation $\boldsymbol{\delta}\mathbf{A}$ in $\mathbf{A}$ is small enough, the constant M above is close to unity. From (5.7), therefore, the overall change $\|\boldsymbol{\delta}\mathbf{x}\|/\|\mathbf{x}\|$ in the unknowns due to small changes in $\mathbf{A}$ and $\mathbf{b}$ will be small if $c(\mathbf{A}) = \|\mathbf{A}\|\,\|\mathbf{A}^{-1}\|$ is not too large. This means that a moderate $c(\mathbf{A})$ implies that the equations are well-conditioned. However, *the converse is not true*; a large $c(\mathbf{A}) = \|\mathbf{A}\|\,\|\mathbf{A}^{-1}\|$ does *not* imply that the equations are ill-conditioned. (**Key Theorem 5.9** gives only an upper bound for $\|\boldsymbol{\delta}\mathbf{x}\|/\|\mathbf{x}\|$, and this upper bound is not necessarily a realistic estimate.) We illustrate this by an example.

EXERCISE 5.26. Let $\mathbf{A}$ be the matrix

$$\mathbf{A} = \begin{bmatrix} 1 & k \\ 0 & 1 \end{bmatrix}$$

so that

$$\mathbf{A}^{-1} = \begin{bmatrix} 1 & -k \\ 0 & 1 \end{bmatrix}.$$

In either the norm $\|\cdot\|_1$ or the norm $\|\cdot\|_\infty$, $\|\mathbf{A}\| = \|\mathbf{A}^{-1}\| = 1 + k$ for $k \geq 0$, so that the condition number $c(\mathbf{A}) = (1 + k)^2$ which is large for large k. However if we consider the system of equations $\mathbf{A}\mathbf{x} = \mathbf{b}$ with

$$\mathbf{b} = \begin{bmatrix} 1 \\ 1 \end{bmatrix}$$

the solution is

$$\mathbf{x} = \begin{bmatrix} 1 - k \\ 1 \end{bmatrix},$$

while if we perturb only **b** via nonzero δ_1, δ_2 to

$$\mathbf{b} + \boldsymbol{\delta}\mathbf{b} = \begin{bmatrix} 1 + \delta_1 \\ 1 + \delta_2 \end{bmatrix}$$

we find the change $\boldsymbol{\delta}\mathbf{x}$ in the solution to be

$$\boldsymbol{\delta}\mathbf{x} = \begin{bmatrix} \delta_1 - k\,\delta_2 \\ \delta_2 \end{bmatrix}$$

so that for large k for either $\|\cdot\|_1$ or $\|\cdot\|_\infty$ we have

$$\frac{\|\boldsymbol{\delta}\mathbf{x}\|}{\|\mathbf{x}\|} \leq 2\frac{\|\boldsymbol{\delta}\mathbf{b}\|}{\|\mathbf{b}\|}$$

so that the solution is very well-conditioned.

Although a large condition number $c(\mathbf{A})$ does not indicate that the solution to $\mathbf{Ax} = \mathbf{b}$ is ill-conditioned for *every* **b**, it is true that it is ill-conditioned for *some* **b**.

EXERCISE 5.27. Carry out the calculations in Ex. 5.26 with the **b** there replaced by

$$\mathbf{b} = \begin{bmatrix} 1 \\ -\frac{1}{k} \end{bmatrix}$$

and show that the solution *is* ill-conditioned.

EXERCISE 5.28. Show that **Key Theorem 5.9** is true when $\|\cdot\|$ is *any* operator norm.

The significance of the condition number can be seen from another viewpoint as well. If **A** is nonsingular, then we know from **Key Theorem 5.8** that $\mathbf{A} + \mathbf{R}$ is nonsingular for all **R** with $\|\mathbf{R}\| < \|\mathbf{A}^{-1}\|^{-1}$, that is, with

$$\frac{\|\mathbf{R}\|}{\|\mathbf{A}\|} < \frac{1}{\|\mathbf{A}\|\,\|\mathbf{A}^{-1}\|}.$$

Therefore $\mathbf{A} + \boldsymbol{\delta}\mathbf{A}$ is nonsingular whenever

$$\frac{\|\boldsymbol{\delta}\mathbf{A}\|}{\|\mathbf{A}\|} < \frac{1}{c(\mathbf{A})};$$

stated differently, if $\mathbf{A} + \boldsymbol{\delta}\mathbf{A}$ *is* singular then

$$\frac{\|\boldsymbol{\delta}\mathbf{A}\|}{\|\mathbf{A}\|} \geq \frac{1}{c(\mathbf{A})},$$

giving a lower bound on $c(\mathbf{A})$. In fact it can be shown that

$$\frac{1}{c(\mathbf{A})} = \min\left\{\frac{\|\boldsymbol{\delta}\mathbf{A}\|}{\|\mathbf{A}\|}\,\middle|\, \mathbf{A} + \boldsymbol{\delta}\mathbf{A} \text{ is singular}\right\}.$$

Therefore the condition number of $\mathbf{A}$ is large if and only if there is a singular matrix $\mathbf{A} + \boldsymbol{\delta}\mathbf{A}$ nearby in the sense that $\|\boldsymbol{\delta}\mathbf{A}\|/\|\mathbf{A}\|$ is small. (See Ex. 5.45.)

EXERCISE 5.29. In the analysis of chemical reactions it turns out that the number of really independent factors in a process is just the rank of a certain matrix $\mathbf{C}$ made up from observable data based on the concentrations of the chemicals in the reaction. Suppose that we have three chemicals and that we measure the concentration matrix $\mathbf{C}$ to be

$$\mathbf{C} = \begin{bmatrix} 1.02 & 2.03 & 4.20 \\ 0.25 & 0.51 & 1.06 \\ 1.74 & 3.46 & 7.17 \end{bmatrix}$$

where each of our measurements may be subject to experimental errors of at most 0.015. To determine the rank of $\mathbf{C}$, we start to reduce it to row-echelon form by operating in the first column; we obtain

$$\begin{bmatrix} 1.00 & 1.99 & 4.12 \\ 0 & 0.013 & -0.03 \\ 0 & 0.003 & 0.005 \end{bmatrix}.$$

If we accept these numbers we can go on to find that the rank is three; however, given the fact that our numbers contain experimental errors, these last two rows look suspiciously like zeros. In fact, the slightly perturbed matrix

$$\mathbf{C}' = \begin{bmatrix} 1.02 & 2.03 & 4.20 \\ 0.26 & 0.517451\ldots & 1.070588\ldots \\ 1.74 & 3.462941\ldots & 7.164706\ldots \end{bmatrix}$$

reduces exactly to

$$\begin{bmatrix} 1.00 & 1.99 & 4.12 \\ 0 & 0 & 0 \\ 0 & 0 & 0 \end{bmatrix}$$

Since the largest change in any element from $\mathbf{C}$ to $\mathbf{C}'$ is $0.010588\ldots < 0.015$, the bound on our experimental error, the data in $\mathbf{C}'$ is just as good as that in $\mathbf{C}$

for which the rank was three rather than one. We probably should conclude that there is only one independent factor in the reaction. From the viewpoint of condition number, since $\|\mathbf{C}\|_\infty = 12.37$ while $\mathbf{C}'$ is singular and

$$\|\mathbf{C} - \mathbf{C}'\|_\infty = 0.028\ldots$$

we know that the condition number $c(\mathbf{C})$ for the norm $\|\cdot\|_\infty$ is at least $12.37/0.028 \geq 441$. In fact, by perturbing only the third row of $\mathbf{C}$, we can find an even closer singular matrix and deduce that $c(\mathbf{C}) \geq 1502$; verify these lower bounds on $c(\mathbf{C})$.

EXERCISE 5.30. Find the condition number of the general 2×2 matrix for the norm $\|\cdot\|_1$.

EXERCISE 5.31. Find the condition number of the general 2×2 matrix for the norm $\|\cdot\|_\infty$.

EXERCISE 5.32. State and prove a version of **Key Theorem 5.9** with the hypothesis that $\|(\boldsymbol{\delta}\mathbf{A})\mathbf{A}^{-1}\| < 1$ replaced by the hypothesis that $\|\mathbf{A}^{-1}\,\boldsymbol{\delta}\mathbf{A}\| < 1$.

EXERCISE 5.33. For the norm $\|\cdot\|_1$, show that $c(\mathbf{A}) \geq 600$ where

$$\mathbf{A} = \begin{bmatrix} 1.1 & 2.1 & 3.1 \\ 1.0 & -1.0 & 2.0 \\ 0.2 & 3.3 & 1.4 \end{bmatrix}.$$

MISCELLANEOUS EXERCISES 5

EXERCISE 5.34. Prove that the transformations in the examples (1)–(4) in Section 5.1 are actually linear transformations.

EXERCISE 5.35. Suppose that A is a linear transformation from the vector space V to the vector space W and that B is a linear transformation from W to the vector space Z. Prove that BA is a linear transformation from V to Z.

EXERCISE 5.36. Let A be a linear transformation from V to W and let V and W have the bases $\mathbf{v}_1, \ldots, \mathbf{v}_n$ and $\mathbf{w}_1, \ldots, \mathbf{w}_m$, respectively (always writing the basis vectors in precisely the same order). Let $\mathbf{A}$ be the $m \times n$ matrix representing this transformation with respect to these bases. Prove that $\mathbf{v}$ is in the null space of A if and only if

$$\mathbf{v} = \sum_{i=1}^{n} x_i \mathbf{v}_i$$

where $\mathbf{x} = [x_i]$ satisfies $\mathbf{Ax} = \mathbf{0}$. Indicate how to find a basis for the null space of A.

EXERCISE 5.37. Assume as in Ex. 5.36. Prove that $\mathbf{w}$ is in the exact range space of A if and only if

$$\mathbf{w} = \sum_{j=1}^{m} y_j \mathbf{w}_j$$

where $\mathbf{y} = [y_j]$ is in the column space of $\mathbf{A}$. Indicate how to find a basis for the exact range space of A.

EXERCISE 5.38. For the proof of Theorem 5.4 explain precisely how we handle the possible renumbering of the unknowns. (See Ex. 5.9.)

EXERCISE 5.39. Under the additional hypothesis that W is finite dimensional, prove Corollary 5.1. (See Exs. 5.36 and 5.37.)

EXERCISE 5.40. Prove Corollary 5.1 essentially by replacing W by W_0, the exact range space of A, and showing W_0 to be finite-dimensional.

EXERCISE 5.41. By using norms, show that the matrices $\mathbf{A}$ described in Section 2.2 on Markov processes cannot possibly have the sequence of powers $\mathbf{A}^i$ become unbounded.

EXERCISE 5.42. Let $\mathbf{P}$ be an $n \times n$ matrix with $\|\mathbf{P}\|_\infty < 1$. Prove that the infinite series $\mathbf{I} - \mathbf{P} + \mathbf{P}^2 - \mathbf{P}^3 + \mathbf{P}^4 - \mathbf{P}^5 \ldots$ converges and that the sum equals $(\mathbf{I} + \mathbf{P})^{-1}$ which must exist. This is another approach to the proof of **Key Lemma 5.1.**

EXERCISE 5.43. Let $\mathbf{A}$ be an $n \times n$ nonsingular matrix and let $c(\mathbf{A})$ be its condition number. Prove that there exists *some* $\mathbf{b}$ and some perturbation $\boldsymbol{\delta}\mathbf{b}$ for which $\|\boldsymbol{\delta}\mathbf{x}\|/\|\mathbf{x}\|$ approximately equals $c(\mathbf{A})[\|\boldsymbol{\delta}\mathbf{b}\|/\|\mathbf{b}\|]$ where $\mathbf{x}$ solves $\mathbf{Ax} = \mathbf{b}$ and $\mathbf{x} + \boldsymbol{\delta}\mathbf{x}$ solves $\mathbf{A}(\mathbf{x} + \boldsymbol{\delta}\mathbf{x}) = \mathbf{b} + \boldsymbol{\delta}\mathbf{b}$.

EXERCISE 5.44. Show, by exhibiting a small perturbation $\boldsymbol{\delta}\mathbf{b}$ causing a large change $\boldsymbol{\delta}\mathbf{x}$, that the equations

$$\begin{aligned} x_1 + x_2 &= 2.0000 \\ 1.00001\, x_1 + x_2 &= 2.00001 \end{aligned}$$

are ill-conditioned. For $\|\cdot\|_\infty$, compute the condition number $c(\mathbf{A})$.

EXERCISE 5.45. For the matrix $\mathbf{A}$ of Ex. 5.26, verify that

$$\frac{1}{c(\mathbf{A})} = \min\left\{\frac{\|\boldsymbol{\delta}\mathbf{A}\|_\infty}{\|\mathbf{A}\|_\infty} \,\middle|\, \mathbf{A} + \boldsymbol{\delta}\mathbf{A} \text{ is singular}\right\}$$

by considering

$$\boldsymbol{\delta}\mathbf{A} = \begin{bmatrix} \dfrac{-1}{k+1} & 0 \\ \dfrac{1}{k+1} & 0 \end{bmatrix}.$$

EXERCISE 5.46. Find a basis for the null space of the usual transformation A generated

(a) by the matrix

$$[1 \quad 1 \quad -1 \quad -1],$$

(b) by the matrix

$$\begin{bmatrix} 1 & -1 & 0 & 0 \\ 0 & 1 & -1 & 0 \\ 1 & -2 & 1 & 0 \end{bmatrix}.$$

EXERCISE 5.47. Prove that the exact range space of a linear transformation is indeed a linear subspace.

EXERCISE 5.48. Prove that the null space of a linear transformation is indeed a linear subspace.

EXERCISE 5.49. Let V and W be vector spaces over the same field F, and let L be the set of all linear transformations from V to W. Prove that L is a vector space over F.

CHAPTER SIX

PRACTICAL SOLUTION OF SYSTEMS OF EQUATIONS

This chapter briefly discusses some practical computational matters concerning Gauss elimination in Theorem 6.1, as well as the notion of a determinant in Theorems 6.5 and 6.9. Theorems 6.2, 6.3, and 6.4 are important from a practical point of view and are only slightly less fundamental.

6.1 INTRODUCTION

Many problems in the applied sciences involve the inversion of $n \times n$ matrices or the solution of n linear equations in n unknowns (see Chapter 2). If n is large (greater than three or four, say) or the entries in the matrix are not integers with only a few digits, the solution of such problems by hand becomes rather tedious and unpleasant; in practice these problems are usually solved by automatic programs on electronic digital computers, perhaps with human guidance to simplify the problem from the outset. The systematic procedure introduced in Section 3.2 provides an excellent tool for determining the nature of the set of solutions and for calculating solutions. We can make this procedure the basis for the desired computational method, if proper precautions are used. Before we consider how to do this, let us make some comments on the nature of the systems of equations which arise in practice so that the reader can see the criteria we use to measure the success of any proposed practical procedures.

We visualize the following when considering the solution of systems of linear equations:

1. The real problem that we want to solve has somehow been modeled mathematically. We suppose that one step in the solution of the mathematical model requires the solution of the set of equations
$$\mathbf{A}^*\mathbf{x}^* = \mathbf{b}^*, \tag{6.1}$$
where $\mathbf{A}^*$ and $\mathbf{b}^*$ are matrices whose elements are precisely *defined* but usually not precisely *known*.
2. Instead of the ideal system (6.1), we usually actually have at hand a set of equations
$$\mathbf{Ax} = \mathbf{b}, \tag{6.2}$$
where typically we have $\mathbf{A} \neq \mathbf{A}^*$ and $\mathbf{b} \neq \mathbf{b}^*$. The errors in the data $\mathbf{A}$ and $\mathbf{b}$ are almost always at least partly due to rounding errors and to binary-decimal conversion when numbers are stored in the computer; quite commonly the data also contain experimental error caused by our inability to measure $\mathbf{A}^*$, $\mathbf{b}^*$ accurately.
3. The system (6.2) of course has an *exact* solution, that is, one for which $\mathbf{Ax} - \mathbf{b}$ is exactly zero provided that the computations involved in evaluating $\mathbf{Ax} - \mathbf{b}$ are done exactly, but the best that we can hope to do on a digital computer or desk calculator or probably even by hand is to obtain an *approximate* solution $\mathbf{x}'$, not generally equal to the exact solution $\mathbf{x}$ to (6.2).

Our practical methods for solving equations will of course be able to deal only with the approximate data $\mathbf{A}$, $\mathbf{b}$ in (6.2). When we try to evaluate how well our method does, that is, how accurate an approximation $\mathbf{x}'$ it produces, we can only mean accuracy with respect to the exact solution $\mathbf{x}$ *for the data given us in* (6.2). We will try to guarantee that our procedures make the error $\mathbf{x} - \mathbf{x}'$ "small."

The applied scientist of course really wants the answer $\mathbf{x}'$ we give to be an accurate approximation to the ideal solution $\mathbf{x}^*$; the errors $\mathbf{x}' - \mathbf{x}$ are of no concern in this regard. Fortunately in Section 5.5 we developed ways of comparing the change $\mathbf{x}^* - \mathbf{x}$ with the changes $\mathbf{A}^* - \mathbf{A}$, $\mathbf{b}^* - \mathbf{b}$, so that we can bound $\mathbf{x}^* - \mathbf{x}$ if the scientist can bound the errors $\mathbf{A}^* - \mathbf{A}$, $\mathbf{b}^* - \mathbf{b}$ in the data; combining these bounds on $\mathbf{x}^* - \mathbf{x}$ with our bounds on $\mathbf{x} - \mathbf{x}'$ that will be obtained in the analysis of our practical procedure, we can give the scientist the desired bounds on the error $\mathbf{x}' - \mathbf{x}^*$ between the approximate and ideal solutions. If we can show that the errors $\mathbf{x} - \mathbf{x}'$ caused by our computing method are no worse than the errors $\mathbf{x}^* - \mathbf{x}$ caused by the inherent rounding and/or experimental errors in the data, then no one can blame our process for producing errors, because the approximate solution $\mathbf{x}'$ that we obtained is no more or less reliable than is the exact solution $\mathbf{x}$ to the experimental data. Of course, as we saw in Section 5.5, if the condition number of $\mathbf{A}^*$ (or of $\mathbf{A}$) is large, then this exact solution $\mathbf{x}$ and our approximate solution $\mathbf{x}'$ can be very unreliable indeed, but this is the fault of the *problem*, not of our computing method. It is all of these considerations that justify our concern over controling the error $\mathbf{x} - \mathbf{x}'$.

EXERCISE 6.1. Suppose that the ideal system of equations is

$$\begin{aligned} x_1^* + x_2^* &= 2 \\ x_1^* + 1.00001x_2^* &= 2.00001, \end{aligned}$$

that the experimentally obtained system is

$$\begin{aligned} x_1 + x_2 &= 2 \\ x_1 + 1.00001x_2 &= 1.99999, \end{aligned}$$

and that some solution method generates $x_1' = 2$, $x_2' = 0$ while the ideal solution is $x_1^* = x_2^* = 1$. Are the numerical answers x_1' and x_2' as accurate as one could *reasonably* ask and expect? Why or why not?

EXERCISE 6.2. Suppose that we create a solution method for approximately solving the experimentally obtained equations (6.2), and that our method produces an $\mathbf{x}'$ which can be proved to be the exact solution of yet a third system $\mathbf{A}'\mathbf{x}' = \mathbf{b}'$, where $||\mathbf{A}' - \mathbf{A}|| \leq ||\mathbf{A} - \mathbf{A}^*||$ and $||\mathbf{b}' - \mathbf{b}|| \leq ||\mathbf{b} - \mathbf{b}^*||$. Is the method a good one? Why or why not?

6.2 GAUSS ELIMINATION

Let us consider the solution of k sets of n simultaneous linear equations in n unknowns, with the same coefficient matrix $\mathbf{A}$ but different right-hand sides $\mathbf{b}_j$:

$$\mathbf{A}\mathbf{x}_j = \mathbf{b}_j \qquad (j = 1, 2, \ldots, k). \tag{6.3}$$

This includes the problem of matrix inversion for which $k = n$ and the $\mathbf{b}_j$ are the unit column matrices $\mathbf{e}_j$.

The method of Gauss elimination has already been described in Sections 3.2 and 3.4. We simplify the situation by assuming that $\mathbf{A}$ is nonsingular, so that our equations always have a unique solution. Suppose that we consider one set of equations, $\mathbf{Ax} = \mathbf{b}$ [compare (3.32)],

$$\begin{aligned} a_{11}x_1 + a_{12}x_2 + \cdots + a_{1n}x_n &= b_1 \\ a_{21}x_1 + a_{22}x_2 + \cdots + a_{2n}x_n &= b_2 \\ &\cdots \\ a_{n1}x_1 + a_{n2}x_2 + \cdots + a_{nn}x_n &= b_n. \end{aligned} \tag{6.4}$$

Suppose that a_{11} is an acceptable pivot; theoretically we only need a_{11} to be nonzero, but in Section 6.3 we will see that it is wiser to be more cautious in choosing a pivot. If a_{11} is not acceptable, we interchange the first equation with a later one to obtain a coefficient for x_1 that is acceptable as a pivot, assuming at least one of the coefficients a_{i1} is nonzero. Throughout this section we assume that we can always find at least one nonzero pivot when needed. In Chapter 3 (e.g., Section 3.4) the next step was to divide the first equation by a_{11}. In this chapter we use a procedure that is slightly different but mathematically equivalent. We divide the coefficients of x_1 in the last $n - 1$ equations by a_{11} to obtain multiplying factors

$$m_{i1} = \frac{a_{i1}}{a_{11}} \qquad (i = 2, \ldots, n).$$

We subtract m_{i1} times the first equation from the ith equation for $i = 2, \ldots, n$ to obtain the first derived system [compare (3.33)],

$$\begin{aligned} a_{11}x_1 + a_{12}x_2 + \cdots + a_{1n}x_n &= b_1 \\ a_{22}^{(2)}x_2 + \cdots + a_{2n}^{(2)}x_n &= b_2^{(2)} \\ &\cdots \\ a_{n2}^{(2)}x_2 + \cdots + a_{nn}^{(2)}x_n &= b_n^{(2)}, \end{aligned}$$

where

$$a_{ij}^{(2)} = a_{ij} - m_{i1}a_{1j}, \qquad b_i^{(2)} = b_i - m_{i1}b_1 \qquad (i, j = 1, \ldots, n).$$

If $a_{22}^{(2)}$ is unacceptable as a pivot, we again interchange one of the *later* equations with the second equation so as to obtain an acceptable pivot in the (2, 2) position, assuming that this is possible. Having found an $a_{22}^{(2)}$ that is acceptable, we divide the coefficients of x_2 in the last $n - 2$ equations by $a_{22}^{(2)}$, obtaining multipliers

$$m_{i2} = \frac{a_{i2}^{(2)}}{a_{22}^{(2)}} \qquad (i = 3, \ldots, n)$$

We subtract m_{i2} times the second equation from the ith equation for $i = 3, \ldots, n$, eliminating the coefficients of x_2 in the ith equation for

$i = 3, \ldots, n$. Proceeding similarly for each column, always eliminating only *below* the main diagonal, we eventually obtain [compare (3.47)]

$$\begin{aligned} a_{11}x_1 + a_{12}x_2 + \cdots + a_{1n}x_n &= b_1 \\ a_{22}^{(2)}x_2 + \cdots + a_{2n}^{(2)}x_n &= b_2^{(2)} \\ \cdots \qquad & \\ a_{nn}^{(n)}x_n &= b_n^{(n)}. \end{aligned} \tag{6.5}$$

The unknowns are easily found from these equations by back-substitution.

EXERCISE 6.3. Show that the equations obtained by the procedure in Section 3.4 are simply (6.5) divided by the diagonal elements $a_{rr}^{(r)}$. (There is no special reason why we have used the slightly different sequence of operations described here, in preference to the method of Section 3.4. In the row-echelon form it is convenient to reduce the pivot elements to unity so that the key columns are unit vectors. In numerical work on a computer the magnitudes of the pivots are important, as we see later, and these are displayed explicitly in (6.5). On the other hand, we could keep track of the sizes of the $a_{rr}^{(r)}$ even though we divided through by them to obtain the row-echelon form.)

The procedure described above can be summarized by saying that we produce $n - 1$ sets of equations equivalent to $\mathbf{Ax} = \mathbf{b}$, say

$$\mathbf{A}^{(r)}\mathbf{x} = \mathbf{b}^{(r)} \qquad (r = 2, \ldots, n),$$

where the original equations $\mathbf{Ax} = \mathbf{b}$ can be considered to be the case $r = 1$. The final matrix $\mathbf{A}^{(n)}$ is upper triangular.

The general equations for calculating the elements of $\mathbf{A}^{(r+1)}$, $\mathbf{b}^{(r+1)}$ from those of $\mathbf{A}^{(r)}$, $\mathbf{b}^{(r)}$ are:

$$m_{ir} = \frac{a_{ir}^{(r)}}{a_{rr}^{(r)}} \qquad (i = r + 1, \ldots, n), \tag{6.6}$$

$$a_{ij}^{(r+1)} = a_{ij}^{(r)} - m_{ir}a_{rj}^{(r)} \qquad (i, j = r + 1, \ldots, n), \tag{6.7}$$

$$b_i^{(r+1)} = b_i^{(r)} - m_{ir}b_r^{(r)} \qquad (i = r + 1, \ldots, n), \tag{6.8}$$

where, for $r = 1$, we must take $a_{ij}^{(1)} = a_{ij}$, $b_i^{(1)} = b_i$.

In order to obtain some idea of the amount of work involved in Gauss elimination both for solving equations and for inverting matrices, we count the number of additions and multiplications required. (For simplicity, additions and subtractions are lumped together as "additions," multiplications and divisions are lumped together as "multiplications.") Consider the solution of k sets of equations, as in (6.3). The calculation of the elements of

$\mathbf{A}^{(r+1)}$, $\mathbf{b}^{(r+1)}$ from $\mathbf{A}^{(r)}$, $\mathbf{b}^{(r)}$ involves the following:

1. Computations of the m_{ir} from (6.6), involving $n - r$ divisions.
2. Computation of the $a_{ij}^{(r+1)}$ from (6.7), involving $(n - r)^2$ multiplications and additions.
3. Computation of k sets $b_i^{(r+1)}$ from (6.8), involving $k(n - r)$ multiplications and additions.

The forward elimination to obtain (6.5) therefore requires

$$\sum_{r=1}^{n-1} \{(n - r)^2 + (k + 1)(n - r)\} = n\left\{\left(\tfrac{1}{3}n^2 - \frac{1}{3}\right) + \tfrac{1}{2}k(n - 1)\right\}$$

multiplications,

$$\sum_{r=1}^{n-1} \{(n - r)^2 + k(n - r)\} = n\left\{\left(\tfrac{1}{3}n^2 - \tfrac{1}{2}n + \frac{1}{6}\right) + \tfrac{1}{2}k(n - 1)\right\}$$

additions,

where we have used the standard results

$$\sum_{s=1}^{m} s = \tfrac{1}{2}m(m + 1), \qquad \sum_{s=1}^{m} s^2 = \tfrac{1}{6}m(m + 1)(2m + 1).$$

In the solution of (6.5) by back-substitution, the determination of x_r involves $n - r$ multiplications and additions, and one division. Regarding the division as equivalent to a multiplication, k back-substitutions involve

$$k(1 + 2 + \cdots + n) = \tfrac{1}{2}kn(n + 1) \quad \text{multiplications,}$$
$$k(1 + 2 + \cdots + n - 1) = \tfrac{1}{2}kn(n - 1) \quad \text{additions.}$$

The overall totals for the complete solution of k sets of equations are:

$$\begin{aligned} &n\left(\tfrac{1}{3}n^2 - \frac{1}{3} + kn\right) && \text{multiplications,} \\ &n\left\{\tfrac{1}{3}n^2 - \tfrac{1}{2}n + \frac{1}{6} + k(n - 1)\right\} && \text{additions.} \end{aligned} \tag{6.9}$$

When inverting $\mathbf{A}$, we start with n unit vectors $\mathbf{e}_j$ instead of $\mathbf{b}_j$ on the right of (6.3). By carefully remembering where there are zeros and ones on the right and then avoiding multiplications by ones and zeros and additions of zeros, we can easily reduce the total work for the forward elimination to

$$\tfrac{1}{2}n^2(n - 1) \quad \text{multiplications,}$$
$$\tfrac{1}{2}n(n - 1)^2 \quad \text{additions.}$$

By also being cautious in the back-substitution, we finally arrive at the following overall totals required for the entire inversion:

$$\begin{aligned} n^3 &\quad \text{multiplications,} \\ n^3 - 2n^2 + n &\quad \text{additions.} \end{aligned} \tag{6.10}$$

The important conclusions to be drawn from the numbers in (6.9) and (6.10) for Gauss elimination are:

1. The numbers of multiplications and additions required to solve a single set of equations are of order $\frac{1}{3}n^3$.
2. The numbers of multiplications and additions required to invert a matrix are of order n^3.

This tells us then that:

1. The amount of work is proportional to n^3, so that if we double the number of equations we have to do eight times the amount of work. The amount of work involved rapidly becomes substantial as the size of the system increases.
2. About three times the amount of work is involved when inverting a matrix as compared with solving a single set of equations, not n times, as we might expect. The ratio is essentially independent of the size of the system.

For n very large at all, these operation counts seem rather discouraging since n^3 will be extremely large. However, one tends to forget that multiplying two matrices of order n together requires n^3 multiplications and additions. Hence the amount of work required to form $\mathbf{A}^{-1}$ is of the same order of magnitude as the amount of work required to form $\mathbf{A}^2$ or $\mathbf{A}^T\mathbf{A}$. Moreover, it is known that no method using entire row and entire column operations can require less than this amount of work. In addition, we forget just how rapidly computers can operate to implement Gauss elimination. By hand it is laborious to solve five equations in five unknowns. On present day computers it is quite practical to solve 100 equations in 100 unknowns, with all 10^4 coefficients nonzero. The arithmetic operations take about 10 seconds and the cost is less than a dollar. This is why digital computers are revolutionizing the way in which we formulate problems in the applied sciences. However, we would still hesitate before tackling a system with $n = 1000$ unless most of the coefficients were zero.

EXERCISE 6.4. Show that the Gauss-Jordan procedure (Section 3.2) requires roughly $\frac{1}{2}n^3$ multiplications and additions to solve a single set of equations, so that the work required is 50% more than for Gauss elimination. Show however, that inversion can be performed with approximately n^3 multiplications and additions so that the Gauss-Jordan and Gauss procedures involve comparable amounts of work for inversion.

EXERCISE 6.5. Describe how to take advantage of the ones and zeros when inverting a matrix; derive the operation counts (6.10).

EXERCISE 6.6. Show, no matter what the size n of each system and no matter what the number k of such systems, that it is *always* more efficient to use Gauss elimination on (6.3) than to compute $\mathbf{A}^{-1}$ as above and then multiply $\mathbf{A}^{-1}$ times the k different right-hand sides $\mathbf{b}_j$.

6.3 THE CHOICE OF PIVOTS IN ELIMINATION

If exact arithmetic is used, the same solution to a system of equations will be computed no matter what sequence of pivots is used to eliminate what variables in what order; this is essentially due to the uniqueness of the row-echelon form (**Key Theorem 3.4**). When the calculations are performed on a digital computer, desk calculator, or even by hand, we usually carry only a finite number of digits to represent each number and we therefore make errors by such approximations; since different such errors would be made by using a different sequence of pivots, in practice the computed solution and hence its accuracy depends on our choice of pivots. Again, this is because of our use of finite precision arithmetic; we describe this, as it appears in modern computers, in a little more detail.

To deal with numbers like $\frac{1}{3} = 0.333\ldots$ or even with numbers requiring a very large finite number of digits in their decimal representation, in practice we must round this number to some moderate number of digits we can manage. (To *round* a number *to t digits*, retain the first t digits of the number, ignoring leading zeros, and temporarily discard the remainder; if the discarded part is less than a half unit in the last place retained, leave the last digit of the retained part unchanged; if it is greater than half a unit, increase the last digit of the retained part by one; if it is exactly half a unit, make the last digit of the retained part even by increasing it by one if necessary.) Computers usually represent numbers as the product of a *fractional part a* having t digits and a *scale factor* 10^b where $0.1 \leq |a| < 1$ (unless $a = 0$), so that the decimal representation of a starts out with the first nonzero digit appearing immediately after the decimal point; in this *floating-point* arithmetic, all rounding is thus performed on the t-digit fractional part.

To illustrate that the choice of pivots is important when working with finite precision, we suppose that we are dealing with two-digit floating-point and need to solve the equations

$$\begin{aligned} x_1 - x_2 &= 0 \\ 0.01x_1 + x_2 &= 1, \end{aligned} \tag{6.11}$$

whose exact solution is $x_1 = x_2 = 1/1.01 = 0.990099 \ldots$. The best approximate solution we could conceivably hope to get with two digits would be $x_1' = x_2' = 0.99$; let's see how well we can do.

If we do not interchange equations and thus pivot on the (1, 1) element, then subtracting 0.01 times the first equation from the second yields

$$\begin{aligned} x_1 \qquad - x_2 &= 0 \\ (1 + 0.01)x_2 &= 1.0. \end{aligned}$$

Assuming, as is usually true, that our arithmetic operates so as to give the correctly rounded result of the true arithmetic result, $1 + 0.01$ would be evaluated as the rounded version of 1.01, namely 1.0. Thus the *computed* simplified version of (6.11) is

$$\begin{aligned} x_1 - x_2 &= 0 \\ 1.0x_2 &= 1.0 \end{aligned} \tag{6.12}$$

from which back-substitution yields $x_1' = x_2' = 1.0$, not a bad approximation to the best approximate answer of 0.99.

If we instead pivot on the (2, 1) element, then it is just as though we interchange equations to obtain

$$\begin{aligned} 0.01x_1 + x_2 &= 1 \\ x_1 - x_2 &= 0. \end{aligned}$$

Subtracting 100 times the first equation from the second yields

$$\begin{aligned} 0.01x_1 + \qquad\qquad x_2 &= 1 \\ -(1 + 100)x_2 &= -100 \end{aligned}$$

and $1 + 100 = 101$ is computed to two figures as 100. Thus the *computed* simplified version of (6.11) is

$$\begin{aligned} 0.01x_1 + \quad x_2 &= 1 \\ -100x_2 &= -100 \end{aligned} \tag{6.13}$$

from which back-substitution yields $x_2' = 1.0$, $x_1' = 0.0$, an absolutely terrible result. This certainly illustrates that different pivot strategies can have radically different results; we study the example further to see what actually happened.

Consider again pivoting on the (1, 1) element in (6.11), leading to the computed system (6.12). Had the second equation in (6.11) been instead $0.01x_1 + 0.99x_2 = 1$, the *exact* arithmetic on the modified system would have produced (6.12). Therefore we can say that the effect of using two-digit arithmetic on (6.11) while pivoting on the (1, 1) element is exactly the same as if we used exact arithmetic on the *slightly* perturbed problem

$$\begin{aligned} x_1 - \quad\quad x_2 &= 0 \\ 0.01x_1 + 0.99x_2 &= 1. \end{aligned}$$

Since the matrix $\mathbf{A}$ for the system (6.11) is well-conditioned (for example, $\|\mathbf{A}\|_\infty \|\mathbf{A}^{-1}\|_\infty \leq 4$), this small perturbation in $\mathbf{A}$ produces a small perturbation in the solution, explaining why we obtained the good results $x_1' = x_2' = 1.0$.

On the other hand, consider again pivoting on the (2, 1) element in (6.11), leading to the computed system (6.13). We want to replace $x_1 - x_2 = 0$ by $x_1 - \alpha x_2 = 0$ and see how to choose α so that exact arithmetic would produce (6.13). Obviously, exact arithmetic on

$$\begin{aligned} 0.01x_1 + x_2 &= 1 \\ x_1 - \alpha x_2 &= 0 \end{aligned}$$

yields

$$\begin{aligned} 0.01x_1 + \quad\quad\quad\quad x_2 &= 1 \\ -(\alpha + 100)x_2 &= -100; \end{aligned}$$

for this to be the same as (6.13), we must take $\alpha = 0$. Therefore we can say that the effect of using two-digit arithmetic on (6.11) while pivoting on the (2, 1) element is exactly the same as if we used exact arithmetic on the *greatly* perturbed problem

$$\begin{aligned} x_1 \quad\quad &= 0 \\ 0.01x_1 + x_2 &= 1. \end{aligned}$$

Even though $\mathbf{A}$ is well-conditioned, this large perturbation in $\mathbf{A}$ is almost certain to lead to a large perturbation in the solution, as indeed it did.

The argument that we used above, showing that the *computed* solution is the *true* solution for a *perturbed* problem, is called *backward* (or *inverse*) *error analysis* and is very fundamental in numerical analysis. Since we often must consider our data to be inaccurate because of experimental errors and errors caused by rounding ideal data $\mathbf{A}^*, \mathbf{b}^*$ to $\mathbf{A}, \mathbf{b}$ during storage in the computer (recall Section 6.1), if we can show by backward error analysis that our computing method exactly solves a very slightly perturbed problem with perturbations of about the size of the inaccuracies in the data, then our computed solution is as good as we can ask.

Although the preceding inverse error analysis was illuminating, it still did not explain *why* the (2, 1) pivot caused problems while the (1, 1) pivot did not. The reader may have subconsciously gained the impression that the reason why the first solution was satisfactory whereas the second was unsatisfactory is connected with the fact that the (1, 1) pivot 1 used first is much larger than the (2, 1) pivot 10^{-2}. It is easy to show that this is not the reason when we are working in floating-point. Suppose that we *rescale* (6.11) by multiplying the first equation by 10^{-2}, the second by 10^{2}, and setting $x_1 = 10^2 z_1$, $x_2 = 10^{-2} z_2$. The equations become

$$\begin{aligned} z_1 - 10^{-4} z_2 &= 0 \\ 10^2 z_1 + \qquad z_2 &= 10^2. \end{aligned} \tag{6.14}$$

If we pivot on the "large" coefficient 10^2, we find, working with two digits in floating-point, $z_2' = 10^2$, $z_1' = 0$, which gives $x_2' = 1$, $x_1' = 0$, that is, the same unsatisfactory solution found before.

EXERCISE 6.7. Show that any of the other three pivots in (6.14), including the "small" pivot 10^{-4}, gives satisfactory approximations for x_1, x_2 when working in floating-point with two digits.

We presented the two versions (6.11) and (6.14) above of the same problem to show that a pivot rule seeking large pivots might not be useful in all cases. Another way to view the above comparison is to say that choosing large pivots is the correct strategy only if *the equations are "properly" scaled.* By *scaling* here we mean multiplying any equation by a nonzero constant or replacing any unknown by a new one which is a nonzero multiple of the old one. The best computer programs typically first attempt to scale the equations in some helpful fashion before proceeding with Gauss elimination; the actual elimination process usually selects pivots by one of two strategies:

1. *Partial pivoting,* in which the unknowns are eliminated in their natural order $x_1, x_2, \ldots,$ and at the rth stage the pivot is taken to be that coefficient of x_r of largest absolute value in the remaining $n - r + 1$ equations.
2. *Complete pivoting,* in which, at the rth stage, we select as pivot the coefficient of largest absolute value of all the $n - r + 1$ unknowns in the remaining $n - r + 1$ equations.

Experience seems to indicate that it is sufficient to use partial pivoting in practice; the theoretical advantages of complete pivoting tend to be outweighed by the bookkeeping required for its implementation. The situation in practice as regards choice of pivots is not as bad as it might seem from this discussion. After all, millions of large linear systems have been solved successfully on digital computers. The situation seems to be that most systems that arise in practice have a built-in "natural" scaling that prevents a disastrous choice of pivots. In surveying problems we do not measure some distances in inches and others in miles. (On the other hand, when distances and angles occur together, it is advisable to scale so that coefficients are comparable in size.) The other factor that should be mentioned is that it is not necessary to find the very *best* choice of pivots–the essential thing is to avoid very *bad* choices of pivots. The problem of finding an automatic method for scaling (so that partial pivoting will avoid a bad choice of pivots) is becoming more acute as the size of problems is increasing and the intervention of human beings in complicated calculations is lessening. Although it is not sufficient to *guarantee* that good pivots will result, a commonly accepted and reasonably successful procedure is to scale both the rows and the columns of **A**, perhaps in order to make the largest element in each row and in each column of the order of unity in magnitude or in order to make the magnitudes of all elements in **A** range between 0.1 and 1.0. It seems to be very difficult however to find a good scaling strategy based only on the entries a_{ij} of **A** itself.

EXERCISE 6.8. Show that neither partial nor complete pivoting is satisfactory for

$$\begin{aligned} 2x_1 + x_2 + x_3 &= 1 \\ x_1 + \epsilon x_2 + \epsilon x_3 &= 2\epsilon \\ x_1 + \epsilon x_2 - \epsilon x_3 &= \epsilon \end{aligned}$$

when ϵ is very small and known accurately. Show that rescaling via $z_1 = x_1/\epsilon$, $z_2 = x_2$, $z_3 = x_3$ and dividing the second and third equations by ϵ so as to obtain

$$\begin{aligned} 2\epsilon z_1 + z_2 + z_3 &= 1 \\ z_1 + z_2 + z_3 &= 2 \\ z_1 + z_2 - z_3 &= 1 \end{aligned}$$

allows both methods to work satisfactorily.

Fortunately, the lesson learned from studying (6.11) is not the only motivation for using partial or complete pivoting after scaling; this process can be further justified in terms of the method of inverse error analysis illustrated earlier. That is, we can show that the computed solution $\mathbf{x}'$ exactly solves a slightly perturbed system of equations, where the size of the perturbations is related to the size of the pivots and thus to the pivotal strategy.

We note that a "small" pivot means a "large" *multiplier* m_{ir} in (6.6) which means, because of (6.7), that the elements $a_{ij}^{(r+1)}$ in $\mathbf{A}^{(r+1)}$ can be much larger than those $a_{ij}^{(r)}$ in $\mathbf{A}^{(r)}$; it turns out that our bounds for the perturbations in $\mathbf{A}$ depend explicitly on the size of these elements $a_{ij}^{(r)}$, so we define

$$\rho = \max_{i,j,r} |\, a_{ij}^{(r)} \,|, \tag{6.15}$$

noting that ρ can be computed during the elimination process. A detailed presentation leading to the following theorem may be found, for example, in Forsythe and Moler [57].

THEOREM 6.1 Suppose that Gauss elimination with partial or complete pivoting is used on a computer having t-digit floating-point arithmetic to the base β (*typically* $\beta = 2$ *or* 16, *although we have spoken of* $\beta = 10$ *earlier*) *to solve the system* $\mathbf{Ax} = \mathbf{b}$, *where the elements of the* $n \times n$ *matrix* $\mathbf{A}$ *and the* $n \times 1$ *matrix* $\mathbf{b}$ *are floating point numbers. The numerically obtained solution* $\mathbf{x}'$ *is the exact solution of* $(\mathbf{A} + \boldsymbol{\delta}\mathbf{A})\mathbf{x}' = \mathbf{b}$, *where*

$$\|\boldsymbol{\delta}\mathbf{A}\|_\infty \le 0.505(n^3 + 3n^2)\rho\|\mathbf{A}\|_\infty \beta^{1-t},$$

where ρ *is defined by* (6.15).

If partial pivoting is used and if $\mathbf{A} = [a_{ij}]$ is scaled so that $|a_{ij}| \le 1$, it is easy to see that $\rho \le 2^{n-1}$; this bound can actually be attained (Ex. 6.14). If complete pivoting is used and again $|a_{ij}| \le 1$, then it can be shown that $\rho \le 1.8n^{(\log n)/4}$ and it is widely believed that the bound $\rho \le n$ is valid for real matrices. In actual practice it usually happens that

$$\|\boldsymbol{\delta}\mathbf{A}\|_\infty \le 0.5n\|\mathbf{A}\|_\infty \beta^{1-t}.$$

Since the simple process of storing an ideal matrix $\mathbf{A}^*$ into the computer would often cause a perturbation of this size, Theorem 6.1 says that Gauss elimination provides as good a solution as we can reasonably ask.

All of the preceding discussion of practical techniques for solving systems of linear equations was intended solely to give the reader an appreciation of the practical difficulties, of the type of analysis needed to understand these difficulties, and of the actual methods used to try to avoid these difficulties; most certainly we did *not* try to arm the reader well enough to attack the problem of writing a good computer program for linear equations. Such programs are very complex; their development takes much time and expertise. Most computer manufacturers provide fairly acceptable programs for this with their machines, and often the reader's computer center can provide improved programs; the very best programs have been described in the open literature and can be copied (see for example Forsythe and Moler [57]) or can be purchased from suppliers of mathematical subroutines. The reader who understands what we have presented is equipped to understand the fundamental principles behind the best programs and should demand that such top quality programs are readily available for general use.

EXERCISE 6.9. Theorem 6.1 says that $\mathbf{Ax}' = \mathbf{b} + \mathbf{r}$ where $\mathbf{r} = -\boldsymbol{\delta}\mathbf{Ax}'$ is small, so that $\mathbf{x}'$ nearly solves $\mathbf{Ax} = \mathbf{b}$. The fact, however, that we have an $\mathbf{x}'$ for which the *residual* $\mathbf{r} = \mathbf{Ax}' - \mathbf{b}$ is small does not *necessarily* mean that $\mathbf{x}' - \mathbf{A}^{-1}\mathbf{b}$ is small. To see this, find the residual for

$$0.89x_1 + 0.53x_2 = 0.36$$
$$0.47x_1 + 0.28x_2 = 0.19$$

when $x_1' = 0.47$ and $x_2' = -0.11$, noticing that the true solution is $x_1 = 1$, $x_2 = -1$. Explain this in terms of the condition of the system.

EXERCISE 6.10. As in Ex. 6.9, suppose that we have an $\mathbf{x}'$ for which $\mathbf{r} = \mathbf{Ax}' - \mathbf{b}$ is small. Find a bound for $\|\mathbf{x}' - \mathbf{A}^{-1}\mathbf{b}\|$ in terms of $\|\mathbf{r}\|$ and the condition number of $\mathbf{A}$.

EXERCISE 6.11. If $x_0 = 0.343169 \cdot 10^0$, $y_0 = 0.341946 \cdot 10^0$, and these are rounded to four figures, giving numbers denoted by x and y, show by performing the calculations that $x - y$ is more than 5% different from $x_0 - y_0$, even though x, y are individually only about 0.01% different from x_0, y_0.

EXERCISE 6.12. Show that floating-point multiplication is not necessarily associative by computing $(0.54 \times 2.0) \times 0.56$ and $0.54 \times (2.0 \times 0.56)$ in two-figure decimal arithmetic. Make up a similar example for addition.

EXERCISE 6.13. Using two-figure arithmetic, use Gauss elimination to solve

$$0.98x_1 + 0.43x_2 = 0.91$$
$$-0.61x_1 + 0.23x_2 = 0.48$$

and compare the answer with the true solution $x_1 = 0.005946\ldots$, $x_2 = 2.102727\ldots$. Use inverse error analysis to find a perturbed system solved exactly by the numerical solution. Observe that the resulting small *absolute* (but large *relative*) change in the solution may be unacceptable because one component of the solution is very small.

EXERCISE 6.14. Show that, if the system

$$\begin{bmatrix} 1 & 0 & 0 & 0 & 1 \\ 1 & 1 & 0 & 0 & -1 \\ -1 & 1 & 1 & 0 & 1 \\ 1 & -1 & 1 & 1 & -1 \\ -1 & 1 & -1 & 1 & 1 \end{bmatrix} \begin{bmatrix} x_1 \\ x_2 \\ x_3 \\ x_4 \\ x_5 \end{bmatrix} = \begin{bmatrix} 1 \\ -1 \\ 1 \\ -1 \\ 1 \end{bmatrix}$$

is solved by partial pivoting, the last diagonal element $a_{55}^{(5)}$ is 16. In the case of a matrix of order n with the same structure, show that the last diagonal element $a_{nn}^{(n)}$ would be 2^{n-1}. Show that if $a_{nn} = a_{nn}^{(1)}$ is altered by an amount δ, then $a_{nn}^{(n)}$ changes to $2^{n-1} + \delta$. Why does this tell us that it will be difficult to solve equations of this type (for large n) on a computer, using partial pivoting? Show that the difficulty disappears if we use complete pivoting.

EXERCISE 6.15. Suppose we are solving a set of equations for which we expect (for example, on physical grounds) that the answers will be of order unity. Suppose that we use a computer program that assumes that the equations have a unique solution. Show that if the equations are actually inconsistent, we should expect the computer to print out very large numbers as the solution, whereas if the equations have an infinite number of solutions, we should expect the computer to print one of the infinite number in which the numbers are of reasonable size. (In this latter case we may be misled into thinking that the equations have a unique solution.) [See (3.47), and **Key Theorem 3.6** and its proof.]

6.4 SOLVING SLIGHTLY MODIFIED SYSTEMS

In many applications it is common to have to solve several sets of systems of equations $\mathbf{Ax} = \mathbf{b}$ where each successive system is a simple modification of a preceding one. For example, in design studies we may want to see the effect of varying some parameter in our mathematical model, and this may lead to a change in $\mathbf{b}$ or a simple change in $\mathbf{A}$. It is usually possible (and much cheaper) to take advantage of this fact in some way when solving the later system.

We first consider changes in the right hand sides $\mathbf{b}$. As we saw in Section 6.2 [see (6.3)], if we know that we want to solve with several right hand sides $\mathbf{b}_1, \ldots, \mathbf{b}_k$, then Gauss elimination can be so organized as to provide the solutions $\mathbf{x}_1, \ldots, \mathbf{x}_k$ simultaneously. Often, however, we may not know what new design to try until we see the results of the old one; in our setting, this means that $\mathbf{b}_{i+1}$ may not be known until after $\mathbf{x}_i$ is computed. Thus we cannot proceed precisely as in Section 6.2. It should be clear however that most of the work in Gauss elimination comes from reducing the matrix $\mathbf{A}$ to a triangular form, not in manipulating $\mathbf{b}$, and that this triangular reduction in no way involves the particular right-hand side. Therefore, if we can only *remember* what operations need to be performed on a new right-hand side, then back-substitution on this reduced form would provide the new solution; fortunately, precisely this information is contained in the *multipliers* m_{ir} [see (6.6)] for $r = 1, \ldots, n-1$ and $i = r+1, \ldots, n$. If we merely retain these multipliers (for example, by storing them in our computer where the eliminated coefficients a_{ir} used to be) then any new right-hand side $\mathbf{b}$ can be appropriately reduced by applying (6.8) for $r = 1, 2, \ldots, n-1$ starting with $\mathbf{b}^{(1)} = \mathbf{b}$ at a cost of n^2 multiplications and $n^2 - n$ additions; since the back-substitution requires $\frac{1}{2}(n^2 + n)$ multiplications and $\frac{1}{2}(n^2 - n)$ additions, we find that *the solution for each new right-hand side can be calculated at a cost of* $\frac{3}{2}n^2 + \frac{1}{2}n$ *multiplications and* $\frac{3}{2}(n^2 - n)$ *additions for each right-hand side.* This should be compared with the roughly $\frac{1}{3}n^3$ operations required for the first right-hand side. We observe (see Ex. 6.6) that it is less work to solve for k different right-hand sides in this fashion than to compute $\mathbf{A}^{-1}$ and multiply it times each of the k right-hand sides. For computational purposes it is often convenient to describe in a somewhat different manner the above important idea of saving the multipliers. Suppose, for simplicity, that no row interchanges occur in the reduction of the $n \times n$ matrix $\mathbf{A}$ to an upper-triangular matrix $\mathbf{U}$ by Gauss elimination; use the multipliers m_{ir} of (6.6) to define the matrix $\mathbf{L}$ via

$$\mathbf{L} = \begin{bmatrix} 1 & 0 & 0 & 0 & 0 & \cdot & \cdot & 0 & 0 \\ m_{21} & 1 & 0 & 0 & 0 & \cdot & \cdot & 0 & 0 \\ m_{31} & m_{32} & 1 & 0 & 0 & \cdot & \cdot & 0 & 0 \\ \cdot & \cdot & \cdot & & \cdot & & & & \\ \cdot & \cdot & \cdot & & & & \cdot & & \\ \cdot & \cdot & \cdot & & & & & \cdot & \\ m_{n1} & m_{n2} & \cdot & \cdot & \cdot & & & m_{n,n-1} & 1 \end{bmatrix}.$$

Then one can show {see Ex. 6.54 and Forsythe-Moler [57]} that, in fact,

$$\mathbf{A} = \mathbf{LU}.$$

The idea of using the multipliers again on any new $\mathbf{b}$ for which one wants to solve $\mathbf{Ax} = \mathbf{b}$ can now be implemented by solving the two systems

$$\mathbf{Ly} = \mathbf{b}$$

and

$$\mathbf{Ux} = \mathbf{y},$$

the first of which merely represents the appropriate processing of $\mathbf{b}$ by the multipliers; since both $\mathbf{L}$ and $\mathbf{U}$ are triangular, these solutions are cheap to obtain. If interchanges are used in the elimination process, then the result can be shown to be equivalent to factoring $\mathbf{A}$ as

$$\mathbf{A} = \mathbf{PLU}$$

where $\mathbf{L}$ and $\mathbf{U}$ are respectively lower and upper triangular matrices computed exactly as before, while $\mathbf{P}$ is a so-called *permutation* matrix obtained by interchanging (permuting) certain rows of the identity matrix $\mathbf{I}$. These representations of $\mathbf{A}$ are known as *LU-decompositions.*

As a concrete example, consider solving $\mathbf{Ax} = \mathbf{b}$ for

$$\mathbf{A} = \begin{bmatrix} 2 & 1 & -2 \\ 4 & -1 & 2 \\ 2 & -1 & 1 \end{bmatrix}, \qquad \mathbf{b} = \begin{bmatrix} 1 \\ 5 \\ 2 \end{bmatrix},$$

so that the augmented matrix is

$$\begin{bmatrix} 2 & 1 & -2 & 1 \\ 4 & -1 & 2 & 5 \\ 2 & -1 & 1 & 2 \end{bmatrix}.$$

Using the (1, 1) element as a pivot to eliminate in the first column, we compute multipliers $m_{21} = 2$ and $m_{31} = 1$ to obtain the reduced matrix

$$\begin{bmatrix} 2 & 1 & -2 & 1 \\ 0 & -3 & 6 & 3 \\ 0 & -2 & 3 & 1 \end{bmatrix}.$$

Using the (2, 2) element as a pivot to eliminate in the second column, we compute the multiplier $m_{23} = \frac{2}{3}$ to obtain the reduced matrix

$$\begin{bmatrix} 2 & 1 & -2 & 1 \\ 0 & -3 & 6 & 3 \\ 0 & 0 & -1 & -1 \end{bmatrix}.$$

In the notation introduced above, we therefore have

$$\mathbf{L} = \begin{bmatrix} 1 & 0 & 0 \\ 2 & 1 & 0 \\ 1 & \frac{2}{3} & 1 \end{bmatrix}, \qquad \mathbf{U} = \begin{bmatrix} 2 & 1 & -2 \\ 0 & -3 & 6 \\ 0 & 0 & -1 \end{bmatrix},$$

and the reader can easily check that $\mathbf{LU}$ equals our original $\mathbf{A}$. Moreover, the processed version $[1 \quad 3 \quad -1]^T$ of our original $\mathbf{b} = [1 \quad 5 \quad 2]^T$ can easily be seen to equal $\mathbf{L}^{-1}\mathbf{b}$ as asserted.

We now turn our attention to the case in which not $\mathbf{b}$ but $\mathbf{A}$ changes in some fairly simple way. If we imagine changing the role in a model played by a particular variable or equation, we might need to see what happens to the solution of a system of equations $\mathbf{Ax} = \mathbf{b}$ when one column or one row of $\mathbf{A}$ is changed. It is easy to describe this situation algebraically. If we want to replace the ith column $\mathbf{c}_i$ of $\mathbf{A}$ by the $n \times 1$ column vector $\mathbf{c}'_i$, we only need add to $\mathbf{A}$ the $n \times n$ matrix $(\mathbf{c}'_i - \mathbf{c}_i)\mathbf{e}_i^T$, where $\mathbf{e}_i$ is the ith unit column vector;

if we want to replace the ith row $\mathbf{r}_i$ of $\mathbf{A}$ by the $1 \times n$ row vector $\mathbf{r}'_i$, we only need add to $\mathbf{A}$ the $n \times n$ matrix $\mathbf{e}_i(\mathbf{r}'_i - \mathbf{r}_i)$.

EXERCISE 6.16. Let

$$\mathbf{A} = \begin{bmatrix} 1 & 2 & 1 \\ 3 & 1 & 2 \\ 0 & -2 & 4 \end{bmatrix}, \qquad \mathbf{d} = \begin{bmatrix} 2 \\ 1 \\ 1 \end{bmatrix}.$$

Then

$$\mathbf{A} + \mathbf{d}\mathbf{e}_1^T = \begin{bmatrix} 1 & 2 & -1 \\ 3 & 1 & 2 \\ 0 & -2 & 4 \end{bmatrix} + \begin{bmatrix} 2 \\ 1 \\ 1 \end{bmatrix} [1 \quad 0 \quad 0]$$

$$= \begin{bmatrix} 1 & 2 & -1 \\ 3 & 1 & 2 \\ 0 & -2 & 4 \end{bmatrix} + \begin{bmatrix} 2 & 0 & 0 \\ 1 & 0 & 0 \\ 1 & 0 & 0 \end{bmatrix} = \begin{bmatrix} 3 & 2 & -1 \\ 4 & 1 & 2 \\ 1 & -2 & 4 \end{bmatrix}$$

while

$$\mathbf{A} + \mathbf{e}_2\mathbf{d}^T = \begin{bmatrix} 1 & 2 & -1 \\ 3 & 1 & 2 \\ 0 & -2 & 4 \end{bmatrix} + \begin{bmatrix} 0 \\ 1 \\ 0 \end{bmatrix} [2 \quad 1 \quad 1] = \begin{bmatrix} 1 & 2 & -1 \\ 5 & 2 & 3 \\ 0 & -2 & 4 \end{bmatrix}.$$

More generally we can consider adding to $\mathbf{A}$ a matrix constructed from a few column vectors and a few row vectors, say of the form $\mathbf{CR}$ where $\mathbf{C}$ is $n \times p$ and $\mathbf{R}$ is $p \times n$ and we consider p to be much less than n. The following result describes the inverse of the new matrix.

THEOREM 6.2 Suppose that $\mathbf{A}$ *is* $n \times n$ *and nonsingular, that* $\mathbf{C}$ *is* $n \times p$ *and* $\mathbf{R}$ *is* $p \times n$ *with* $p \leq n$, *and that the* $p \times p$ *matrix* $\mathbf{K} \equiv \mathbf{I} + \mathbf{R}\mathbf{A}^{-1}\mathbf{C}$ *is nonsingular. Then* $\mathbf{A}' = \mathbf{A} + \mathbf{CR}$ *is nonsingular and*

$$(\mathbf{A}')^{-1} = \mathbf{A}^{-1} - \mathbf{A}^{-1}\mathbf{C}\mathbf{K}^{-1}\mathbf{R}\mathbf{A}^{-1}.$$

Proof: We simply show that

$$\mathbf{A}'[\mathbf{A}^{-1} - \mathbf{A}^{-1}\mathbf{C}\mathbf{K}^{-1}\mathbf{R}\mathbf{A}^{-1}] = \mathbf{I} = [\mathbf{A}^{-1} - \mathbf{A}^{-1}\mathbf{C}\mathbf{K}^{-1}\mathbf{R}\mathbf{A}^{-1}]\mathbf{A}',$$

proving that $\mathbf{A}'$ is nonsingular and has the stated inverse; we show the first equality, from which the second follows as in Ex. 3.53 or by direct calculation. We have

$$\begin{aligned} \mathbf{A}'[\mathbf{A}^{-1} - \mathbf{A}^{-1}\mathbf{C}\mathbf{K}^{-1}\mathbf{R}\mathbf{A}^{-1}] &= [\mathbf{A} + \mathbf{CR}][\mathbf{A}^{-1} - \mathbf{A}^{-1}\mathbf{C}\mathbf{K}^{-1}\mathbf{R}\mathbf{A}^{-1}] \\ &= \mathbf{A}\mathbf{A}^{-1} - \mathbf{A}\mathbf{A}^{-1}\mathbf{C}\mathbf{K}^{-1}\mathbf{R}\mathbf{A}^{-1} + \mathbf{C}\mathbf{R}\mathbf{A}^{-1} - \mathbf{C}\mathbf{R}\mathbf{A}^{-1}\mathbf{C}\mathbf{K}^{-1}\mathbf{R}\mathbf{A}^{-1} \\ &= \mathbf{I} - \mathbf{C}\mathbf{K}^{-1}\mathbf{R}\mathbf{A}^{-1} + \mathbf{C}\mathbf{R}\mathbf{A}^{-1} - \mathbf{C}\mathbf{R}\mathbf{A}^{-1}\mathbf{C}\mathbf{K}^{-1}\mathbf{R}\mathbf{A}^{-1} \\ &= \mathbf{I} + \mathbf{C}[-(\mathbf{I} + \mathbf{R}\mathbf{A}^{-1}\mathbf{C})\mathbf{K}^{-1} + \mathbf{I}]\mathbf{R}\mathbf{A}^{-1} \\ &= \mathbf{I} + \mathbf{C}[-\mathbf{K}\mathbf{K}^{-1} + \mathbf{I}]\mathbf{R}\mathbf{A}^{-1} = \mathbf{I}, \end{aligned}$$

as desired.

We remark that, given the nonsingularity of $\mathbf{A}$, $\mathbf{A}'$ is nonsingular if and only if $\mathbf{K}$ is nonsingular.

COROLLARY 6.1. Under the hypotheses of Theorem 6.2, let $\mathbf{x}$ *be the unique solution of* $\mathbf{Ax} = \mathbf{b}$ *and let* $\mathbf{x}'$ *be the unique solution of* $\mathbf{A}'\mathbf{x}' = \mathbf{b}$. *Then*

$$\mathbf{x}' = \mathbf{x} - \mathbf{A}^{-1}\mathbf{C}\mathbf{K}^{-1}\mathbf{R}\mathbf{x}.$$

We now give some examples of the use of this theorem.

EXERCISE 6.17. Find the inverse, if it exists, of the matrix $\mathbf{A}'$ formed by adding λ to the (i, j) element of a matrix $\mathbf{A}$.

SOLUTION: In the notation of Theorem 6.2 we take $p = 1$, $\mathbf{C} = \lambda\mathbf{e}_i$, $\mathbf{R} = \mathbf{e}_j^T$, so that $\mathbf{K} = [1 + \lambda\mathbf{e}_j^T\mathbf{A}^{-1}\mathbf{e}_i]$ is 1×1 and equals $[1 + \lambda\alpha_{ji}]$ where α_{ji} is the (j, i) element of $\mathbf{A}^{-1}$. Therefore, if $\lambda \neq -1/\alpha_{ji}$, $\mathbf{K}$ is nonsingular and hence $\mathbf{A}'$ is nonsingular and

$$(\mathbf{A}')^{-1} = \mathbf{A}^{-1} - \frac{\lambda}{1 + \lambda\alpha_{ji}}\mathbf{c}_i\mathbf{r}_j^T$$

where $\mathbf{c}_i$ is the ith column and $\mathbf{r}_j^T$ the jth row of $\mathbf{A}^{-1}$.

EXERCISE 6.18. Let $\mathbf{c}$ be an $n \times 1$ matrix and $\mathbf{r}^T$ be a $1 \times n$ matrix. Discuss the inversion of $\mathbf{I} + \mathbf{c}\mathbf{r}^T$.

SOLUTION: In the notation of Theorem 6.2 we take $\mathbf{A} = \mathbf{I}$, $p = 1$, $\mathbf{C} = \mathbf{c}$, $\mathbf{R} = \mathbf{r}^T$, so that $\mathbf{K} = [1 + \mathbf{r}^T\mathbf{c}]$ is 1×1. Therefore if $\mathbf{r}^T\mathbf{c} \neq -1$, $\mathbf{K}$ is nonsingular and

$$(\mathbf{I} + \mathbf{c}\mathbf{r}^T)^{-1} = \mathbf{I} - \frac{1}{1 + \mathbf{r}^T\mathbf{c}}\mathbf{c}\mathbf{r}^T.$$

If we had to rely on inspection of $\mathbf{A}'$ in order to express it in a form suitable for Theorem 6.2, the applications of that theorem would be very limited. However there are certain classes of problems in which one can see that the overall system can be broken down into several subsystems, the interconnections between subsystems being small in number compared with the internal connections within the subsystems. (Many of the problems in mechanical structures, electrical networks, economic systems, etc., are of this type.) In this case, we can use our knowledge of the physical structure of the system in order to write $\mathbf{A}'$ in a suitable form. A typical example might lead to an $n \times n$ matrix $\mathbf{A}'$ of the form

$$\mathbf{A}' = \begin{array}{c} \\ (r) \\ (s) \end{array}\!\!\begin{array}{c} \begin{array}{ccc} (m) & (p) & (q) \end{array} \\ \left[\begin{array}{c|c|c} \mathbf{B} & \mathbf{D} & \mathbf{0} \\ \hline \mathbf{0} & \mathbf{E} & \mathbf{F} \end{array}\right] \end{array} \tag{6.16}$$

where $\mathbf{B}$ is $r \times m$, $\mathbf{D}$ is $r \times p$, $\mathbf{E}$ is $s \times p$, and $\mathbf{F}$ is $s \times q$, with

$$m + p + q = r + s = n,$$

and r, s, m, and q are roughly $\frac{1}{2}n$ and p is small compared to n. Here the overall system almost splits up into two subsystems of large numbers (m and q) of variables with only a small number (p) of variables providing links between the subsystems. There are a variety of ways in which to split $\mathbf{A}'$; suppose that we choose one which separates the p columns of

$$\begin{bmatrix} \mathbf{D} \\ \mathbf{E} \end{bmatrix}$$

into p_1 on the left and p_2 on the right, giving

$$\mathbf{A}' = \begin{array}{c} \\ (r) \\ (s) \end{array}\begin{matrix} (m) & (p_1) & (p_2) & (q) \\ \begin{bmatrix} \mathbf{B} & \mathbf{D}_1 & \mathbf{D}_2 & \mathbf{0} \\ \hline \mathbf{0} & \mathbf{E}_1 & \mathbf{E}_2 & \mathbf{F} \end{bmatrix} \end{matrix}.$$

We suppose that $m + p_1 = r$ and $q + p_2 = s$ so that

$$\mathbf{G} = [\mathbf{B} \quad \mathbf{D}_1] \quad \text{and} \quad \mathbf{L} = [\mathbf{E}_2 \quad \mathbf{F}]$$

are $r \times r$ and $s \times s$ respectively. We can then write (6.16) as

$$\mathbf{A}' = \begin{array}{c} \\ (r) \\ (s) \end{array}\begin{matrix} (r) & (p_2) & (q) \\ \begin{bmatrix} \mathbf{G} & \mathbf{D}_2 & \mathbf{0} \\ \hline \mathbf{0} & \mathbf{E}_1 & \mathbf{L} \end{bmatrix} \\ (m) \quad (p_1) \quad\quad (s) \end{matrix} \qquad (6.17)$$

where $\mathbf{G}$ is $r \times r$ (square), $\mathbf{L}$ is $s \times s$ (square), $\mathbf{D}_2$ is $r \times p_2$, and $\mathbf{E}_1$ is $s \times p_1$, with $r + s = n$, r and s both roughly $\frac{1}{2}n$, and both p_1 and p_2 are small compared to n. Then $\mathbf{A}' = \mathbf{A} + \mathbf{M}$ with (in the same partitioning)

$$\mathbf{A} = \begin{bmatrix} \mathbf{G} & \mathbf{0} & \mathbf{0} \\ \hline \mathbf{0} & \mathbf{0} & \mathbf{L} \end{bmatrix}, \quad \mathbf{M} = \begin{bmatrix} \mathbf{0} & \mathbf{D}_2 & \mathbf{0} \\ \hline \mathbf{0} & \mathbf{E}_1 & \mathbf{0} \end{bmatrix}.$$

If $\mathbf{G}$ and $\mathbf{L}$ are nonsingular, then

$$\mathbf{A}^{-1} = \begin{bmatrix} \mathbf{G}^{-1} & \mathbf{0} & \mathbf{0} \\ \hline \mathbf{0} & \mathbf{0} & \mathbf{L}^{-1} \end{bmatrix};$$

$\mathbf{M}$ can be written as $\mathbf{CR}$ for the application of Theorem 6.2 with

$$\mathbf{C} = \begin{array}{c} \\ (r) \\ (s) \end{array}\begin{matrix} (p_1) & (p_2) \\ \begin{bmatrix} \mathbf{0} & \mathbf{D}_2 \\ \hline \mathbf{E}_1 & \mathbf{0} \end{bmatrix} \end{matrix}, \quad \mathbf{R} = (p_1 + p_2)\begin{matrix} (r - p_1) & (p_1 + p_2) & (s - p_2) \\ \begin{bmatrix} \mathbf{0} & \mathbf{I} & \mathbf{0} \end{bmatrix} \end{matrix}$$

EXERCISE 6.19. Suppose that $n = 9$ and that $\mathbf{A}'$ has the structure

$$\mathbf{A}' = \left[\begin{array}{ccc|cc|cccc} \times & \times & \times & \times & \times & 0 & 0 & 0 & 0 \\ \times & \times & \times & \times & \times & 0 & 0 & 0 & 0 \\ \times & \times & \times & \times & \times & 0 & 0 & 0 & 0 \\ \times & \times & \times & \times & \times & 0 & 0 & 0 & 0 \\ \hline 0 & 0 & 0 & \times & \times & \times & \times & \times & \times \\ 0 & 0 & 0 & \times & \times & \times & \times & \times & \times \\ 0 & 0 & 0 & \times & \times & \times & \times & \times & \times \\ 0 & 0 & 0 & \times & \times & \times & \times & \times & \times \\ 0 & 0 & 0 & \times & \times & \times & \times & \times & \times \end{array}\right]$$

where the dashed lines indicate the partition in (6.16) and $m = 3, p = 2, q = 4, r = 4$, and $s = 5$. We can partition this in the form of (6.17) by means of $p_1 = p_2 = 1$ as

$$\mathbf{A}' = \left[\begin{array}{ccccccccc} \times & \times & \times & \times & \times & 0 & 0 & 0 & 0 \\ \times & \times & \times & \times & \times & 0 & 0 & 0 & 0 \\ \times & \times & \times & \times & \times & 0 & 0 & 0 & 0 \\ \times & \times & \times & \times & \times & 0 & 0 & 0 & 0 \\ \hline 0 & 0 & 0 & \times & \times & \times & \times & \times & \times \\ 0 & 0 & 0 & \times & \times & \times & \times & \times & \times \\ 0 & 0 & 0 & \times & \times & \times & \times & \times & \times \\ 0 & 0 & 0 & \times & \times & \times & \times & \times & \times \\ 0 & 0 & 0 & \times & \times & \times & \times & \times & \times \end{array}\right]$$

The matrix $\mathbf{K}$ of Theorem 6.2 would then be 2×2.

Various special techniques essentially based on Theorem 6.2 have been developed in a wide variety of applications areas. The same basic idea leads, for example, to the method of *tearing* in network analysis {see Kron [96]}, to *capacitance matrix* methods in the direct numerical solution of certain differential equations {see Hockney [59] or Buzbee et al [54]}, to various *up-date* methods for the constrained or unconstrained minimization of non-linear functions {see Murray [100]}, and to certain implementations of the *simplex* method for linear programming {see Simonnard [108]}.

EXERCISE 6.20. Verify that $\frac{3}{2}n^2 + \frac{1}{2}n$ multiplications and $\frac{3}{2}(n^2 - n)$ additions are needed to solve $\mathbf{Ax} = \mathbf{b}$ when a new $\mathbf{b}$ is presented and all the multipliers m_{ir} of (6.6) have been saved.

EXERCISE 6.21. In the notation of Theorem 6.2, show that if both $\mathbf{A}$ and $\mathbf{A}'$ are nonsingular then $\mathbf{K}$ is nonsingular.

EXERCISE 6.22. Explain what computational advantage can be gained via Corollary 6.1.

EXERCISE 6.23. Let **A** be an $n \times n$ matrix and **r** be a $1 \times n$ matrix. Discuss the inversion of $\mathbf{A} + \mathbf{e}_i\mathbf{r}$ and of $\mathbf{A} + \mathbf{r}^T\mathbf{e}_i^T$ by Theorem 6.2.

EXERCISE 6.24. Describe the procedure that would result in Ex. 6.19 if we partitioned $\mathbf{A}'$ in the form of (6.17) with $r = 5$ and $s = 4$.

EXERCISE 6.25. Use the partitioned approach of (6.17) with $r = s = 2$ and $p_1 = p_2 = 1$ to invert

$$\left[\begin{array}{cc|c|c} 3 & 5 & 0 & 1 \\ 1 & 2 & 1 & 0 \\ \hline 0 & 1 & 2 & 3 \\ 0 & 0 & 1 & 2 \end{array}\right].$$

6.5 DETERMINANTS

Many readers may have learned some years ago how to solve systems of linear equations by determinants and may have wondered why we have not mentioned them yet. In fact, for practical and efficient computation of solutions of systems larger than about 3×3, determinants are useless, but they are useful as a conceptual or descriptive device for representing solutions and inverses. For this reason, but *not* for computational purposes, we now take up determinants briefly.

We are going to define a single number associated with an $n \times n$ matrix $\mathbf{A} = [a_{ij}]$, called the *determinant* of the matrix, and denoted by det **A** or $|\mathbf{A}|$. (The vertical lines here have nothing to do with the absolute value or modulus of a real or complex number.) The determinant of order n will be defined in terms of determinants of order $n - 1$, together with the statement that, if we are considering a 1×1 matrix consisting of a single element, $\mathbf{A} = [a_{11}]$, then

$$|\mathbf{A}| = \det \mathbf{A} = a_{11}. \tag{6.18}$$

This means that second-order determinants are defined in terms of the first-order determinant just defined. Third-order determinants are defined in terms of second-order determinants, and so on. To be precise, we introduce the following definitions.

DEFINITION 6.1. The determinant of the $(n - 1) \times (n - 1)$ matrix formed by omitting the ith row and the jth column of the $n \times n$ matrix **A** is called the *minor* of a_{ij}, and is denoted by M_{ij}. The number

$$A_{ij} = (-1)^{i+j} M_{ij}, \tag{6.19}$$

is called the *cofactor* of a_{ij}.

The signs $(-1)^{i+j}$ form a checkerboard pattern:

$$\begin{bmatrix} + & - & + & \cdots \\ - & + & - & \cdots \\ + & - & + & \cdots \\ & \cdots & & \end{bmatrix}.$$

DEFINITION 6.2. The *determinant* of an $n \times n$ matrix $\mathbf{A}$ is defined by the expression

$$|\mathbf{A}| = \det \mathbf{A} = \sum_{j=1}^{n} a_{1j} A_{1j}, \tag{6.20}$$

in conjunction with the definition of det A for $n = 1$ given in (6.18). In sentence form, the determinant of A is the sum of the products of the elements in the first row of the matrix times their respective cofactors.

As a simple example, consider

$$\begin{aligned} |\mathbf{A}| = \begin{vmatrix} a_{11} & a_{12} \\ a_{21} & a_{22} \end{vmatrix} &= a_{11} A_{11} + a_{12} A_{12} \\ &= a_{11} \det[a_{22}] - a_{12} \det[a_{21}] \\ &= a_{11} a_{22} - a_{12} a_{21}. \end{aligned}$$

Similarly,

$$\begin{aligned} \begin{vmatrix} a_{11} & a_{12} & a_{13} \\ a_{21} & a_{22} & a_{23} \\ a_{31} & a_{32} & a_{33} \end{vmatrix} &= a_{11} A_{11} + a_{12} A_{12} + a_{13} A_{13} \\ &= a_{11} \begin{vmatrix} a_{22} & a_{23} \\ a_{32} & a_{33} \end{vmatrix} - a_{12} \begin{vmatrix} a_{21} & a_{23} \\ a_{31} & a_{33} \end{vmatrix} \\ &\quad + a_{13} \begin{vmatrix} a_{21} & a_{22} \\ a_{31} & a_{32} \end{vmatrix} \\ &= a_{11} a_{22} a_{33} - a_{11} a_{32} a_{23} - a_{12} a_{21} a_{33} + a_{12} a_{31} a_{23} \\ &\quad + a_{13} a_{21} a_{32} - a_{13} a_{31} a_{22}. \end{aligned} \tag{6.21}$$

If we choose any term in this expansion, say $a_{12} a_{31} a_{23}$, we see that there is one element from each of the three rows of $\mathbf{A}$, and one from each of the three columns. No two elements in the product lie in the same row of **A**, and no two elements lie in the same column. It is clear from the definition that the same rule holds for a determinant of order n. In the general case, the determinant is the sum of all possible products of n elements chosen from the matrix, with appropriate signs, where, in each product, no two elements belong to the same row, and no two elements belong to the same column. A determinant is often defined in this way, with a suitable specification of the signs of the products, but we shall not continue this method of approach.

The first row of the matrix plays a special role in (6.20) of Definition 6.2. The first important result we require is that the value of the determinant can be found by expanding in terms of *any* row or *any* column. Thus, in addition to (6.21) we have the equivalent expansions:

$$\begin{aligned}\det \mathbf{A} &= a_{21}A_{21} + a_{22}A_{22} + a_{23}A_{23} \\ &= a_{11}A_{11} + a_{21}A_{21} + a_{31}A_{31} \\ &= a_{13}A_{13} + a_{23}A_{23} + a_{33}A_{33},\end{aligned}$$

and so on.

THEOREM 6.3. The sum of the products of the elements in any row (or column) of a square matrix with their corresponding cofactors equals the determinant of the matrix. In symbols,

$$\det \mathbf{A} = \sum_{j=1}^{n} a_{pj}A_{pj} = \sum_{i=1}^{n} a_{iq}A_{iq} \qquad (\text{all } p,q). \tag{6.22}$$

Proof: The proof is more complex than instructive and is merely outlined in Ex. 6.32.

Since we eventually want to use determinants to discuss solutions of systems of equations and since we presently solve systems by applying elementary row operations, it is not surprising that we investigate the effect of elementary row operations on the determinant of a matrix; the following theorem describes these effects, among others.

THEOREM 6.4.

(i) *The determinant of the transpose of a matrix is equal to the determinant of the matrix:*

$$|\mathbf{A}^T| = |\mathbf{A}|. \tag{6.23}$$

(ii) *If all the elements of any row or any column of a matrix are zero, then the determinant of the matrix is zero.*

(iii) *If the elements of one row or one column of a matrix are multiplied by a constant c, then the determinant is multiplied by c. If* **A** *is* $n \times n$, *then*

$$|c\mathbf{A}| = c^n|\mathbf{A}|. \tag{6.24}$$

(iv) *If* **A** *and* **B** *differ only in their kth columns, then* det **A** + det **B** = det **C** *where* **C** *is a matrix whose columns are the same as those of* **A** (*or* **B**) *except that the kth column of* **C** *is the sum of the kth columns of* **A** *and* **B**. *In symbols, partitioning in columns,*

$$\det[\mathbf{c}_1, \ldots, \mathbf{c}_k, \ldots, \mathbf{c}_n] + \det[\mathbf{c}_1, \ldots, \boldsymbol{\gamma}_k, \ldots, \mathbf{c}_n]$$
$$= \det[\mathbf{c}_1, \ldots, \mathbf{c}_k + \boldsymbol{\gamma}_k, \ldots, \mathbf{c}_n]. \quad (6.25)$$

A similar result holds for rows.

(v) *If a determinant has two equal rows* (*or columns*), *then its value is zero. If any row* (*or column*) *of a matrix is a multiple of any other row* (*or column*), *then its determinant is zero.*

(vi) *The value of a determinant is unchanged if a multiple of one row* (*or column*) *is added to another row* (*or column*). *For* $j \neq q$,

$$\begin{vmatrix} a_{11} & \ldots & a_{1j} & \ldots & a_{1q} & \ldots & a_{1n} \\ & & & \ldots & & & \\ a_{n1} & \ldots & a_{nj} & \ldots & a_{nq} & \ldots & a_{nn} \end{vmatrix} = \begin{vmatrix} a_{11} & \ldots & a_{1j} + ca_{1q} & \ldots & a_{1q} & \ldots & a_{1n} \\ & & & \ldots & & & \\ a_{n1} & \ldots & a_{nj} + ca_{nq} & \ldots & a_{nq} & \ldots & a_{nn} \end{vmatrix}.$$

(vii) *If two rows* (*or columns*) *of a matrix are interchanged, then the determinant is multiplied by* (−1).

Proof: To prove (i) we note that expansion of $\mathbf{A}^T$ by its first row is the same as expanding **A** by its first column, so that the result follows from Theorem 6.3. Part (ii) also follows from Theorem 6.3 on expanding the determinant by the row or column containing the zero elements. The first part of (iii) follows similarly on expanding by the row or column whose elements are multiplied by c. The last part follows on noting that $c\mathbf{A}$ is $[ca_{ij}]$ by definition, and $|c\mathbf{A}|$ can be found in terms of $|\mathbf{A}|$ by applying the first part of (iii) to each of the n rows of $c\mathbf{A}$. To prove (iv), expand the determinant on the right of (6.25) by the jth column, and the result is obvious. The first part of (v) requires induction on n, the order of the matrix. Certainly (v) holds for $n = 2$ since

$$\begin{vmatrix} a & b \\ a & b \end{vmatrix} = ab - ab = 0.$$

Once (v) holds for $n - 1$ then it holds for n by using Theorem 6.3 and expanding along one of the nonidentical rows, since each cofactor vanishes according to the inductive hypothesis. The last part of (v) follows on combining this result and the first part of (iii). Part (vi) follows from (iv) and (v) since, if the final determinant is expanded by (iv), one of the determinants in the sum is zero by the second part of (v). For part (vii), suppose we want to interchange the ith and jth rows $\mathbf{r}_i$ and $\mathbf{r}_j$ of **A**, and for any $1 \times n$ vectors **u** and **v** let [**u**, **v**] denote the determinant of the matrix resulting from replacing the ith row of **A** by **u** and the jth row of **A** by **v**; in this notation, we must prove $[\mathbf{r}_i, \mathbf{r}_j] = -[\mathbf{r}_j, \mathbf{r}_i]$. By using (vi) three consecutive times and then using the first part of (iii) we obtain $[\mathbf{r}_i, \mathbf{r}_j] = [\mathbf{r}_i - \mathbf{r}_j, \mathbf{r}_j] = [\mathbf{r}_i - \mathbf{r}_j, \mathbf{r}_i] = [-\mathbf{r}_j, \mathbf{r}_i] = -[\mathbf{r}_j, \mathbf{r}_i]$ as required.

There is a great deal of material in the above theorem and perhaps the best way to illustrate some of the implications is to work a numerical example. Part (vi) of the theorem is the main tool used in the practical evaluation of determinants when the elements are whole numbers or contain symbols. Consider

$$|\mathbf{A}| = \begin{vmatrix} 2 & -3 & 2 & 5 \\ 1 & -1 & 1 & 2 \\ 3 & 2 & 2 & 1 \\ 1 & 1 & -3 & -1 \end{vmatrix}.$$

We reduce the first element in the first, third, and fourth rows to zero by subtracting suitable multiples of the second row from each of these other rows. (The second row was chosen since the numbers in it are very simple. In particular, the first element of the second row is unity.) By subtracting twice the second row from the first, and so on, we obtain from (vi) of Theorem 6.4 that

$$|\mathbf{A}| = \begin{vmatrix} 0 & -1 & 0 & 1 \\ 1 & -1 & 1 & 2 \\ 0 & 5 & -1 & -5 \\ 0 & 2 & -4 & -3 \end{vmatrix}. \tag{6.26}$$

It is important to remember that the row that we are adding to the other rows must itself remain unchanged. We now expand (6.26) by the first column. This is the point of reducing all the elements in the first column to zero except one, since the expansion is now simple:

$$|\mathbf{A}| = -\begin{vmatrix} -1 & 0 & 1 \\ 5 & -1 & -5 \\ 2 & -4 & -3 \end{vmatrix}.$$

On adding the last column to the first, we see from (vi) of Theorem 6.4 that

$$|\mathbf{A}| = -\begin{vmatrix} 0 & 0 & 1 \\ 0 & -1 & -5 \\ -1 & -4 & -3 \end{vmatrix} = +\begin{vmatrix} 0 & 1 \\ -1 & -4 \end{vmatrix} = +1,$$

where we have now expanded by the last element in the first column.

In this method of evaluating determinants we use essentially two results:

1. We use part (vi) of Theorem 6.4, which says that the value of a determinant is unchanged if a multiple of one row is added to another. By repeated application of this result we reduce all the elements except one in some row or column to zero. (If the elements of the determinant are integers, there is usually considerable scope for ingenuity in doing this in such a way that the arithmetic is as simple as possible.)
2. Having carried out (1), Theorem 6.3 tells us that if all the elements in the ith row (or the jth column) of $\mathbf{A}$ are zero except a_{ij}, then

$$|\mathbf{A}| = a_{ij}A_{ij} = (-1)^{i+j}a_{ij}M_{ij},$$

where M_{ij} is the minor corresponding to a_{ij}.

The operations described in (iii), (vi), and (vii) are just the elementary row operations of Definition 3.1 which can be implemented in terms of the elementary matrices of Definition 3.3. Since each elementary matrix results from applying its corresponding row operation to the unit matrix $\mathbf{I}$ whose determinant clearly equals unity, we trivially deduce the values of the determinants of the elementary matrices from Theorem 6.4. In the notation of Definition 3.3 we have

$$\begin{aligned} \det \mathbf{E}_{pq} &= -1 \\ \det \mathbf{E}_p(c) &= c \\ \det \mathbf{E}_{pq}(c) &= 1. \end{aligned} \tag{6.27}$$

Recalling again from Theorem 3.1 that row operations are equivalent to premultiplication by elementary matrices and comparing Theorem 6.4 with (6.27), we immediately obtain

$$\det \mathbf{EA} = \det \mathbf{E} \det \mathbf{A} \tag{6.28}$$

for every elementary matrix $\mathbf{E}$. This leads directly to two important theorems.

THEOREM 6.5. An $n \times n$ matrix $\mathbf{A}$ is singular if and only if det $\mathbf{A} = 0$.

Proof: From Theorem 3.1 and the definition of row-echelon form (see Definition 3.2) we know that there is a sequence $\mathbf{E}^{(1)}, \ldots, \mathbf{E}^{(r)}$ of elementary matrices such that $\mathbf{A} = \mathbf{E}^{(1)} \ldots \mathbf{E}^{(r)}\mathbf{V}$ and $\mathbf{V}$ is the row-echelon form for $\mathbf{A}$. Then

$$\det \mathbf{A} = \det \mathbf{E}^{(1)} \cdots \det \mathbf{E}^{(r)} \det \mathbf{V}$$

so that, from (6.27), det $\mathbf{A} = 0$ if and only if det $\mathbf{V} = 0$. If $\mathbf{A}$ is nonsingular then rank $(\mathbf{A}) = n$ and $\mathbf{V} = \mathbf{I}$ so that det $\mathbf{A} \neq 0$; if $\mathbf{A}$ is singular then rank $(\mathbf{A}) < n$ and the last row of $\mathbf{V}$ is zero so that det $\mathbf{V}$ vanishes because of Theorem 6.4(ii) and hence so also does det $\mathbf{A}$ as required.

THEOREM 6.6. If $\mathbf{A}$ and $\mathbf{B}$ are $n \times n$ matrices then det $\mathbf{AB}$ = det $\mathbf{A}$ det $\mathbf{B}$.

Proof: If $\mathbf{A}$ is nonsingular then as in the preceding proof $\mathbf{A} = \mathbf{E}^{(1)} \dots \mathbf{E}^{(r)}$ for a sequence of elementary matrices and hence by repeated application of (6.28) we obtain $\det \mathbf{AB} = \det \mathbf{E}^{(1)} \dots \mathbf{E}^{(r)}\mathbf{B} = \det \mathbf{E}^{(1)} \dots \det \mathbf{E}^{(r)} \det \mathbf{B} = \det \mathbf{A} \det \mathbf{B}$ as desired. On the other hand, if $\mathbf{A}$ is singular then so is $\mathbf{A}^T$ since it has the same rank as $\mathbf{A}$ and hence there exists an $\mathbf{x} \neq \mathbf{0}$ such that $\mathbf{A}^T\mathbf{x} = \mathbf{0}$; therefore $\mathbf{B}^T\mathbf{A}^T\mathbf{x} = \mathbf{0}$ and $\mathbf{B}^T\mathbf{A}^T$ is also singular. By Theorem 6.4(i) and Theorem 6.5 we then have $\det \mathbf{A} \det \mathbf{B} = 0 \cdot \det \mathbf{B} = 0 = \det \mathbf{B}^T\mathbf{A}^T = \det \mathbf{AB}$ as desired.

EXERCISE 6.26. Suppose that $\mathbf{A}$ is an $n \times n$ matrix such that in each row and in each column there is precisely one nonzero element, which is unity. Show that $\det \mathbf{A} = \pm 1$.

EXERCISE 6.27. Show that

$$\det \begin{bmatrix} 2 & 0 & 3 \\ 10 & 1 & 17 \\ 7 & 12 & -4 \end{bmatrix} = -77.$$

Verify that the same result is obtained by expanding by *any* row or column.

EXERCISE 6.28. Verify that the determinant of the following matrix is 60.

$$\begin{bmatrix} 1 & 1 & 3 & 0 & 2 \\ 3 & 1 & 0 & 1 & 2 \\ 0 & 1 & 3 & 0 & 2 \\ 4 & -2 & 3 & 1 & 0 \\ 5 & 1 & 0 & 0 & 6 \end{bmatrix}.$$

EXERCISE 6.29. Verify that, if $\theta_1 + \theta_2 + \theta_3 = 3\pi/2$,

$$\begin{vmatrix} 1 & \sin\theta_1 & \sin\theta_2 \\ \sin\theta_1 & 1 & \sin\theta_3 \\ \sin\theta_2 & \sin\theta_3 & 1 \end{vmatrix} = 0.$$

EXERCISE 6.30. If

$$k_p = \begin{vmatrix} a_1 & 1 & 0 & \dots & 0 \\ 1 & a_2 & 1 & \dots & 0 \\ 0 & 1 & a_3 & \dots & 0 \\ & & \dots & & \\ 0 & 0 & 0 & \dots & a_p \end{vmatrix},$$

show that

$$k_p = a_p k_{p-1} - k_{p-2} \qquad (p \geq 3).$$

If $a_i = 2\cos\theta$ for all i, deduce that $k_n = \operatorname{cosec}\theta \sin(n+1)\theta$.

EXERCISE 6.31. Evaluate the *Vandermonde determinant:*

$$\begin{vmatrix} 1 & x_1 & \cdots & x_1^{n-1} \\ 1 & x_2 & \cdots & x_2^{n-1} \\ & & \cdots & \\ 1 & x_n & \cdots & x_n^{n-1} \end{vmatrix} = \prod_{i<j} (x_j - x_i).$$

EXERCISE 6.32. Prove Theorem 6.3 as follows. First prove it for $n = 2$ and then use induction on n. To prove it for n, take three steps:

(a) expansion by first row = expansion by first column;
(b) expansion by first row = expansion by any row; and
(c) expansion by first column = expansion by any column.

For (a), show that the term multiplying $a_{i1}a_{1j}$ is the same in either expansion. For (b), use the pth row and show that the term multiplying $a_{1j}a_{pk}$ is the same in either expansion. For (c), apply (b) to $\mathbf{A}^T$.

EXERCISE 6.33. Prove that (6.27) holds and that $\det \mathbf{I} = 1$.

EXERCISE 6.34. Prove that (6.28) holds.

EXERCISE 6.35. If $\mathbf{A}$ is nonsingular, use Theorem 6.6 to prove that $\det \mathbf{A}^{-1} = (\det \mathbf{A})^{-1}$.

6.6 DETERMINANTAL REPRESENTATION OF SOLUTIONS AND INVERSES

We next explain how the solution of a set of n simultaneous linear equations in n unknowns can be expressed in terms of determinants. The following theorem will play a central role:

THEOREM 6.7. Let $\mathbf{A}$ *be an* $n \times n$ *matrix. Then*

$$\sum_{i=1}^{n} a_{ij}A_{ip} = \begin{cases} \det \mathbf{A} & (j = p), \\ 0 & (j \neq p). \end{cases} \tag{6.29}$$

Proof: We have already met the part $j = p$ of (6.29) in Theorem 6.3, which states that the sum of the products of the elements in the jth column of a matrix times their cofactors equals the determinant of the matrix. The remaining part $j \neq p$ of (6.29) states that the sum of the products of the elements in the jth column of $\mathbf{A}$ times the cofactors of the corresponding elements in *another* column of $\mathbf{A}$ is zero. This result is almost obvious since it is the expansion by the pth column of a matrix $\mathbf{B}$ whose pth column is the same as its jth:

$$\mathbf{B} = \begin{bmatrix} a_{11} & \dots & a_{1j} & \dots & a_{1j} & \dots & a_{1n} \\ a_{21} & \dots & a_{2j} & \dots & a_{2j} & \dots & a_{2n} \\ & & & \dots & & & \\ a_{n1} & \dots & a_{nj} & \dots & a_{nj} & \dots & a_{nn} \end{bmatrix}.$$

Hence, the value of the sum in (6.29) when $j \neq p$ is equal to the determinant of a matrix with two identical columns, and this is zero by Theorem 6.4 (v).

EXERCISE 6.36. As a numerical illustration, consider

$$|\mathbf{A}| = \begin{vmatrix} 2 & -3 & 2 & 5 \\ 1 & -1 & 1 & 2 \\ 3 & 2 & 2 & 1 \\ 1 & 1 & -3 & -1 \end{vmatrix}.$$

By straightforward evaluation of 3×3 determinants we find, for example,

$$A_{13} = -5, \qquad A_{23} = 13, \qquad A_{33} = -1, \qquad A_{43} = 0.$$

We verify (6.29):

$$\begin{aligned} 2(-5) + 1(13) + 3(-1) + 1(0) &= 0 \\ -3(-5) - 1(13) + 2(-1) + 1(0) &= 0 \\ 2(-5) + 1(13) + 2(--1) - 3(0) &= 1 \\ 5(-5) + 2(13) + 1(-1) - 1(0) &= 0. \end{aligned}$$

We now use Theorem 6.7 to obtain *Cramer's rule* for representing the solution of a system of n equations in n unknowns. For simplicity, we first examine the special case where $n = 3$:

$$\begin{aligned} a_{11}x_1 + a_{12}x_2 + a_{13}x_3 &= b_1 \\ a_{21}x_1 + a_{22}x_2 + a_{23}x_3 &= b_2 \\ a_{31}x_1 + a_{32}x_2 + a_{33}x_3 &= b_3. \end{aligned}$$

We multiply the first equation by A_{11}, the second by A_{21}, the third by A_{31} and add. The coefficients of x_2 and x_3 vanish because of Theorem 6.7 and we find

$$(a_{11}A_{11} + a_{21}A_{21} + a_{31}A_{31})x_1 = (b_1A_{11} + b_aA_{21} + b_3A_{31})$$

or

$$\begin{vmatrix} a_{11} & a_{12} & a_{13} \\ a_{21} & a_{22} & a_{23} \\ a_{31} & a_{32} & a_{33} \end{vmatrix} x_1 = \begin{vmatrix} b_1 & a_{12} & a_{13} \\ b_2 & a_{22} & a_{23} \\ b_3 & a_{32} & a_{33} \end{vmatrix}.$$

This introduces the following theorem:

THEOREM 6.8 (Cramer's Rule). If det $\mathbf{A} \neq 0$, *then the solution* $\mathbf{x} = [x_i]$ *of the system of linear equations* $\mathbf{Ax} = \mathbf{b}$ *is given by*

$$x_i = \frac{\Delta_i}{\Delta} \qquad (i = 1, \ldots, n) \tag{6.30}$$

where $\Delta = \det \mathbf{A}$ *and* $\Delta_i = \det \mathbf{A}^{(i)}$, *where* $\mathbf{A}^{(i)}$ *is the matrix obtained by replacing the ith column of* $\mathbf{A}$ *by* $\mathbf{b}$.

Proof: The equations $\mathbf{Ax} = \mathbf{b}$ are

$$\sum_{j=1}^{n} a_{ij}x_j = b_i \qquad (i = 1, 2, \ldots, n).$$

We multiply the ith equation by A_{ip}, and sum over i:

$$\sum_{i=1}^{n} A_{ip} \sum_{j=1}^{n} a_{ij}x_j = \sum_{i=1}^{n} A_{ip}b_i.$$

On interchanging orders of summation

$$\sum_{j=1}^{n} \left\{\sum_{i=1}^{n} a_{ij}A_{ip}\right\} x_j = \sum_{i=1}^{n} A_{ip}b_i.$$

We have just shown in Theorem 6.7 that the inner sum on the left is zero unless $j = p$, in which case it has the value det $\mathbf{A}$. Hence,

$$(\det \mathbf{A})x_p = \sum_{i=1}^{n} A_{ip}b_i.$$

The expression on the right is the expansion, by the pth column, of the determinant of the matrix given by replacing the pth column of $\mathbf{A}$ by $\mathbf{b}$. This matrix is $\mathbf{A}^{(p)}$ in the notation used in the theorem, so that

$$x_p = \frac{\det \mathbf{A}^{(p)}}{\det \mathbf{A}},$$

as desired.

We next express the inverse matrix in terms of the cofactors of the elements of the original matrix. Suppose that $\mathbf{A}^{-1}$ is defined by the equation $\mathbf{AA}^{-1} = \mathbf{I}$, and that the columns of $\mathbf{A}^{-1}$ are denoted by $\boldsymbol{\alpha}_1, \boldsymbol{\alpha}_2, \ldots, \boldsymbol{\alpha}_n$, the ith element of $\boldsymbol{\alpha}_j$ being α_{ij}:

$$\mathbf{A}\boldsymbol{\alpha}_j = \mathbf{e}_j,$$

where $\mathbf{e}_j$ is the unit column matrix whose jth element is unity, and all other elements are zero. Hence, from Cramer's rule,

$$\alpha_{ij} = \frac{\Delta_{ij}}{\Delta},$$

where $\Delta = |\mathbf{A}|$ and Δ_{ij} is the determinant obtained by replacing the ith *column* of $\mathbf{A}$ by $\mathbf{e}_j$, that is, Δ_{ij} is the determinant of a matrix with unity in the (j, i) position, and zeros in the other positions in the ith column. Hence, Δ_{ij} is simply A_{ij}, the cofactor of the element a_{ij} in $\mathbf{A}$, so that

$$\alpha_{ij} = \frac{A_{ji}}{\Delta}.$$

This result states that the inverse matrix is $(1/\Delta)$ times the transpose of the matrix whose elements are the cofactors of the elements of the original matrix. We restate this representation of $\mathbf{A}^{-1}$ formally.

THEOREM 6.9 *If A is nonsingular, then $\mathbf{A}^{-1}$ is given by*

$$\mathbf{A}^{-1} = \frac{1}{\det \mathbf{A}} \operatorname{adj} \mathbf{A}$$

where the adjoint of $\mathbf{A}$, *denoted* adj $\mathbf{A}$, *is*

$$\operatorname{adj} \mathbf{A} = [A_{ij}]^T$$

and A_{ij} is the cofactor of a_{ij} in $\mathbf{A}$.

We cannot emphasize enough that Theorem 6.8 and Theorem 6.9 give representations which are *useful theoretically* but are *not useful numerically on large systems.* Since an $n \times n$ determinant is defined in terms of n different $(n - 1) \times (n - 1)$ determinants, it is easy to see that the effort required to evaluate a general $n \times n$ determinant grows like $n! = n(n - 1)(n - 2) \ldots 3 \cdot 2 \cdot 1$; therefore the determinantal solution of equations or inversion of matrices requires effort like $n!$ compared with n^3 for Gauss elimination. It is painfully clear from Table 6.1 why Gauss elimination is preferable as n increases.

Table 6.1

$n =$	1	2	3	4	5	6	7	10
$n^3 =$	1	8	27	64	125	216	343	1000
$n! =$	1	2	6	24	120	720	5040	3,628,800

EXERCISE 6.37. Use Theorem 6.8 to solve (3.1).

EXERCISE 6.38. Use Theorem 6.9 to invert the matrix formed from the first four columns of (3.17) [see (3.22)].

EXERCISE 6.39. Exactly how many multiplications and additions are required to use the definition to evaluate the general determinant that is: (a) 2×2; (b) 3×3; (c) 4×4; (d) $n \times n$.

EXERCISE 6.40. Use Theorem 6.8 to study the condition of the solution $\mathbf{x}$ to $\mathbf{Ax} = \mathbf{b}$ when the ith component b_i of $\mathbf{b}$ is subject to perturbations so that $\mathbf{b}$ is perturbed to $\mathbf{b} + \epsilon\mathbf{e}_i$ for small ϵ.

EXERCISE 6.41. $\epsilon\mathbf{I}$ for small ϵ is an example of an extremely well-conditioned matrix whose determinant is small (so one might think it ill-conditioned because its determinant is nearly zero). Show by example a matrix $\mathbf{A}$ with $\det \mathbf{A} = 1$ and yet $\mathbf{A}$ is ill-conditioned. Conclude that the determinant's size is not necessarily a reliable indication of condition.

MISCELLANEOUS EXERCISES 6

EXERCISE 6.42. Show that the straight line through two points (x_1, y_1) and (x_2, y_2) is given by the equation

$$\begin{vmatrix} x & y & 1 \\ x_1 & y_1 & 1 \\ x_2 & y_2 & 1 \end{vmatrix} = 0.$$

Deduce the familiar formula

$$\frac{x - x_1}{x_1 - x_2} = \frac{y - y_1}{y_1 - y_2}$$

for such a line.

EXERCISE 6.43. Show that the area of the triangle whose vertices are the points (x_1, y_1), (x_2, y_2), (x_3, y_3) is given by the formula

$$\mathbf{A} = \frac{1}{2}\begin{vmatrix} x_1 & y_1 & 1 \\ x_2 & y_2 & 1 \\ x_3 & y_3 & 1 \end{vmatrix},$$

if the vertices are numbered 1, 2, 3 in the counterclockwise direction. How must this expression be altered if the vertices are numbered clockwise?

EXERCISE 6.44. Show that the equation of the circle through three given points (x_i, y_i), $i = 1, 2, 3$, is

$$\begin{vmatrix} x^2+y^2 & x & y & 1 \\ x_1^2+y_1^2 & x_1 & y_1 & 1 \\ x_2^2+y_2^2 & x_2 & y_2 & 1 \\ x_3^2+y_3^2 & x_3 & y_3 & 1 \end{vmatrix} = 0.$$

EXERCISE 6.45. If $\mathbf{A}$ is skewsymmetric (i.e., $\mathbf{A}^T = -\mathbf{A}$), prove that

$$\det \mathbf{A} = (-1)^n \det \mathbf{A}.$$

Deduce that the determinant of a skewsymmetric matrix of odd order is zero.

EXERCISE 6.46. If $\mathbf{A}$ is hermitian (i.e., $\mathbf{A} = \bar{\mathbf{A}}^T = \mathbf{A}^H$), prove that $\det \mathbf{A}$ is real. If $\mathbf{A}$ is a general square matrix, prove that $\det \mathbf{A}^H\mathbf{A}$ is real.

EXERCISE 6.47. If $\mathbf{A}$ is singular, prove that

$$\mathbf{A} \operatorname{adj} \mathbf{A} = \mathbf{0}.$$

EXERCISE 6.48. Prove that

$$\det (\operatorname{adj} \mathbf{A}) = (\det \mathbf{A})^{n-1}.$$

EXERCISE 6.49. If $\mathbf{A}, \mathbf{D}$ are square matrices of orders m, n, respectively, show that

$$\det \begin{bmatrix} \mathbf{A} & \mathbf{0} \\ \mathbf{C} & \mathbf{D} \end{bmatrix} = \det \mathbf{A} \det \mathbf{D}.$$

EXERCISE 6.50. If $\mathbf{A}, \mathbf{D}$ are nonsingular matrices of orders m, n, and $\mathbf{B}, \mathbf{C}$ are $m \times n$, $n \times m$, respectively, show that

$$\det \mathbf{A} \det (\mathbf{D} + \mathbf{C}\mathbf{A}^{-1}\mathbf{B}) = \det \mathbf{D} \det (\mathbf{A} + \mathbf{B}\mathbf{D}^{-1}\mathbf{C}) = \det \begin{bmatrix} \mathbf{A} & \mathbf{B} \\ -\mathbf{C} & \mathbf{D} \end{bmatrix}.$$

EXERCISE 6.51. Show that if $\mathbf{u}, \mathbf{v}$ are $m \times 1$ column vectors,

$$\det (\mathbf{I} + \mathbf{u}\mathbf{v}^T) = 1 + \mathbf{v}^T\mathbf{u}.$$

EXERCISE 6.52. If $\mathbf{D}$ is a diagonal matrix of order m whose ith diagonal element is d_i, and $\mathbf{u}$ is an $m \times 1$ column matrix, show that

$$\det (\mathbf{D} + \mathbf{u}\mathbf{u}^T) = \det \mathbf{D}\left\{1 + \sum_{i=1}^{m} \frac{u_i^2}{d_i}\right\}.$$

Deduce that

$$\begin{vmatrix} d_1+1 & 1 & \cdots & 1 \\ 1 & d_2+1 & \cdots & 1 \\ & \cdots & & \\ 1 & 1 & \cdots & d_m+1 \end{vmatrix} = d_1 d_2 \ldots d_m \left\{1 + \sum_{i=1}^{m} \frac{1}{d_i}\right\}.$$

EXERCISE 6.53. Show that if $\mathbf{A} = [a_{ij}]$ with $a_{ii} = a$, $i = 1, \ldots, n$, and $a_{ij} = b$ $(i \neq j)$, then

$$\det \mathbf{A} = (a - b)^{n-1}[a + (n - 1)b].$$

EXERCISE 6.54. Show that Gauss elimination for $\mathbf{Ax} = \mathbf{b}$ without interchanges can be described by the LU-decomposition $\mathbf{A} = \mathbf{LU}$ by arguing as follows. The first step in the elimination is to subtract m_{21} times the first row of $[\mathbf{A}, \mathbf{b}]$ from the second row; use Theorem 3.1 to show that this is equivalent to multiplying $[\mathbf{A}, \mathbf{b}]$ on the left by the elementary matrix $\mathbf{E}_{21}(-m_{12})$. More generally, show that the final reduced form $[\mathbf{U}, \mathbf{b}']$ can be obtained as

$$[\mathbf{U}, \mathbf{b}'] = \mathbf{E}_{n,n-1}(-m_{n,n-1}) \ldots \mathbf{E}_{n,2}(-m_{n2}) \ldots \mathbf{E}_{32}(-m_{32})\mathbf{E}_{n1}(-m_{n1}) \ldots \mathbf{E}_{21}(-m_{21}).$$

Deduce that

$$\mathbf{L}[\mathbf{U}, \mathbf{b}'] = [\mathbf{A}, \mathbf{b}]$$

where $\mathbf{L}$ is the product of the inverses of the elementary matrices above; show finally from Theorem 3.1 and Theorem 3.2 that $\mathbf{L}$ has the form asserted in Section 6.4.

EXERCISE 6.55. Find the LU-decomposition of the matrix

$$\mathbf{A} = \begin{bmatrix} 1 & 2 & -6 \\ 2 & 4 & 3 \\ -3 & 4 & 1 \end{bmatrix}$$

by Gauss elimination and verify that in fact $\mathbf{A} = \mathbf{LU}$.

EXERCISE 6.56. Use Gauss elimination on the matrix

$$\mathbf{A} = \begin{bmatrix} 2 & -1 & 3 & 4 \\ 2 & -1 & 4 & 2 \\ -4 & 2 & -7 & -7 \\ -2 & 1 & -1 & -1 \end{bmatrix}$$

to find its LU-decomposition. After reducing the first column you will find that the second column need not be further reduced, so that you may take $m_{32} = m_{42} = 0$. Verify that $\mathbf{A} = \mathbf{LU}$.

EXERCISE 6.57. Find the *LU*-decomposition of the matrix

$$\mathbf{A} = \begin{bmatrix} 2 & 2 \\ 4 & -1 \end{bmatrix}$$

directly by writing $\mathbf{A} = \mathbf{LU}$ with

$$\mathbf{L} = \begin{bmatrix} 1 & 0 \\ l_{21} & 1 \end{bmatrix}, \qquad \mathbf{U} = \begin{bmatrix} u_{11} & u_{12} \\ 0 & u_{22} \end{bmatrix}$$

and solving the resulting equations for $u_{11}, u_{12}, l_{21}, u_{22}$ in precisely the order just listed. The generalization of this procedure to arbitrary $n \times n$ matrices is very useful; it is usually known as the *Crout* method.

EXERCISE 6.58. To show that the *LU*-decomposition is not unique if **A** is singular, show that

$$\begin{bmatrix} 0 & 1 \\ 0 & 1 \end{bmatrix} = \begin{bmatrix} 1 & 0 \\ l & 1 \end{bmatrix} \begin{bmatrix} 0 & 1 \\ 0 & 1 - l \end{bmatrix}$$

for all numbers l. (See Ex. 6.60).

EXERCISE 6.59. If $\mathbf{A} = \mathbf{LU}$ is the *LU*-decomposition of **A**, show that det $\mathbf{A} = \det \mathbf{U} = u_{11}u_{22} \ldots u_{nn}$, so that **A** is nonsingular if and only if **U** is nonsingular.

EXERCISE 6.60. Let **A** be nonsingular and let $\mathbf{A} = \mathbf{L}_1\mathbf{U}_1 = \mathbf{L}_2\mathbf{U}_2$ be two *LU*-decompositions of **A**. By using Ex. 6.59 conclude that $\mathbf{L}_2^{-1}\mathbf{L}_1 = \mathbf{U}_2\mathbf{U}_1^{-1}$; from this deduce that $\mathbf{L}_2^{-1}\mathbf{L}_1$ is both lower and upper triangular, hence diagonal. Finally show that $\mathbf{L}_2^{-1}\mathbf{L}_1$ (and hence $\mathbf{U}_2\mathbf{U}_1^{-1}$) equals **I**, so that the *LU*-decomposition of a nonsingular matrix is unique (see Ex. 6.58). Note that it is an essential part of this result that **L** be defined to have unit diagonal elements.

CHAPTER SEVEN

LINEAR PROGRAMMING

This chapter applies matrix methods to the solution of problems in linear programming, perhaps the largest area outside the physical sciences in which matrices arise. The most important theoretical facts concerning linear programs and their dual programs are given in **Key Theorems 7.2** and **7.5,** while **Key Theorems 7.4** and **7.6** are fundamental to the understanding of the more computational aspects of such programs.

7.1 THE SIMPLEX METHOD: AN EXAMPLE

In a wide variety of economic, political, social, and scientific operations situations often arise where we wish to maximize or minimize some quantity which is a measure of the efficiency of an activity. This quantity may be, for example, the total output over a given period of time, or the cost of the operation. Optimization problems of this type are known as (*mathematical*) *programming problems*. We will concern ourselves here with a special but important class of programming problems involving only *linear* equations and inequalities. We will confine our attention to a method for solving these linear programming problems known as the *simplex method*. In Section 2.5 we considered, in some detail, the problem of allocating a plant's resources between the manufacture of two different products so as to maximize profits; it will be helpful to read Section 2.5 again at this point. Here we merely restate the linear programming problem [see (2.38), (2.39)] that arose from our model: maximize

$$M = 40x_1 + 60x_2 \tag{7.1}$$

where x_1 and x_2 must satisfy the constraints

$$\begin{aligned} 2x_1 + x_2 &\leq 70 \\ x_1 + x_2 &\leq 40 \\ x_1 + 3x_2 &\leq 90 \\ x_1 &\geq 0 \\ x_2 &\geq 0. \end{aligned} \tag{7.2}$$

For this simple case, we were able to use graphical arguments to obtain the solution $x_1 = 15$, $x_2 = 25$, resulting in $M = 2100$; we now want to present a more systematic procedure which extends easily to much larger linear programs. This procedure is called the *simplex method.*

We introduce the simplex method by solving the specific problem formulated in (7.1)–(7.2). The first step is to introduce *slack* variables x_3, x_4, x_5 in order to convert the first three inequalities in (7.2) into equalities:

$$\begin{aligned} 2x_1 + x_2 + x_3 &= 70 \\ x_1 + x_2 + x_4 &= 40 \\ x_1 + 3x_2 + x_5 &= 90. \end{aligned} \tag{7.3}$$

Because of the signs of the inequalities, the slack variables must be nonnegative, so that

$$x_j \geq 0 \qquad (j = 1, 2, \ldots, 5). \tag{7.4}$$

The problem is to find x_j, $j = 1, \ldots, 5$, that satisfy (7.3) and (7.4) and also maximize M in (7.1).

Equations (7.3) constitute a system of three equations in five unknowns with rank 3, so that if the values of two of the variables are assigned arbitrarily, then, in general, we can solve the system for the remaining variables. An easy way of assigning values to two of the variables is simply to set these equal to zero. The resulting solution is called a *basic solution*, and the variables *other* than those that are set equal to zero are known as *basic variables.* Thus if we set $x_1 = x_2 = 0$ in (7.3) we obtain the basic solution

$$x_3 = 70, \qquad x_4 = 40, \qquad x_5 = 90, \tag{7.5}$$

in terms of the basic variables x_3, x_4, x_5. It is convenient to introduce the following terminology:

Any solution of (7.3) that also satisfies (7.4) is called a *feasible solution.*

A feasible solution that also maximizes (7.1) is called an *optimal feasible solution.*

The fundamental existence theorem (which we discuss later) states that *whenever there exists an optimal feasible solution, there exists one which is also basic.* This provides the motivation for the simplex method, which we now describe.

We start with the basic solution (7.5), which is also clearly a feasible solution. The nonbasic variables x_1, x_2 are both zero. The value of M corresponding to this basic solution is $M = 0$. To increase M we can make either x_1 or x_2 assume a positive value. We choose x_2 for this purpose since $M = 40x_1 + 60x_2$ so that increasing x_2 by one unit increases M by 60, but increasing x_1 by one unit increases M by only 40 units. We therefore increase x_2, *keeping* x_1 *equal to* 0. This change in the value of x_2 increases M, but it forces us to reduce x_3, x_4, x_5 in order to preserve the equalities in (7.3). Since we are keeping $x_1 = 0$, these equations give

$$\begin{aligned} x_3 &= 70 - x_2 \\ x_4 &= 40 - x_2 \\ x_5 &= 90 - 3x_2. \end{aligned} \tag{7.6}$$

Since x_3, x_4, x_5 must be nonnegative, we cannot increase x_2 indefinitely. In fact, the first equation means that x_2 cannot be greater than 70, the second implies $x_2 \leq 40$, and the third $x_2 \leq 30$. We must clearly take the *smallest* of these numbers, namely 30, as the maximum permissible value of x_2. We now have a new basic feasible solution, using (7.6):

$$x_1 = 0, \qquad x_2 = 30, \qquad x_3 = 40, \qquad x_4 = 10, \qquad x_5 = 0, \tag{7.7}$$

and the new basic (nonzero) variables are x_2, x_3, x_4. The value of M is now

$$M = 40x_1 + 60x_2 = 1800,$$

which is greater than the previous value of zero.

Notice that we were able to write down (7.6) easily because only one of the basic variables x_3, x_4, x_5 occurred in each of Equations (7.3). We therefore make sure that each of our new basic variables x_2, x_3, x_4 occurs in only one equation. By inspection of (7.3) it is apparent that this can be done by using the third equation in (7.3) to eliminate x_2 from the first and second. This gives

$$\begin{aligned} \tfrac{5}{3}x_1 \quad + x_3 \quad - \tfrac{1}{3}x_5 &= 40 \\ \tfrac{2}{3}x_1 \quad + x_4 - \tfrac{1}{3}x_5 &= 10 \\ \tfrac{1}{3}x_1 + x_2 \quad + \tfrac{1}{3}x_5 &= 30. \end{aligned} \tag{7.8}$$

We also express M in terms of the *nonbasic* variables x_1, x_5. Eliminating x_2 between (7.1) and the third equation in (7.3), note that

$$M = 1800 + 20x_1 - 20x_5. \tag{7.9}$$

We are still working with the basic feasible solution (7.7) with $x_1 = x_5 = 0$ so that M is still 1800 as it should be, but the advantage of the new form (7.9) over the old form (7.1) is that it shows us what happens if we increase x_1 or x_5. Looking at (7.9) we see that (following a procedure similar to that used previously when we varied x_2) we must now vary x_1, *keeping* x_5 *equal to zero.* Equations (7.8) give [compare (7.6)]

$$\begin{aligned} x_3 &= 40 - \tfrac{5}{3}x_1 \\ x_4 &= 10 - \tfrac{2}{3}x_1 \\ x_5 &= 30 - \tfrac{1}{3}x_1. \end{aligned} \tag{7.10}$$

Since x_3, x_4, x_2 cannot be negative these imply that x_1 cannot be greater than 24, 15, 90, respectively, and we must choose 15 as the maximum permissible value of x_1. This leads to the new basic feasible solution:

$$x_1 = 15, \qquad x_2 = 25, \qquad x_3 = 15, \qquad x_4 = 0, \qquad x_5 = 0, \tag{7.11}$$

The new basic variables are x_1, x_2, x_3 and, using the previous line of reasoning, we arrange for each of these to appear in only one equation. This is done by eliminating x_1 from the first and third equations in (7.8) by means of the second equation, giving

$$\begin{aligned} x_3 - 2.5x_4 + 0.5x_5 &= 15 \\ x_1 \quad + 1.5x_4 - 0.5x_5 &= 15 \\ x_2 \quad - 0.5x_4 + 0.5x_5 &= 25. \end{aligned} \tag{7.12}$$

Eliminating the new basic variable x_1 between (7.9) and the second equation in (7.12) so as to express M in terms of the nonbasic variables, we see that

$$M = 2100 - 30x_4 - 10x_5. \tag{7.13}$$

The only way we can alter x_4 and x_5 is to make them positive, but this would decrease M. Hence no further improvement is possible. We have obtained the required optimal feasible solution (7.11). From (7.13), the corresponding value of M is 2100. This is of course the same solution derived by graphical methods in Section 2.5.

EXERCISE 7.1. Use the simplex method to maximize $M = x_1 + 2x_2$ subject to the constraints

$$\begin{aligned} x_1 \qquad &\leq 25 \\ x_1 + x_2 &\leq 30 \\ -x_1 + x_2 &\leq 10 \\ x_1 \qquad &\geq 0 \\ x_2 &\geq 0. \end{aligned}$$

It is not too difficult to develop a better understanding of what is actually happening in the simplex method. First we note that the optimal solution x_1^*, x_2^* should not be strictly inside the set of points satisfying the constraints (7.2), that is, at least one inequality in (7.2) should be satisfied as an equality at x_1^*, x_2^*; for if not, we could add some small positive multiple of [40, 60] to $[x_1^*, x_2^*]$ obtaining $x_1^* + \epsilon \cdot 40$ and $x_2^* + \epsilon \cdot 60$ so that the new point satisfies (7.2) but $40(x_1^* + \epsilon \cdot 40) + 60(x_2^* + \epsilon \cdot 60)$ would be greater than the maximum $40x_1^* + 60x_2^*$, a contradiction. Therefore x_1^*, x_2^* must be on the boundary of the region described by (7.2) and sketched in Figure 2.6 and again below in Figure 7.1. Similarly we can argue with respect to the edges of this set and finally conclude that a solution must be at a vertex, of which there are only finitely many (namely five). In Figure 7.1 we label these vertices V_1, V_2, V_3, V_4, V_5; the values of M at these points are 0 at V_1, 1800 at V_2, 2100 at V_3,

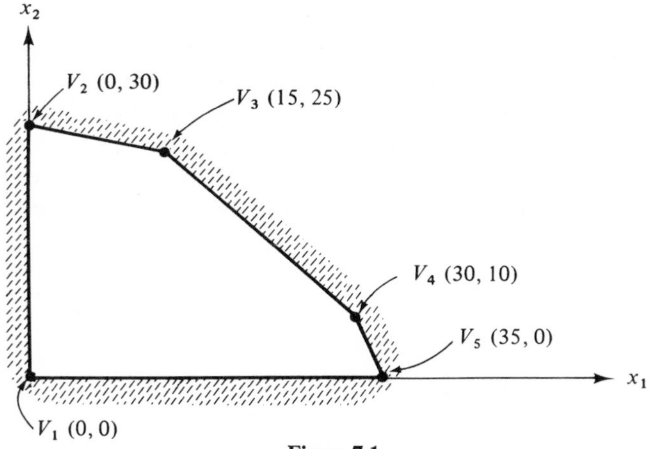

Figure 7.1

1800 at V_4, and 1400 at V_5. Instead of randomly examining the values of M at each of the vertices, the simplex method moves from vertex to vertex along a connecting edge so as to increase the value of M and to increase it as much as possible; thus we chose to move from V_1 to V_2 rather than to V_5. This picture should at least help you visualize *what* is happening, if not precisely *how*.

EXERCISE 7.2. Describe the results of Ex. 7.1 geometrically as above.

It is important to interpret the procedure for solving (7.1), (7.2) in terms of row operations on matrices. We begin by writing (7.3), (7.1) in the form of an array:

$$\begin{array}{c} M \\ 0 \\ 0 \\ 0 \\ \\ 1 \end{array} \quad \begin{array}{cccccc} x_1 & x_2 & x_3 & x_4 & x_5 & b \end{array} \\ \left[\begin{array}{cccccc} 2 & 1 & 1 & 0 & 0 & 70 \\ 1 & 1 & 0 & 1 & 0 & 40 \\ 1 & 3 & 0 & 0 & 1 & 90 \\ \cdots & \cdots & \cdots & \cdots & \cdots & \cdots \\ -40 & -60 & 0 & 0 & 0 & 0 \end{array}\right], \tag{7.14}$$

where the last line must be interpreted as

$$M - 40x_1 - 60x_2 = 0,$$

and we will not carry along the column for M, since it is not changed in subsequent steps.

We started our previous analysis by deciding to change x_2 because a unit change in x_2 would produce a greater change in M than a unit change in x_1 [see (7.1)]. In terms of the matrix (7.14) we choose the column with the *most negative* number in the last row. The step following (7.6) consists of dividng the first three elements in the last column of (7.14) by the corresponding elements in the second column, and choosing the row corresponding to the *smallest* of these three positive numbers. This is the third row. The next step in the previous analysis was to eliminate x_2 from the first and second equations in (7.3) and from (7.1). In matrix terms, we pivot on the (3, 2) element in (7.14) and reduce the first, second, and fourth elements in the second column to zero, as in the Gauss-Jordan procedure. This gives [compare (7.8), (7.9)]

$$\left[\begin{array}{cccccc} \frac{5}{3} & 0 & 1 & 0 & -\frac{1}{3} & 40 \\ \frac{2}{3} & 0 & 0 & 1 & -\frac{1}{3} & 10 \\ \frac{1}{3} & 1 & 0 & 0 & \frac{1}{3} & 30 \\ \cdots & \cdots & \cdots & \cdots & \cdots & \cdots \\ -20 & 0 & 0 & 0 & 20 & 1800 \end{array}\right]. \tag{7.15}$$

The first three rows, written out in longhand as separate equations, are (7.8). The last row of (7.15) must be interpreted as

$$M - 20x_1 + 20x_5 = 1800,$$

which is (7.9).

In exactly the same way we now find the column that contains the most negative number in the last row of (7.15), in this case the first. We next divide each of the first three elements in the last column of (7.15) by the corresponding element in the first column, and choose the smallest positive number among the results. This is the second. [Compare (7.10) and the statement following.] This leads us to pivot on the (2, 1) element in (7.15), and the standard Gauss-Jordan procedure then yields

$$\begin{bmatrix} 0 & 0 & 1 & -2.5 & 0.5 & 15 \\ 1 & 0 & 0 & 1.5 & -0.5 & 15 \\ 0 & 1 & 0 & -0.5 & 0.5 & 25 \\ \cdots & \cdots & \cdots & \cdots & \cdots & \cdots \\ 0 & 0 & 0 & 30 & 10 & 2100 \end{bmatrix}. \tag{7.16}$$

The first three rows give (7.12), and the last row must be interpreted as

$$M + 30x_4 + 10x_5 = 2100,$$

which is (7.13). The elements in the last row of (7.16) are all positive, and this indicates that M cannot be increased further. We have obtained the same solution as before, by row operations on matrices.

EXERCISE 7.3. Use the matrix methods just described to solve the linear program in Ex. 7.1.

EXERCISE 7.4. The linear program in Ex. 2.13 can be rewritten as:

$$\begin{aligned} \text{maximize} \quad & -70y_1 - 40y_2 - 90y_3 \\ \text{subject to} \quad & -2y_1 - y_2 - y_3 + y_4 = -40 \\ & -y_1 - y_2 - 3y_3 + y_5 = -60 \\ & y_i \geq 0 \quad \text{for} \quad i = 1, 2, 3, 4, 5. \end{aligned}$$

Starting with $y_1 = 12$ and $y_3 = 16$ as the basic variables, so that

$$y_2 = y_4 = y_5 = 0,$$

use the simplex method to solve this linear program.

EXERCISE 7.5. Use the simplex method to

$$\text{maximize} \qquad x_1 + 2x_2$$

$$\begin{aligned}\text{subject to}\quad -x_1 + x_2 &\leq 10\\ x_2 &\leq 20\\ x_1 + x_2 &\leq 60\\ x_1 \qquad &\leq 50\\ x_1 \qquad &\geq 0\\ x_2 &\geq 0.\end{aligned}$$

7.2 THE GENERAL LINEAR PROGRAM

In Section 2.5 we considered a more general linear program (2.41)–(2.42) than that of the preceding specific example; to do this let us recall some notation.

DEFINITION 7.1. A matrix $\mathbf{P}$ is said to be *greater than* $\mathbf{Q}$, written $\mathbf{P} > \mathbf{Q}$, when $\mathbf{P}$ and $\mathbf{Q}$ have the same numbers of rows and columns, and each element of $\mathbf{P}$ is greater than the corresponding element of $\mathbf{Q}$. Similar definitions hold for $\geq$, $<$, and $\leq$. If $\mathbf{P} > \mathbf{0}$, we say that $\mathbf{P}$ is *positive*. If $\mathbf{P} \geq \mathbf{0}$, we say that $\mathbf{P}$ is *non-negative*.

We then consider the general linear program analogous to that of (7.1)–(7.2) to be:

$$\text{maximize} \qquad M = \mathbf{c}^T\mathbf{x} \tag{7.17}$$

$$\text{over those } \mathbf{x} \text{ satisfying} \quad \mathbf{A}\mathbf{x} \leq \mathbf{b}, \qquad \mathbf{x} \geq \mathbf{0}, \tag{7.18}$$

where $\mathbf{x}$ is $n \times 1$, $\mathbf{c}$ is $n \times 1$, $\mathbf{A}$ is $m \times n$, and $\mathbf{b}$ is $m \times 1$. (Strictly speaking $\mathbf{c}^T\mathbf{x}$ is a 1×1 matrix, but we adopt the convention in this chapter that $\mathbf{c}^T\mathbf{x}$ denotes the single element in that matrix.)

EXERCISE 7.6. Find the matrices $\mathbf{A}$, $\mathbf{c}$, and $\mathbf{b}$ for each of the linear programs in (7.1)–(7.2), Exs. 7.1, 7.4, and 7.5.

The first step in our method of solving (7.17)–(7.18) is to introduce slack variables $x_{n+1}, \ldots, x_{n+m}$ so as to convert the inequalities $\mathbf{A}\mathbf{x} \leq \mathbf{b}$ into equalities. This of course *extends* $\mathbf{x}$ from an $n \times 1$ matrix into an $(n + m) \times 1$ matrix which we denote by $\mathbf{x}_e$ (the subscript "e" is for "*extended*") and similarly extends $\mathbf{A}$ from an $m \times n$ matrix into an $m \times (n + m)$ matrix which we denote by $\mathbf{A}_e$ and extends $\mathbf{c}$ from an $n \times 1$ matrix to an $(n + m) \times 1$ matrix denoted by $\mathbf{c}_e$. We thus replace the linear program in (7.17)–(7.18) by:

$$\text{maximize} \qquad M = \mathbf{c}_e^T\mathbf{x}_e + u \tag{7.19}$$

$$\text{subject to} \quad \mathbf{A}_e\mathbf{x}_e = \mathbf{b}, \qquad \mathbf{x}_e \geq \mathbf{0} \tag{7.20}$$

where u is a scalar constant, $\mathbf{x}_e$ is $q \times 1$, $\mathbf{c}_e$ is $q \times 1$, $\mathbf{A}_e$ is $m \times q$, and $\mathbf{b}$ is $m \times 1$; we introduced the constant u because the simplex method creates such constants, and we introduced the dimension q to allow us to consider *general* linear programs of the form of (7.19), (7.20), not just those which arise by adding slack variables in (7.17), (7.18). *We henceforth consider* (7.19), (7.20) *as our general linear program*; when this in fact arises from (7.17), (7.18) we must recall that $\mathbf{A}_e = [\mathbf{A}, \mathbf{I}_m]$, $\mathbf{c}_e^T = [\mathbf{c}^T, \mathbf{0}^T]$, $q = m + n$, and $u = 0$. Let us now study properties of the general linear program.

DEFINITION 7.2. Any vector $\mathbf{x}_e$ satisfying (7.20) is said to be a *feasible vector;* a feasible vector that maximizes M in (7.19) is said to be an *optimal feasible vector.*

In the simplex method described in Section 7.1, row operations were performed on the 3×5 matrix in the upper left of (7.14). At each stage, three of the columns of the resulting matrix were the unit vectors [see (7.15), (7.16)]. This implies that the corresponding 3×3 matrices in the original matrix $\mathbf{A}_e$ in (7.14) were nonsingular. In the general theory it is convenient to assume that *every* $m \times m$ submatrix of the augmented matrix $[\mathbf{A}_e, \mathbf{b}]$ is nonsingular. This appears to be a sweeping assumption but it makes the theory much simpler. Special devices must be used in both the theory and the practical solution of linear programming problems when this assumption is not true. The assumption is valid in most of the problems encountered in practice.

DEFINITION 7.3. A linear programming problem (7.19), (7.20) is said to be *nondegenerate* if every $m \times m$ submatrix selected from the $m \times (q + 1)$ augmented matrix $[\mathbf{A}_e, \mathbf{b}]$ is nonsingular. Otherwise the problem is said to be *degenerate.*

We now give a precise definition of the basic variables we introduced in the last section. A solution of the equation $\mathbf{A}_e\mathbf{x}_e = \mathbf{b}$ can be obtained by choosing a nonsingular $m \times m$ submatrix, say $\mathbf{K}$, from $\mathbf{A}_e$, solving the equations $\mathbf{Kz} = \mathbf{b}$, and setting $x_i = 0$ unless i corresponds to one of the columns of $\mathbf{A}_e$ selected for $\mathbf{K}$, in which case x_i is given the corresponding value in $\mathbf{z}$. Solutions of this type, in which at least $q - m$ elements are zero, are particularly important in linear programming, and they are given a special name.

DEFINITION 7.4. The variables associated with the columns of the matrix $\mathbf{K}$ chosen in the last paragraph are called *basic* variables. The other variables are called *nonbasic.* A solution in which the nonbasic variables are zero is called a *basic* solution of the equations.

How one decides *which* columns to choose for $\mathbf{K}$ at any moment will be clarified later.

Given a basic solution of the equations we will later need to know which variables are actually the basic variables. Of course the $q - m$ nonbasic variables are zero, but if some of the basic variables are also zero then we cannot be certain which zero variables are nonbasic and which are basic. However, if any of the basic variables $\mathbf{z}$ solving $\mathbf{Kz} = \mathbf{b}$, as above, are zero, $\mathbf{b}$ would be written as a linear combination of $m - 1$ columns of $\mathbf{K}$ (and thus of $\mathbf{A}_e$) since $\mathbf{Kz} = z_1\mathbf{k}_1 + \cdots + z_m\mathbf{k}_m$ where $\mathbf{z} = [z_i]$ and $\mathbf{K} = [\mathbf{k}_1, \ldots, \mathbf{k}_m]$; this would imply that the problem is degenerate. We state this more formally.

THEOREM 7.1. *If a problem is nondegenerate, then the basic variables in any basic solution are precisely the nonzero variables.*

Recall that in the example of the previous section we constructed a sequence of basic feasible vectors in our attempt to find an optimal feasible vector; we justified this geometrically for that example by arguing that there had to be an optimal solution at a vertex of the set of feasible vectors. The more general version of this fundamental result asserts that if there is an optimal feasible vector at all, then there is a *basic* optimal feasible vector; this basic solution corresponds to a vertex of the set of feasible solutions, as before. Equivalently, this result asserts that given a feasible vector $\mathbf{x}_e$ there is a basic feasible vector with no smaller a value of M. To show this we proceed much as we did in the simplex method described in the material preceding Figure 7.1: we move away from $\mathbf{x}_e$, increasing M, until at $\mathbf{z}$ we hit a boundary surface of the feasible set; we then move from $\mathbf{z}$ along this surface, increasing M, until at $\mathbf{w}$ we hit its boundary; continuing in this fashion we ultimately reach a vertex, which corresponds to a basic feasible vector. To describe this process algebraically, we recall [see the first assention of **Key Theorem 3.7**] that two feasible vectors, as solutions of the same system of equations, must differ by an element of the null space of (the linear transformation defined by) $\mathbf{A}_e$. Thus we must move from $\mathbf{x}_e$ along an element of this null space until at $\mathbf{z}$ some new component z_i of the feasible vector $\mathbf{z}$ becomes zero; then we must move from $\mathbf{z}$ along another element of the null space, keeping the zero components at zero, until at $\mathbf{w}$ still another component w_j becomes zero; continuing in this fashion we eventually produce a feasible vector with $q - m$ components zero, that is, a basic feasible vector. Since the proof for the general result is rather complex, we simply illustrate the argument with a numerical example.

Consider the detailed example (7.1), (7.3), (7.4) of the preceding section with $\mathbf{A}_e$ given by (7.14); we start with the feasible vector $\mathbf{x}_e$:

$$x_1 = 20, \qquad x_2 = 15, \qquad x_3 = 15, \qquad x_4 = 5, \qquad x_5 = 25.$$

Any element of the null space of $\mathbf{A}_e$ can be represented in terms of a basis for this null space; a basis for this space computed as in Theorem 5.2 is formed by the vectors

$$\mathbf{v}_1 = \begin{bmatrix} -3 \\ 1 \\ 5 \\ 2 \\ 0 \end{bmatrix}, \qquad \mathbf{v}_2 = \begin{bmatrix} 1 \\ -1 \\ -1 \\ 0 \\ 2 \end{bmatrix}.$$

New feasible vectors $\mathbf{z} = \mathbf{x}_e + \alpha\mathbf{v}_1 + \beta\mathbf{v}_2$ have components

$$z_1 = 20 - 3\alpha + \beta, \qquad z_2 = 15 + \alpha - \beta, \qquad z_3 = 15 + 5\alpha - \beta,$$
$$z_4 = 5 + 2\alpha, \qquad z_5 = 25 + 2\beta$$

and the associated value of M is $40z_1 + 60z_2 = 1700 - 60\alpha - 20\beta$. In order for the new feasible vector to increase M we need $-60\alpha - 20\beta \geq 0$, that is

$$\beta \leq -3\alpha,$$

and we of course must keep all $z_i \geq 0$. If for convenience we let $\beta = -3\alpha$ then M does not change and we require

$$z_1 = 20 - 6\alpha \geq 0, \qquad z_2 = 15 + 4\alpha \geq 0, \qquad z_3 = 15 + 8\alpha \geq 0,$$
$$z_4 = 5 + 2\alpha \geq 0, \qquad z_5 = 25 - 6\alpha \geq 0,$$

which forces

$$-\frac{15}{8} \leq \alpha \leq \frac{10}{3}.$$

Choosing $\alpha = \frac{10}{3}$ yields $\beta = -10$ and the new feasible vector

$$z_1 = 0, \qquad z_2 = \frac{85}{3}, \qquad z_3 = \frac{125}{3}, \qquad z_4 = \frac{35}{3}, \qquad z_5 = 5$$

having one more zero component than before in $\mathbf{x}_e$. We next move to a feasible vector $\mathbf{w} = \mathbf{z} + \alpha\mathbf{v}_1 + \beta\mathbf{v}_2$ for new values α and β. We find

$$w_1 = -3\alpha + \beta, \qquad w_2 = \frac{85}{3} + \alpha - \beta, \qquad w_3 = \frac{125}{3} + 5\alpha - \beta,$$
$$w_4 = \frac{35}{3} + 2\alpha, \qquad w_5 = 5 + 2\beta$$

and since we want to keep $w_1 = 0$ (since $z_1 = 0$) we must have $\beta = 3\alpha$. The nonnegativity condition then gives

$$w_1 = 0, \qquad w_2 = \frac{85}{3} - 2\alpha \geq 0, \qquad w_3 = \frac{125}{3} + 2\alpha \geq 0,$$
$$w_4 = \frac{35}{3} + 2\alpha \geq 0, \qquad w_3 = 5 + 6\alpha \geq 0$$

so that

$$-\frac{5}{6} \leq \alpha \leq \frac{85}{6},$$

while the condition that M at $\mathbf{w}$ exceed $M = 1700$ at $\mathbf{z}$ is just

$$40w_1 + 60w_2 \geq 1700,$$

so that

$$-120\alpha \geq 0$$

and α must be negative. Choosing $\alpha = -5/6$ yields the new feasible vector

$$w_1 = 0, \qquad w_2 = 30, \qquad w_3 = 40, \qquad w_4 = 10, \qquad w_5 = 0$$

having two zero components and a value for M of 1800. This is the required basic feasible vector with an increased value of M. In terms of Figure 7.1, we moved from $x_1 = 20$, $x_2 = 15$ (strictly inside the region) to $x_1 = 0$, $x_2 = \frac{85}{3}$ (on the line from V_1 to V_2) and then to $x_1 = 0$, $x_2 = 30$ (the vertex V_2). This argument can be generalized to prove the following **key** result, fundamental because it justifies considering only basic feasible vectors as in the simplex method.

- KEY THEOREM 7.2. *If the linear program described by* (7.19), (7.20) *is nondegenerate and has a feasible vector with a value of, say, M_1 for M, then either the value of M can be made as large as desired by feasible vectors* (*so that no maximum value of M exists*) *or there exists a basic feasible vector that provides a maximum value for M* (*in other words, a basic optimal feasible vector exists*).

EXERCISE 7.7. Explicitly find the matrices $\mathbf{A}_e$ and $\mathbf{c}_e$ for each of the linear programs in (7.1)–(7.2), Exs. 7.1, 7.4, and 7.5.

EXERCISE 7.8. Exhibit a linear program of the form of (7.19), (7.20) that does *not* arise by adding slack variables for (7.17), (7.18). Make your example from a "real world" situation.

EXERCISE 7.9. Provide the details for our sketched proof of Theorem 7.1.

EXERCISE 7.10. Give an example of a degenerate linear program and a basic solution with a zero basic variable.

EXERCISE 7.11. Determine whether or not the linear programs in (7.1)–(7.2), Exs. 7.1, 7.4, and 7.5 are degenerate.

EXERCISE 7.12. Use our argument leading to **Key Theorem 7.2** to find a basic feasible vector for the linear program (extended with slack variables) of Ex. 7.5 for which the value of M exceeds that at the feasible point x_e:

$$x_1 = 5, \qquad x_2 = 5, \qquad x_3 = 10, \qquad x_4 = 15, \qquad x_5 = 50, \qquad x_6 = 45.$$

EXERCISE 7.13. Generalize our argument to prove **Key Theorem 7.2**.

7.3 THE SIMPLEX METHOD IN GENERAL

We now know, thanks to **Key Theorem 7.2**, that we can restrict ourselves to basic feasible vectors as candidates for optimal vectors. This is precisely what the simplex method does. Let us study what happens when the simplex method is applied to the general linear program (7.19), (7.20). The method creates the *tableau*

$$\mathbf{T} = \begin{bmatrix} \mathbf{A}_e & \mathbf{b} \\ -\mathbf{c}_e^T & u \end{bmatrix} \tag{7.21}$$

[see (7.14)] and then executes a sequence of elementary row operations on $\mathbf{T}$ [see the calculations leading to (7.15) and (7.16)], except that the only operation allowed on the last row is to add to it linear combinations of the earlier rows, which means that $[\mathbf{A}_e, \mathbf{b}]$ itself is transformed by elementary row operations into a new matrix $[\mathbf{A}'_e, \mathbf{b}']$. According to Theorem 3.5 the vectors satisfying (7.20) (feasible vectors) are precisely those satisfying $\mathbf{A}'_e\,\mathbf{x}_e = \mathbf{b}'$, $\mathbf{x}_e \geq \mathbf{0}$. Similarly a new $[-\mathbf{c}_e'^T, u']$ produced by these operations must be of the form $[-\mathbf{c}_e^T, u] + \mathbf{z}^T[\mathbf{A}_e, \mathbf{b}]$ so that whenever $\mathbf{x}_e$ solves (7.20) we have

$$\begin{aligned} u' + \mathbf{c}_e'^T\mathbf{x}_e &= u + \mathbf{z}^T\mathbf{b} + (\mathbf{c}_e^T - \mathbf{z}^T\mathbf{A}_e)\mathbf{x}_e = u + \mathbf{c}_e^T\mathbf{x}_e + \mathbf{z}^T(\mathbf{b} - \mathbf{A}_e\mathbf{x}_e) \\ &= u + \mathbf{c}_e^T\mathbf{x}_e. \end{aligned}$$

This proves the following theorem, which states that the successive tableaus generated by the simplex method all represent equivalent linear programs. (Compare Theorem 3.5 for solutions of equations.)

THEOREM 7.3. Suppose that elementary row operations are performed on the matrix (tableau)

$$\mathbf{T} = \begin{bmatrix} \mathbf{A}_e & \mathbf{b} \\ -\mathbf{c}_e^T & u \end{bmatrix}$$

in any way, except that the only operation that is allowed on the last row is to add to it sums of multiples of the other rows. Denote the result by

$$\begin{bmatrix} \mathbf{A}'_e & \mathbf{b}' \\ -(\mathbf{c}'_e)^T & u' \end{bmatrix}.$$

Then the following linear programming problem is completely equivalent to (7.19), (7.20):

$$\textit{maximize} \quad M' = u' + (\mathbf{c}'_e)^T\mathbf{x}_e,$$

$$\textit{subject to} \quad \mathbf{A}'_e\mathbf{x}_e = \mathbf{b}', \qquad \mathbf{x}_e \geq \mathbf{0}.$$

EXERCISE 7.14. Write out the linear programs represented by the tableaus (7.15) and (7.16) and explicitly verify as outlined above that these are equivalent to (7.1), (7.3), (7.4).

The simplex method generates a sequence of tableaus $\mathbf{T}_r$ having the form of (7.21) but with the elements having been modified by row operations; each tableau corresponds to a basic feasible vector. Suppose that, at the rth stage, a basic feasible solution has been found for which, by renumbering the variables if necessary (which can be done without loss of generality), we have made the basic variables be $x_1, x_2, \ldots, x_m$. Also, suppose that, as in the concrete example in Section 7.1, we have performed row operations in the tableau $\mathbf{T}_r$ so that the columns of $\mathbf{T}_r$ corresponding to the basic variables are unit vectors. Explicitly, $\mathbf{T}_r$ will have the form

$$\begin{bmatrix} 1 & 0 & \ldots & 0 & a_{1,m+1} & \ldots & a_{1q} & b_1 \\ 0 & 1 & \ldots & 0 & a_{2,m+1} & \ldots & a_{2q} & b_2 \\ & & & & \ldots & & & \\ 0 & 0 & \ldots & 1 & a_{m,m+1} & \ldots & a_{mq} & b_m \\ 0 & 0 & \ldots & 0 & -c_{m+1} & \ldots & -c_q & u_r \end{bmatrix}. \tag{7.22}$$

Also,

$$M = u_r + c_{m+1}x_{m+1} + \cdots + c_qx_q. \tag{7.23}$$

This expression for M is valid whatever the values of $x_1, \ldots, x_q$. For the basic solution available at this stage we have $x_{m+1} = \cdots = x_q = 0$, so that the estimate of the maximum at this stage is $M = u_r$. We wish to find out whether we can modify the values of $x_{m+1}, \ldots, x_q$ so as to increase M. The only way that we are allowed to change these variables is to make their values positive. If all the c_j, $j = m + 1, \ldots, q$, are negative, then from (7.23) it is clear that any increase in the values of $x_{m+1}, \ldots, x_q$ can only decrease M. In this case, it is *not* possible to improve the basic feasible solution that we are working with at this stage. If at least one c_j is positive, say c_s, we can increase our value of M by increasing x_s. From **Key Theorem 7.2** we know that we can restrict our attention to basic feasible solutions. We will maintain all the variables $x_{m+1}, \ldots, x_q$ at the value zero, except x_s, and try to find a basic feasible solution in which $x_s \neq 0$ and one of $x_1, \ldots, x_m$ is zero. The equations represented by the tableau $\mathbf{T}_r$ are, then,

$$x_i + a_{is}x_s = b_i \qquad (i = 1, \ldots, m),$$

or

$$x_i = b_i - a_{is}x_s \qquad (i = 1, \ldots, m). \tag{7.24}$$

The b_i are known to be positive since $x_i = b_i$ ($i = 1$ to m), $x_i = 0$ ($i = m + 1$ to q) is assumed to be a basic feasible solution. If x_s is increased from zero, there are two possibilities:

1. If $a_{is} \leq 0$, for all i with $1 \leq i \leq m$, then from (7.24) it is clear that x_i is always positive for $i = 1, \ldots, m$. This means that

$$(x_1, \ldots, x_m, 0, \ldots, 0, x_s, 0, \ldots, 0) \tag{7.25}$$

is a feasible solution for all $x_s > 0$.

2. If $a_{is} > 0$, for some i with $1 \leq i \leq m$, then from (7.24), if $x_s > b_i/a_{is}$, the value of x_i is negative, which is not permissible. The value of x_s must lie between 0 and b_i/a_{is} for each i. In order to maximize M, we wish to choose x_s as large as possible, so that the optimum value of x_s is given by the *smallest* of the ratios

$$\frac{a_{is}}{b_i} \qquad (i = 1, \ldots, m;\ a_{is} > 0). \tag{7.26}$$

These considerations give the following result:

● *KEY THEOREM 7.4. Consider a nondegenerate linear programming problem that has been transformed into the form* (7.22), (7.23).

(i) *If* $c_j < 0$ *for all* $j = m + 1, \ldots, q$, *then the basic feasible solution* $(x_1, \ldots, x_m, 0, \ldots, 0)$ *is the unique solution of the linear program, that is, it is an optimal basic feasible vector and it is the only optimal feasible vector.*

(ii) *If one of the* c_j *is greater than zero, say* c_s, *then there are two possibilities:*

(a) *If* $a_{is} \leq 0$ *for all* $i = 1, \ldots, m$, *then feasible solutions exist such that* M *can be made as large as desired and the linear program has no solution.*

(b) *If* $a_{is} > 0$ *for at least one* i, *a new basic solution can be found such that the value of* M *is strictly increased.*

(iii) *If* $c_j \leq 0$ *for all* j *while* $c_j = 0$ *for at least one value of* j, *say* s, *corresponding to a nonbasic variable, then* M *attains its maximum value for at least two different* $\mathbf{x}$. *If two such optimal solutions are denoted by* $\mathbf{x}^{(1)}$ *and* $\mathbf{x}^{(2)}$, *then*

$$\mathbf{x} = \alpha\mathbf{x}^{(1)} + (1 - \alpha)\mathbf{x}^{(2)} \qquad (0 \leq \alpha \leq 1)$$

is also an optimal solution.

(iv) *If procedure* (ii) (b) *is repeated, then either the situation in* (ii) (a) *is reached after a finite number of steps, in which case M can be made as large as desired, or a situation is reached, after a finite number of steps, in which* $c_j \leq 0$ *for all j, and M is finite, and a solution to the linear program has been found.*

We leave the proof of this main theorem on the simplex method to the exercises; first we emphasize some portions of the theorem, illustrating the great power of the simplex method.

1. If the values M on the feasible set are bounded above, then there exists a solution to the linear program.
2. If a solution to the linear program exists, then the simplex method will locate a solution in a finite number of steps.
3. If the values M on the feasible set are unbounded so that no solution to the linear program exists, then this fact will be discovered by the simplex method in a finite number of steps.

EXERCISE 7.15. Under the hypotheses of **Key Theorem 7.4**(i), show first that the value of M for the basic feasible vector $(x_1, \ldots, x_m, 0, \ldots, 0)$ cannot be increased by making any of the $x_{m+1}, \ldots, x_q$ nonzero. Use the fact that the problem is nondegenerate to deduce that the present basic feasible vector is the only feasible vector with $x_{m+1} = \cdots = x_q = 0$. Finally conclude that **Key Theorem 7.4**(i) is valid.

EXERCISE 7.16. Under the hypotheses of **Key Theorem 7.4**(ii)(a), use (7.25) to prove that M can be made arbitrarily large on the feasible set.

EXERCISE 7.17. Prove **Key Theorem 7.4**(ii)(b), using the nondegeneracy of the problem to show that $\mathbf{b} > \mathbf{0}$ and then using x_s as defined before (7.26).

EXERCISE 7.18. Prove **Key Theorem 7.4**(iii) by modifying Ex. 7.15 so as to construct a different feasible vector with the same value for M.

EXERCISE 7.19. Prove **Key Theorem 7.4**(iv) by showing that there are only a finite number of basic feasible vectors and using the fact that (ii)(b) provides a strict increase in the value of M.

Since the preceding theory may have obscured the basic simplicity of the calculations involved in the simplex method, we summarize, one final time, the steps involved in implementing the location of a new basic solution as required in (ii)(b) of **Key Theorem 7.4**; the process is precisely the one we used on the specific example in Section 7.1.

1. Find the most negative element $-c_s$ among $-c_{m+1}, \ldots, -c_q$ in the row of the array (7.22). (We could choose any negative element but it is natural to pick the most negative one since a unit change in the cor-

responding variable will cause the largest resulting change in M. The choice of the most negative element is not necessarily the best.)

2. Calculate the ratios b_i/a_{is} for the positive a_{is}. Choose the smallest of these ratios.
3. Pivot on the element a_{is} corresponding to the minimum ratio in (2), thus changing to zero all the elements in the sth column of the tableau except of course for the pivot a_{is} itself.

There is one aspect of the simplex method that seems to be quite inefficient. To determine which variable to insert into the basis [step (1) above] we only need to know the $q - m$ numbers $c_{m+1}, \ldots, c_q$; having decided to insert, say, variable x_s, to determine next which variable to delete from the basis [step (2) above] we only need to know the $2m$ numbers $a_{1s}, \ldots, a_{ms}, b_1, \ldots, b_m$. Thus to select our pivot we only need to have these $m + q$ numbers available even though to get the present tableau from its predecessor we performed row operations generating $q - m + 1$ columns for a total of $m(q - m + 1)$ numbers. It would clearly be more efficient if we could first compute just the numbers $c_{m+1}, \ldots, c_q$, then perform step (1) above, then compute only the sth column $a_{1s}, \ldots, a_{ms}$ and $b_1, \ldots, b_m$ needed for step (2), and then perform step (2). A way of implementing this is the *revised simplex method,* whose workings we now roughly indicate.

Suppose that our original tableau is written in partitioned form as

$$\mathbf{T} = \begin{bmatrix} \mathbf{a}_1 & \mathbf{a}_2 & \cdots & \mathbf{a}_m & \mathbf{a}_{m+1} & \cdots & \mathbf{a}_q & \vdots & \mathbf{b} \\ \cdots & \cdots & \cdots & \cdots & \cdots & \cdots & \cdots & \cdots & \cdots \\ & & & -\mathbf{c}^T & & & & \vdots & u \end{bmatrix}$$

where the $\mathbf{a}_i$ for $i = 1, \ldots, q$ and $\mathbf{b}$ are $m \times 1$ matrices and $\mathbf{c}$ is $q \times 1$; suppose that the basic variables are $x_1, \ldots, x_m$. Our first step then with the simplex method would be to perform row operations (restricted as in Theorem 7.3) so as to reduce the vectors $\mathbf{a}_1, \ldots, \mathbf{a}_m$ to the unit column vectors $\mathbf{e}_1, \ldots, \mathbf{e}_m$ and to reduce the first m components of $-\mathbf{c}^T$ to zeros. By **Key Theorem 3.3** this is equivalent to premultiplying $\mathbf{T}$ by a nonsingular matrix to accomplish these goals; because of the restriction on the row operations, this matrix must have the form

$$\begin{bmatrix} \mathbf{F} & \mathbf{0} \\ \mathbf{v}^T & 1 \end{bmatrix}$$

where $\mathbf{F}$ is $m \times m$ and $\mathbf{v}$ is $m \times 1$. If we partition the tableau $\mathbf{T}$ by letting

$$\mathbf{B} = [\mathbf{a}_1, \ldots, \mathbf{a}_m], \qquad \mathbf{D} = [\mathbf{a}_{m+1}, \ldots, \mathbf{a}_q],$$
$$\mathbf{c}_1^T = [c_1, \ldots, c_m], \qquad \mathbf{c}_2^T = [c_{m+1}, \ldots, c_q],$$

then the requirements on the effects of the row operations are expressed as

$$\left[\begin{array}{c|c} \mathbf{F} & \mathbf{0} \\ \hline \mathbf{v}^T & 1 \end{array}\right]\left[\begin{array}{c|c|c} \mathbf{B} & \mathbf{D} & \mathbf{b} \\ \hline -\mathbf{c}_1^T & -\mathbf{c}_2^T & u \end{array}\right] = \left[\begin{array}{c|c|c} \mathbf{I} & \mathbf{D}' & \mathbf{b}' \\ \hline \mathbf{0}^T & -\mathbf{c}_2'^T & u' \end{array}\right].$$

This tells us what **F** and **v** must be, namely

$$\mathbf{F} = \mathbf{B}^{-1}, \qquad \mathbf{v} = \mathbf{F}^T\mathbf{c}_1, \qquad \mathbf{D}' = \mathbf{F}\mathbf{D}, \qquad \mathbf{b}' = \mathbf{F}\mathbf{b}.$$

Thus, if we have at our disposal (or can compute) $\mathbf{B}^{-1}$, we can compute the vector $-\mathbf{c}_2'^T$ we need to determine which variable to insert in the basis since

$$-\mathbf{c}_2'^T = -\mathbf{c}_2^T + \mathbf{v}^T\mathbf{D}.$$

We can then determine which variable to delete from the basis by examining the column of $\mathbf{D}'$ just chosen which can be computed by multiplying **F** times the corresponding column of **D**. In this way, *once we have* $\mathbf{B}^{-1} = \mathbf{F}$, *we can determine what variables to exchange in the basis without actually changing the original tableau.* To continue this process we need to exchange the variables (which corresponds to *replacing precisely one column of* **B** *by a column from* **D**), and then to compute the inverse of the newly created matrix $\mathbf{B}'$ of basic columns. Since $\mathbf{B}'$ differs from **B** in only one column, however, we can efficiently compute $\mathbf{B}'^{-1}$ as described in Section 6.4. In this way the revised simplex method maintains only the original tableau and easily updates sequentially the inverses of $m \times m$ matrices formed from the columns in the tableau corresponding to the basic variables. When the number m of constraints is small compared with the number q of variables, the revised simplex method provides a considerable gain in efficiency over the standard simplex method; most computer programs use the revised simplex method with modifications to handle numerical difficulties.

EXERCISE 7.20. Solve the linear program (7.1), (7.2) by the revised simplex method.

SOLUTION: Our tableau is

$$\begin{bmatrix} 2 & 1 & 1 & 0 & 0 & 70 \\ 1 & 1 & 0 & 1 & 0 & 40 \\ 1 & 3 & 0 & 0 & 1 & 90 \\ -40 & -60 & 0 & 0 & 0 & 0 \end{bmatrix}$$

and we start with the basic variables x_3, x_4, x_5. Therefore our matrix **B** is just formed by columns 3, 4, and 5, so that $\mathbf{B} = \mathbf{I}$ and there is no elimination to perform at this step. We decide as usual (see Section 7.1) to replace x_5 in the basis by x_2, so that the basic variables are just x_2, x_3, x_4 and our new matrix **B** equals

$$\begin{bmatrix} 1 & 1 & 0 \\ 1 & 0 & 1 \\ 3 & 0 & 0 \end{bmatrix}$$

so that $\mathbf{F} = \mathbf{B}^{-1}$ is simply

$$\begin{bmatrix} 0 & 0 & \frac{1}{3} \\ 1 & 0 & -\frac{1}{3} \\ 0 & 1 & -\frac{1}{3} \end{bmatrix}.$$

The vector $-\mathbf{c}_1^T$ (under the basic columns) equals $[-60 \quad 0 \quad 0]$ so that $\mathbf{v}^T = \mathbf{c}_1^T\mathbf{F} = [0 \quad 0 \quad 20]$; since $-\mathbf{c}_2^T$ (under the nonbasic columns) equals $[-40 \quad 0]$, we find that

$$-\mathbf{c}_2'^T = [-40 \quad 0] + [0 \quad 0 \quad 20]\begin{bmatrix} 2 & 0 \\ 1 & 0 \\ 1 & 1 \end{bmatrix} = [-20 \quad 20]$$

just as in (7.15). As usual, we decide to add x_1 to the basis; to determine which variable to delete, we must compute

$$\begin{bmatrix} 0 & 0 & \frac{1}{3} \\ 1 & 0 & -\frac{1}{3} \\ 0 & 1 & -\frac{1}{3} \end{bmatrix}\begin{bmatrix} 2 \\ 1 \\ 1 \end{bmatrix} = \begin{bmatrix} \frac{1}{3} \\ \frac{5}{3} \\ \frac{2}{3} \end{bmatrix}$$

and

$$\begin{bmatrix} 0 & 0 & \frac{1}{3} \\ 1 & 0 & -\frac{1}{3} \\ 0 & 1 & -\frac{1}{3} \end{bmatrix}\begin{bmatrix} 70 \\ 40 \\ 90 \end{bmatrix} = \begin{bmatrix} 30 \\ 40 \\ 10 \end{bmatrix}$$

and by examining $30/\frac{1}{3}$, $40/\frac{5}{3}$, $10/\frac{2}{3}$ we decide to delete the third one of our basic variables, namely x_4. We have thus found a new basis x_1, x_2, x_3 and our new **B** equals

$$\begin{bmatrix} 2 & 1 & 1 \\ 1 & 1 & 0 \\ 1 & 3 & 0 \end{bmatrix}$$

so that $\mathbf{F} = \mathbf{B}^{-1}$ equals

$$\begin{bmatrix} 0 & \frac{3}{2} & -\frac{1}{2} \\ 0 & -\frac{1}{2} & \frac{1}{2} \\ 1 & -\frac{5}{2} & \frac{1}{2} \end{bmatrix}.$$

From $-\mathbf{c}_1^T = [-40 \quad -60 \quad 0]$ we find $\mathbf{v}^T = [0 \quad 30 \quad 10]$ and hence

$$-\mathbf{c}_2^{T'} = [0 \quad 0] + [0 \quad 30 \quad 10]\begin{bmatrix} 0 & 0 \\ 1 & 0 \\ 0 & 1 \end{bmatrix} = [30 \quad 10]$$

just as in (7.16). Since $-\mathbf{c}_2'^T$ contains no negative values, our solution has been found. To find it, we compute $\mathbf{B}^{-1}\mathbf{b}$ which yields $x_1 = 15$, $x_2 = 25$, $x_3 = 15$ as before.

In this example we wrote out everything in English, so it seemed less efficient than our earlier method; nonetheless, it is a useful and efficient modification of the standard simplex method.

EXERCISE 7.21. Use the revised simplex method to solve Ex. 7.1.

EXERCISE 7.22. Maximize $-4x_1 - x_2 + x_3 - 2x_4$

$$\begin{aligned} \text{subject to} \quad 3x_1 - 3x_2 + x_3 &= 3 \\ 6x_2 - 2x_3 + x_4 &= 2 \\ x_i \geq 0 \ (i = 1, 2, 3, 4). \end{aligned}$$

EXERCISE 7.23. Maximize $x_1 + 4x_2 + 3x_3$

$$\begin{aligned} \text{subject to} \quad 3x_1 + 2x_2 + x_3 &\leq 4 \\ x_1 + 5x_2 + 4x_3 &\leq 14 \\ x_i \geq 0 \quad (i = 1, 2, 3). \end{aligned}$$

7.4 COMPLICATIONS FOR THE SIMPLEX METHOD

In the preceding sections we have often assumed, especially in the concrete examples, that no unpleasant difficulties would arise to complicate matters. Our theorems were for nondegenerate programs, our examples had one and only one solution, and it was always easy in our examples to find an initial feasible vector that was basic. In practice, of course, life is not always so uncomplicated. In this section we describe a number of more complex problems to illustrate the variety of situations that can arise.

1. No Obvious Basic Feasible Solution

We were fortunate in the detailed example of Section 7.1 that the basic feasible vector of (7.5) was readily available. In the more general situation represented in (7.18), a basic feasible vector is always readily available if $\mathbf{b} \geq \mathbf{0}$, for when we add slack variables to create the extended form of (7.20), we can let the original variables vanish while the slack variables equal the corresponding components of $\mathbf{b}$, giving us our basic feasible vector. If $\mathbf{b}$ is not nonnegative, some ingenuity is often required to find a basic feasible vector. The trick is to define a new linear program, for which an initial basic feasible vector is *easily* found, and whose solution gives us the desired initial basic feasible vector for the *original* linear program. We illustrate the idea by a concrete example.

Consider the linear program

$$\text{maximize} \quad 40x_1 + 60x_2 \tag{7.27}$$

$$\text{subject to} \quad 2x_1 + x_2 \leq 70, \qquad x_1 + x_2 \geq 40, \qquad x_1 + 3x_2 \leq 90, \qquad x_i \geq 0 \text{ (for } i = 1, 2, 3). \tag{7.28}$$

The point $x_1 = x_2 = 0$ does not satisfy the constraints because of the requirement $x_1 + x_2 \geq 40$ [compare (7.2)], so we must find a different initial vector. If, as usual, we introduce slack variables, the offending inequality becomes $x_1 + x_2 - x_4 = 40$ with $x_4 \geq 0$; our inability to let $x_1 = x_2 = 0$ for the inequality shows up here in the fact that $x_1 = x_2 = 0$ yields $x_4 = -40$ in violation of $x_4 \geq 0$.

The clever trick is to introduce another variable x_6, a so-called *artificial variable*, and then to replace x_4 in the equation $x_1 + x_2 - x_4 = 0$ by $x_4 - x_6$. That is, we write $x_1 + x_2 \geq 40$ as $x_1 + x_2 - x_4 + x_6 = 40$ with $x_4 \geq 0$ and $x_6 \geq 0$; now we can let $x_1 = x_2 = 0$ since we can easily have $x_6 - x_4 = 40$ while both x_4 and x_6 are nonnegative. We therefore consider the extended constraints

$$\begin{aligned} 2x_1 + x_2 + x_3 \qquad\qquad\qquad &= 70 \\ x_1 + x_2 \qquad - x_4 \qquad + x_6 &= 40 \\ x_1 + 3x_2 \qquad\qquad + x_5 \qquad &= 90 \end{aligned} \tag{7.29}$$

for which it is easy to find a basic feasible vector, for example,

$$x_1 = x_2 = x_4 = 0, \qquad x_3 = 70, \qquad x_6 = 40, \qquad x_5 = 90.$$

If we can somehow start from this vector and eventually find another basic feasible vector for (7.29) but for which the sixth component x_6 is zero, then the first five components of this vector will form a basic feasible vector for the orignal linear program with slack variables, namely for

$$\begin{aligned} 2x_1 + x_2 + x_3 \qquad\qquad &= 70 \\ x_1 + x_2 \qquad - x_4 \qquad &= 40 \\ x_1 + 3x_2 \qquad\qquad + x_5 &= 90 \\ x_i \geq 0 \qquad (\text{for } i = 1, \ldots, 5). \end{aligned} \tag{7.30}$$

Conversely, any basic feasible vector for (7.30) generates one for (7.29) by adding a sixth component x_6 with $x_6 = 0$. Therefore the constraints we really care about, namely (7.30), have a basic feasible vector if and only if (7.29) has a basic feasible vector for which $x_6 = 0$; since of course $x_6 \geq 0$, we are really trying to *minimize* x_6 (or maximize $-x_6$) subject to (7.29) and ask whether or not the optimal value is zero. Practically, we attack the linear program

$$\begin{aligned} &\text{maximize} \quad M = -x_6 \\ &\text{subject to} \quad (7.29) \text{ and } x_i \geq 0 \text{ (for } i = 1, \ldots, 6). \end{aligned}$$

Since $x_6 \geq 0$, there are two possibilities:

(1) The maximum value is in fact 0. This means that we can find a basic feasible solution of (7.29) in which $x_6 = 0$, that is, we can find a basic feasible solution of (7.30).

(2) The maximum value of M is less than 0. This means that no basic feasible solution of (7.30) exists because, if such a solution did exist, then we could find a basic feasible solution of (7.29) with $x_6 = 0$, that is, $M = 0$, and we have just assumed that this is not possible.

We apply the simplex method to maximize $-x_6$ subject to (7.29); the corresponding tableau is

$$\begin{bmatrix} 2 & 1 & 1 & 0 & 0 & 0 & 70 \\ 1 & 1 & 0 & -1 & 0 & 1 & 40 \\ 1 & 3 & 0 & 0 & 1 & 0 & 90 \\ 0 & 0 & 0 & 0 & 0 & 1 & 0 \end{bmatrix}. \tag{7.31}$$

There is an obvious basic feasible solution $x_1 = x_2 = x_4 = 0$, $x_3 = 70$, $x_5 = 90$, $x_6 = 40$, but this array is not quite analogous to those considered previously since the elements in the last row of (7.31) corresponding to the basic feasible variables x_3, x_5, and x_6 are not all zero; this is because we have not yet performed the Gauss-Jordan elimination in the columns corresponding to the basic variables, as indicated by the presence of *two* 1s in the sixth column. Pivoting on the (2, 6) element leads to the tableau

$$\begin{bmatrix} 2 & 1 & 1 & 0 & 0 & 0 & 70 \\ 1 & 1 & 0 & -1 & 0 & 1 & 40 \\ 1 & 3 & 0 & 0 & 1 & 0 & 90 \\ -1 & -1 & 0 & 1 & 0 & 0 & -40 \end{bmatrix} \tag{7.32}$$

which is the standard tableau. If we proceed from here with the standard simplex method we eventually obtain the tableau

$$\begin{bmatrix} 0 & 0 & 1 & 2.5 & 0.5 & -2.5 & 15 \\ 1 & 0 & 0 & -1.5 & -0.5 & 1.5 & 15 \\ 0 & 1 & 0 & 0.5 & 0.5 & 0.5 & 25 \\ 0 & 0 & 0 & 0 & 0 & 0 & 0 \end{bmatrix}, \tag{7.33}$$

and the simplex method halts because of the lack of negative numbers in the last row. Since the optimal value for this linear program is zero [see the (4, 7) element], we deduce that we have a basic feasible vector for the original program (7.30), namely

$$x_1 = 15, \qquad x_2 = 25, \qquad x_3 = 15, \qquad x_4 = 0, \qquad x_5 = 0.$$

In order to use this basic feasible solution as a starting point for maximizing $M = 40x_1 + 60x_2$, we must start with the array

$$\begin{bmatrix} 2 & 1 & 1 & 0 & 0 & 70 \\ 1 & 1 & 0 & -1 & 0 & 40 \\ 1 & 3 & 0 & 0 & 1 & 90 \\ -40 & -60 & 0 & 0 & 0 & 0 \end{bmatrix} \tag{7.34}$$

and modify it to reflect the fact that the basic variables are x_1, x_2, and x_3, that is, we must perform row operations (Gauss-Jordan elimination) to change the first three columns into unit column vectors. But, except for the presence of an added column in (7.31), this is precisely what we just finished doing in transforming (7.31) into (7.33), and the first three rows and columns, which determine the row operations, are identical in (7.34) and (7.33). Hence we could obtain the required array from (7.34) by performing Gauss-Jordan elimination to make the resulting first three columns identical with those of (7.33). More conveniently, we can do this at exactly the same time that the reduction of (7.31) to (7.33) is accomplished. This can be achieved by carrying along the last row in (7.31) (with one extra element inserted to compensate for the missing column) as an extra row in (7.34); thus, we start with

$$\begin{bmatrix} 2 & 1 & 1 & 0 & 0 & 0 & 70 \\ 1 & 1 & 0 & -1 & 0 & 1 & 40 \\ 1 & 3 & 0 & 0 & 1 & 0 & 90 \\ -40 & -60 & 0 & 0 & 0 & 0 & 0 \\ 0 & 0 & 0 & 0 & 0 & 1 & 0 \end{bmatrix} \tag{7.35}$$

instead of (7.31). Reducing this, using the same pivots as for (7.31), we find

$$\begin{bmatrix} 0 & 0 & 1 & 2.5 & 0.5 & -2.5 & 15 \\ 1 & 0 & 0 & -1.5 & -0.5 & 1.5 & 15 \\ 0 & 1 & 0 & 0.5 & 0.5 & 0.5 & 25 \\ 0 & 0 & 0 & -30 & 10 & 0 & 2100 \\ 0 & 0 & 0 & 0 & 0 & 0 & 0 \end{bmatrix}. \tag{7.36}$$

If we delete the sixth column because it corresponds to x_6 and the fifth row because it represents the artificial problem we had to solve to find a basic feasible vector for (7.30), we obtain the tableau

$$\begin{bmatrix} 0 & 0 & 1 & 2.5 & 0.5 & 15 \\ 1 & 0 & 0 & -1.5 & -0.5 & 15 \\ 0 & 1 & 0 & 0.5 & 0.5 & 25 \\ 0 & 0 & 0 & -30 & 10 & 2100 \end{bmatrix} \tag{7.37}$$

as our initial tableau corresponding to the basic feasible vector with $x_1 = 15$, $x_2 = 25$, $x_3 = 15$, $x_4 = 0$, $x_5 = 0$ to use the simplex method to solve (7.27), (7.28). It is easy to complete the simplex method, arriving eventually at a maximum value of $M = 2280$ generated by $x_1 = 24$, $x_2 = 22$, $x_3 = 0$, $x_4 = 6$, $x_5 = 0$.

EXERCISE 7.24. Verify that (7.31) leads to (7.32) and thence to (7.33) via the simplex method.

EXERCISE 7.25. Verify that (7.35) reduces to (7.36).

EXERCISE 7.26. Solve the linear program described by (7.34) using the simplex method starting from $x_1 = 15$, $x_2 = 25$, $x_3 = 15$, $x_4 = x_5 = 0$; obtain the same solution starting from (7.37).

EXERCISE 7.27. Use the above method, as in (7.35), to

$$\begin{aligned} \text{maximize} \quad & x_1 + x_2 \\ \text{subject to} \quad & -x_1 + x_2 \leq 10 \\ & x_1 + x_2 \geq 5 \\ & 2x_1 + x_2 \leq 40 \\ & x_1 \geq 0, \quad x_2 \geq 0. \end{aligned}$$

The above device of using an auxiliary linear program to obtain an initial basic feasible vector can of course be used quite generally; we merely have to introduce one artificial variable for each constraint violated by the vector zero. We then create and solve an auxiliary linear program with an objective function which is maximized only when the artificial variables all vanish; there is of course some freedom in choosing this function.

EXERCISE 7.28. Find a basic feasible vector for

$$\begin{aligned} x_1 - \quad x_2 &\geq 2 \\ x_1 + 0.2x_2 &\geq 4 \\ 3x_1 - \quad x_2 &\geq 14 \\ x_1 \geq 0, \quad x_2 &\geq 0. \end{aligned}$$

SOLUTION: By adding slack variables x_3, x_4, x_5 we transform this to

$$\begin{aligned} x_1 - x_2 - x_3 \qquad\qquad &= 2 \\ x_1 + 0.2x_2 \qquad - x_4 \qquad &= 4 \\ 3x_1 - x_2 \qquad\qquad - x_5 &= 14 \\ x_i \geq 0 \text{ (for } i = 1, \ldots, 5). \end{aligned} \tag{7.38}$$

Since we cannot simply let $x_1 = x_2 = 0$ while keeping all five $x_i \geq 0$, we add artificial variables x_6, x_7, x_8, resulting in

$$\begin{aligned} x_1 - x_2 - x_3 \qquad\qquad + x_6 \qquad\qquad &= 2 \\ x_1 + 0.2x_2 \qquad - x_4 \qquad\qquad + x_7 \qquad &= 4 \\ 3x_1 - x_2 \qquad\qquad - x_5 \qquad\qquad + x_8 &= 14 \\ x_1 \geq 0 \text{ (for } i = 1, \ldots, 8). \end{aligned} \tag{7.39}$$

If we try to eliminate the nonzero values of the artificial variables by choosing to maximize $-(x_6 + x_7 + 0.1x_8)$, we arrive at the basic feasible vector for (7.38) given by $x_1 = 6$, $x_2 = 4$, $x_3 = 0$, $x_4 = 2.8$, $x_5 = 0$; on the other hand, if we choose to maximize $-(x_6 + x_7 + 0.3x_8)$, we arrive at the basic feasible vector for (7.38) given by $x_1 = \frac{14}{3}$, $x_2 = 0$, $x_3 = \frac{8}{3}$, $x_4 = \frac{2}{3}$, $x_5 = 0$. Thus different artificial objective functions can produce different basic feasible vectors for the extended problem (7.38) and even for the original problem in the (x_1, x_2)– variables only.

EXERCISE 7.29. Use the simplex method to maximize $-(x_6 + x_7 + 0.1x_8)$ subject to (7.39) to obtain the basic feasible vector for (7.38) found first in Ex. 7.28.

EXERCISE 7.30. Use the simplex method to maximize $-(x_6 + x_7 + 0.3x_8)$ subject to (7.39) to obtain the basic feasible vector for (7.38) found second in Ex. 7.28.

EXERCISE 7.31. Find a basic feasible vector for

$$\begin{aligned} x_1 + 2x_2 &\geq 3 \\ 10x_1 + x_2 &\geq 11 \\ 4x_1 + 3x_2 &\leq 33 \\ x_1 \geq 0, \qquad x_2 &\geq 0. \end{aligned}$$

2. Degeneracy

So far we have always assumed that the linear programming problem was nondegenerate (see Definition 7.3). Consider the constraints

$$2x_1 + x_2 \leq 70, \qquad x_1 + x_2 \leq 40, \qquad x_1 + 3x_2 \leq 120.$$

The corresponding initial array to maximize $40x_1 + 60x_2$ is

$$\begin{bmatrix} 2 & 1 & 1 & 0 & 0 & 70 \\ 1 & 1 & 0 & 1 & 0 & 40 \\ 1 & 3 & 0 & 0 & 1 & 120 \\ -40 & -60 & 0 & 0 & 0 & 0 \end{bmatrix}. \tag{7.40}$$

The -60 in the last row leads us to examine the ratios $70/1 = 70$, $40/1 = 40$, $120/3 = 40$. There is now a choice of pivoting on either the (2, 2) or the (3, 2) element. If we pivot on the (3, 2) element, this gives the array

$$\begin{bmatrix} \frac{5}{3} & 0 & 1 & 0 & -\frac{1}{3} & 30 \\ \frac{2}{3} & 0 & 0 & 1 & -\frac{1}{3} & 0 \\ \frac{1}{3} & 1 & 0 & 0 & \frac{1}{3} & 40 \\ -20 & 0 & 0 & 0 & 20 & 2400 \end{bmatrix}.$$

The basic variable x_4 is zero–i.e., the problem is degenerate. If we next pivot on the element $\frac{2}{3}$ in the first column, we obtain

$$\begin{bmatrix} 0 & 0 & 1 & -2.5 & 0.5 & 30 \\ 1 & 0 & 0 & 1.5 & -0.5 & 0 \\ 0 & 1 & 0 & -0.5 & 0.5 & 40 \\ 0 & 0 & 0 & 30 & 10 & 2400 \end{bmatrix}.$$

The procedure now terminates and the required maximum is 2400. Two points should be noted:

1. In going from the first array to the second, a choice of pivots was available.
2. In going from the second array to the third, the value of M was not changed.

The proof that either the simplex method will terminate or M is unbounded is based on the fact that a change of basic variables in a nondegenerate problem always increased M. This guarantees that the basic solution found at each stage is always different from that found at previous stages. If the problem is degenerate, this proof breaks down, and in fact the result is not true. The basic solution found at one stage may be identical with a basic solution found at a previous stage, in which case the procedure can cycle indefinitely.

Graphically, the situation in the above concrete example is illustrated in Figure 7.2. Two lines PQ, PR respresenting inequality constraints go through one point P. This means that if the basic solution we are considering corresponds to the point P, then the basis can consist of either PQ and QR or PR and QR. At the point P we can shift from one basis to another without chang-

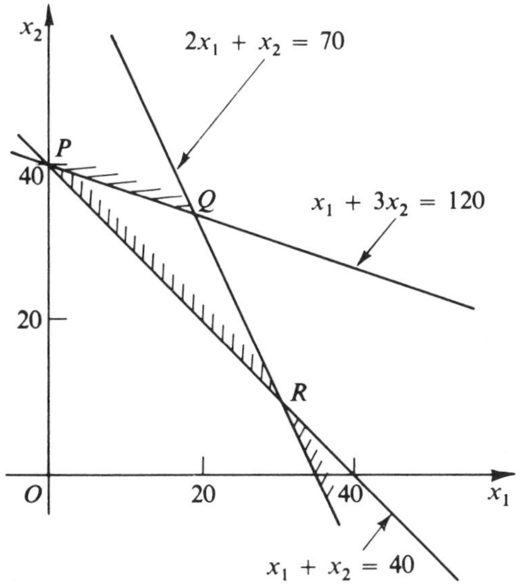

Figure 7.2. Degeneracy.

ing the value of the quantity $\mathbf{c}^T\mathbf{x}$ that we wish to maximize. The possibility arises that we might, at one step in the simplex method, change basis from PR, QR to PQ, QR and then at the next step change from PQ, QR to PR, QR. If this were to happen, the procedure would cycle indefinitely. No difficulty of this type occurred in the above example, but the difficulty might occur in complicated examples. Cycling would be avoided if, when at the point P, we ensured that the next step in the simplex method involved a move away from P to an adjacent point R on the polygon such that this increased $\mathbf{c}^T\mathbf{x}$. However, cycling seems to be rare in practice and will not be discussed further here.

EXERCISE 7.32. Verify our computations with the simplex method starting with (7.40).

EXERCISE 7.33. Complete the application of the simplex method when we first pivot on the (2,2) rather than the (3,2) element in (7.40).

EXERCISE 7.34. Try to create an example in which cycling occurs.

3. Contradictory Constraints

Suppose that we consider

$$2x_1 + x_2 \geq 70, \qquad x_1 + x_2 \leq 40, \qquad x_1 + 3x_2 \geq 90. \qquad (7.41)$$

The situation is illustrated in Figure 7.3, from which we see that there is no point (x_1, x_2) that satisfies all three conditions–i.e., the constraints are contradictory. Algebraically, if we insert slack variables to convert (7.41) into

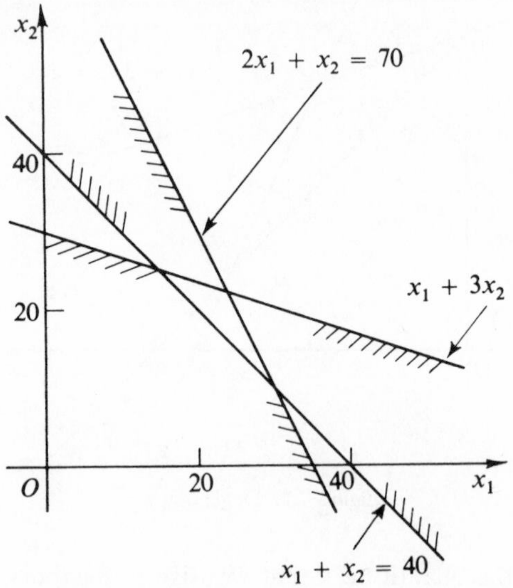

Figure 7.3. Contradictory constraints.

equalities, there is no obvious feasible vector. In attempting to find a feasible solution, we introduce two artificial variables and first maximize

$$M = -x_6 - x_7.$$

This means that we start from the array

$$\begin{bmatrix} 2 & 1 & -1 & 0 & 0 & 1 & 0 & 70 \\ 1 & 1 & 0 & 1 & 0 & 0 & 0 & 40 \\ 1 & 3 & 0 & 0 & -1 & 0 & 1 & 90 \\ 0 & 0 & 0 & 0 & 0 & 1 & 1 & 0 \end{bmatrix}. \tag{7.42}$$

An obvious basic feasible vector is

$$x_4 = 40, \quad x_6 = 70, \quad x_7 = 90, \quad x_1 = x_2 = x_3 = x_5 = 0.$$

We therefore reduce the elements in the sixth and seventh columns of the last row of (7.42) to zero and then apply the simplex method in the usual way. This eventually gives the array

$$\begin{bmatrix} 0 & 0 & -1 & -2.5 & -0.5 & 1 & 0.5 & 15 \\ 1 & 0 & 0 & 1.5 & 0.5 & 0 & -0.5 & 15 \\ 0 & 1 & 0 & -0.5 & -0.5 & 0 & 0.5 & 25 \\ 0 & 0 & 1 & 2.5 & 0.5 & 0 & 0.5 & -15 \end{bmatrix}.$$

Since there are no negative numbers in the first seven elements of the last row, we know that we have maximized $-x_6 - x_7$. Hence, the maximum value of M is -15, corresponding to $x_1 = 15$, $x_2 = 25$, $x_6 = 15$, and all the other variables zero. We are now in situation (2) preceding (7.31) and no basic feasible solution of (7.41) exists. From a slightly different point of view, the reason for our difficulties is that the feasible solution obtained contains one of the artificial variables, namely x_6.

EXERCISE 7.35. Verify our computations (immediately above) by applying the simplex method, starting with (7.42).

EXERCISE 7.36. Use the simplex method to show that there are no feasible vectors for

$$\begin{aligned} x_1 + x_2 \qquad &\geq 3 \\ x_1 + 2x_2 + x_3 &\geq 4 \\ 2x_1 + x_2 + x_3 &\leq 2 \\ x_1 \geq 0, \quad x_2 \geq 0, \quad & x_3 \geq 0. \end{aligned}$$

4. No Maximum Exists

Consider the constraints

$$2x_1 + x_2 \geq 70, \qquad x_1 + x_2 \geq 40, \qquad x_1 + 3x_2 \geq 90.$$

It is obvious both graphically and algebraically that any sufficiently large values of x_1, x_2 satisfy these constraints so that $M = 40x_1 + 60x_2$ can be made as large as desired. If we introduce slack and artificial variables, and perform the usual pivoting operations we obtain the array

$$\begin{bmatrix} 1 & 0 & -1 & 1 & 0 & 30 \\ 5 & 0 & -3 & 0 & 1 & 120 \\ 2 & 1 & -1 & 0 & 0 & 70 \\ 80 & 0 & -60 & 0 & 0 & 4200 \end{bmatrix}. \tag{7.43}$$

The elements in the third column are all negative, and this is the situation in **Key Theorem 7.4**(ii)(a), which shows that M can be made as large as desired and hence that no solution exists.

EXERCISE 7.37. Verify that the simplex method leads to (7.43).

EXERCISE 7.38. Use the simplex method to show that $M = 3x_1 - x_2$ does not achieve a maximum value subject to

$$\begin{aligned} -2x_1 + x_2 &\le 1 \\ x_1 - 2x_2 &\le 2 \\ x_1 \ge 0, \quad x_2 &\ge 0. \end{aligned}$$

EXERCISE 7.39. Use the simplex method to show that $M = 3x_2 - x_1$ does not achieve a maximum value subject to

$$\begin{aligned} x_1 \qquad &\le 10 \\ x_1 + x_2 &\ge 5 \\ x_1 - x_2 &\le 5 \\ x_1 \ge 0, \quad x_2 &\ge 0. \end{aligned}$$

5. Nonunique Optimal Solutions

Suppose that, instead of maximizing M defined in (7.1), we wish to maximize

$$M = 40x_1 + 40x_2$$

subject to (7.2). Graphically, it is clear that the corresponding family of lines, for various M, is parallel to the line $x_1 + x_2 = 40$ in Figure 2.6, which is one of the boundaries of the polygonal region. Any point on the line PQ in the figure will give the maximum value of M. The maximum value itself is unique but an infinite number of points (x_1, x_2) will give this maximum value.

We now investigate what happens if we solve this problem by the simplex method. The initial tableau is

$$\begin{bmatrix} 2 & 1 & 1 & 0 & 0 & 70 \\ 1 & 1 & 0 & 1 & 0 & 40 \\ 1 & 3 & 0 & 0 & 1 & 90 \\ -40 & -40 & 0 & 0 & 0 & 0 \end{bmatrix}.$$

Since both nonzero numbers in the last row are -40, we can introduce either x_1 or x_2 into the basic solution. If we choose to introduce x_2, then the rule described previously indicates that we should pivot on the element 3 in the third row, second column. In the resulting tableau, the standard rule indicates that we should pivot on the element in the second row, first column. This gives the tableau

$$\begin{bmatrix} 0 & 0 & 1 & -2.5 & 0.5 & 15 \\ 1 & 0 & 0 & 1.5 & -0.5 & 15 \\ 0 & 1 & 0 & -0.5 & 0.5 & 25 \\ 0 & 0 & 0 & 40 & 0 & 1600 \end{bmatrix}. \tag{7.44}$$

which yields the basic optimal feasible vector

$$x_1 = 15, \quad x_2 = 25, \quad x_3 = 15, \quad x_4 = 0, \quad x_5 = 0. \tag{7.45}$$

Also,

$$M = 1600 - 40x_4.$$

Since x_5 does not appear in this equation and positive numbers appear in the fifth column in (7.44), we can obtain a second feasible solution from (7.44), without changing the value of M. By pivoting on the fifth element of the first row of (7.44), we obtain

$$\begin{bmatrix} 0 & 0 & 2 & -5 & 1 & 30 \\ 1 & 0 & 1 & -1 & 0 & 30 \\ 0 & 1 & 0 & 2 & 0 & 10 \\ 0 & 0 & 0 & 40 & 0 & 1600 \end{bmatrix},$$

giving a second basic optimal feasible vector

$$x_1 = 30, \quad x_2 = 10, \quad x_3 = 0, \quad x_4 = 0, \quad x_5 = 30. \tag{7.46}$$

Both optimal vectors (7.45), (7.46) of course yield $M = 1600$. This illustrates **Key Theorem 7.4**(iii), where $c_5 = 0$; we note that for $0 \leq \alpha \leq 1$, another optimal vector is given by

$$\begin{aligned} &x_1 = 30 - 15\alpha, \quad && x_2 = 10 + 15\alpha, \quad && x_3 = 15\alpha, \\ &x_4 = 0, && x_5 = 30 - 30\alpha. \end{aligned} \tag{7.47}$$

EXERCISE 7.40. Verify that the simplex method leads to (7.44) if we first pivot on the (3, 2) element of the original tableau.

EXERCISE 7.41. Maximize $40x_1 + 40x_2$ subject to (7.2) using the simplex method and first pivoting on the (1, 1) element of the original tableau.

EXERCISE 7.42. Verify that the vectors given by (7.47) for $0 \leq \alpha \leq 1$ satisfy the relevant constraints and yield a value of M that is optimal.

7.5 DUALITY

We return for a moment to our original concrete example (7.1), (7.2) and consider it geometrically by examining Figure 2.6. In that figure we wish to maximize $40x_1 + 60x_2$ over the region $ORPQS$; this corresponds to moving the lines described by the equations $40x_1 + 60x_2 = M$ for various constants M (see the dashed lines in Figure 2.6) as far as possible to the "northeast," thus locating P as the maximizing point. Geometrically, we see that the "northeasterly" perpendicular to the line $40x_1 + 60x_2 = 2100$ through the optimal vector P lies in the angle between the "northeasterly" perpendiculars to the lines RP and PQ at P; this says that the first perpendicular can be written as a nonnegative linear combination of the latter two perpendiculars. In this case, since the mentioned perpendiculars are in the directions

$$\begin{bmatrix} 40 \\ 60 \end{bmatrix}, \quad \begin{bmatrix} 1 \\ 3 \end{bmatrix}, \quad \text{and} \quad \begin{bmatrix} 1 \\ 1 \end{bmatrix},$$

respectively, we have

$$\begin{bmatrix} 40 \\ 60 \end{bmatrix} = 10\begin{bmatrix} 1 \\ 3 \end{bmatrix} + 30\begin{bmatrix} 1 \\ 1 \end{bmatrix}.$$

It should be noted that the vector

$$\begin{bmatrix} 40 \\ 60 \end{bmatrix}$$

is just the vector $\mathbf{c}$ appearing in our function $M = \mathbf{c}^T\mathbf{x}$ to be maximized, while the vectors

$$\begin{bmatrix} 1 \\ 3 \end{bmatrix}, \quad \begin{bmatrix} 1 \\ 1 \end{bmatrix}$$

are just the (transposes of the) rows of the matrix $\mathbf{A}$ in $\mathbf{Ax} \leq \mathbf{b}$ which are satisfied as *equalities* at the optimal vector. This situation occurs in general (see Ex. 7.59); that is, if $\mathbf{x}^*$ maximizes $\mathbf{c}^T\mathbf{x}$ subject to $\mathbf{Ax} \leq \mathbf{b}$ and $\mathbf{x} \geq \mathbf{0}$ where $\mathbf{A}$ is $m \times n$, then $\mathbf{c}^T$ can be written as a nonnegative linear combination of those rows of $\mathbf{A}$ corresponding to inequalities actually satisfied as equalities at the solution $\mathbf{x}^*$ (the so-called *active* constraints at $\mathbf{x}^*$). We explore briefly what this means.

Since $\mathbf{c}^T$ is a nonnegative linear combination of the rows of $\mathbf{A}$ corresponding to the active constraints at $\mathbf{x}^*$, there is an $m \times 1$ vector $\mathbf{y}^* \geq \mathbf{0}$ such that

$$\mathbf{c}^T = \mathbf{y}^{*T}\mathbf{A}, \quad \text{that is} \quad \mathbf{A}^T\mathbf{y}^* = \mathbf{c},$$

and such that those components of $\mathbf{y}^*$ multiplying rows of $\mathbf{A}$ corresponding to *inactive* constraints at $\mathbf{x}^*$ must vanish. Suppose now that $\mathbf{y}$ is any other $m \times 1$ vector satisfying the *dual constraints*

$$\mathbf{A}^T\mathbf{y} \geq \mathbf{c}, \qquad \mathbf{y} \geq \mathbf{0}. \tag{7.48}$$

Since $\mathbf{A}\mathbf{x}^* \leq \mathbf{b}$ and $\mathbf{y} \geq \mathbf{0}$ it follows that

$$\mathbf{y}^T\mathbf{b} \geq \mathbf{y}^T\mathbf{A}\mathbf{x}^*.$$

Since $\mathbf{y}^T\mathbf{A} \geq \mathbf{c}^T$ and $\mathbf{x}^* \geq \mathbf{0}$, it further follows that

$$\mathbf{y}^T\mathbf{b} \geq \mathbf{y}^T\mathbf{A}\mathbf{x}^* \geq \mathbf{c}^T\mathbf{x}^*. \tag{7.49}$$

Furthermore, since $\mathbf{y}^*$ vanishes in the components corresponding to inactive rows while in the other components the inequalities $\mathbf{A}\mathbf{x}^* \leq \mathbf{b}$ are satisfied as equalities, it follows that

$$\mathbf{y}^{*T}\mathbf{b} = \mathbf{y}^{*T}\mathbf{A}\mathbf{x}^* = \mathbf{c}^T\mathbf{x}^*. \tag{7.50}$$

But then (7.48) and (7.49) together state that

$$\mathbf{y}^T\mathbf{b} \geq \mathbf{y}^{*T}\mathbf{b} = \mathbf{c}^T\mathbf{x}^*$$

whenever $\mathbf{y}$ satisfies (7.48). That is, $\mathbf{y}^*$ *solves the linear program*

$$\begin{aligned} &\text{minimize} \quad \mathbf{b}^T\mathbf{y} \\ &\text{subject to} \quad \mathbf{A}^T\mathbf{y} \geq \mathbf{c}, \quad \mathbf{y} \geq \mathbf{0} \end{aligned} \tag{7.51}$$

and the minimal value of this program (7.51) *equals the maximal value of the original program*

$$\begin{aligned} &\text{maximize} \quad \mathbf{c}^T\mathbf{x} \\ &\text{subject to} \quad \mathbf{A}\mathbf{x} \leq \mathbf{b}, \qquad \mathbf{x} \geq \mathbf{0}. \end{aligned} \tag{7.52}$$

The original linear program (7.52) is called the *primal* problem, and the new linear program (7.51) we have just defined is called the *dual* problem. The relationships between these two programs are very important both theoretically and computationally. Although we motivated the dual program geometrically, once we have defined these two programs we can proceed to explore their relationships from basic principles.

EXERCISE 7.43. Find the dual of the linear program (7.1), (7.2).

SOLUTION: The matrices **A**, **b**, and **c** for this program were found in (2.40). The dual can immediately be written down as

$$\begin{aligned} &\text{minimize} \quad 70y_1 + 40y_2 + 90y_3 \\ &\text{subject to} \quad 2y_1 + y_2 + y_3 \geq 40 \\ &\qquad\qquad y_1 + y_2 + 3y_3 \geq 60 \\ &y_1 \geq 0, \quad y_2 \geq 0, \quad y_3 \geq 0. \end{aligned} \tag{7.53}$$

EXERCISE 7.44. Find the dual of the linear program (7.53).

SOLUTION: We have to write this in our standard format of *maximization* and "$\leq$" constraints; this yields

$$\begin{aligned} &\text{maximize} \quad -70y_1 - 40y_2 - 90y_3 \\ &\text{subject to} \quad -2y_1 - y_2 - y_3 \leq -40 \\ &\qquad\qquad -y_1 - y_2 - 3y_3 \leq -60 \\ &y_1 \geq 0, \quad y_2 \geq 0, \quad y_3 \geq 0. \end{aligned}$$

We then write down the dual:

$$\begin{aligned} &\text{minimize} \quad -40z_1 - 60z_2 \\ &\text{subject to} \quad -2z_1 - z_2 \geq -70 \\ &\qquad\qquad -z_1 - z_2 \geq -40 \\ &\qquad\qquad -z_1 - 3z_2 \geq -90 \end{aligned}$$

which we discover is equivalent to (7.1), (7.2); the dual of the dual turned out to be the primal!

We first demonstrate the **key** theoretical relationships between the primal and dual linear programs.

- *KEY THEOREM 7.5*

(i) *The dual of the dual is the primal.*
(ii) *If* **x** *satisfies the primal constraints* (7.52) *and* **y** *satisfies the dual constraints* (7.51), *then* $\mathbf{c}^T\mathbf{x} \leq \mathbf{b}^T\mathbf{y}$.
(iii) *If* $\bar{\mathbf{x}}$ *satisfies the primal constraints* (7.52) *and* $\bar{\mathbf{y}}$ *satisfies the dual constraints* (7.51) *and* $\mathbf{c}^T\bar{\mathbf{x}} = \mathbf{b}^T\bar{\mathbf{y}}$, *then* $\bar{\mathbf{x}}$ *and* $\bar{\mathbf{y}}$ *are optimal vectors for the primal and dual, respectively.*
(iv) *If both the primal and the dual are nondegenerate and have feasible vectors, then they both have optimal feasible vectors and the optimal values are equal.*
(v) *If either the primal or the dual fails to have a feasible vector, then both programs fail to have optimal vectors.*

Proof: Part (i) follows trivially from the definitions, just as in Ex. 7.44. For part (ii), just as in our arguments motivating our original definition of the dual, we have $\mathbf{c}^T\mathbf{x} \leq (\mathbf{y}^T\mathbf{A})\mathbf{x} = \mathbf{y}^T(\mathbf{A}\mathbf{x}) \leq \mathbf{y}^T\mathbf{b}$ as asserted. To prove (iii), note from (ii) that for any primal feasible $\mathbf{x}$ we have $\mathbf{c}^T\mathbf{x} \leq \mathbf{b}^T\bar{\mathbf{y}} = \mathbf{c}^T\bar{\mathbf{x}}$ and hence $\bar{\mathbf{x}}$ is optimal; the argument for $\bar{\mathbf{y}}$ is similar. To prove (iv), let $\mathbf{x}_0$ and $\mathbf{y}_0$ be the guaranteed primal and dual feasible vectors. From (ii), we have $\mathbf{c}^T\mathbf{x} \leq \mathbf{b}^T\mathbf{y}_0$ for all primal feasible $\mathbf{x}$, so that $M = \mathbf{c}^T\mathbf{x}$ is *not* unbounded above; since also the primal feasible set is nonempty ($\mathbf{x}_0$ is in it), by **Key Theorem 7.4**(iv) the primal has an optimal vector. Using (i), it also follows that the dual has an optimal vector. The fact that the optimal values must be equal will be proved in the next theorem when we construct an optimal dual solution from an optimal primal solution. To prove (v), if, say, the primal problem has an optimal vector, then the construction that we give in the next theorem creates an optimal vector for the dual problem as well; by (i), we conclude that either *both* or *neither* programs have optimal vectors (and hence nonempty feasible sets), completing the proof.

Twice in the above proof we had to appeal to the next theorem, which gives a construction for an optimal vector for the dual (respectively, primal) in terms of an available optimal vector for the primal (respectively, dual). Although delaying this theorem makes for an awkward proof, the computational impact of the next theorem is so great that the result needs to be stated independently. Before we proceed with this however, we give some examples to clarify **Key Theorem 7.5** and its use.

EXERCISE 7.45. Use duality theory to demonstrate that there are no feasible vectors for (7.41).

SOLUTION: Suppose that we want to maximize $c_1x_1 + c_2x_2$ subject to (7.41). By **Key Theorem 7.5(ii)**, if the dual is unbounded below, the primal can have no feasible points. The dual in this case is

$$\begin{aligned} &\text{minimize} \quad -70y_1 + 40y_2 - 90y_3 \\ &\text{subject to} \quad -2y_1 + y_2 - y_3 \geq c_1 \\ &\qquad\qquad\quad -y_1 + y_2 - 3y_3 \geq c_2, \\ &y_1 \geq 0, \quad y_2 \geq 0, \quad y_3 \geq 0. \end{aligned}$$

If we let $y_1 = c_2 - c_1 + 2\alpha$, $y_2 = 2c_2 - c_1 + 5\alpha$, $y_3 = \alpha$ and let α tend to plus infinity, we see that the constraints are satisfied and $-70y_1 + 40y_2 - 90y_3$ has the value $30c_1 + 10c_2 - 3\alpha$, which is unbounded below. Thus the dual has feasible vectors and is unbounded below, so the primal has no feasible vectors.

EXERCISE 7.46. Use the simplex method to solve the dual in Ex. 7.45, treating c_1 and c_2 as parameters; show that we arrive at the situation indicating an unbounded objective function, and that we generate the vectors $\mathbf{y}$ already found in Ex. 7.45, that make the objective function unbounded.

EXERCISE 7.47. Use duality theory to demonstrate that the program in (7.43) is unbounded above.

SOLUTION: The problem is to maximize $40x_1 + 60x_2$ subject to $2x_1 + x_2 \geq 40$, $x_1 + 3x_2 \geq 90$, $x_1 \geq 0$, $x_2 \geq 0$. By **Key Theorem 7.5**(iv), (v), since feasible vectors clearly exist, what we must show is that the *dual* has no feasible vector. The dual constraints are just

$$\begin{aligned} -2y_1 - y_2 - y_3 &\geq 40 \\ -y_1 - y_2 - 3y_3 &\geq 60 \\ y_1 \geq 0, \quad y_2 \geq 0, \quad y_3 &\geq 0 \end{aligned}$$

which is clearly impossible to satisfy.

Now we show how to obtain an optimal vector for the dual from an optimal vector for the primal; first we work with our familiar concrete example (7.1), (7.2),

$$\begin{aligned} \text{maximize} \quad & 40x_1 + 60x_2 \\ \text{subject to} \quad & 2x_1 + x_2 \leq 70 \\ & x_1 + x_2 \leq 40 \\ & x_1 + 3x_2 \leq 90 \\ & x_1 \geq 0, \quad x_2 \geq 0, \end{aligned} \tag{7.54}$$

whose dual we found in Ex. 7.43 to be equivalent to

$$\begin{aligned} \text{maximize} \quad & -70y_1 - 40y_2 - 90y_3 \\ \text{subject to} \quad & -2y_1 - y_2 - y_3 \leq -40 \\ & -y_1 - y_2 - 3y_3 \leq -60 \\ & y_1 \geq 0, \quad y_2 \geq 0, \quad y_3 \geq 0. \end{aligned} \tag{7.55}$$

Applying the simplex method (see Section 7.1) to (7.54) we terminate with the tableau of (7.16), namely

$$\begin{bmatrix} 0 & 0 & 1 & -2.5 & 0.5 & 15 \\ 1 & 0 & 0 & 1.5 & -0.5 & 15 \\ 0 & 1 & 0 & -0.5 & 0.5 & 25 \\ 0 & 0 & 0 & 30 & 10 & 2100 \end{bmatrix}. \tag{7.56}$$

If we add slack variables y_4, y_5 to (7.55) and then subtract artificial variables y_6, y_7 and maximize $-(y_6 + y_7)$, we obtain the feasible vector $y_1 = 12$,

$y_2 = 0$, $y_3 = 16$, $y_4 = y_5 = 0$ for the extended problem. Applying the simplex method again finally leads to the tableau

$$\begin{bmatrix} 2.5 & 1 & 0 & -1.5 & 0.5 & 30 \\ -0.5 & 0 & 1 & 0.5 & -0.5 & 10 \\ 15 & 0 & 0 & 15 & 25 & -2100 \end{bmatrix}. \tag{7.57}$$

The same numbers appear in the two terminal tableaus (7.56), (7.57), but in different patterns; while the numbers in the last row (column) of (7.56) appear in the last column (row) of (7.57), the correspondence between the remaining numbers is not so obvious. To understand this correspondence better, we consider the general problem of solving

$$\begin{aligned} &\text{maximize} \quad \mathbf{c}^T\mathbf{x} \\ &\text{subject to} \quad \mathbf{Ax} \leq \mathbf{b}, \qquad \mathbf{x} \geq \mathbf{0} \end{aligned} \tag{7.58}$$

by the simplex method. As described in Section 7.2 we add slack variables and begin the simplex method with the initial tableau

$$\mathbf{T} = \begin{bmatrix} \mathbf{A} & \mathbf{I} & \mathbf{b} \\ -\mathbf{c}^T & \mathbf{0}^T & 0 \end{bmatrix}. \tag{7.59}$$

It was apparent from our discussion of the revised simplex method in Section 7.2 that the row operations we use in this method to transform $\mathbf{T}$ into the final tableau are equivalent to premultiplying $\mathbf{T}$ by

$$\left[\begin{array}{c|c} \mathbf{F} & \mathbf{0} \\ \hline \mathbf{v}^T & 1 \end{array}\right].$$

If the final tableau obtained is partitioned as in (7.59), then this tableau can be written

$$\begin{bmatrix} \mathbf{P}_1 & \mathbf{P}_2 & \mathbf{x}_B \\ \mathbf{d}_1^T & \mathbf{d}_2^T & M_0 \end{bmatrix} = \begin{bmatrix} \mathbf{F} & \mathbf{0} \\ \mathbf{v}^T & 1 \end{bmatrix} \begin{bmatrix} \mathbf{A} & \mathbf{I} & \mathbf{b} \\ -\mathbf{c}^T & \mathbf{0}^T & 0 \end{bmatrix}; \tag{7.60}$$

hence

$$\mathbf{d}_1^T = \mathbf{v}^T\mathbf{A} - \mathbf{c}^T \tag{7.61}$$

$$\mathbf{d}_2^T = \mathbf{v}^T \tag{7.62}$$

$$M_0 = \mathbf{v}^T\mathbf{b}. \tag{7.63}$$

If this is the terminal tableau in the simplex method and we have found an optimal solution, then $\mathbf{d}_1^T \geq \mathbf{0}$, $\mathbf{d}_2^T \geq \mathbf{0}$, M_0 is the maximum value of $\mathbf{c}^T\mathbf{x}$, and $\mathbf{x}_B$ contains the values of the optimal basic variables. Since $\mathbf{d}_1^T \geq \mathbf{0}$, (7.61) tells us that

$$\mathbf{A}^T\mathbf{v} \geq \mathbf{c}.$$

Since $\mathbf{d}_2^T \geq \mathbf{0}$, (7.62) tells us that

$$\mathbf{v} \geq \mathbf{0}.$$

Therefore $\mathbf{v}$ is a feasible vector for the dual problem

$$\begin{aligned} &\text{minimize} \quad \mathbf{b}^T\mathbf{y} \\ &\text{subject to} \quad \mathbf{A}^T\mathbf{y} \geq \mathbf{c} \end{aligned} \tag{7.64}$$

associated with the primal (7.58). Moreover, since $\mathbf{v}^T\mathbf{b} = M_0$ from (7.63) while M_0 is the maximum value of $\mathbf{c}^T\mathbf{x}$, **Key Theorem 7.5**(iii) states that $\mathbf{v}$ is in fact an optimal feasible vector for the dual problem. Thus we have proved that an optimal feasible vector for the dual problem can be read off immediately from the last row of the simplex tableau corresponding to an optimal basic feasible vector for the primal; likewise, by **Key Theorem 7.5**(i), we can solve the primal via the dual's final tableau. We restate this **key** result.

- *KEY THEOREM 7.6. Let the nondegenerate primal linear program* (7.58) *have an optimal feasible vector* $\bar{\mathbf{x}}$. *Then the dual linear program* (7.64) *also has an optimal feasible vector* $\bar{\mathbf{y}}$ *which can be generated as follows. For any optimal basic feasible solution* $\mathbf{x}^*$ *of the primal problem, let* (7.60) *describe the final tableau corresponding to* $\mathbf{x}^*$, *so that the basic variables in* $\mathbf{x}^*$ *are just given by* $\mathbf{x}_B$. *Then* $\mathbf{y} = \mathbf{d}_2$ *is an optimal feasible vector for the dual linear program. That is, the numbers in the last row of the final tableau for the primal problem, in the columns corresponding to the slack variables in the original tableau, form an optimal vector for the dual problem.*

Before considering examples, we emphasize the *computational* importance of **Key Theorem 7.6.** Suppose that we have very many constraints but only a few variables, so that $m \gg n$ where $\mathbf{A}$ is $m \times n$. Then, in applying the standard simplex method to the primal problem, each step will require elimination in a large number m of rows; similarly, the revised simplex method requires the updating of large $m \times m$ matrices. If, however, we attack the dual problem instead, since $\mathbf{A}^T$ is $n \times m$ we have few constraints and many variables; we need only eliminate in n rows or update small $n \times n$ matrices in applying the simplex method to the dual problem. Once we have obtained an optimal basic feasible vector for the dual in this way, we use **Key Theorem 7.6** to read off an optimal feasible vector for the primal at no extra cost. Thus solving the dual can be much more efficient than solving the primal if $m \gg n$ (see Ex. 7.50).

EXERCISE 7.48.

$$\begin{aligned} \text{Minimize} \quad & y_1 + y_2 \\ \text{subject to} \quad & 2y_1 + y_2 \geq 2 \\ & y_1 + 2y_2 \geq 2 \\ & 6y_1 + y_2 \geq 3 \\ & y_1 \geq 0, \quad y_2 \geq 0. \end{aligned}$$

SOLUTION: This is the dual of:

$$\begin{aligned} \text{maximize} \quad & 2x_1 + 2x_2 + 3x_3 \\ \text{subject to} \quad & 2x_1 + x_2 + 6x_3 \leq 1 \\ & x_1 + 2x_2 + x_3 \leq 1 \\ & x_1 \geq 0, \quad x_2 \geq 0, \quad x_3 \geq 0, \end{aligned}$$

and vice versa. Adding slack variables x_4, x_5, the initial tableau is

$$\begin{bmatrix} 2 & 1 & 6 & 1 & 0 & 1 \\ 1 & 2 & 1 & 0 & 1 & 1 \\ -2 & -2 & -3 & 0 & 0 & 0 \end{bmatrix}.$$

We pivot on the (1, 3) element and produce the tableau

$$\begin{bmatrix} \frac{1}{3} & \frac{1}{6} & 1 & \frac{1}{6} & 0 & \frac{1}{6} \\ \frac{2}{3} & \frac{11}{6} & 0 & -\frac{1}{6} & 1 & \frac{5}{6} \\ -1 & -\frac{3}{2} & 0 & \frac{1}{2} & 0 & \frac{1}{2} \end{bmatrix}.$$

We pivot on the (2, 2) element and produce the tableau

$$\begin{bmatrix} \frac{3}{11} & 0 & 1 & \frac{2}{11} & -\frac{1}{11} & \frac{1}{11} \\ \frac{4}{11} & 1 & 0 & -\frac{1}{11} & \frac{6}{11} & \frac{5}{11} \\ -\frac{5}{11} & 0 & \frac{1}{2} & \frac{4}{11} & \frac{9}{11} & \frac{13}{11} \end{bmatrix}.$$

We pivot on the (1, 1) element and produce the tableau

$$\begin{bmatrix} 1 & 0 & \frac{11}{3} & \frac{2}{3} & -\frac{1}{3} & \frac{1}{3} \\ 0 & 1 & -\frac{4}{3} & -\frac{1}{3} & \frac{2}{3} & \frac{1}{3} \\ 0 & 0 & \frac{13}{6} & \frac{2}{3} & \frac{2}{3} & \frac{4}{3} \end{bmatrix}$$

so that the solution in the **x** variables is $x_1 = \frac{1}{3}$, $x_2 = \frac{1}{3}$, $x_3 = 0$, while the solution in the **y** variables comes from the fourth and fifth elements in the last row, namely $y_1 = y_2 = \frac{2}{3}$.

EXERCISE 7.49.

$$\begin{aligned} \text{Maximize} \quad & x_1 + x_2 \\ \text{subject to} \quad & -x_1 + x_2 \le 10 \\ & x_1 + 2x_2 \le 50 \\ & 5x_1 + x_2 \le 160 \\ & x_2 \le 15 \\ & x_1 \ge 0, \quad x_2 \ge 0. \end{aligned}$$

SOLUTION: Since there are only two variables but four constraints we attack the dual:

$$\begin{aligned} \text{minimize} \quad & 10y_1 + 50y_2 + 160y_3 + 15y_4 \\ \text{subject to} \quad & -y_1 + y_2 + 5y_3 \ge 1 \\ & y_1 + 2y_2 + y_3 + y_4 \ge 1 \\ & y_i \ge 0 \quad (\text{for } i = 1, \ldots, 4). \end{aligned}$$

If we subtract slack variables y_5, y_6, introduce artificial variables y_7, y_8 and maximize $-(y_7 + y_8)$ we obtain a feasible vector for the extended problem; applying the simplex method yields the final tableau

$$\begin{bmatrix} -\frac{1}{3} & 0 & 1 & -\frac{1}{9} & -\frac{2}{9} & \frac{1}{9} & \frac{1}{9} \\ \frac{2}{3} & 0 & 0 & \frac{5}{9} & \frac{1}{9} & -\frac{5}{9} & \frac{4}{9} \\ 30 & 0 & 0 & 5 & 30 & 10 & -40 \end{bmatrix}.$$

Since the slack variables were y_5, y_6, we read off the solution from columns five and six of the bottom row; an optimal feasible vector for the primal problem is $x_1 = 30$, $x_2 = 10$, and the corresponding optimal value is 40.

EXERCISE 7.50. Suppose that $\mathbf{A}$ is $m \times n$, so that the tableau for the linear program with the constraints $\mathbf{Ax} \le \mathbf{b}$ will be $(m + 1) \times (m + n + 1)$ after inserting slack variables. Find the number of multiplication-divisions and addition-subtractions necessary in one pivot step of the simplex method, that is, in performing row operations to reduce a column to a unit column vector. Explain why it is better to solve the dual than the primal when $m \gg n$.

It would take us too far afield to illustrate in more detail why the primal-dual relationship is both important and useful. We have already noted that it is often more efficient to solve one problem rather than the other. Moreover a basic feasible solution may be obvious in one case but not in the other. Also, if $\mathbf{A}$ is $m \times n$, then basic solutions of the primal and dual have, respectively, m and n nonzero variables. Other things being equal, it will be simpler to solve the problem for which the basic solution has the smaller number of basic variables. Finally, if we have solved a given linear programming problem and wish to add an extra constraint, it would seem to be necessary to start the solution from the beginning, because feasible solutions of the original system do not necessarily satisfy the additional constraint. However, in the dual it is only necessary to add an extra *variable*. In this case we can start from the previous solution, initially setting the additional variable in the dual equal to zero.

EXERCISE 7.51. Use the simplex method to solve (7.55) so as to obtain (7.57).

EXERCISE 7.52. Show that (7.61)–(7.63) are valid. Find formulas for $\mathbf{P}_1$, $\mathbf{P}_2$, and $\mathbf{x}_B$ in (7.60).

EXERCISE 7.53. Solve the original linear program of Ex. 7.48 in the **y** variables by the simplex method.

EXERCISE 7.54. Describe the nature of the linear program:

$$\begin{aligned} \text{minimize} \quad & -x_1 + 2x_2 \\ \text{subject to} \quad & -5x_1 + x_2 \geq 2 \\ & 4x_1 - x_2 \geq 3 \\ & x_1 \geq 0, \quad x_2 \geq 0 \end{aligned}$$

by solving its dual.

EXERCISE 7.55.

$$\begin{aligned} \text{Maximize} \quad & x_1 - x_2 \\ \text{subject to} \quad & -2x_1 + x_2 \leq 2 \\ & x_1 - 2x_2 \leq 1 \\ & x_1 + x_2 \leq 4 \\ & x_1 \geq 0, \quad x_2 \geq 0 \end{aligned}$$

by solving its dual.

MISCELLANEOUS EXERCISES 7

EXERCISE 7.56. By rewriting the condition $\mathbf{Bx} = \mathbf{d}$ as $\mathbf{Bx} \leq \mathbf{d}$ and $-\mathbf{Bx} \leq -\mathbf{d}$, find the dual of: maximize $\mathbf{c}^T\mathbf{x}$ subject to $\mathbf{Bx} = \mathbf{d}$, $\mathbf{x} \geq 0$.

EXERCISE 7.57. Show that the problem of maximizing

$$\tfrac{3}{4}x_1 - 150x_2 + \tfrac{1}{50}x_3 - 6x_4$$

subject to

$$\begin{aligned} \tfrac{1}{4}x_1 - 60x_2 - \tfrac{1}{25}x_3 + 9x_4 &\leq 0 \\ \tfrac{1}{2}x_1 - 90x_2 - \tfrac{1}{50}x_3 + 3x_4 &\leq 0 \\ x_3 &\leq 1 \end{aligned}$$

$$x_1 \geq 0, \quad x_2 \geq 0, \quad x_3 \geq 0, \quad x_4 \geq 0,$$

is degenerate. When the simplex method is applied, there will be "ties" in the column of the tableau in deciding which row to use in the elimination process; if ties are broken by using the row with the least row index, then the simplex method cycles and never converges. Demonstrate this by applying the method.

EXERCISE 7.58. Suppose that to the problem (7.1), (7.2) we add the constraint $4x_1 + x_2 \leq 80$. Use duality as indicated at the very end of Section 7.5 to solve the resulting problem.

EXERCISE 7.59. Let $\mathbf{x}^*$ solve the primal (7.52) and $\mathbf{y}^*$ solve the dual (7.51) of a nondegenerate problem, and let $\mathbf{r} = \mathbf{b} - \mathbf{A}\mathbf{x}^*$ and $\mathbf{p} = \mathbf{A}^T\mathbf{y}^* - \mathbf{c}$ be the corresponding slack values. Prove that $\mathbf{p}^T\mathbf{x}^* = \mathbf{r}^T\mathbf{y}^* = 0$; deduce that $\mathbf{c}$ is a nonnegative combination of the rows of $\mathbf{A}$ corresponding to constraints active at $\mathbf{x}^*$.

EXERCISE 7.60. Consider the linear program

$$\text{minimize} \quad \mathbf{c}^T\mathbf{x}$$

$$\text{subject to} \quad \mathbf{A}\mathbf{x} = \mathbf{b}, \quad \mathbf{x} \geq \mathbf{0}$$

where $\mathbf{A}$ is $m \times n$ and $(n/2) < m < n$ and $\mathbf{A}$ has full rank m. Show how to reduce this to a *smaller* problem of size $(n - m) \times n$ as follows. Show that we can find a nonsingular $n \times n$ matrix $\mathbf{S}$ such that

$$\mathbf{A}\mathbf{S} = [\mathbf{B} \quad \mathbf{0}] \tag{7.65}$$

where $\mathbf{B}$ is $m \times m$ and nonsingular. Let $\mathbf{y} = \mathbf{S}^{-1}\mathbf{x}$ and partition the matrices so that

$$\mathbf{y} = \mathbf{S}^{-1}\mathbf{x} = \begin{bmatrix} \mathbf{y}_1 \\ \mathbf{y}_2 \end{bmatrix}, \qquad \mathbf{S} = [\mathbf{S}_1, \mathbf{S}_2]$$

where $\mathbf{y}_1$ is in $\mathbb{R}^m$, $\mathbf{y}_2$ is in $\mathbb{R}^{n-m}$, $\mathbf{S}_1$ is $n \times m$, and $\mathbf{S}_2$ is $n \times (n - m)$. Show then that $\mathbf{A}\mathbf{x} = \mathbf{b}$ and $\mathbf{x} \geq \mathbf{0}$ if and only if $\mathbf{y}_1 = \mathbf{B}^{-1}\mathbf{b}$ and $\mathbf{S}_1\mathbf{y}_1 + \mathbf{S}_2\mathbf{y}_2 \geq \mathbf{0}$. Deduce that the primal linear program is equivalent to

$$\text{minimize} \quad \mathbf{c}^T\mathbf{S}_2\mathbf{y}_2$$

$$\text{subject to} \quad \mathbf{S}_2\mathbf{y}_2 \geq -\mathbf{S}_1\mathbf{B}^{-1}\mathbf{b}$$

whose dual is

$$\text{maximize} \quad -(\mathbf{S}_1\mathbf{B}^{-1}\mathbf{b})^T\mathbf{x}_2$$

$$\text{subject to} \quad \mathbf{S}_2^T\mathbf{x}_2 = \mathbf{S}_2^T\mathbf{c}, \qquad \mathbf{x}_2 \geq \mathbf{0}$$

and the matrix of the equality constraints is now only $(n - m) \times n$. (This technique was communicated to us by Dr. David M. Gay.)

EXERCISE 7.61. Use the method of Ex. 7.60 to transform the problem of minimizing the function (7.27) of five variables subject to the three equality constraints (7.30) to one with only two equality constraints on five variables.

EXERCISE 7.62. A university library is open 24 hours per day, and each librarian works a steady eight-hour shift beginning at 12 midnight, 4 a.m., 8 a.m., 12 noon, 4 p.m., or 8 p.m. To handle the demands for service, the library requires the following numbers of librarians on hand during various time periods: 3 from midnight to 3:59 a.m.; 2 from 4 a.m. to 7:59 a.m.; 10 from 8 a.m. to 11:59 a.m.; 14 from noon to 3:59 p.m.; 8 from 4 p.m. to 7:59 p.m.; and 10 from 8 p.m. to 11:59 p.m. Let $x_1, x_2, \ldots, x_6$ denote the number of persons to start their eight-hour shift at midnight, 4 a.m., . . . , 8 p.m., respectively, and pose as a linear program the problem of minimizing the total number of librarians used to operate the library. Show that an optimal solution is given by $x_1 = 2, x_2 = 0, x_3 = 14, x_4 = 0, x_5 = 8, x_6 = 2$, once the system is in operation.

EXERCISE 7.63. It is assumed that the number N of cell divisions in a certain organism in each time period is approximately $N_0 + bp$ where p is the amount of a certain growth stimulus that is added and where N_0 and b are model parameters to be determined. An experiment is performed using three different amounts $p_1 = 1, p_2 = 3, p_3 = 6$ of the stimulus, and it is found that the respective number of cell divisions is $N_1 = 40, N_2 = 102, N_3 = 190$. To determine the model parameters N_0 and b so as to minimize the maximum error in the model over the three experiments, that is to minimize $\max\{|N_0 + b - 40|, |N_0 + 3b - 102|, |N_0 + 6b - 190|\}$, introduce an additional variable ϵ and solve a linear program whose constraints are equivalent to $\max\{|N_0 + b - 40|, |N_0 + 3b - 102|, |N_0 + 6b - 190|\} \leq \epsilon$.

EXERCISE 7.64. Write an essay on connections between linear programming and the theory of games. Possible references are Glicksman [82] and Vajda [109].

EXERCISE 7.65. Write an essay on applications of linear programming in economics. Possible references are Dorfman, Samuelson, and Solow [76], and Gale [77].

CHAPTER EIGHT

EIGENVALUES AND EIGENVECTORS: AN OVERVIEW

This chapter introduces new concepts that will prove to be extremely useful in simplifying the study of complicated systems. Since the subject is quite complex, this chapter merely motivates and introduces the main concepts and develops some of the basic tools, especially those needed in the more detailed subsequent two chapters. **Key Theorem 8.2** gives the simplest properties of eigensystems, while Section 8.3 presents some important alternative viewpoints for studying the structure of eigensystems; **Key Theorem 8.3** interrelates these viewpoints, which are then explored further in the following two sections, culminating in a preview of the two **key** results from the next two chapters on the structure of eigensystems.

8.1 INTRODUCTION

We have already noted that applied problems often reduce to the study of linear transformations and their effects. The linear transformation might, for example, describe the evolution of some complicated system from one point in time to the next; elementary examples of this appeared in Sections 2.2 and 2.3. More generally, the state of the system at any time might be described by the variables $\mathbf{x}$ in some vector space V, while the linear transformation A

transforms the state $\mathbf{x}$ in V into the state $A(\mathbf{x})$ in V. Since V is often of quite high dimension for real systems, it becomes difficult to understand how the system works. To help us to understand the system we often seek much smaller subsystems, that is, certain sets of variables or combinations of variables, such that the resulting transformed state can also be expressed in terms of those variables or combinations. Algebraically, this corresponds to finding a low-dimensional subspace V_0 of V such that $A(\mathbf{x})$ is in V_0 whenever $\mathbf{x}$ is in V_0; we could then study the simpler problem of the nature of the transformation on this smaller space. Ideally we might even find such a subspace of dimension unity. This, in fact, means that we would try to find a nonzero vector $\mathbf{x}$ (an *eigenvector*) and a scalar λ (an *eigenvalue*) for which $A(\mathbf{x}) = \lambda\mathbf{x}$ so that A maps the one-dimensional subspace V_0 spanned by $\mathbf{x}$ into itself. This motivates the study of *eigensystems* (that is, sets of pairs of eigenvectors and eigenvalues) as related to *invariant subspaces* V_0 of a linear transformation A.

For an example of eigensystems in this context we consider in detail the linear transformation A of $\mathbb{R}^3$ into $\mathbb{R}^3$ defined by the matrix $\mathbf{A}$ in (2.10) of Section 2.2, describing the change during one month in the division of a market among three milk producers. To find an eigenvector $\mathbf{x}$ associated with an eigenvalue λ we need to choose λ so that the equation $A(\mathbf{x}) = \lambda\mathbf{x}$ has a nonzero solution $\mathbf{x}$; since this means that $(\mathbf{A} - \lambda\mathbf{I})\mathbf{x} = \mathbf{0}$ while $\mathbf{x} \neq \mathbf{0}$, we know that the matrix $\mathbf{A} - \lambda\mathbf{I}$ given explicitly by

$$\mathbf{A} - \lambda\mathbf{I} = \begin{bmatrix} 0.8 - \lambda & 0.2 & 0.1 \\ 0.1 & 0.7 - \lambda & 0.3 \\ 0.1 & 0.1 & 0.6 - \lambda \end{bmatrix}$$

must be singular. By Theorem 6.5, therefore, this is equivalent to

$$\det(\mathbf{A} - \lambda\mathbf{I}) = 0.$$

Direct calculation gives

$$\det(\mathbf{A} - \lambda\mathbf{I}) = -\lambda^3 + 2.1\lambda^2 - 1.4\lambda + 0.3 = -(\lambda - 0.5)(\lambda - 0.6)(\lambda - 1.0)$$

so that the eigenvalues λ must be

$$\lambda = 0.5, \qquad \lambda = 0.6, \qquad \lambda = 1.0.$$

To find the associated eigenvectors, consider, for example, the equation $(\mathbf{A} - \lambda\mathbf{I})\mathbf{x} = \mathbf{0}$ for $\lambda = 0.5$. In terms of the components x_1, x_2, x_3 of $\mathbf{x}$, this is

$$\begin{aligned} 0.3x_1 + 0.2x_2 + 0.1x_3 &= 0 \\ 0.1x_1 + 0.2x_2 + 0.3x_3 &= 0 \\ 0.1x_1 + 0.1x_2 + 0.1x_3 &= 0. \end{aligned}$$

Our usual techniques for linear systems of equations reduce this to

$$
\begin{aligned}
x_1 + 0x_2 - x_3 &= 0 \\
0x_1 + x_2 + 2x_3 &= 0 \\
0x_1 + 0x_2 + 0x_3 &= 0;
\end{aligned}
$$

hence we can let x_3 equal an arbitrary α, $x_1 = \alpha$, and $x_2 = -2\alpha$, so that

$$
\mathbf{x} = \alpha \begin{bmatrix} 1 \\ -2 \\ 1 \end{bmatrix}
$$

is an eigenvector associated with the eigenvalue $\lambda = 0.5$ for every nonzero α. By similar methods we find that arbitrary multiples of the vectors

$$
\begin{bmatrix} 1 \\ -1 \\ 0 \end{bmatrix}, \quad \begin{bmatrix} 0.45 \\ 0.35 \\ 0.20 \end{bmatrix} \tag{8.1}
$$

are eigenvectors associated with the eigenvalues $\lambda = 0.6$ and $\lambda = 1.0$, respectively. Since the components x_1, x_2, x_3 represent market shares, and negative shares make no sense, only the last eigenvector actually is physically meaningful. Since in this last case $\mathbf{Ax} = \lambda\mathbf{x}$ with $\lambda = 1$, we deduce that the market division described by the shares 45%, 35%, 20% is *stable* in the sense that once this division is achieved it will remain constant throughout time.

EXERCISE 8.1. Verify that the vectors in (8.1) are indeed eigenvectors as asserted.

EXERCISE 8.2. Consider again the preceding example leading to (8.1). When $\mathbf{A} - \lambda\mathbf{I}$ is singular there also exists a nonzero vector $\mathbf{y}$ such that $\mathbf{y}^T(\mathbf{A} - \lambda\mathbf{I}) = \mathbf{0}$. Thus for any market distribution $\mathbf{x}$, the new market distribution $\mathbf{Ax}$ satisfies

$$
\mathbf{y}^T(\mathbf{Ax}) = (\mathbf{y}^T\mathbf{A})\mathbf{x} = (\lambda\mathbf{y}^T)\mathbf{x} = \lambda(\mathbf{y}^T\mathbf{x})
$$

which tells how a certain linear combination (namely $\mathbf{y}^T\mathbf{x}$) of the variables x_1, x_2, x_3 behaves under the transformation described by $\mathbf{A}$. For $\lambda = 0.5$, show that the associated (*left-eigenvector*) $\mathbf{y}$ can be taken to be $y_1 = 1, y_2 = 1, y_3 = -4$, so that, after the transformation, $x_1 + x_2 - 4x_3$ equals 0.5 times its value before the transformation. For $\lambda = 0.6$, find $y_1 = 3$, $y_2 = -1$, $y_3 = -5$, so that, after the transformation, $3x_1 - x_2 - 5x_3$ equals 0.6 times its value before the transformation. For $\lambda = 1.0$, find $y_1 = 1$, $y_2 = 1$, $y_3 = 1$ so that $x_1 + x_2 + x_3$ remains constant under the transformation, just as noted in (2.9) by more direct means.

We have so far discussed how eigensystems arise in the study of linear transformations. They also arise very commonly in the physical sciences when we study vibrations; we consider such an example in some detail.

First of all, recall the differential equation for simple harmonic motion. If a particle of mass m moves in a straight line and is attracted to a point in the line by a force which is proportional to the distance X of the particle from the point, then the equation of motion of the particle is given by

$$m\frac{d^2X}{dt^2} + pX = 0, \tag{8.2}$$

where p is a positive constant of proportionality. To solve this equation we set, in the usual way,

$$X = xe^{i\omega t}$$

where ω is the angular frequency of the vibration and x is a constant representing the amplitude of vibration. Equation (8.2) gives

$$(-m\omega^2 + p)xe^{i\omega t} = 0. \tag{8.3}$$

Since the particle is assumed to be moving, $x \neq 0$. Also, $e^{i\omega t}$ is nonzero for all t. Hence (8.3) implies that

$$m\omega^2 = p \quad \text{or} \quad \omega = \left(\frac{p}{m}\right)^{1/2}.$$

This means physically that there is only one frequency of free vibration.

We now discuss a more complicated example. Consider three particles, each of mass m, placed at positions l, $3l$, $5l$, respectively, along an elastic string of length $6l$, as in Figure 8.1(a). The string is fixed at both ends and is under a tension T. Suppose that the particles execute small transverse vibrations under no external forces, the displacements of the three particles in a direction perpendicular to the equilibrium line of the string being X_1, X_2, X_3, respectively.

The forces on bead 1 are shown in Figure 8.1(b). The resultant force perpendicular to the line of equilibrium is

$$T(\sin\theta_2 - \sin\theta_1).$$

We assume that the displacements are small so that the tension T can be taken to be constant, and $\sin\theta_1$, $\sin\theta_2$ can be approximated by $\tan\theta_1$, $\tan\theta_2$ as shown in Figure 8.1(c). Newton's second law of motion (that the mass times the acceleration is equal to the force on the particle) gives, for bead 1,

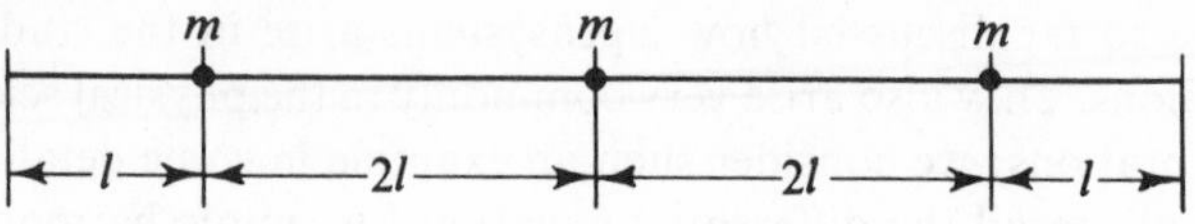

(a) Three beads on a string

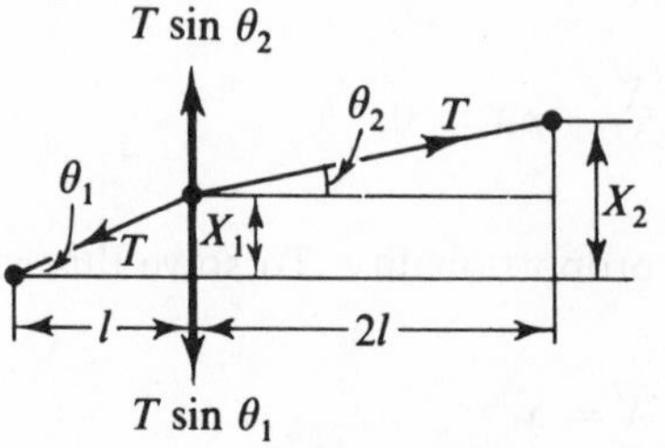

(b) The forces on bead 1

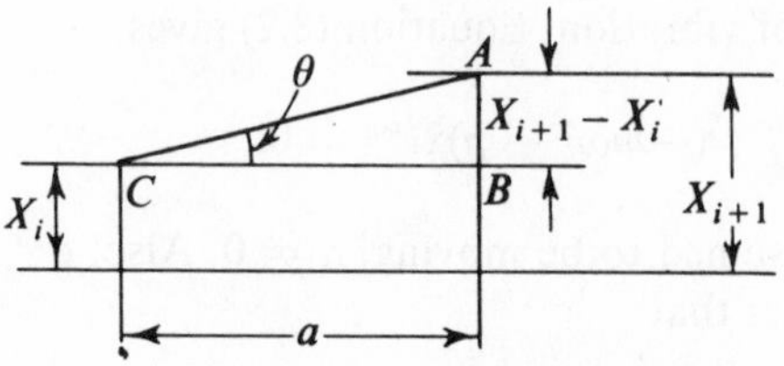

(c) The approximation for $\sin\theta$:
$$\sin\theta = \frac{AB}{AC} \approx \frac{AB}{BC} = \frac{X_{i+1} - X_i}{a}$$

Figure 8.1. Vibration of beads perpendicular to a string.

on using the above results,

$$m\frac{d^2X_1}{dt^2} = -\frac{TX_1}{l} + \frac{T(X_2 - X_1)}{2l}.$$

Similarly, for the motion of beads 2, 3, we obtain

$$m\frac{d^2X_2}{dt^2} = -\frac{T(X_2 - X_1)}{2l} + \frac{T(X_3 - X_2)}{2l},$$

$$m\frac{d^2X_3}{dt^2} = -\frac{T(X_3 - X_2)}{2l} - \frac{TX_3}{l}.$$

We now assume that all quantities vary sinusoidally with time with the same frequency, and so we set

$$X_r = x_r e^{i\omega t} \qquad (r = 1, 2, 3). \tag{8.4}$$

Then the above equations become

$$\begin{aligned} (3-\lambda)x_1 - \quad x_2 \qquad\qquad\qquad & \\ -x_1 + (2-\lambda)x_2 - \quad x_3 & \\ - \quad x_2 + (3-\lambda)x_3 &= 0, \end{aligned} \tag{8.5}$$

where $\lambda = 2\omega^2 ml/T$; this states that λ must be an eigenvalue and $x = [x_1 \quad x_2 \quad x_3]^T$ its associated eigenvector for the matrix

$$\begin{bmatrix} 3 & -1 & 0 \\ -1 & 2 & -1 \\ 0 & -1 & 3 \end{bmatrix}.$$

As we saw before, nonzero solutions $\mathbf{x}$ will exist precisely when the determinant of coefficients in (8.5) is zero, that is, when

$$\det \begin{bmatrix} 3-\lambda & -1 & 0 \\ -1 & 2-\lambda & -1 \\ 0 & -1 & 3-\lambda \end{bmatrix} = 0. \tag{8.6}$$

This determinant is easily evaluated so that (8.6) becomes

$$-\lambda^3 + 8\lambda^2 - 19\lambda + 12 = 0,$$

whose roots are

$$\lambda = 1, \lambda = 3, \lambda = 4.$$

For $\lambda = 1$, (8.5) becomes

$$\begin{aligned} 2x_1 - x_2 \qquad &= 0 \\ -x_1 + x_2 - x_3 &= 0 \\ - x_2 + 2x_3 &= 0 \end{aligned}$$

which is easily solved to give

$$\lambda = 1, \mathbf{x} = \begin{bmatrix} x_1 \\ x_2 \\ x_3 \end{bmatrix} = \alpha \begin{bmatrix} 1 \\ 2 \\ 1 \end{bmatrix} \text{ for arbitrary } \alpha \neq 0. \tag{8.7}$$

For $\lambda = 3$, (8.5) yields

$$\lambda = 3, \mathbf{x} = \begin{bmatrix} x_1 \\ x_2 \\ x_3 \end{bmatrix} = \beta \begin{bmatrix} 1 \\ 0 \\ -1 \end{bmatrix} \text{ for arbitrary } \beta \neq 0. \tag{8.8}$$

For $\lambda = 4$, (8.5) yields

$$\lambda = 4,\ \mathbf{x} = \begin{bmatrix} x_1 \\ x_2 \\ x_3 \end{bmatrix} = \gamma \begin{bmatrix} 1 \\ -1 \\ 1 \end{bmatrix} \text{ for arbitrary } \gamma \neq 0. \tag{8.9}$$

The physical meaning of the above mathematical results on eigensystems is the following. Corresponding to $\lambda = 1$, for example, there is a free vibration with angular frequency ω given by $\omega^2 = \frac{1}{2}T/ml$, and corresponding to this frequency of vibration there is a mode of oscillation given by (8.4) and (8.5) such that the ratios $x_1 : x_2 : x_3$ are $1 : 2 : 1$. Similarly for $\lambda = 3$ and $\lambda = 4$. The modes of vibration are illustrated graphically in Figure 8.2. These three frequencies and modes of vibration are the only ones that can exist.

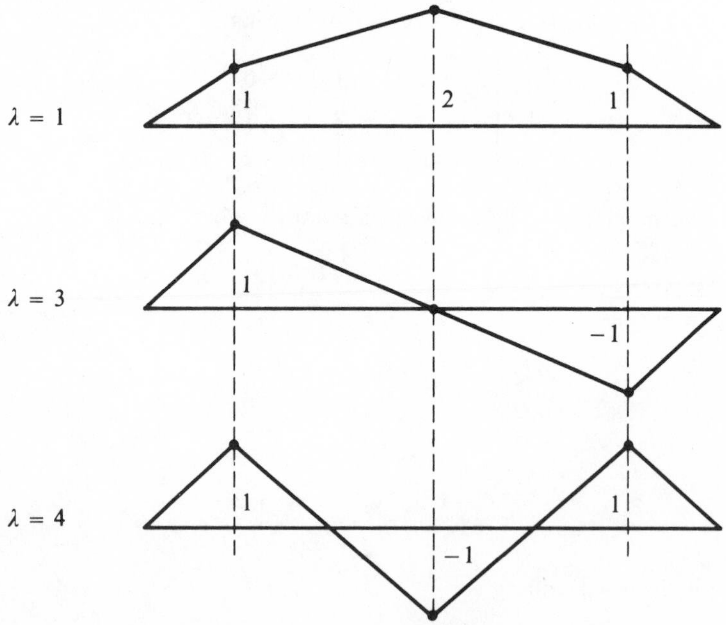

Figure 8.2. Modes of vibration.

EXERCISE 8.3. Verify the calculation of the eigenvalues and eigenvectors of the matrix in the preceding example on vibrations.

EXERCISE 8.4. Verify that the determinant in (8.6) has the value as given in the text.

EXERCISE 8.5. Find the eigenvalues and eigenvectors of the matrices

$$\begin{bmatrix} 2 & 2 \\ 1 & 3 \end{bmatrix}, \quad \begin{bmatrix} 4 & -20 & -10 \\ -2 & 10 & 4 \\ 6 & -30 & -13 \end{bmatrix}.$$

EXERCISE 8.6. Obtain the determinantal equation corresponding to (8.6) for the case of n particles of masses m_i ($i = 1$ to n) at arbitrary positions along a string with fixed ends.

8.2 DEFINITIONS AND BASIC PROPERTIES

We express the ideas in Section 8.1 more formally. Generalizing (8.5), we consider a system of n homogeneous equations in n unknowns:

$$\begin{aligned} (a_{11} - \lambda)x_1 + \quad a_{12}x_2 + \cdots + \quad a_{1n}x_n &= 0 \\ a_{21}x_1 + (a_{22} - \lambda)x_2 + \cdots + \quad a_{2n}x_n &= 0 \\ \cdots \\ a_{n1}x_1 + \quad a_{n2}x_2 + \cdots + (a_{nn} - \lambda)x_n &= 0. \end{aligned} \tag{8.10}$$

In matrix notation, these are

$$(\mathbf{A} - \lambda\mathbf{I})\mathbf{x} = \mathbf{0} \tag{8.11}$$

or

$$\mathbf{A}\mathbf{x} = \lambda\mathbf{x}. \tag{8.12}$$

The homogeneous set of equations possesses only the trivial solution $\mathbf{x} = \mathbf{0}$, unless the determinant of coefficients is zero:

$$f(\lambda) \equiv \det(\mathbf{A} - \lambda\mathbf{I}) = \begin{vmatrix} a_{11} - \lambda & a_{12} & \cdots & a_{1n} \\ a_{21} & a_{22} - \lambda & \cdots & a_{2n} \\ & \cdots & & \\ a_{n1} & a_{n2} & \cdots & a_{nn} - \lambda \end{vmatrix} = 0. \tag{8.13}$$

The function f defined in (8.13) is actually a polynomial of exact degree n, as we will show in Theorem 8.1; it is called the *characteristic* (or *secular*) *polynomial* of $\mathbf{A}$, while (8.13) is called the *characteristic* (or *secular*) *equation* corresponding to $\mathbf{A}$. The roots of this equation are special values of λ for which the simultaneous equations (8.10) possess nonzero solutions. They are called the *eigenvalues* of $\mathbf{A}$, and will be denoted by λ_i ($i = 1, 2, \ldots, n$). The λ_i are sometimes called latent roots, characteristic roots, or proper values. Corresponding to each of the λ_i there will be a solution of (8.10) of the form $\alpha\mathbf{x}_i$, where $\mathbf{x}_i$ is a nonzero vector, and α is an arbitrary constant. These solutions are called the *eigenvectors* (or latent vectors, characteristic vectors, or proper vectors). No confusion should arise from the use of $\mathbf{x}_1, \ldots, \mathbf{x}_n$ to denote the n eigenvectors, and $x_1, \ldots, x_n$ to denote the elements of a given eigenvector. We recap all this in a definition; see also Definition 10.3.

DEFINITION 8.1. The polynomial $f(\lambda) = \det(\mathbf{A} - \lambda\mathbf{I})$ is called the *characteristic polynomial* and the equation $f(\lambda) = 0$ is called the *characteristic equation* of the $n \times n$ matrix $\mathbf{A}$. The *eigenvalues* of $\mathbf{A}$ are the scalars λ for which $\mathbf{Ax} = \lambda\mathbf{x}$ possesses nonzero solutions. The corresponding nonzero solutions $\mathbf{x}$ are the *eigenvectors* of $\mathbf{A}$. The eigenvalues and eigenvectors together are called the *eigensystem* of $\mathbf{A}$.

We obtain the following basic formula directly from this definition of eigenvalues and eigenvectors:

$$\mathbf{Ax}_i = \lambda_i\mathbf{x}_i \qquad (i = 1, \ldots, n). \tag{8.14}$$

Some useful results concerning the characteristic polynomial are summarized in the next theorem. Recall the definition of the *trace*, tr $\mathbf{A}$, of a square matrix $\mathbf{A}$, which is simply the sum of the diagonal elements (see Ex. 1.58).

THEOREM 8.1. The characteristic polynomial of a square matrix $\mathbf{A}$ *of order n is a polynomial of exact degree n with leading coefficient* $(-1)^n$ *and constant term* det $\mathbf{A}$. *The coefficient of* λ^{n-1} *is* $(-1)^{n-1}$ *tr* $\mathbf{A}$. *There are n eigenvalues, and if these are* $\lambda_1, \lambda_2, \ldots, \lambda_n$, *then*

$$\sum_{i=1}^{n} \lambda_i = \sum_{i=1}^{n} a_{ii} = \operatorname{tr} \mathbf{A} \tag{8.15}$$

$$\lambda_1\lambda_2 \ldots \lambda_n = \det \mathbf{A}. \tag{8.16}$$

Proof: If we expand $\det(\mathbf{A} - \lambda\mathbf{I})$ in terms of elements in the first row, we see that

$$f(\lambda) = \det(\mathbf{A} - \lambda\mathbf{I}) = (a_{11} - \lambda)B_{11} + \sum_{j=2}^{n} a_{1j}B_{1j}, \tag{8.17}$$

where B_{ij} is the cofactor of the (i, j) element in $\mathbf{A} - \lambda\mathbf{I}$. There are only $n - 2$ elements $a_{ii} - \lambda$ involving λ in the B_{1j} for $j = 2, \ldots, n$, so that the largest power of λ that can be obtained by expansion of these is λ^{n-2}. Hence (8.17) gives

$$f(\lambda) = (a_{11} - \lambda)B_{11} + \{\text{terms of degree } n - 2 \text{ or less in } \lambda\}. \tag{8.18}$$

The same argument can be applied to B_{11}, and by repetition we see that

$$\begin{aligned} f(\lambda) &= (a_{11} - \lambda)(a_{22} - \lambda) \ldots (a_{nn} - \lambda) + \{\text{terms of degree } n - 2 \text{ or less in } \lambda\} \\ &= (-1)^n\lambda^n + (-1)^{n-1}\lambda^{n-1} \sum_{i=1}^{n} a_{ii} + \{\text{terms of degree } n - 2 \text{ or less in } \lambda\}. \end{aligned} \tag{8.19}$$

Hence the characteristic polynomial is of degree n, and the coefficients of λ^n and λ^{n-1} agree with those stated in the theorem. To see that the constant term in $f(\lambda)$ is det **A** we set $\lambda = 0$ in the definition $f(\lambda) = \det(\mathbf{A} - \lambda\mathbf{I})$. The λ_i are the roots of the characteristic polynomial, so that

$$f(\lambda) = \det(\mathbf{A} - \lambda\mathbf{I}) = (\lambda_1 - \lambda)(\lambda_2 - \lambda)\cdots(\lambda_n - \lambda)$$
$$= (-1)^n\lambda^n + (-1)^{n-1}\lambda^{n-1}\sum_{i=1}^{n}\lambda_i + \cdots + \lambda_1\lambda_2\ldots\lambda_n \tag{8.20}$$

A comparison of this result with (8.19) gives (8.15). To obtain (8.16) we simply set $\lambda = 0$ in (8.20).

Since the eigenvalues of a matrix of order n are the roots of a polynomial of degree n, we know that there will be n eigenvalues, but that they need not be distinct in the general case; we must take into account their *multiplicities* in order to count n of them. As a simple example, consider the case in which **A** is the unit matrix **I**. Then the characteristic polynomial is given by

$$\det(\mathbf{A} - \lambda\mathbf{I}) = \det[(1 - \lambda)\mathbf{I}] = (1 - \lambda)^n \det\mathbf{I} = (1 - \lambda)^n$$

and the eigenvalues of **A** are $\lambda = 1$ repeated n times; we say that $\lambda = 1$ has *multiplicity* equal to n. A slightly less trivial example is provided by the matrix

$$\mathbf{A} = \begin{bmatrix} 3 & 0 & 0 \\ 0 & 3 & 0 \\ 0 & 0 & 4 \end{bmatrix}$$

whose characteristic polynomial is easily evaluated as

$$\det(\mathbf{A} - \lambda\mathbf{I}) = \det\begin{bmatrix} 3-\lambda & 0 & 0 \\ 0 & 3-\lambda & 0 \\ 0 & 0 & 4-\lambda \end{bmatrix} = (3 - \lambda)^2(4 - \lambda).$$

We have a total of three eigenvalues, but we must count $\lambda = 3$ twice and $\lambda = 4$ once; we say that $\lambda = 3$ has multiplicity equal to two while the multiplicity of $\lambda = 4$ is one.

Since we know now that there are exactly n eigen*values*, the next natural question is: "How many eigen*vectors* are there?" Care must be taken in phrasing this question precisely, because if $\mathbf{x}_i$ is an eigenvector of **A** associated with λ_i then so is $\alpha\mathbf{x}_i$ for any arbitrary nonzero scalar α since

$$\mathbf{A}(\alpha\mathbf{x}_i) = \alpha\mathbf{A}\mathbf{x}_i = \alpha(\lambda_i\mathbf{x}_i) = \lambda_i(\alpha\mathbf{x}_i).$$

What we probably mean to ask is how many "really different" eigenvectors there are, that is, how many eigenvectors can be found such that each is linearly independent of the others; we will use the language in exactly this way, so that if we say "there are only three eigenvectors" we mean "there are only three eigenvectors such that each is linearly independent of the others, that is, such that they form an independent set."

This question of the number of eigenvectors is a deep and subtle one; at this point we can only give some partial answers. We will show that there is always *at least one* eigenvector associated with each *distinct* eigenvalue, and that eigenvectors associated with *different* eigenvalues are linearly independent of each other; thus there are at least as many eigenvectors as there are distinct eigenvalues. For example, suppose that a matrix **A** of order six has eigenvalues $\lambda = 10$ of multiplicity three, $\lambda = -7$ of multiplicity two, and $\lambda = 3$ of multiplicity one, so that the totality of eigenvalues is 10, 10, 10, -7, -7, 3, while the distinct eigenvalues are 10, -7, 3; then we can conclude that there are at least three eigenvectors: at least one for $\lambda = 10$, at least one for $\lambda = -7$, and at least one for $\lambda = 3$. The difficult question is how many eigenvectors can be associated with an eigenvalue λ_i of multiplicity m_i greater than unity; once we have developed some machinery in Section 8.4, we can easily show that the number of eigenvectors cannot exceed the multiplicity m_i of λ_i (see Ex. 8.81), but this is not the central issue. We must postpone answering this subtle question until later, but we can already show something about the set of all eigenvectors of **A** associated with one given eigenvalue λ_i: this set generates a linear subspace V_0 with the property that $\mathbf{A}\mathbf{v}_0$ is in V_0 whenever $\mathbf{v}_0$ is in V_0; such *invariant* subspaces will be important in our later analysis of the number of eigenvectors.

As an example of invariant subspaces, consider the matrix

$$\mathbf{A} = \begin{bmatrix} 7 & 0 & 0 & 0 \\ 0 & 7 & 0 & 0 \\ 0 & 0 & 4 & 1 \\ 0 & 0 & 0 & 4 \end{bmatrix}$$

whose characteristic polynomial clearly is

$$\det(\mathbf{A} - \lambda\mathbf{I}) = (7 - \lambda)^2(4 - \lambda)^2$$

so that the eigenvalues are 7, 7, 4, 4. To find eigenvectors corresponding to $\lambda = 7$ we write $(\mathbf{A} - 7\mathbf{I})\mathbf{x} = \mathbf{0}$, that is,

$$\begin{bmatrix} 0 & 0 & 0 & 0 \\ 0 & 0 & 0 & 0 \\ 0 & 0 & -3 & 1 \\ 0 & 0 & 0 & -3 \end{bmatrix} \begin{bmatrix} x_1 \\ x_2 \\ x_3 \\ x_4 \end{bmatrix} = \begin{bmatrix} 0 \\ 0 \\ 0 \\ 0 \end{bmatrix}$$

from which we find that any vector **x** from the two-dimensional subspace

$$V_0 = \left\{ \begin{bmatrix} \alpha \\ \beta \\ 0 \\ 0 \end{bmatrix} \middle| \alpha, \beta \text{ arbitrary scalars} \right\}$$

will satisfy $\mathbf{Ax} = 7\mathbf{x}$. Thus we have found the space V_0 of all the eigenvectors associated with $\lambda = 7$; clearly V_0 is an invariant subspace since $\mathbf{Av}_0 = 7\mathbf{v}_0$ is in V_0 for every $\mathbf{v}_0$ in V_0. Any of an infinite number of sets of two eigenvectors could be described as "the" eigenvectors associated with $\lambda = 7$ and generating the space V_0 of all such eigenvectors, such as

$$\left\{ \begin{bmatrix} 1 \\ 0 \\ 0 \\ 0 \end{bmatrix}, \begin{bmatrix} 0 \\ 1 \\ 0 \\ 0 \end{bmatrix} \right\}, \left\{ \begin{bmatrix} 1 \\ 2 \\ 0 \\ 0 \end{bmatrix}, \begin{bmatrix} 3 \\ 0 \\ 0 \\ 0 \end{bmatrix} \right\}, \left\{ \begin{bmatrix} -1 \\ 3 \\ 0 \\ 0 \end{bmatrix}, \begin{bmatrix} 2 \\ -2 \\ 0 \\ 0 \end{bmatrix} \right\}.$$

As a warning of things to come, we note that the subspace of vectors of the form

$$\begin{bmatrix} 0 \\ 0 \\ \alpha \\ \beta \end{bmatrix}$$

is also an invariant subspace of **A**, since

$$\mathbf{A} \begin{bmatrix} 0 \\ 0 \\ \alpha \\ \beta \end{bmatrix} = \begin{bmatrix} 0 \\ 0 \\ 4\alpha + \beta \\ 4\beta \end{bmatrix}$$

is of the same form, although it is *not* the case that such a vector is an *eigenvector* (unless $\beta = 0$), unlike the situation with V_0 above. More examples of invariant subspaces appear in Exs. 10.4–10.7.

Before rigorously proving the assertions we are making, we formally define some of the concepts we introduced.

DEFINITION 8.2. If the n roots of the characteristic equation associated with an $n \times n$ matrix **A** are such that the value of a particular root λ_i is repeated precisely k times, then λ_i is said to be an eigenvalue of *multiplicity* k; if λ_i occurs only once, that is, if its multiplicity is unity, then λ_i is said to be a *simple* eigenvalue. Any linear subspace V_0 of $\mathbb{C}^n$ for which $\mathbf{A}\mathbf{v}_0$ is in V_0 whenever $\mathbf{v}_0$ is in V_0 is called an *invariant subspace* of $\mathbf{A}$.

• *KEY THEOREM 8.2. Let* $\mathbf{A}$ *be an* $n \times n$ *matrix.*

(i) *There exists at least one eigenvector associated with each distinct value for an eigenvalue* λ. *If* $\mathbf{A}$ *and* λ *are real then the eigenvector may be taken to be real.*
(ii) *The set of all eigenvectors corresponding to a given eigenvalue forms an invariant subspace (if we add the zero vector* $\mathbf{0}$ *to the set).*
(iii) *If* $\lambda_1, \ldots, \lambda_s$ *is a collection of distinct eigenvalues and if* $\mathbf{x}_1, \ldots, \mathbf{x}_s$ *form a set of associated eigenvectors, then* $\{\mathbf{x}_1, \ldots, \mathbf{x}_s\}$ *is linearly independent.*
(iv) *If* λ *is an eigenvalue of* $\mathbf{A}$ *and* $\|\mathbf{A}\|$ *is an operator norm of* $\mathbf{A}$ *induced by a vector norm* $\|\cdot\|$, *then* $|\lambda| \leq \|\mathbf{A}\|$.

Proof: If λ is an eigenvalue, then $\det(\mathbf{A} - \lambda\mathbf{I}) = 0$ and hence by Theorem 6.5 $\mathbf{A} - \lambda\mathbf{I}$ is singular and there exists a nonzero vector $\mathbf{x}$ such that $(\mathbf{A} - \lambda\mathbf{I})\mathbf{x} = \mathbf{0}$. If $\mathbf{x} = \mathbf{u} + i\mathbf{v}$ where $i = \sqrt{-1}$ and $\mathbf{u}, \mathbf{v}, \mathbf{A}$, and λ are all real, then from $\mathbf{A}(\mathbf{u} + i\mathbf{v}) = \lambda(\mathbf{u} + i\mathbf{v})$ it follows that $\mathbf{A}\mathbf{u} = \lambda\mathbf{u}$ and $\mathbf{A}\mathbf{v} = \lambda\mathbf{v}$. Since at least one of $\mathbf{u}$ and $\mathbf{v}$ must be nonzero we can choose the eigenvector to be real; thus (i) is proved.

For (ii), if $\mathbf{u}$ and $\mathbf{w}$ are two eigenvectors associated with the eigenvalue λ of $\mathbf{A}$, then of course

$$\mathbf{A}\mathbf{u} = \lambda\mathbf{u} \quad \text{and} \quad \mathbf{A}\mathbf{w} = \lambda\mathbf{w}$$

so that $\mathbf{A}(\alpha\mathbf{u} + \beta\mathbf{w}) = \alpha\mathbf{A}\mathbf{u} + \beta\mathbf{A}\mathbf{w} = \alpha\lambda\mathbf{u} + \beta\lambda\mathbf{w} = \lambda(\alpha\mathbf{u} + \beta\mathbf{w})$ for scalars α and β, which says that $\alpha\mathbf{u} + \beta\mathbf{w}$ is also an eigenvector for λ or is zero. Thus the set is a linear subspace. Since any $\mathbf{u}$ in the subspace satisfies $\mathbf{A}\mathbf{u} = \lambda\mathbf{u}$ and $\lambda\mathbf{u}$ is also in the subspace, we have an invariant subspace.

For (iii), suppose that the set is linearly dependent. By Theorem 4.1(iii) we can let p be the smallest integer such that $\mathbf{x}_p$ can be written as a linear combination of $\mathbf{x}_1, \ldots, \mathbf{x}_{p-1}$, say

$$\mathbf{x}_p = \alpha_1\mathbf{x}_1 + \cdots + \alpha_{p-1}\mathbf{x}_{p-1},$$

where not all of $\alpha_1, \ldots, \alpha_{p-1}$ can be zero since $\mathbf{x}_p \neq \mathbf{0}$. Then

$$\lambda_p\mathbf{x} = \mathbf{A}\mathbf{x}_p = \alpha_1\mathbf{A}\mathbf{x}_1 + \cdots + \alpha_{p-1}\mathbf{A}\mathbf{x}_{p-1} = \alpha_1\lambda_1\mathbf{x}_1 + \cdots + \alpha_{p-1}\lambda_{p-1}\mathbf{x}_{p-1}.$$

Also, of course,

$$\lambda_p\mathbf{x}_p = \lambda_p(\alpha_1\mathbf{x}_1 + \cdots + \alpha_{p-1}\mathbf{x}_{p-1}).$$

Subtracting these two representations of $\lambda_p\mathbf{x}_p$ yields

$$\mathbf{0} = \alpha_1(\lambda_1 - \lambda_p)\mathbf{x}_1 + \cdots + \alpha_{p-1}(\lambda_{p-1} - \lambda_p)\mathbf{x}_{p-1}.$$

Since $\lambda_i - \lambda_p \neq 0$ for $i = 1, \ldots, p-1$ and not all of $\alpha_1, \ldots, \alpha_{p-1}$ can be zero, this equality states that $\{\mathbf{x}_1, \ldots, \mathbf{x}_{p-1}\}$ is linearly dependent. By Theorem

4.1(iii) again, one of $\mathbf{x}_1, \ldots, \mathbf{x}_{p-1}$ can be written as a linear combination of the preceding vectors, in contradiction to the fact that p was the smallest such integer.

For (iv), since the eigenvector $\mathbf{x}$ associated with λ has $\|\mathbf{x}\| \neq 0$, we have $|\lambda| = \|\mathbf{Ax}\|/\|\mathbf{x}\| \leq \|\mathbf{A}\|$ by the definition of $\|\mathbf{A}\|$. This completes the proof.

COROLLARY 8.1. If an $n \times n$ matrix $\mathbf{A}$ has n distinct eigenvalues, then there exists a linearly independent set of n eigenvectors, one associated with each eigenvalue, and any eigenvector of $\mathbf{A}$ is a multiple of one of these n eigenvectors.

Proof: By **Key Theorem 8.2**(i), (iii) the set $\{\mathbf{x}_1, \ldots, \mathbf{x}_n\}$ of eigenvectors is linearly independent and thus spans the space by Theorem 4.6. Hence if $\mathbf{v}$ is any eigenvector corresponding, say, to λ_1 for convenience, there exist $\alpha_1, \ldots, \alpha_n$ not all zero for which $\mathbf{v} = \alpha_1\mathbf{x}_1 + \cdots + \alpha_n\mathbf{x}_n$. By **Key Theorem 8.2**(ii), $\mathbf{v} - \alpha_1\mathbf{x}_1$ is either zero as asserted or an eigenvector associated with λ_1; in this latter case, by **Key Theorem 8.2**(iii), $\{\mathbf{v} - \alpha_1\mathbf{x}_1, \mathbf{x}_2, \ldots, \mathbf{x}_n\}$ would be linearly independent so that $\alpha_2 = \cdots = \alpha_n = 0$ and $\mathbf{v} = \alpha_1\mathbf{x}_1$ as asserted.

Key Theorem 8.2 does not, of course, give the whole story. If there are n distinct eigenvalues then we get a linearly independent set of n eigenvectors, the maximum possible number in our n dimensional space. If we have only k distinct values $\lambda_1, \ldots, \lambda_k$, with λ_i having multiplicity m_i, from **Key Theorem 8.2** we know only that we can find a linearly independent set of k eigenvectors, *one* for each distinct eigenvalue. Since $m_1 + \cdots + m_k = n$, we might hope to get an independent set with the maximum number n of vectors in $\mathbb{R}^n$ or $\mathbb{C}^n$ by adding $m_i - 1$ more eigenvectors for each distinct eigenvalue. In other words, we might hope to get a linearly independent set of n eigenvectors where each distinct eigenvalue λ_i is associated with m_i of the eigenvectors. To do this, of course, the invariant subspace of all eigenvectors associated with λ_i must have dimension (at least) m_i since it would contain a linearly independent set of m_i vectors. Unfortunately, this is not true; the subspace of eigenvectors associated with an eigenvalue of multiplicity m may have dimension *anywhere* from 1 through m. It is this possibility that complicates the discussion of the "number" of eigenvectors we should expect for an $n \times n$ matrix.

EXERCISE 8.7. To illustrate briefly how this can happen, we simply consider the matrix

$$\mathbf{A} = \begin{bmatrix} 2 & 1 \\ 0 & 2 \end{bmatrix}. \tag{8.21}$$

Since

$$\det(\mathbf{A} - \lambda\mathbf{I}) = (2 - \lambda)^2,$$

it is clear that $\lambda = 2$ is the only distinct eigenvalue of $\mathbf{A}$ and $\lambda = 2$ has multiplicity two. If we are to find associated eigenvectors $\mathbf{x} = [x_1, x_2]^T$ by solving

$(\mathbf{A} - 2\mathbf{I})\mathbf{x} = \mathbf{0}$, we must solve

$$\begin{aligned} (2-2)x_1 \qquad\quad + x_2 &= 0 \\ 0x_1 + (2-2)x_2 &= 0 \end{aligned}$$

whose only solution obviously has

$$x_1 = z(\text{arbitrary}), \qquad x_2 = 0,$$

so that

$$\mathbf{x} = z\mathbf{e}_1 = z\begin{bmatrix} 1 \\ 0 \end{bmatrix}.$$

Thus, despite the fact that $\lambda = 2$ has multiplicity two in Ex. 8.7, the invariant subspace of eigenvectors associated with $\lambda = 2$ has dimension equal to only unity; that is, the double eigenvalue has only a single eigenvector associated with it. We will see later that our applications of matrices are much simpler when the $n \times n$ matrix in question has a full set of eigenvectors, that is, has an independent set of n eigenvectors. For this reason it is very important for us to identify large classes of matrices for which this does occur and to understand more fully what happens when we fail to have a full set of eigenvectors. Although we will describe briefly some of these results in the remainder of this chapter, much of the following two chapters will be required to derive and present a more detailed analysis.

EXERCISE 8.8. Show that the matrix

$$\begin{bmatrix} 2 & -1 & 0 \\ -1 & 2 & -1 \\ 0 & -1 & 2 \end{bmatrix}$$

has eigenvalues 2, $2 \pm \sqrt{2}$, and find the corresponding eigenvectors.

EXERCISE 8.9. Find a linearly independent set of two eigenvectors of the matrix

$$\begin{bmatrix} 2 & 2 & -6 \\ 2 & -1 & -3 \\ -2 & -1 & 1 \end{bmatrix}$$

corresponding to the eigenvalue $\lambda = -2$. Find the other eigenvalue and eigenvector.

EXERCISE 8.10. Prove that $\lambda = 0$ is an eigenvalue of a matrix $\mathbf{A}$ if and only if $\mathbf{A}$ is singular.

EXERCISE 8.11. Let **A** have the eigenvalues λ_i. Prove:

(a) The transpose of **A** has the same eigenvalues as **A**.
(b) The matrix $k\mathbf{A}$ has the eigenvalues $k\lambda_i$.
(c) The matrix $\mathbf{A}^p$, where p is a positive integer, has the eigenvalues λ_i^p.
(d) If **A** is nonsingular, $\mathbf{A}^{-1}$ has the eigenvalues $1/\lambda_i$.
(e) The matrix $\mathbf{A} + k\mathbf{I}$ has the eigenvalues $\lambda_i + k$.

EXERCISE 8.12. If $f(x)$ is a polynomial in x, then $f(\mathbf{A})$ denotes the matrix obtained by replacing x by the (square) matrix **A**. If λ is an eigenvalue of **A**, show that $f(\lambda)$ is an eigenvalue of $f(\mathbf{A})$.

EXERCISE 8.13. Find the characteristic polynomial f of the matrix **A** in (8.21); show that $f(\mathbf{A})$ is the zero matrix. Do the same things for the matrices in Ex. 8.5.

EXERCISE 8.14. If **A** is a real $n \times n$ matrix, show that the eigenvalues of **A** are real or complex conjugate in pairs. Also, if n is odd, show that **A** has at least one real eigenvalue.

EXERCISE 8.15. Find all three eigenvalues and as many eigenvectors as possible for the matrix

$$\mathbf{A} = \begin{bmatrix} 7 & 1 & 2 \\ -1 & 7 & 0 \\ 1 & -1 & 6 \end{bmatrix}.$$

8.3 EIGENSYSTEMS, DECOMPOSITIONS, AND TRANSFORMATIONS

In the preceding section we mentioned the importance of understanding the *structure* of an eigensystem, including the number of eigenvectors associated with a given eigenvalue. In this section we will describe two other ways of viewing this same question, one of which is manipulative in that it decomposes a given matrix into a special product of other matrices, and the other of which is geometric in that it gives a geometrically transparent description of a linear transformation described by a given matrix. The three viewpoints together make it easier to understand the structure of eigensystems.

Recall the matrix

$$\mathbf{A} = \begin{bmatrix} 3 & -1 & 0 \\ -1 & 2 & -1 \\ 0 & -1 & 3 \end{bmatrix} \tag{8.22}$$

studied in relation to the vibration phenomenon in Section 8.1; we found its eigenvalues to be

$$\lambda_1 = 1, \qquad \lambda_2 = 3, \qquad \lambda_3 = 4 \tag{8.23}$$

with associated eigenvectors

$$\mathbf{x}_1 = \begin{bmatrix} 1 \\ 2 \\ 1 \end{bmatrix}, \qquad \mathbf{x}_2 = \begin{bmatrix} 1 \\ 0 \\ -1 \end{bmatrix}, \qquad \mathbf{x}_3 = \begin{bmatrix} 1 \\ -1 \\ 1 \end{bmatrix}. \tag{8.24}$$

We write the three equations $\mathbf{A}\mathbf{x}_1 = \lambda_1\mathbf{x}_1$, $\mathbf{A}\mathbf{x}_2 = \lambda_2\mathbf{x}_2$, $\mathbf{A}\mathbf{x}_3 = \lambda_3\mathbf{x}_3$ in the block form

$$\begin{aligned} \mathbf{A}[\mathbf{x}_1 \quad \mathbf{x}_2 \quad \mathbf{x}_3] &= [\lambda_1\mathbf{x}_1 \quad \lambda_2\mathbf{x}_2 \quad \lambda_3\mathbf{x}_3] \\ &= [\mathbf{x}_1 \quad \mathbf{x}_2 \quad \mathbf{x}_3]\begin{bmatrix} \lambda_1 & 0 & 0 \\ 0 & \lambda_2 & 0 \\ 0 & 0 & \lambda_3 \end{bmatrix}. \end{aligned} \tag{8.25}$$

To write (8.25) in more compact form, we let

$$\mathbf{P} = [\mathbf{x}_1 \quad \mathbf{x}_2 \quad \mathbf{x}_3]$$

and

$$\mathbf{\Lambda} = \begin{bmatrix} \lambda_1 & 0 & 0 \\ 0 & \lambda_2 & 0 \\ 0 & 0 & \lambda_3 \end{bmatrix}.$$

We can then rewrite (8.25) as

$$\mathbf{AP} = \mathbf{P\Lambda}. \tag{8.26}$$

As a numerical check, the reader can easily verify that

$$\begin{aligned} \begin{bmatrix} 3 & -1 & 0 \\ -1 & 2 & -1 \\ 0 & -1 & 3 \end{bmatrix}\begin{bmatrix} 1 & 1 & 1 \\ 2 & 0 & -1 \\ 1 & -1 & 1 \end{bmatrix} &= \begin{bmatrix} 1 & 3 & 4 \\ 2 & 0 & -4 \\ 1 & -3 & 4 \end{bmatrix} \\ &= \begin{bmatrix} 1 & 1 & 1 \\ 2 & 0 & -1 \\ 1 & -1 & 1 \end{bmatrix}\begin{bmatrix} 1 & 0 & 0 \\ 0 & 3 & 0 \\ 0 & 0 & 4 \end{bmatrix}. \end{aligned}$$

Since the set of columns of **P** is linearly independent, **P** is nonsingular; (8.26) can then be formulated as

$$\mathbf{A} = \mathbf{P\Lambda P}^{-1}. \tag{8.27}$$

Thus the existence of a linearly independent set of three eigenvectors of **A** allows us to *decompose* **A** via (8.27) into a special kind of product. We will discuss such special products from a different viewpoint in the following two sections; for the present, we merely want to observe that such a decomposition is *equivalent* to the existence of a full set of eigenvectors, as indicated in the preceding example.

- KEY THEOREM 8.3. *A given $n \times n$ matrix $\mathbf{A}$ has a linearly independent set of n eigenvectors if and only if there exists a nonsingular $n \times n$ matrix $\mathbf{P}$ and a diagonal matrix $\mathbf{\Lambda}$ for which*

$$\mathbf{A} = \mathbf{P}\,\mathbf{\Lambda}\,\mathbf{P}^{-1}, \tag{8.28}$$

$$\mathbf{\Lambda} = \mathbf{P}^{-1}\mathbf{A}\mathbf{P}. \tag{8.29}$$

The columns $\mathbf{p}_1, \ldots, \mathbf{p}_n$ of $\mathbf{P} = [\mathbf{p}_1, \ldots, \mathbf{p}_n]$ may be taken as the eigenvectors of $\mathbf{A}$ associated respectively with the eigenvalues λ_i, where λ_i is the (i, i) element of the diagonal matrix $\mathbf{\Lambda}$.

Proof: We proceed just as in the example preceding this theorem. If **A** has a linearly independent set of n eigenvectors $\mathbf{p}_1, \ldots, \mathbf{p}_n$ associated with the eigenvalues $\lambda_1, \ldots, \lambda_n$, respectively, then the matrix

$$\mathbf{P} = [\mathbf{p}_1, \ldots, \mathbf{p}_n]$$

is nonsingular. From the equations $\mathbf{A}\mathbf{p}_i = \lambda_i\mathbf{p}_i$ for $1 \leq i \leq n$, we obtain

$$\mathbf{A}\mathbf{P} = [\mathbf{A}\mathbf{p}_1, \ldots, \mathbf{A}\mathbf{p}_n] = [\lambda_1\mathbf{p}_1, \ldots, \lambda_n\mathbf{p}_n] = \mathbf{P}\mathbf{\Lambda} \tag{8.30}$$

where $\mathbf{\Lambda}$ is the diagonal matrix

$$\mathbf{\Lambda} = \begin{bmatrix} \lambda_1 & 0 & \cdots & 0 \\ 0 & \lambda_2 & \cdots & 0 \\ & & \cdots & \\ 0 & 0 & \cdots & \lambda_n \end{bmatrix}.$$

Since **P** is nonsingular, (8.30) immediately yields (8.28) and (8.29). Conversely, if (8.28) and (8.29) hold then multiplication by **P** gives

$$\mathbf{A}\mathbf{P} = \mathbf{P}\mathbf{\Lambda}$$

which, because of the structure of **P** and $\mathbf{\Lambda}$, yields the desired equations

$$\mathbf{A}\mathbf{p}_i = \lambda_i\mathbf{p}_i \qquad (\text{for } 1 \leq i \leq n).$$

Since **P** is nonsingular, the eigenvectors $\mathbf{p}_1, \ldots, \mathbf{p}_n$ form an independent set and the proof is complete.

EXERCISE 8.16. Illustrate **Key Theorem 8.3** numerically, using

$$\mathbf{A} = \begin{bmatrix} 1 & 1 & 0 \\ 0 & 2 & 1 \\ 0 & 0 & 3 \end{bmatrix}.$$

SOLUTION: The eigenvalues are the diagonal elements $\lambda_1 = 1, \lambda_2 = 2, \lambda_3 = 3$. The corresponding eigenvectors, written as the columns of a matrix, give

$$\mathbf{P} = \begin{bmatrix} 1 & 1 & 1 \\ 0 & 1 & 2 \\ 0 & 0 & 2 \end{bmatrix}.$$

A straightforward computation leads to

$$\mathbf{P}^{-1} = \tfrac{1}{2}\begin{bmatrix} 2 & -2 & 1 \\ 0 & 2 & -2 \\ 0 & 0 & 1 \end{bmatrix}.$$

We can check that

$$\mathbf{P}^{-1}\mathbf{AP} = \begin{bmatrix} 1 & 0 & 0 \\ 0 & 2 & 0 \\ 0 & 0 & 3 \end{bmatrix}.$$

EXERCISE 8.17. Illustrate **Key Theorem 8.3** as in Ex. 8.16, using the matrix (2.10) of Section 2.2 whose eigensystem was described near the beginning of Section 8.1.

The decomposition (8.28) of **Key Theorem 8.3** is sufficient to describe the structure of the eigensystem from the viewpoint of a manipulative decomposition; the reader may well ask why, then, we also presented the representation (8.29). This is because it is a natural one when we view the structure of eigensystems from our remaining viewpoint, namely the geometrical one involving transformations; we now turn to this approach.

Consider once again the matrix **A** of (8.22), whose eigensystem is given by (8.23) and (8.24) or equivalently by the decomposition (8.27). We now view **A** as defining a linear transformation A from $\mathbb{R}^3$ to $\mathbb{R}^3$, so that the vector **x** with components x_1, x_2, x_3 is transformed by **A** into $\mathbf{y} = A(\mathbf{x}) = \mathbf{Ax}$ with components

$$y_1 = 3x_1 - x_2, \qquad y_2 = -x_1 + 2x_2 - x_3, \qquad y_3 = -x_2 + 3x_3.$$

Since the set of eigenvectors $\mathbf{x}_1, \mathbf{x}_2, \mathbf{x}_3$ of (8.24) is linearly independent, it can be used as a basis for $\mathbb{R}^3$. If we express a physical vector **x** as a sum of multiples of these basis vectors via $\mathbf{x} = x_1'\mathbf{x}_1 + x_2'\mathbf{x}_2 + x_3'\mathbf{x}_3$, then the numbers

x_1', x_2', x_3' can be thought of as "coordinates" of $\mathbf{x}$ with respect to this basis. If as before we let

$$\mathbf{P} = [\mathbf{x}_1 \quad \mathbf{x}_2 \quad \mathbf{x}_3]$$

then

$$\mathbf{x} = x_1'\mathbf{x}_1 + x_2'\mathbf{x}_2 + x_3'\mathbf{x}_3 = \mathbf{P}\mathbf{x}' \quad \text{where} \quad \mathbf{x}' = \begin{bmatrix} x_1' \\ x_2' \\ x_3' \end{bmatrix}. \tag{8.31}$$

Since $\mathbf{P}$ is nonsingular, we also have

$$\mathbf{x}' = \mathbf{P}^{-1}\mathbf{x}. \tag{8.32}$$

The equations (8.31) and (8.32) algebraically describe the *change of basis* from $\mathbf{e}_1, \mathbf{e}_2, \mathbf{e}_3$ to $\mathbf{x}_1, \mathbf{x}_2, \mathbf{x}_3$: a physical vector $\mathbf{x}$ whose coordinates with respect to the basis vectors $\mathbf{e}_1, \mathbf{e}_2, \mathbf{e}_3$ are x_1, x_2, x_3 can be equivalently described by the coordinates x_1', x_2', x_3' with respect to the basis vectors $\mathbf{x}_1, \mathbf{x}_2, \mathbf{x}_3$, and these two sets of coordinates are related by (8.31), (8.32).

EXERCISE 8.18. Consider the point labeled w in Figure 8.3. With respect to the standard basis ("axes") it has coordinates $x_1 = 3$, $x_2 = 1$, while if we use the basis ("axes") vectors $\mathbf{x}_1$, $\mathbf{x}_2$ with $\mathbf{x}_1 = [2, 3]^T$, $\mathbf{x}_2 = [-1, 2]^T$, then the coordinates are $x_1' = 1$, $x_2' = -1$. Verify that (8.31), (8.32) are valid.

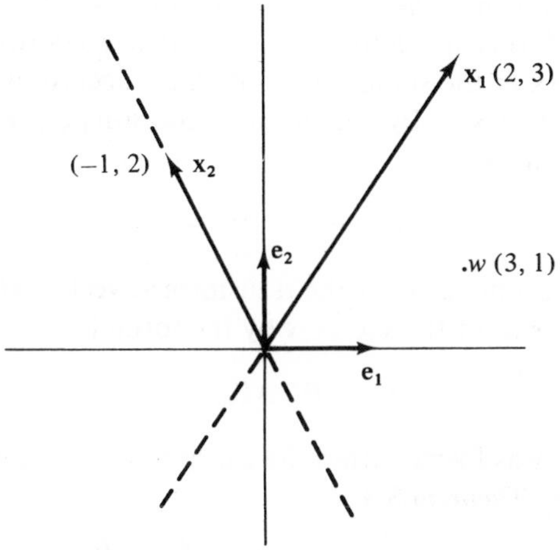

Figure 8.3. Change of basis.

The relationships (8.31), (8.32), of course, do not depend on our having used precisely the three vectors of (8.24) as a basis; for completeness we state the general result.

THEOREM 8.4. Let the vectors $\mathbf{x}_1, \ldots, \mathbf{x}_n$ *form a basis for* $\mathbb{R}^n$ *(or* $\mathbb{C}^n$*), and suppose that a vector*

$$\mathbf{x} = \begin{bmatrix} x_1 \\ \cdot \\ \cdot \\ \cdot \\ x_n \end{bmatrix}$$

has components $x'_1, \ldots, x'_n$ *with respect to that basis in the sense that*

$$\mathbf{x} = x'_1\mathbf{x}_1 + \cdots + x'_n\mathbf{x}_n.$$

Then if

$$\mathbf{x}' = \begin{bmatrix} x'_1 \\ \cdot \\ \cdot \\ \cdot \\ x'_n \end{bmatrix}, \qquad \mathbf{P} = [\mathbf{x}_1, \ldots, \mathbf{x}_n],$$

the coordinates $\mathbf{x}$ *and* $\mathbf{x}'$ *are related by* (8.31), (8.32), *that is,*

$$\mathbf{x} = \mathbf{P}\mathbf{x}', \qquad \mathbf{x}' = \mathbf{P}^{-1}\,\mathbf{x}.$$

We continue now with our discussion of the matrix (8.22) which led to our discovery of (8.31); recall that we are interested in the transformation A defined by $\mathbf{A}$. Since we have decided to represent our vectors in $\mathbb{R}^3$ with respect to the basis vectors $\mathbf{x}_1, \mathbf{x}_2, \mathbf{x}_3$ of (8.24), it is reasonable to ask how the transformation A modifies coordinates with respect to this basis. We know that A transforms the vector $\mathbf{x} = \mathbf{P}\mathbf{x}'$ with new coordinates $\mathbf{x}' = \mathbf{P}^{-1}\mathbf{x}$ into $\mathbf{y} = A(\mathbf{x}) = \mathbf{A}\mathbf{x}$ in the standard coordinates; according to (8.31), (8.32), $\mathbf{y}$ can be expressed as $\mathbf{y} = \mathbf{P}\mathbf{y}'$ in the new coordinates, where $\mathbf{y}' = \mathbf{P}^{-1}\mathbf{y}$. Since $\mathbf{y} = A(\mathbf{x})$ we have

$$\mathbf{y}' = \mathbf{P}^{-1}\mathbf{y} = \mathbf{P}^{-1}A(\mathbf{x}) = \mathbf{P}^{-1}\mathbf{A}\mathbf{x} = \mathbf{P}^{-1}\mathbf{A}\mathbf{P}\mathbf{x}',$$

so that the new coordinates $\mathbf{y}'$ of the transformed vector $A(\mathbf{x})$ are related to the new coordinates $\mathbf{x}'$ of the vector $\mathbf{x}$ by the formula

$$\mathbf{y}' = \mathbf{P}^{-1}\mathbf{A}\mathbf{P}\mathbf{x}'. \tag{8.33}$$

Since the matrix $\mathbf{P}$ was formed from the eigenvectors of $\mathbf{A}$, however, according to (8.29) of **Key Theorem 8.3**

$$\mathbf{P}^{-1}\mathbf{A}\mathbf{P} = \mathbf{\Lambda} = \begin{bmatrix} 1 & 0 & 0 \\ 0 & 3 & 0 \\ 0 & 0 & 4 \end{bmatrix}$$

is diagonal, so that according to (8.33) the linear transformation A is described with respect to the basis vectors $\mathbf{x}_1, \mathbf{x}_2, \mathbf{x}_3$ by the diagonal matrix $\mathbf{\Lambda} = \mathbf{P}^{-1}\mathbf{A}\mathbf{P}$. Such a transformation of the matrix $\mathbf{A}$ into $\mathbf{\Lambda} = \mathbf{P}^{-1}\mathbf{A}\mathbf{P}$ is called a *similarity transformation* and will be studied in detail in the next two sections. In summary, our reason for stating (8.29) in **Key Theorem 8.3** is that it tells us that *if a linear transformation A is represented by the $n \times n$ matrix* $\mathbf{A}$ *with respect to the usual basis* $\mathbf{e}_1, \ldots, \mathbf{e}_n$ *and if* $\mathbf{A}$ *has an independent set of n eigenvectors* $\mathbf{x}_1, \ldots, \mathbf{x}_n$, *then* $\mathbf{A}$ *is represented by a diagonal matrix with respect to the basis vectors* $\mathbf{x}_1, \ldots, \mathbf{x}_n$.

EXERCISE 8.19. Let the linear transformation $\mathbf{y} = A(\mathbf{x})$ from $\mathbb{R}^2$ into $\mathbb{R}^2$ be described by

$$y_1 = 2x_1 + x_2$$
$$y_2 = x_1 + 2x_2.$$

Find the matrix representing A when vectors in $\mathbb{R}^2$ are represented in terms of the basis vectors

$$\begin{bmatrix} 2 \\ 1 \end{bmatrix}, \quad \begin{bmatrix} 1 \\ 1 \end{bmatrix}$$

rather than the usual $\mathbf{e}_1$, $\mathbf{e}_2$, respectively.

EXERCISE 8.20. Find a basis of vectors $\mathbf{x}_1$, $\mathbf{x}_2$ with respect to which the linear transformation A of Ex. 8.19 will be represented by a diagonal matrix.

We now know how to state the results of **Key Theorem 8.3** from three different equivalent viewpoints: (i) the $n \times n$ matrix $\mathbf{A}$ has a linearly independent set of n eigenvectors; (ii) the $n \times n$ matrix $\mathbf{A}$ can be decomposed as $\mathbf{A} = \mathbf{P}\mathbf{\Lambda}\mathbf{P}^{-1}$ where $\mathbf{\Lambda}$ is an $n \times n$ diagonal matrix; (iii) the transformation A of $\mathbb{R}^n$ into $\mathbb{R}^n$ (or $\mathbb{C}^n$ into $\mathbb{C}^n$) defined by $\mathbf{A}$ via $A(\mathbf{x}) = \mathbf{A}\mathbf{x}$ can be represented by a diagonal matrix $\mathbf{\Lambda}$ with respect to an appropriately chosen basis. Similarly, when $\mathbf{A}$ does *not* have a full set of eigenvectors, it may be helpful to describe the structure of the eigensystem from one of these other two viewpoints by describing how "simple" a $\mathbf{\Lambda}$ can be found in a decomposition of $\mathbf{A}$ or in a representation of the linear transformation defined by $\mathbf{A}$. It is important to grasp the equivalence of these three viewpoints; we depict the relationships in Figure 8.4.

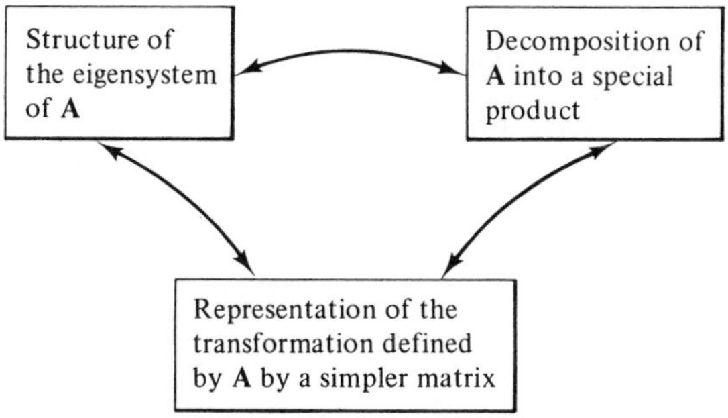

Figure 8.4. Equivalent viewpoints.

EXERCISE 8.21. Decompose the matrix

$$\mathbf{A} = \begin{bmatrix} 2 & 1 \\ 1 & 2 \end{bmatrix}$$

of Ex. 8.19 in the form (8.28) of **Key Theorem 8.3.**

EXERCISE 8.22. Find the eigensystem of the matrix **A** in Ex. 8.21.

EXERCISE 8.23. Explain why the matrix **A** of (8.21) in Ex. 8.7 cannot be decomposed in the form (8.28) of **Key Theorem 8.3** for diagonal $\mathbf{\Lambda}$.

EXERCISE 8.24. Let the linear transformation A of $\mathbb{R}^2$ into $\mathbb{R}^2$ be defined by $A(\mathbf{x}) = \mathbf{Ax}$ where **A** is the matrix (8.21) in Ex. 8.7. Explain why there does *not* exist a basis with respect to which A has a diagonal matrix representation.

8.4 SIMILARITY TRANSFORMATIONS

We just observed, in the study of linear transformation from $\mathbb{R}^n$ to $\mathbb{R}^n$ (or $\mathbb{C}^n$ to $\mathbb{C}^n$), that when a new basis $\{\mathbf{p}_1, \ldots, \mathbf{p}_n\}$ is introduced for representing the geometric vectors in $\mathbb{R}^n$ (or $\mathbb{C}^n$) it is natural to consider the transformation of a given matrix **A** into the matrix $\mathbf{P}^{-1}\mathbf{AP}$ for the nonsingular matrix $\mathbf{P} = [\mathbf{p}_1, \ldots, \mathbf{p}_n]$.

DEFINITION 8.3. If there exists a nonsingular matrix **P** such that $\mathbf{P}^{-1}\mathbf{AP} = \mathbf{B}$, then the matrix **B** is said to be *similar* to **A**, and we say that **B** is obtained from **A** by means of a *similarity transformation.*

EXERCISE 8.25. Let

$$\mathbf{A} = \begin{bmatrix} 2 & -3 \\ 1 & -1 \end{bmatrix}, \qquad \mathbf{P} = \begin{bmatrix} 2 & 1 \\ 1 & 1 \end{bmatrix}.$$

Find $\mathbf{B} = \mathbf{P}^{-1}\mathbf{AP}$.

SOLUTION: We easily compute

$$\mathbf{P}^{-1} = \begin{bmatrix} 1 & -1 \\ -1 & 2 \end{bmatrix}, \qquad \mathbf{AP} = \begin{bmatrix} 1 & -1 \\ 1 & 0 \end{bmatrix}$$

and hence

$$\mathbf{B} = \begin{bmatrix} 0 & -1 \\ 1 & 1 \end{bmatrix}$$

is similar to **A**.

Similarity transformations have some simple and obvious properties.

THEOREM 8.5. Similarity is an equivalence relation—i.e.: $\mathbf{A}$ *is similar to itself; if* $\mathbf{B}$ *is similar to* $\mathbf{A}$*, then* $\mathbf{A}$ *is similar to* $\mathbf{B}$*; and if* $\mathbf{C}$ *is similar to* $\mathbf{B}$ *and* $\mathbf{B}$ *to* $\mathbf{A}$*, then* $\mathbf{C}$ *is similar to* $\mathbf{A}$.

Proof: Using $\mathbf{P} = \mathbf{I}$, we see immediately that $\mathbf{A}$ is similar to $\mathbf{A}$. If $\mathbf{P}^{-1}\mathbf{AP} = \mathbf{B}$ then $\mathbf{A} = \mathbf{PBP}^{-1} = \mathbf{Q}^{-1}\mathbf{BQ}$, where $\mathbf{Q} = \mathbf{P}^{-1}$, so that $\mathbf{A}$ is similar to $\mathbf{B}$. If $\mathbf{C} = \mathbf{P}^{-1}\mathbf{BP}$, $\mathbf{B} = \mathbf{Q}^{-1}\mathbf{AQ}$, then $\mathbf{C} = (\mathbf{QP})^{-1}\mathbf{A}(\mathbf{QP})$, so that $\mathbf{C}$ is similar to $\mathbf{B}$.

EXERCISE 8.26. Verify in Ex. 8.25 that $\mathbf{A}$ is similar to $\mathbf{B}$, that is, that there exists $\mathbf{Q}$ with $\mathbf{A} = \mathbf{Q}^{-1}\mathbf{BQ}$.

SOLUTION: As in the proof of Theorem 8.5, we take

$$\mathbf{Q} = \mathbf{P}^{-1} = \begin{bmatrix} 1 & -1 \\ -1 & 2 \end{bmatrix},$$

and compute

$$\mathbf{BQ} = \begin{bmatrix} 1 & -2 \\ 0 & 1 \end{bmatrix}, \qquad \mathbf{Q}^{-1}\mathbf{BQ} = \begin{bmatrix} 2 & -3 \\ 1 & -1 \end{bmatrix} = \mathbf{A}$$

as asserted.

EXERCISE 8.27. Either prove or give a counterexample: if $\mathbf{A}$ and $\mathbf{B}$ are $n \times n$ and $\mathbf{A}'$ is similar to $\mathbf{A}$ and $\mathbf{B}'$ is similar to $\mathbf{B}$, then $\mathbf{A}'\mathbf{B}'$ is similar to $\mathbf{AB}$.

THEOREM 8.6.

(i) *Similar matrices have the same characteristic equation and the same eigenvalues.*
(ii) *If* $\mathbf{P}^{-1}\mathbf{AP} = \mathbf{B}$ *and* $\mathbf{x}$ *is an eigenvector of* $\mathbf{A}$ *corresponding to the eigenvalue* λ*, then* $\mathbf{P}^{-1}\mathbf{x}$ *is an eigenvector of* $\mathbf{B}$ *corresponding to* λ.

Proof: Since $\det \mathbf{P}^{-1} \det \mathbf{P} = \det(\mathbf{P}^{-1}\mathbf{P}) = \det \mathbf{I} = 1$, we have

$$\begin{aligned} \det(\mathbf{B} - \lambda\mathbf{I}) &= \det \mathbf{P}^{-1}(\mathbf{A} - \lambda\mathbf{I})\mathbf{P} \\ &= \det \mathbf{P}^{-1} \det(\mathbf{A} - \lambda\mathbf{I}) \det \mathbf{P} = \det(\mathbf{A} - \lambda\mathbf{I}), \end{aligned}$$

which proves (i). If $\mathbf{Ax} = \lambda\mathbf{x}$, we have

$$(\mathbf{P}^{-1}\mathbf{AP})(\mathbf{P}^{-1}\mathbf{x}) = \lambda(\mathbf{P}^{-1}\mathbf{x}),$$

or

$$\mathbf{B}(\mathbf{P}^{-1}\mathbf{x}) = \lambda(\mathbf{P}^{-1}\mathbf{x}),$$

which proves (ii), since $\mathbf{P}^{-1}\mathbf{x}$ cannot be zero because $\mathbf{P}^{-1}$ is nonsingular and $\mathbf{x} \neq \mathbf{0}$.

EXERCISE 8.28. Verify Theorem 8.6(i) for the matrices of Ex. 8.25.

SOLUTION: We compute $\det(\mathbf{A} - \lambda\mathbf{I})$ as

$$\det\begin{bmatrix} 2-\lambda & -3 \\ 1 & -1-\lambda \end{bmatrix} = (2-\lambda)(-1-\lambda) + 3 = \lambda^2 - \lambda + 1,$$

while det $(\mathbf{B} - \lambda\mathbf{I})$ equals

$$\det\begin{bmatrix} -\lambda & -1 \\ 1 & 1-\lambda \end{bmatrix} = -\lambda(1-\lambda) + 1 = \lambda^2 - \lambda + 1$$

just as asserted.

EXERCISE 8.29. Prove that $\det \mathbf{A} = \det \mathbf{B}$ if $\mathbf{A}$ and $\mathbf{B}$ are similar matrices.

EXERCISE 8.30. If $\mathbf{A}$ is similar to a lower triangular matrix $\mathbf{L}$, so that $l_{ij} = 0$ for $j > i$, prove that the eigenvalues of $\mathbf{A}$ are just the diagonal elements l_{ii} of $\mathbf{L}$.

THEOREM 8.7. If a matrix $\mathbf{P}$ *exists such that* $\mathbf{B} = \mathbf{P}^{-1}\mathbf{A}\mathbf{P}$, *then*

(i) *For any positive integer k*

$$\mathbf{B}^k = \mathbf{P}^{-1}\mathbf{A}^k\mathbf{P}.$$

If $\mathbf{A}$ *is nonsingular then so is* $\mathbf{B}$ *and the result above is true for any negative integer k. In particular,* $\mathbf{B}^{-1} = \mathbf{P}^{-1}\mathbf{A}^{-1}\mathbf{P}$.

(ii) *If* $f(x)$ *is a polynomial of degree m,*

$$f(x) = a_0x^m + a_1x^{m-1} + \cdots + a_m,$$

and by $f(\mathbf{A})$ *we mean*

$$f(\mathbf{A}) = a_0\mathbf{A}^m + a_1\mathbf{A}^{m-1} + \cdots + a_m\mathbf{I},$$

then

$$f(\mathbf{B}) = \mathbf{P}^{-1}f(\mathbf{A})\mathbf{P}.$$

Proof: For any positive integer k we have

$$\mathbf{B}^k = (\mathbf{P}^{-1}\mathbf{A}\mathbf{P})(\mathbf{P}^{-1}\mathbf{A}\mathbf{P})(\mathbf{P}^{-1}\mathbf{A}\mathbf{P})\ldots$$

repeated k times. The associative law allows us to remove the parentheses. Since $\mathbf{P}\mathbf{P}^{-1} = \mathbf{I}$, the product collapses to give the required result immediately. If $k = -1$, we have

$$\mathbf{B}^{-1} = (\mathbf{P}^{-1}\mathbf{A}\mathbf{P})^{-1} = (\mathbf{P})^{-1}\mathbf{A}^{-1}(\mathbf{P}^{-1})^{-1} = \mathbf{P}^{-1}\mathbf{A}^{-1}\mathbf{P},$$

so that the result is true for $k = -1$. The method used to prove the result for positive k can now be used to prove it for negative integer k. To prove (ii) we use part (i). We have

$$\begin{aligned} f(\mathbf{B}) &= a_0\mathbf{P}^{-1}\mathbf{A}^m\mathbf{P} + a_1\mathbf{P}^{-1}\mathbf{A}^{m-1}\mathbf{P} + \cdots + a_m\mathbf{P}^{-1}\mathbf{P} \\ &= \mathbf{P}^{-1}(a_0\mathbf{A}^m + a_1\mathbf{A}^{m-1} + \cdots + a_m\mathbf{I})\mathbf{P} = \mathbf{P}^{-1}f(\mathbf{A})\mathbf{P}. \end{aligned}$$

EXERCISE 8.31. Find $\mathbf{A}^k$ for all positive k for the matrix $\mathbf{A}$ of Ex. 8.25.

SOLUTION: We examine instead $\mathbf{B}^k$ since $\mathbf{B}$ seems "simpler" than $\mathbf{A}$. We compute easily

$$\mathbf{B}^2 = \begin{bmatrix} -1 & -1 \\ 1 & 0 \end{bmatrix}, \qquad \mathbf{B}^3 = \begin{bmatrix} -1 & 0 \\ 0 & -1 \end{bmatrix} = -\mathbf{I},$$

so that $\mathbf{B}^4 = \mathbf{B}\mathbf{B}^3 = \mathbf{B}(-\mathbf{I}) = -\mathbf{B}$, $\mathbf{B}^5 = \mathbf{B}^2\mathbf{B}^3 = -\mathbf{B}^2$, $\mathbf{B}^6 = -\mathbf{B}^3 = \mathbf{I}$, etc. In general,

$$\mathbf{B}^{3i} = (-1)^i\mathbf{I}, \qquad \mathbf{B}^{3i+1} = (-1)^i\mathbf{B}, \qquad \mathbf{B}^{3i+2} = (-1)^i\mathbf{B}^2,$$

and therefore

$$\mathbf{A}^{3i} = (-1)^i\mathbf{I}, \qquad \mathbf{A}^{3i+1} = (-1)^i\mathbf{A}, \qquad \mathbf{A}^{3i+2} = (-1)^i\mathbf{A}^2$$

where

$$\mathbf{A}^2 = \begin{bmatrix} 1 & -3 \\ 1 & -2 \end{bmatrix}.$$

EXERCISE 8.32. Let f be a polynomial as in Theorem 8.7, and let $\mathbf{A}$ be as in Ex. 8.25 and Ex. 8.31. Find a simplified expression for $f(\mathbf{A})$.

By restating **Key Theorem 8.3** from Section 8.3, we again recall the relationship between similarity transformations and our study of the structure of eigensystems:

- *RESTATEMENT OF* ***KEY THEOREM 8.3.*** *An $n \times n$ matrix $\mathbf{A}$ has a linearly independent set of n eigenvectors if and only if $\mathbf{A}$ is similar to a diagonal matrix.*

As indicated in Section 8.3, our detailed study of the structure of eigensystems in the next two chapters is equivalent to the study of into how "simple" a matrix $\mathbf{B}$ can a matrix $\mathbf{A}$ be transformed by similarity transformations. We will not answer this in full generality until **Key Theorem 10.2.** A rough preview of this theorem follows.

- *ROUGH PREVIEW OF* ***KEY THEOREM 10.2.*** *Every $n \times n$ matrix $\mathbf{A}$ is similar to a matrix $\mathbf{J}$ having:*

  (i) *the eigenvalues of $\mathbf{A}$ on the principal diagonal of $\mathbf{J}$;*
  (ii) *zeros below the principal diagonal, so that $(\mathbf{J})_{ij} = 0$ for $i \geq j + 1$;*
  (iii) *zeros above the first superdiagonal, so that $(\mathbf{J})_{ij} = 0$ for $j \geq i + 2$.*

In the next section we consider a special class of similarity transformations that are of great importance in applications; matrices arising in a large number of practical situations can actually be diagonalized by such special similarity transformations, known as *unitary transformations.*

EXERCISE 8.33. Use Theorem 8.7 and the result of Ex. 8.16 to compute $\mathbf{A}^k$ for all integers k, where $\mathbf{A}$ is as in Ex. 8.16.

EXERCISE 8.34. Suppose that f is a polynomial and $f(\mathbf{A}) = \mathbf{0}$. Prove that $f(\mathbf{B}) = \mathbf{0}$ if $\mathbf{B}$ is similar to $\mathbf{A}$.

EXERCISE 8.35. Prove that two diagonal matrices are similar if and only if the diagonal elements of one are simply a rearrangement of the diagonal elements of the other.

8.5 UNITARY MATRICES AND TRANSFORMATIONS

In Section 8.3 we motivated our study of similarity transformations $\mathbf{B} = \mathbf{P}^{-1}\mathbf{A}\mathbf{P}$ in Section 8.4 by considering the representation of linear transformations with respect to a changed basis $\{\mathbf{p}_1, \ldots, \mathbf{p}_n\}$. Since in Section 4.7 we discussed the convenience of using an *orthonormal* basis $\{\mathbf{p}_1, \ldots, \mathbf{p}_n\}$ (see Definition 4.14), so that $(\mathbf{p}_i, \mathbf{p}_j) = 0$ if $i \neq j$ while $(\mathbf{p}_i, \mathbf{p}_i) = 1$ where $(\mathbf{x}, \mathbf{y}) = \mathbf{x}^H\mathbf{y}$ is an inner product, it is natural to consider similarity transformations induced by such matrices $\mathbf{P}$. Since $\mathbf{p}_i^H\mathbf{p}_i = 1$ while $\mathbf{p}_i^H\mathbf{p}_j = 0$ if $i \neq j$, if

$$\mathbf{P} = [\mathbf{p}_1, \ldots, \mathbf{p}_n],$$

it follows that $\mathbf{P}^H\mathbf{P} = \mathbf{I}$ and hence $\mathbf{P}^{-1} = \mathbf{P}^H$. Such matrices are important enough to be given a special name.

DEFINITION 8.4. A matrix $\mathbf{P}$ such that $\mathbf{P}^H\mathbf{P} = \mathbf{P}\mathbf{P}^H = \mathbf{I}$ is said to be a *unitary* matrix. As a special case, a real matrix $\mathbf{P}$ such that $\mathbf{P}^T\mathbf{P} = \mathbf{P}\mathbf{P}^T = \mathbf{I}$ is said to be an *orthogonal* matrix.

EXERCISE 8.36. Verify that

$$\mathbf{P} = \frac{1}{2}\begin{bmatrix} 1+i & -1+i \\ 1+i & 1-i \end{bmatrix}$$

is unitary, where $i = \sqrt{-1}$.

The statement that $\mathbf{P}^H\mathbf{P} = \mathbf{I}$ tells us that the columns of $\mathbf{P}$ form an orthonormal set; similarly, $\mathbf{P}\mathbf{P}^H = \mathbf{I}$ states that the rows of $\mathbf{P}$ form an orthonormal set. Unitary matrices have further striking properties.

THEOREM 8.8

(i) *Both the columns and the rows of a unitary* (*or orthogonal*) *matrix form an orthonormal set.*

(ii) *If* $\mathbf{P}$ *is unitary, then* $|\det \mathbf{P}| = 1$.

(iii) *If* $\mathbf{P}$ and $\mathbf{Q}$ *are unitary, then so is* $\mathbf{PQ}$.

(iv) *If* $\mathbf{P}$ *is unitary, then for all* $\mathbf{x}$, $\mathbf{y}$ *we have* $(\mathbf{Px}, \mathbf{Py}) = (\mathbf{x}, \mathbf{y})$, $\|\mathbf{Px}\|_2 = \|\mathbf{x}\|_2$, and $\|\mathbf{P}\|_2 = 1$; *interpreted geometrically, this states that* $\mathbf{P}$ *preserves both angles and lengths when considered as defining a linear transformation.*

(v) *If* λ *is an eigenvalue of the unitary matrix* $\mathbf{P}$, *then* $|\lambda| = 1$.

Proof: We have already considered (i). For (ii), we note that

$$1 = \det \mathbf{I} = \det \mathbf{P}^H\mathbf{P} = (\det \mathbf{P}^H)(\det \mathbf{P}) = |\det \mathbf{P}|^2.$$

For (iii), we merely write

$$(\mathbf{PQ})^H(\mathbf{PQ}) = \mathbf{Q}^H\mathbf{P}^H\mathbf{PQ} = \mathbf{Q}^H\mathbf{Q} = \mathbf{I}$$

as asserted. For (iv), from Theorem 4.10(iii) we obtain

$$(\mathbf{Px}, \mathbf{Py}) = (\mathbf{P}^H\mathbf{Px}, \mathbf{y}) = (\mathbf{x}, \mathbf{y}),$$

and letting $\mathbf{x} = \mathbf{y}$ yields $\|\mathbf{Px}\|_2 = \|\mathbf{x}\|_2$ so that $\|\mathbf{P}\|_2 = \sup \|\mathbf{Px}\|_2/\|\mathbf{x}\|_2 = 1$; we interpret this as preserving length and angle because of Definition 4.11. Finally, for (v) if $\mathbf{Px} = \lambda\mathbf{x}$ then from (iv) we have

$$\|\mathbf{x}\|_2 = \|\mathbf{Px}\|_2 = \|\lambda\mathbf{x}\|_2 = |\lambda|\,\|\mathbf{x}\|_2$$

and, since $\mathbf{x} \neq \mathbf{0}$, we obtain $|\lambda| = 1$.

EXERCISE 8.37. Verify Theorem 8.8(i), (ii), (iv), (v) for the matrix of Ex. 8.36.

EXERCISE 8.38. If $\mathbf{P}$ is unitary, show that $\bar{\mathbf{P}}$, $\mathbf{P}^T$ and $\mathbf{P}^H = \mathbf{P}^{-1}$ are unitary.

EXERCISE 8.39. In Theorem 8.8 we deduced that $\|\mathbf{Px}\|_2 = \|\mathbf{x}\|_2$ *from* $(\mathbf{Px}, \mathbf{Py}) = (\mathbf{x}, \mathbf{y})$. Prove conversely that if $\mathbf{P}$ is a matrix for which

$$\|\mathbf{Px}\|_2 = \|\mathbf{x}\|_2$$

for all $\mathbf{x}$, then $(\mathbf{Px}, \mathbf{Py}) = (\mathbf{x}, \mathbf{y})$ for all $\mathbf{x}$, $\mathbf{y}$.

EXERCISE 8.40. Use Ex. 8.39 to help prove that a matrix $\mathbf{P}$ is unitary if and only if $\|\mathbf{Px}\|_2 = \|\mathbf{x}\|_2$ for all complex vectors $\mathbf{x}$.

Our geometric interpretation of Theorem 8.8(iv) indicates that linear transformations defined by unitary matrices behave very much like simple rotations in space. We now show that simple plane rotations are, in fact, represented by unitary matrices. Suppose as in Figure 8.5 that we represent points P with respect to the standard perpendicular x–y coordinate axes, and that we want to determine the coordinates of P with respect to new X–Y axes obtained by rotating the old axes in the plane through an angle θ. Simple

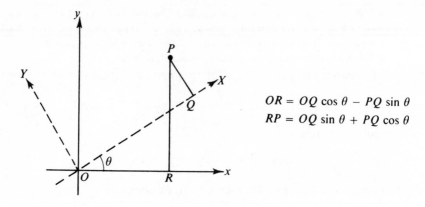

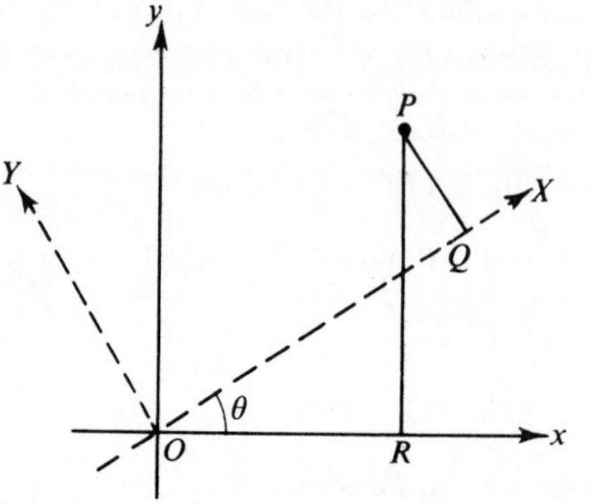

$$OR = OQ\cos\theta - PQ\sin\theta$$
$$RP = OQ\sin\theta + PQ\cos\theta$$

Figure 8.5. Rotation of axes.

trigonometry tells us that

$$OQ = QR\cos\theta + RP\sin\theta$$
$$QP = -OR\sin\theta + RP\cos\theta.$$

If we let

$$\mathbf{x} = \begin{bmatrix} OR \\ RP \end{bmatrix}$$

represent the coordinates with respect to the old axes and

$$\mathbf{x}' = \begin{bmatrix} OQ \\ QP \end{bmatrix}$$

represent the coordinates with respect to the new axes, then we have

$$\mathbf{x}' = \mathbf{P}\mathbf{x}$$

where

$$\mathbf{P} = \begin{bmatrix} \cos\theta & \sin\theta \\ -\sin\theta & \cos\theta \end{bmatrix}$$

is easily seen to be a unitary (in fact, orthogonal) matrix. More generally, in n dimensions we may wish to keep all but two axes of the x_1-x_2-$\cdots$-x_n coordinate system fixed but rotate the x_i and x_j axes through an angle θ from the x_i axis towards the x_j axis in the x_i-x_j plane. This is described in matrix notation by the *plane rotation matrix* $\mathbf{R}_{ij}$ obtained by replacing: (1) the (i, i) element of the unit matrix $\mathbf{I}$ with $\cos\theta$; (2) the (i, j) element of $\mathbf{I}$ with $\sin\theta$; (3) the (j, i) element of $\mathbf{I}$ with $-\sin\theta$; and (4) the (j, j) element of $\mathbf{I}$ with $\cos\theta$. It is trivial to see that $\mathbf{R}_{ij}^T\mathbf{R}_{ij} = \mathbf{I}$, so that $\mathbf{R}_{ij}$ is *orthogonal.*

EXERCISE 8.41. Find the 3 × 3 matrix $\mathbf{R}_{13}$ for $\theta = 45°$.

SOLUTION: For $\theta = 45°$, we have $\sin\theta = \cos\theta = \sqrt{2}/2$. Therefore

$$\mathbf{R}_{13} = \begin{bmatrix} \frac{\sqrt{2}}{2} & 0 & \frac{\sqrt{2}}{2} \\ 0 & 1 & 0 \\ \frac{-\sqrt{2}}{2} & 0 & \frac{\sqrt{2}}{2} \end{bmatrix}.$$

EXERCISE 8.42. Find the 3 × 3 matrix $\mathbf{R}_{21}$ for $\theta = 90°$.

EXERCISE 8.43. Use Exs. 8.41 and 8.42 to find a 3 × 3 matrix representing the combined effects in $\mathbb{R}^3$ of *first* rotating the x_1 axis through 45° toward the x_3 axis in the x_1-x_3 plane, and *then* rotating the x_2 axis through 90° toward the *new* x_1 axis in the x_2-x_1 plane.

We now consider another geometrically simple unitary transformation (see Figure 8.6). Suppose that we wish to transform a real arbitrary vector $\mathbf{u}$

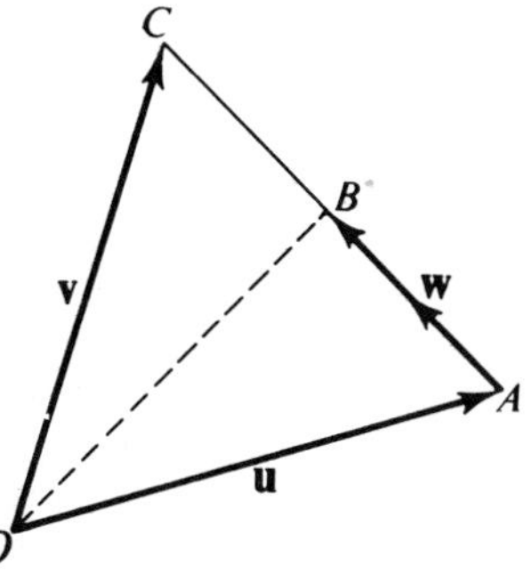

Figure 8.6. Geometrical interpretation of the Householder transformation.

into a second real vector $\mathbf{v}$ of the same length by means of an equation

$$\mathbf{v} = \mathbf{P}\mathbf{u} \tag{8.34}$$

where $\mathbf{P}$ is a square matrix that will depend on $\mathbf{u}$ and $\mathbf{v}$. This can be done by noting that

$$\overrightarrow{OC} = \overrightarrow{OA} + 2\overrightarrow{AB}$$

where B is the midpoint of AC so that $\overrightarrow{OB}$ is perpendicular to $\overrightarrow{AC}$ and $\overrightarrow{AB}$ is *minus* the projection of $\overrightarrow{OA}$ on $\overrightarrow{AC}$. The idea of a projection has already been mentioned in Section 4.7 in connection with Gram-Schmidt orthogonalization. If we introduce

$$\mathbf{w} = \frac{\mathbf{v} - \mathbf{u}}{\|\mathbf{v} - \mathbf{u}\|_2}$$

so that **w** is a unit vector along $\overrightarrow{AC}$, then, remembering a minus because of the directions of the vectors **u** and **w**,

$$\mathbf{v} = \mathbf{u} - 2(\mathbf{w}, \mathbf{u})\mathbf{w}.$$

It is clear that

$$(\mathbf{w}, \mathbf{u})\mathbf{w} = \mathbf{w}(\mathbf{w}, \mathbf{u}) = \mathbf{w}\mathbf{w}^T\mathbf{u},$$

so that the preceding equation gives

$$\mathbf{v} = (\mathbf{I} - 2\mathbf{w}\mathbf{w}^T)\mathbf{u}.$$

Summing up, we wish to consider what we shall call *Householder transformations* which are of the form (8.34) with

$$\mathbf{P} = \mathbf{I} - 2\mathbf{w}\mathbf{w}^T, \qquad \|\mathbf{w}\|_2 = 1.$$

We have derived these results by a geometrical argument, but of course (8.34) is now an algebraic identity which can be verified directly. We have

$$\mathbf{P}\mathbf{u} = (\mathbf{I} - 2\mathbf{w}\mathbf{w}^T)\mathbf{u} = \mathbf{u} - \frac{2(\mathbf{v} - \mathbf{u})(\mathbf{v} - \mathbf{u})^T\mathbf{u}}{\|\mathbf{v} - \mathbf{u}\|_2^2} = \mathbf{u} - \frac{2(\mathbf{v} - \mathbf{u}, \mathbf{u})}{\|\mathbf{v} - \mathbf{u}\|_2^2}(\mathbf{v} - \mathbf{u}).$$

Since $\|\mathbf{u}\|_2 = \|\mathbf{v}\|_2$, we have

$$\|\mathbf{v} - \mathbf{u}\|_2^2 = (\mathbf{v}, \mathbf{v}) - (\mathbf{v}, \mathbf{u}) - (\mathbf{u}, \mathbf{v}) + (\mathbf{u}, \mathbf{u}) = -2(\mathbf{v} - \mathbf{u}, \mathbf{u}),$$

and therefore

$$\mathbf{P}\mathbf{u} = \mathbf{u} + (\mathbf{v} - \mathbf{u}) = \mathbf{v}.$$

Our derivation of the formula for **P** shows that it depends only on **w**, the direction of $\mathbf{v} - \mathbf{u}$. Thus for any other two vectors $\mathbf{u}'$ and $\mathbf{v}'$ for which $\|\mathbf{u}'\|_2 = \|\mathbf{v}'\|_2$ and $\mathbf{v}' - \mathbf{u}'$ is in the same direction as $\mathbf{v} - \mathbf{u}$, we would also have $\mathbf{u}'$ mapped onto $\mathbf{v}'$, that is, $\mathbf{P}\mathbf{u}' = \mathbf{v}'$. This implies that **P** merely *reflects* each point $\mathbf{u}'$ through a plane that is perpendicular to $\mathbf{v} - \mathbf{u}$ and which contains the line through the origin and $\frac{1}{2}(\mathbf{u} + \mathbf{v})$; that is, a Householder transformation is simply a reflection, and our geometrical intuition tells us that this should preserve lengths and angles and hence be a unitary transformation. In fact, for $\mathbf{P} = \mathbf{I} - 2\mathbf{w}\mathbf{w}^T$ with $\|\mathbf{w}\|_2 = 1$, we easily compute $\mathbf{P}^H = \mathbf{P}$ and $\mathbf{P}^H\mathbf{P} = (\mathbf{I} - 2\mathbf{w}\mathbf{w}^T)^2 = \mathbf{I} - 4\mathbf{w}\mathbf{w}^T + 4\mathbf{w}\mathbf{w}^T\mathbf{w}\mathbf{w}^T = \mathbf{I}$.

EXERCISE 8.44. Find the Householder transformation in $\mathbb{R}^2$ which reflects across the line $x_1 = x_2$.

SOLUTION: If we let

$$\mathbf{u} = \begin{bmatrix} 1 \\ 0 \end{bmatrix}, \qquad \mathbf{v} = \begin{bmatrix} 0 \\ 1 \end{bmatrix}$$

then we would reflect as desired. This choice yields

$$\mathbf{w} = \begin{bmatrix} -\frac{\sqrt{2}}{2} \\ \frac{\sqrt{2}}{2} \end{bmatrix}$$

and hence

$$\mathbf{P} = \begin{bmatrix} 1 & 0 \\ 0 & 1 \end{bmatrix} - 2\begin{bmatrix} -\frac{\sqrt{2}}{2} \\ \frac{\sqrt{2}}{2} \end{bmatrix}\begin{bmatrix} -\frac{\sqrt{2}}{2} & \frac{\sqrt{2}}{2} \end{bmatrix} = \begin{bmatrix} 0 & 1 \\ 1 & 0 \end{bmatrix}$$

which simply interchanges the x_1- and x_2-axes, as we would expect.

EXERCISE 8.45. Find the Householder transformation $\mathbb{R}^3$ which results from

$$\mathbf{u} = \begin{bmatrix} 1 \\ 0 \\ 0 \end{bmatrix}, \quad \mathbf{v} = \begin{bmatrix} 0 \\ 1 \\ 0 \end{bmatrix}.$$

Interpret the resulting transformation geometrically.

EXERCISE 8.46. Let $\mathbf{A}$ be an $n \times n$ real matrix. Prove that there exists an orthogonal transformation $\mathbf{H}$ such that the first column of $\mathbf{HA}$ is just a scalar multiple of the unit column vector $\mathbf{e}_1$. Describe how to find such an $\mathbf{H}$.

Recall that we began our discussion of unitary matrices because we wished to consider similarity transformations induced by unitary matrices (that is, by a change to an orthonormal bases); we now return to this topic.

DEFINITION 8.5. If there exists a unitary matrix $\mathbf{P}$ such that $\mathbf{B} = \mathbf{P}^H\mathbf{AP}$, then $\mathbf{A}$ and $\mathbf{B}$ are said to be *unitarily equivalent* and $\mathbf{B}$ is said to have been derived from $\mathbf{A}$ by a *unitary transformation.*

It is trivial to see that unitary equivalence is in fact an equivalence relation. Since unitarily equivalent matrices are also similar matrices (why?), most of our earlier results hold with $\mathbf{P}^{-1}$ replaced by $\mathbf{P}^H$.

THEOREM 8.9

(i) *The results of Theorems 8.5, 8.6, and 8.7 hold with "similar" replaced by "unitarily equivalent" and with $\mathbf{P}^{-1}$ replaced by $\mathbf{P}^H$, where $\mathbf{P}$ is unitary.*

(ii) *If $\mathbf{P}$ and $\mathbf{Q}$ are unitary matrices and $\mathbf{A}$ is a matrix, then*

$$\|\mathbf{A}\|_2 = \|\mathbf{PA}\|_2 = \|\mathbf{AQ}\|_2 = \|\mathbf{PAQ}\|_2,$$

so in particular unitarily equivalent matrices have the same $\|\cdot\|_2$-norm.

Proof: We need only prove (ii); recall that

$$\|\mathbf{B}\|_2 = \sup_{\|\mathbf{x}\|_2=1} \|\mathbf{Bx}\|_2.$$

Therefore

$$\|\mathbf{PA}\|_2 = \sup_{\|\mathbf{x}\|_2=1} \|\mathbf{PAx}\|_2 = \sup_{\|\mathbf{x}\|_2=1} \|\mathbf{Ax}\|_2 = \|\mathbf{A}\|_2$$

where we have used Theorem 8.8(iv). Similarly,

$$\|\mathbf{AQ}\|_2 = \sup_{\|\mathbf{x}\|_2=1} \|\mathbf{AQx}\|_2 = \sup_{\|\mathbf{Qx}\|_2=1} \|\mathbf{AQx}\|_2 = \sup_{\|\mathbf{y}\|_2=1} \|\mathbf{Ay}\|_2 = \|\mathbf{A}\|_2.$$

Finally, from what we have already proved,

$$\|\mathbf{PAQ}\|_2 = \|\mathbf{AQ}\|_2 = \|\mathbf{A}\|_2$$

as asserted.

The properties described in this theorem account for much of the practical importance of unitary matrices and transformations; crudely stated, unitary matrices cannot create or eliminate large numbers or small numbers by accident, hence they are very useful in practical computational procedures.

We again recall that we introduced similarity transformations and their subclass of unitary transformations to provide a particular viewpoint for the study of eigensystems. While **Key Theorem 8.3** described the class of matrices having a full set of eigenvectors as those matrices which are similar to diagonal matrices, this condition cannot be tested simply by looking at the matrix or by performing a simple computation. In Chapter 9 we will show that hermitian and real symmetric matrices (see Definitions 1.7, 1.8) and more generally the so-called *normal* matrices (satisfying $\mathbf{A}^H\mathbf{A} = \mathbf{A}\mathbf{A}^H$) *always* have a full set of eigenvectors and that in fact these eigenvectors can be chosen to be orthonormal so that the similarity transformation of **Key Theorem 8.3** in fact becomes a unitary transformation. We will not prove this until **Key Theorems 9.2** and **9.3**, but we give a preview of this fundamental result now.

- *ROUGH PREVIEW OF **KEY THEOREMS** 9.2 AND 9.3. Every $n \times n$ normal matrix (that is, satisfying $\mathbf{A}^H\mathbf{A} = \mathbf{A}\mathbf{A}^H$), and in particular every hermitian or real symmetric matrix, can be diagonalized by a unitary transformation; equivalently, every such matrix has an orthonormal set of n eigenvectors.*

EXERCISE 8.47. Prove that any matrix **A** which is unitarily equivalent to a real or complex diagonal matrix must be *normal*, that is, **A** satisfies $\mathbf{A}^H\mathbf{A} = \mathbf{A}\mathbf{A}^H$. If $\mathbf{D} = \mathbf{P}^H\mathbf{A}\mathbf{P}$ where **D** is diagonal and **P** is unitary, find the eigenvalues and eigenvectors of **A**.

EXERCISE 8.48. Prove that if **B** is unitarily equivalent to **A**, then $\mathbf{B}^H\mathbf{B}$ is unitarily equivalent to $\mathbf{A}^H\mathbf{A}$.

8.6 THE CONDITION OF EIGENSYSTEMS

We have remarked before, especially in Sections 5.5 and 6.1, that in practice we should think of any matrix **A** arising in a real problem as being only an approximation, perhaps because of measurement errors, to some ideal matrix; it is important to know what effect these inaccuracies in **A** have on its eigenvalues and eigenvectors. As in Section 5.5, we describe the question of the size of perturbations in eigenvalues and eigenvectors caused by perturbations in **A** as being a question of the *condition* of the eigenvalues and eigenvectors of **A** in terms of the data **A**. This problem is far too complex for us to study in any detail here, but we do give some indications of what can be expected in general by examining some special cases; the reader seeking detail should turn to Wilkinson [67] or Stewart [65].

To get an idea of the general case we first consider the very special but illustrative case when **A** is a *diagonal* matrix:

$$\mathbf{A} = \begin{bmatrix} a_{11} & 0 & \dots & 0 \\ 0 & a_{22} & \dots & 0 \\ & & \dots & \\ 0 & 0 & \dots & a_{nn} \end{bmatrix} \tag{8.35}$$

Obviously, in this case the eigenvalues λ_i of **A** are $\lambda_i = a_{ii}$ for $i = 1, \ldots, n$ with the associated eigenvectors being $\mathbf{e}_i$, the unit column vectors; the question is: what happens when **A** is perturbed to $\mathbf{A} + \boldsymbol{\delta}\mathbf{A}$ for some "small" matrix $\boldsymbol{\delta}\mathbf{A}$? The following important theorem allows us to answer this question.

THEOREM 8.10 (Gerschgorin Circle Theorem). *Each (real or complex) eigenvalue λ of an $n \times n$ matrix* **B** *satisfies at least one of the inequalities*

$$|\lambda - b_{ii}| \leq r_i, \quad \text{where} \quad r_i = \sum_{\substack{j=1 \\ j \neq i}}^{n} |b_{ij}| \qquad (i = 1, \ldots, n). \tag{8.36}$$

That is, each eigenvalue lies in at least one of the discs with center b_{ii} and radius r_i in the complex plane. Moreover, if the union of p of the discs (8.36) *is disjoint from the remainder (so that the union does not touch or overlap any of the remaining $n - p$ discs), then there are precisely p eigenvalues of* **B** *in the union of the p discs.*

Proof: If λ is an eigenvalue of **B**, and **x** the corresponding eigenvector, then $\mathbf{Bx} = \lambda\mathbf{x}$, which implies

$$(\lambda - b_{ii})x_i = \sum_{j=1}^{n}{}' b_{ij}x_j \qquad (i = 1, \ldots, n),$$

where the prime indicates that the term $i = j$ in the sum has been omitted Suppose that x_k is the largest element of $\mathbf{x}$. Then $|x_j/x_k| \leq 1$ for all j, and

$$|\lambda - b_{kk}| \leq \sum_{j=1}^{n}{}' |b_{kj}| \left|\frac{x_j}{x_k}\right| \leq \sum_{j=1}^{n}{}' |b_{kj}|.$$

Since this is true for *any* eigenvalue, this proves the first part of the theorem. The proof of the latter part is beyond the scope of this book, but some hints may be found in Exs. 8.77 and 8.78.

EXERCISE 8.49. Estimate the eigenvalues of

$$\mathbf{B} = \begin{bmatrix} 1 & -10^{-5} & 2\cdot10^{-5} \\ 4\cdot10^{-5} & 0.5 & -3\cdot10^{-5} \\ -10^{-5} & 3\cdot10^{-5} & 0.1 \end{bmatrix}.$$

SOLUTION: Direct application of the Gerschgorin theorem shows that the eigenvalues λ_i of $\mathbf{B}$ satisfy

$$|1 - \lambda_1| \leq 3\cdot10^{-5}, \qquad |0.5 - \lambda_2| \leq 7\cdot10^{-5}, \qquad |0.1 - \lambda_3| \leq 4\cdot10^{-5}.$$

Since each of these discs is disjoint from the other two, there is precisely one eigenvalue in each disc. We can easily obtain better estimates, however; see Ex. 8.79.

EXERCISE 8.50. Estimate the eigenvalues of

$$\mathbf{B} = \begin{bmatrix} 1 & 0.1 & 2 \\ 0.01 & 10 & 10 \\ -0.01 & 1 & 100 \end{bmatrix}.$$

SOLUTION: Direct application of the Gerschgorin theorem to $\mathbf{B}$ shows that the three eigenvalues satisfy

$$|\lambda - 1| \leq 2.1, \qquad |\lambda - 10| \leq 10.01, \qquad |\lambda - 100| \leq 1.01.$$

Since the third disc is disjoint from the others there is precisely one eigenvalue λ_3 with $|\lambda_3 - 100| \leq 1.01$. Since the first two discs overlap, however, we can only conclude that two eigenvalues lie somewhere in the union of the two discs. We also know that the eigenvalues of $\mathbf{B}$ and of $\mathbf{B}^T$ are equal. Gerschgorin's theorem applied to $\mathbf{B}^T$ yields the existence of precisely one eigenvalue in each of the three disjoint discs

$$|\lambda_1 - 1| \leq 0.02, \qquad |\lambda_2 - 10| \leq 1.1, \qquad |\lambda_3 - 100| \leq 12.$$

EXERCISE 8.51. Estimate the eigenvalues of

$$\mathbf{B} = \begin{bmatrix} 2 & 0.1 & 0.2 \\ -0.1 & 2 & -0.1 \\ 1 & -1 & 10 \end{bmatrix}.$$

EXERCISE 8.52. Suppose that the $n \times n$ matrix **B** is *strictly diagonally dominant* in the sense that

$$|b_{ii}| > \sum_{\substack{j=1 \\ j \neq i}}^{n} |b_{ij}| \qquad (\text{for } i = 1, \ldots, n).$$

Show that **B** is nonsingular.

We return now to consideration of the result of perturbing the diagonal matrix **A** of (8.35) to $\mathbf{A} + \boldsymbol{\delta}\mathbf{A}$. By the Gerschgorin theorem, the eigenvalues will lie in the discs

$$|a_{ii} + \delta a_{ii} - \lambda| \leq r_i = \sum_{\substack{j=1 \\ j \neq i}}^{n} |\delta a_{ij}| \quad (\text{for } i = 1, \ldots, n).$$

Since the original matrix **A** has as eigenvalues $\lambda_i = a_{ii}$, this says that the eigenvalues of $\mathbf{A} + \boldsymbol{\delta}\mathbf{A}$ are in the discs

$$|\lambda_i - \lambda| \leq |\delta a_{ii}| + r_i \qquad (\text{for } i = 1, \ldots, n).$$

Since we can bound $|\delta a_{ii}| + r_i$ by $||\boldsymbol{\delta}\mathbf{A}||_\infty$, the last inequality above tells us that $|\lambda_i - \lambda| \leq ||\boldsymbol{\delta}\mathbf{A}||_\infty$, that is, *the eigenvalues of a diagonal matrix are well-conditioned.* The situation for eigen*vectors* is more complicated, however, as we show by the following three examples; the eigenvectors are well-conditioned in the first and ill-conditioned in the last two (but for different reasons).

EXERCISE 8.53. Consider the matrix

$$\mathbf{A} = \begin{bmatrix} 1 & 0 \\ 0 & 2 \end{bmatrix},$$

having eigenvectors $\mathbf{e}_1$ and $\mathbf{e}_2$, and the perturbed matrix

$$\mathbf{A} + \boldsymbol{\delta}\mathbf{A} = \begin{bmatrix} 1 & \epsilon \\ 0 & 2 \end{bmatrix}$$

having eigenvectors $\mathbf{e}_1$ and $\epsilon\mathbf{e}_1 + \mathbf{e}_2$. The eigenvectors of **A** are thus perturbed by $||\mathbf{e}_1 - \mathbf{e}_1|| = 0$ and by $||\mathbf{e}_2 - (\mathbf{e}_2 + \epsilon\mathbf{e}_1)|| = ||\epsilon\mathbf{e}_1||$, which is of order ϵ; we see that in this case the eigenvectors are well-conditioned.

EXERCISE 8.54. Consider the matrix

$$\mathbf{A} = \begin{bmatrix} 2+\delta & 0 \\ 0 & 2 \end{bmatrix},$$

where $\delta \neq 0$, having the eigenvectors $\mathbf{e}_1$ and $\mathbf{e}_2$. Consider also the perturbed matrix

$$\mathbf{A} + \boldsymbol{\delta}\mathbf{A} = \begin{bmatrix} 2+\delta & \epsilon \\ 0 & 2 \end{bmatrix}$$

having eigenvectors $\mathbf{e}_1$ and $\mathbf{e}_2 - (\epsilon/\delta)\mathbf{e}_1$. Whether or not the perturbation from $\mathbf{e}_2$ to $\mathbf{e}_2 - (\epsilon/\delta)\mathbf{e}_1$ is small depends not on ϵ alone but on its size relative to δ, the separation between the eigenvalues of $\mathbf{A}$.

EXERCISE 8.55. Consider the matrix

$$\mathbf{A} = \begin{bmatrix} 2 & 0 \\ 0 & 2 \end{bmatrix}$$

having eigenvectors $\mathbf{e}_1$ and $\mathbf{e}_2$, and the perturbed matrix

$$\mathbf{A} + \boldsymbol{\delta}\mathbf{A} = \begin{bmatrix} 2 & \epsilon \\ 0 & 2 \end{bmatrix}$$

having only one eigenvector $\mathbf{e}_1$. Although we "lost" an eigenvector in the perturbation, we might be tempted to call the problem well-conditioned since one vector changed not at all. However, had we chosen as "the" eigenvectors for $\mathbf{A}$ the vectors $\mathbf{e}_1 + \mathbf{e}_2$ and $\mathbf{e}_1 - \mathbf{e}_2$, say, the problem would be revealed as ill-conditioned since the eigenvector of $\mathbf{A} + \boldsymbol{\delta}\mathbf{A}$ would still be $\mathbf{e}_1$, a large jump from either eigenvector of $\mathbf{A}$.

These examples show that we might expect a diagonal matrix with distinct and well-separated eigenvalues to have well-conditioned eigenvectors, while one with repeated or clustered eigenvalues may have ill-conditioned eigen*vectors*; in all cases, however, the eigen*values* were well-conditioned. We now extend our analysis from mere diagonal matrices to the case of $n \times n$ matrices having a linearly independent set containing n eigenvectors. We will base our analysis on the following result.

THEOREM 8.11. Suppose that the $n \times n$ matrix $\mathbf{A}$ has a linearly independent set of n eigenvectors $\mathbf{x}_1, \ldots, \mathbf{x}_n$ associated with the eigenvalues $\lambda_1, \ldots, \lambda_n$, respectively, and suppose that the real or complex number μ and the vector $\mathbf{v}$ with $\|\mathbf{v}\| = 1$ are considered as approximating an eigenvalue and eigenvector of $\mathbf{A}$, where $\|\cdot\|$ denotes any fixed one of the norms $\|\cdot\|_1, \|\cdot\|_2, \|\cdot\|_\infty$. Let $\mathbf{r}$ denote the amount by which μ and $\mathbf{v}$ fail to be part of the eigensystem, that is, let

$$\mathbf{r} = \mathbf{A}\mathbf{v} - \mu\mathbf{v}, \tag{8.37}$$

and let **P** *be the matrix of eigenvectors, that is, let*

$$\mathbf{P} = [\mathbf{x}_1, \ldots, \mathbf{x}_n].$$

Then at least one of the eigenvalues λ_i *of* **A** *satisfies*

$$|\lambda_i - \mu| \leq ||\mathbf{r}|| \, ||\mathbf{P}|| \, ||\mathbf{P}^{-1}||. \tag{8.38}$$

If **P** *is unitary (in particular, if* **A** *is real symmetric or hermitian or normal), then at least one eigenvalue* λ_i *satisfies*

$$|\lambda_i - \mu| \leq ||\mathbf{r}||_2. \tag{8.39}$$

Proof: By **Key Theorem 8.3**, we have $\mathbf{A} = \mathbf{P}\boldsymbol{\Lambda}\mathbf{P}^{-1}$ where $\boldsymbol{\Lambda}$ is the diagonal matrix whose ith diagonal element is λ_i. Hence (8.37) can be rewritten as

$$\mathbf{r} = \mathbf{P}(\boldsymbol{\Lambda} - \mu\mathbf{I})\mathbf{P}^{-1}\mathbf{v}. \tag{8.40}$$

If μ equals one of the eigenvalues λ_i of **A**, then certainly (8.38) holds trivially; if μ is not an eigenvalue, then $\boldsymbol{\Lambda} - \mu\mathbf{I}$ is nonsingular and (8.40) implies that

$$\mathbf{v} = \mathbf{P}(\boldsymbol{\Lambda} - \mu\mathbf{I})^{-1}\mathbf{P}^{-1}\mathbf{r}.$$

Taking norms and recalling that $||\mathbf{v}|| = 1$ we find that

$$1 = ||\mathbf{v}|| \leq ||\mathbf{P}|| \, ||(\boldsymbol{\Lambda} - \mu\mathbf{I})^{-1}|| \, ||\mathbf{P}^{-1}|| \, ||\mathbf{r}||. \tag{8.41}$$

Since $(\boldsymbol{\Lambda} - \mu\mathbf{I})^{-1}$ is a diagonal matrix, in any of the three norms considered it follows that

$$||(\boldsymbol{\Lambda} - \mu\mathbf{I})^{-1}|| = \max_i |\lambda_i - \mu|^{-1} = [\min_i |\lambda_i - \mu|]^{-1}. \tag{8.42}$$

From (8.41) and (8.42) we obtain

$$\min_i |\lambda_i - \mu| \leq ||\mathbf{r}|| \, ||\mathbf{P}|| \, ||\mathbf{P}^{-1}||,$$

and therefore (8.38) holds for some λ_i. If **P** is unitary then so is $\mathbf{P}^{-1}$ and so by Theorem 8.8(iv) we have $||\mathbf{P}||_2 = ||\mathbf{P}^{-1}||_2 = 1$; (8.39) then follows from (8.38) by letting $||\cdot||$ denote $||\cdot||_2$.

EXERCISE 8.56. Apply Theorem 8.11 to the matrix

$$\mathbf{A} = \begin{bmatrix} 2.1 & 1 \\ 1 & 2.1 \end{bmatrix}$$

with $\mu = 1$ and

$$\mathbf{v} = \begin{bmatrix} \frac{\sqrt{2}}{2} \\ -\frac{\sqrt{2}}{2} \end{bmatrix}.$$

We can use Theorem 8.11 to study the condition of eigenvalues as follows. If $\mathbf{A}$, $\mathbf{x}_1, \ldots, \mathbf{x}_n$, and $\lambda_1, \ldots, \lambda_n$ are as in Theorem 8.11 and if μ is an eigenvalue associated with the eigenvector $\mathbf{v}$ with $\|\mathbf{v}\| = 1$ of the perturbed matrix $\mathbf{A} + \delta\mathbf{A}$, then by definition

$$(\mathbf{A} + \delta\mathbf{A})\mathbf{v} = \mu\mathbf{v}$$

which we choose to write as

$$\mathbf{A}\mathbf{v} - \mu\mathbf{v} = \mathbf{r}, \qquad \mathbf{r} = -(\delta\mathbf{A})\mathbf{v}.$$

By Theorem 8.11, one of the eigenvalues λ_i of $\mathbf{A}$ satisfies (8.38), and since $\|\mathbf{r}\| \leq \|\delta\mathbf{A}\|\,\|\mathbf{v}\| = \|\delta\mathbf{A}\|$, we obtain

$$|\lambda_i - \mu| \leq \|\delta\mathbf{A}\|\,\|\mathbf{P}\|\,\|\mathbf{P}^{-1}\|$$

for some λ_i. This proves the following.

THEOREM 8.12. Let $\mathbf{A}$, $\mathbf{x}_1, \ldots, \mathbf{x}_n$, $\lambda_1, \ldots, \lambda_n$, $\mathbf{P}$, *and* $\|\cdot\|$ *be as in Theorem 8.11, and let* μ *be an eigenvalue associated with the eigenvector* $\mathbf{v}$ *of a perturbed matrix* $\mathbf{A} + \delta\mathbf{A}$, *where* $\|\mathbf{v}\| = 1$. *Then at least one eigenvalue* λ_i *of* $\mathbf{A}$ *satisfies*

$$|\lambda_i - \mu| \leq \|\delta\mathbf{A}\|\,\|\mathbf{P}\|\,\|\mathbf{P}^{-1}\|.$$

If $\mathbf{P}$ *is unitary (in particular if* $\mathbf{A}$ *is real symmetric or hermitian or normal) then at least one eigenvalue* λ_i *of* $\mathbf{A}$ *satisfies*

$$|\lambda_i - \mu| \leq \|\delta\mathbf{A}\|_2.$$

This theorem states that the *eigenvalues* of hermitian matrices are always well-conditioned. This statement should be emphasized.

EXERCISE 8.57. Study the behavior of the eigenvalues when

$$\mathbf{A} = \begin{bmatrix} 1 & 10^6 \\ 0 & 2 \end{bmatrix}$$

is perturbed to

$$\mathbf{A}' = \mathbf{A} + \delta\mathbf{A} = \begin{bmatrix} 1 & 10^6 \\ \epsilon & 2 \end{bmatrix}.$$

SOLUTION: The matrix $\mathbf{A}$ has eigenvalues $\lambda_1 = 1$, $\lambda_2 = 2$, and associated eigenvectors $\mathbf{x}_1 = \mathbf{e}_1$, $\mathbf{x}_2 = \mathbf{e}_1 + 10^{-6}\mathbf{e}_2$. Thus we have

$$\mathbf{P} = \begin{bmatrix} 1 & 1 \\ 0 & 10^{-6} \end{bmatrix}, \qquad \mathbf{P}^{-1} = \begin{bmatrix} 1 & -10^6 \\ 0 & 10^6 \end{bmatrix}$$

so that the eigenvalues λ' of $\mathbf{A}'$ lie in the discs

$$|\lambda' - 1| \leq 2(1 + 10^6)\epsilon, \qquad |\lambda' - 2| \leq 2(1 + 10^6)\epsilon,$$

indicating very large perturbations. In fact, for $\epsilon = 0.75 \times 10^{-6}$ the perturbed eigenvalues are precisely $\lambda_1' = 0.5$ and $\lambda_2' = 2.5$, compared with $\lambda_1 = 1$ and $\lambda_2 = 2$.

EXERCISE 8.58. Study the behavior of the eigenvalues when

$$\mathbf{A} = \begin{bmatrix} 1 & \alpha \\ 0 & 2 \end{bmatrix}$$

is perturbed to

$$\mathbf{A}' = \mathbf{A} + \boldsymbol{\delta}\mathbf{A} = \begin{bmatrix} 1 & \alpha \\ \epsilon & 2 \end{bmatrix};$$

your results should depend on the size of α.

A detailed discussion of the condition of the eigen*vectors* is beyond the scope of this text. Recall that our earlier examples (see Exs. 8.53 to 8.55) revealed that the separation of the eigenvalues strongly affects the condition of the eigenvectors. An eigenvector associated with a simple eigenvalue which is well separated from the remaining eigenvalues can be expected to be well-conditioned, while eigenvectors associated with multiple or clustered eigenvalues can be expected to be ill-conditioned. In fact, it can be shown that if $\mathbf{A}$ is as in Theorem 8.11 and Theorem 8.12, if the eigenvalue λ_1 is distinct from the remaining eigenvalues $\lambda_2, \ldots, \lambda_n$, and if $\mathbf{B}$ is a fixed matrix, then for small ϵ there is an eigenvector $\mathbf{x}_1(\epsilon)$ of $\mathbf{A} + \epsilon\mathbf{B}$ near $\mathbf{x}_1$ and given approximately by

$$\mathbf{x}_1(\epsilon) \approx \mathbf{x}_1 + \epsilon\left\{\sum_{i=2}^{n} \frac{(\mathbf{y}_i, \mathbf{B}\mathbf{x}_i)}{(\lambda_1 - \lambda_i)(\mathbf{y}_i, \mathbf{x}_i)}\mathbf{x}_i\right\}$$

where $\mathbf{y}_i$ is the left-eigenvector associated with λ_i satisfying $\mathbf{y}_i^T\mathbf{A} = \lambda_i\mathbf{y}_i^T$; the term in brackets will certainly be large if any of $\lambda_2, \ldots, \lambda_n$ are sufficiently close to λ_1.

MISCELLANEOUS EXERCISES 8

EXERCISE 8.59. Use Theorem 8.9(ii) to prove that $\|\mathbf{P}\|_2 = 1$ for a unitary matrix $\mathbf{P}$.

EXERCISE 8.60. Let $\mathbf{P}$ be an $n \times n$ unitary matrix. Prove that

$$\frac{1}{\sqrt{n}} \leq \|\mathbf{P}\| \leq \sqrt{n},$$

where $\|\cdot\|$ can denote either $\|\cdot\|_\infty$ or $\|\cdot\|_1$.

EXERCISE 8.61. Verify that the matrix

$$\mathbf{A} = \tfrac{1}{2}\begin{bmatrix} 1+i & -1+i \\ 1+i & 1-i \end{bmatrix}$$

is unitary, where $i = \sqrt{-1}$.

EXERCISE 8.62. Find the eigenvalues and eigenvectors of the unitary matrix

$$\mathbf{A} = \begin{bmatrix} \cos\theta & -\sin\theta \\ \sin\theta & \cos\theta \end{bmatrix}.$$

EXERCISE 8.63. Show that the following matrix has the eigenvectors indicated, corresponding to eigenvalues $a + ib$ and $a - ib$, where $i = \sqrt{-1}$:

$$\mathbf{A} = \begin{bmatrix} a & b \\ -b & a \end{bmatrix}, \qquad \mathbf{x}_1 = \begin{bmatrix} 1 \\ i \end{bmatrix}, \qquad \mathbf{x}_2 = \begin{bmatrix} 1 \\ -i \end{bmatrix}.$$

EXERCISE 8.64. Prove that the eigenvalues of a triangular matrix are the diagonal elements and find the corresponding eigenvectors.

EXERCISE 8.65. Show that the $n \times n$ matrix

$$\begin{bmatrix} k & 1 & 0 & \dots & 0 \\ 1 & k & 1 & \dots & 0 \\ 0 & 1 & k & \dots & 0 \\ & & \dots & & \\ 0 & 0 & 0 & \dots & k \end{bmatrix}$$

has eigenvectors $\mathbf{x}_i$ whose jth element is given by $\sin\{ij\pi/(n+1)\}$. Deduce the corresponding eigenvalues.

EXERCISE 8.66. If $\mathbf{A}^2 = \mathbf{I}$, show that $(\mathbf{I} + \mathbf{A})\mathbf{x}$ and $(\mathbf{I} - \mathbf{A})\mathbf{x}$ are eigenvectors of $\mathbf{A}$, corresponding to eigenvalues $1, -1$, respectively, where $\mathbf{x}$ is an arbitrary vector. If

$$\mathbf{A} = \begin{bmatrix} 0 & 0 & 1 & 0 \\ 0 & 0 & 0 & -1 \\ 1 & 0 & 0 & 0 \\ 0 & -1 & 0 & 0 \end{bmatrix},$$

show by inspection of $(\mathbf{I} + \mathbf{A})\mathbf{x}$ and $(\mathbf{I} - \mathbf{A})\mathbf{x}$ that there are two independent eigenvectors corresponding to each of the eigenvalues $1, -1$.

EXERCISE 8.67. If $\mathbf{A}$ is a real square matrix of even order and det $\mathbf{A} < 0$, prove that $\mathbf{A}$ has at least two real eigenvalues.

EXERCISE 8.68. Prove that the characteristic polynomial of

$$\mathbf{A} = \begin{bmatrix} \mathbf{B} & \mathbf{0} \\ \mathbf{0} & \mathbf{C} \end{bmatrix},$$

where $\mathbf{B}$ and $\mathbf{C}$ are square, is the product of the characteristic polynomials of $\mathbf{B}$ and $\mathbf{C}$. Investigate the relationship among the eigenvectors of $\mathbf{A}$, $\mathbf{B}$, $\mathbf{C}$.

EXERCISE 8.69. If $\mathbf{A}$ is $m \times n$ and $\mathbf{B}$ is $n \times m$, and if $\mathbf{ABx} = \lambda\mathbf{x}$ where $\lambda \neq 0$, show that $\mathbf{Bx}$ is an eigenvector of $\mathbf{BA}$ corresponding to the same eigenvalue. Show that $\mathbf{AB}$ and $\mathbf{BA}$ have the same eigenvalues except that the product which is of larger order has $|m - n|$ extra zero eigenvalues.

EXERCISE 8.70. Show that every polynomial equation is the characteristic equation of some matrix. More specifically, show that

$$(-1)^n \det \begin{bmatrix} -a_1 - \lambda & -a_2 & \dots & -a_{n-1} & -a_n \\ 1 & -\lambda & \dots & 0 & 0 \\ 0 & 1 & \dots & 0 & 0 \\ & & \dots & & \\ 0 & 0 & \dots & 1 & -\lambda \end{bmatrix} = \lambda^n + a_1\lambda^{n-1} + \cdots + a_n.$$

EXERCISE 8.71. If the rank of $\mathbf{A}$ is r, show that at least $n - r$ eigenvalues of $\mathbf{A}$ are zero. Give an example to show that more than $n - r$ eigenvalues may be zero.

EXERCISE 8.72. Let $\mathbf{P}$ denote the *permutation matrix*

$$\begin{bmatrix} 0 & 0 & \dots & 0 & 1 \\ 1 & 0 & \dots & 0 & 0 \\ & & \dots & & \\ 0 & 0 & \dots & 1 & 0 \end{bmatrix}$$

for which $a_{i+1,i} = 1$, $a_{1n} = 1$, and all other elements are zero. Prove that the characteristic polynomial is $\lambda^n - 1 = 0$. If $\mu = \exp(2\pi i/n)$ with $i = \sqrt{-1}$, show that the eigenvectors are given by

$$\mathbf{x}_r = [\mu^r, \mu^{2r}, \dots, \mu^{nr}]^T, \qquad (r = 1, \dots, n)$$

corresponding to eigenvalues μ^{n-r}, respectively. Prove that these are orthogonal. Prove that

$$\mathbf{Pe}_i = \mathbf{e}_{i+1}, \qquad (i = 1, \dots, n - 1); \quad \mathbf{Pe}_n = \mathbf{e}_1.$$

Deduce that if $\mathbf{A}$ is the *circulant*

$$\mathbf{A} = \begin{bmatrix} c_0 & c_{n-1} & \cdots & c_1 \\ c_1 & c_0 & \cdots & c_2 \\ & & \cdots & \\ c_{n-1} & c_{n-2} & \cdots & c_0 \end{bmatrix},$$

then $\mathbf{A} = f(\mathbf{P})$ where $f(x) = c_0 + c_1x + \cdots + c_{n-1}x^{n-1}$. Deduce that the eigenvalues of $\mathbf{A}$ are given by $f(\mu^{n-r})$, $r = 1, \ldots, n$, with corresponding eigenvectors $\mathbf{x}_r$, defined above.

EXERCISE 8.73. Two masses are suspended by weightless springs as shown in

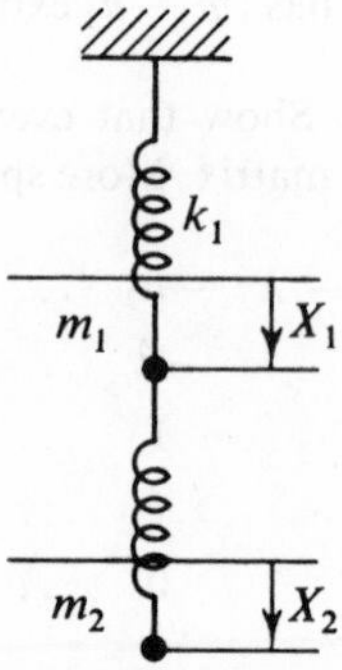

Figure 8.7. Masses on springs for Ex. 8.73.

Figure 8.7. Show that the motion is governed by the equations

$$m_1\frac{d^2X_1}{dt^2} = -k_1X_1 + k_2(X_2 - X_1),$$

$$m_2\frac{d^2X_2}{dt^2} = -k_2(X_2 - X_1),$$

where k_1, k_2 are the spring constants. Show that the free vibrations of this system may be deduced from the solutions of an eigenvalue problem $\mathbf{Ax} = \lambda\mathbf{x}$, where

$$\mathbf{A} = \begin{bmatrix} \dfrac{(k_1 + k_2)}{m_1} & \dfrac{-k_2}{m_1} \\ \dfrac{-k_2}{m_2} & \dfrac{k_2}{m_2} \end{bmatrix}.$$

Find the frequencies and modes of free vibration if $m_1 = m_2$ and $k_1 = \frac{3}{2}k_2$.

EXERCISE 8.74. Establish the equations of motion for the linear triatomic molecule, shown diagrammatically in Figure 8.8 as two masses m attached by

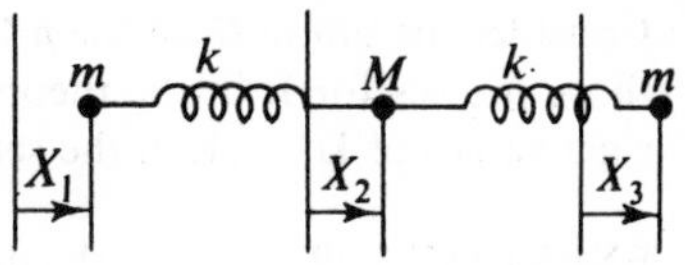

Figure 8.8. Linear triatomic molecule for Ex. 8.74.

springs to a central mass M:

$$m\frac{d^2X_1}{dt^2} = -k(X_1 - X_2),$$

$$M\frac{d^2X_2}{dt^2} = k(X_1 - 2X_2 + X_3),$$

$$m\frac{d^2X_3}{dt^2} = -k(X_3 - X_2).$$

Deduce that there are three frequencies of free vibration given by

$$\omega_1 = 0, \qquad \omega_2 = \left(\frac{k}{m}\right)^{1/2}, \qquad \omega_3 = \left\{\frac{k(m+M)}{mM}\right\}^{1/2}.$$

What is the meaning of $\omega_1 = 0$? Find the modes of vibration.

EXERCISE 8.75. Show that the matrix

$$\begin{bmatrix} 2\sqrt{-1} & 1 \\ 1 & 0 \end{bmatrix}$$

has only one eigenvector, implying that *complex* symmetric matrices need not have a full set of eigenvectors as do *real* symmetric matrices.

EXERCISE 8.76. Show how to derive the eigensystem of the hermitian matrix $\mathbf{A} + \mathbf{B}\sqrt{-1}$ from that of the real symmetric matrix

$$\begin{bmatrix} \mathbf{A} & -\mathbf{B} \\ \mathbf{B} & \mathbf{A} \end{bmatrix}.$$

EXERCISE 8.77. Assuming that the roots of a polynomial are continuous functions of the coefficients of the polynomial, prove that the eigenvalues of a matrix are continuous functions of the elements of the matrix.

EXERCISE 8.78. Prove the last portion of Theorem 8.10 as follows. Consider the matrices $\mathbf{D} + t\mathbf{E}$ for $0 \le t \le 1$, where

$$\mathbf{D} = \begin{bmatrix} b_{11} & 0 & \dots & 0 \\ 0 & b_{22} & \dots & 0 \\ & \dots & & \\ 0 & 0 & \dots & b_{nn} \end{bmatrix}, \qquad \mathbf{E} = \mathbf{B} - \mathbf{D}.$$

Consider the union G_t of the p Gerschgorin discs for $\mathbf{D} + t\mathbf{E}$ corresponding to those p discs for $\mathbf{B}$ in the theorem. Show that for each t there are precisely p eigenvalues of $\mathbf{D} + t\mathbf{E}$ in the set G_t.

EXERCISE 8.79. Provide the details to improve the bounds on the eigenvalues in Ex. 8.49 as follows. Let

$$\mathbf{S} = \begin{bmatrix} \alpha & 0 & 0 \\ 0 & 1 & 0 \\ 0 & 0 & 1 \end{bmatrix}$$

with $\alpha > 0$, and compute $\mathbf{S}^{-1}\mathbf{B}\mathbf{S}$. Choose α as large as possible without the Gerschgorin circle about $b_{11} = 1$ overlapping the other two circles, and then write down the Gerschgorin inequality for λ_1 derived from $\mathbf{S}^{-1}\mathbf{B}\mathbf{S}$; $\alpha = 10^4$, for example, gives $|1 - \lambda_1| \leq 3.10^{-9}$. Use analogous matrices $\mathbf{S}$ to obtain sharper bounds on λ_2 and λ_3 also.

EXERCISE 8.80. Use Theorem 8.10 to show that if an $n \times n$ matrix $\mathbf{A}$ has each of its n Gerschgorin discs disjoint from the remainder, then each disc contains precisely one eigenvalue; if *in addition* $\mathbf{A}$ is real, show that all of its eigenvalues are real.

EXERCISE 8.81. Suppose that the $n \times n$ matrix $\mathbf{A}$ has an eigenvalue λ_i of multiplicity m_i; show that the number of eigenvectors associated with λ_i and independent of one another cannot exceed m_i. *Hint:* Let $\{\mathbf{x}_1, \ldots, \mathbf{x}_p\}$ be a linearly independent set of eigenvectors of $\mathbf{A}$ associated with λ_i. Add vectors $\mathbf{x}_{p+1}, \ldots, \mathbf{x}_n$ to this set so that $\mathbf{S} = [\mathbf{x}_1, \ldots, \mathbf{x}_n]$ is nonsingular, and show that

$$\mathbf{A}' = \mathbf{S}^{-1}\mathbf{A}\mathbf{S} = \begin{bmatrix} \lambda_i \mathbf{I}_p & \mathbf{B} \\ \mathbf{0} & \mathbf{C} \end{bmatrix}$$

where $\mathbf{B}$ is $p \times (n - p)$, $\mathbf{0}$ is an $(n - p) \times p$ null matrix, and $\mathbf{C}$ is

$$(n - p) \times (n - p).$$

By considering the characteristic polynomial of $\mathbf{A}'$ and using Theorem 8.6, deduce that $m_i \geq p$.

CHAPTER NINE

UNITARY TRANSFORMATIONS, EIGENSYSTEMS, AND APPLICATIONS

This chapter details precisely what can be accomplished in the simplification of matrices (or linear transformations) using the unitary matrices and transformations introduced in Section 8.5. **Key Theorems 9.2** and **9.3** characterize normal matrices as those diagonalizable by unitary transformations (having an orthonormal set of eigenvectors). These findings describe the result of changes of basis from the standard one $\{\mathbf{e}_1, \ldots, \mathbf{e}_n\}$ to a single other orthonormal basis; from this viewpoint the results are extended in **Key Theorems 9.6** and **9.7** on the QR and singular-value decompositions. As examples of their usefulness, the decompositions are applied to analyze the important practical problems of least-squares solutions of systems of linear equations (**Key Theorem 9.8** and **Key Corollary 9.1**) and of the recognition of important factors influencing experimental results.

9.1 INTRODUCTION

The important question of the structure of the eigensystem of a matrix **A** is equivalent to questions concerning decompositions of **A** and changes of basis to simplify the representation of the standard linear transformation defined by **A** (see Figure 8.4). In particular, Section 8.5 emphasized the advantage of using unitary matrices in such decompositions, since this is equivalent to using computationally convenient orthonormal bases to represent vector spaces. In this chapter we present a number of decompositions using unitary matrices (equivalently, changes of orthonormal bases) which are very important not only for the analysis of the structure of eigensystems (Section 9.2) but for other applications as well (Sections 9.5, 9.6, 10.5, 10.6, 10.7).

9.2 SCHUR DECOMPOSITION OR CANONICAL FORM

If an $n \times n$ matrix $\mathbf{A}$ is considered as defining a linear transformation A from $\mathbb{R}^n$ to $\mathbb{R}^n$ (or $\mathbb{C}^n$ to $\mathbb{C}^n$) via $A(\mathbf{x}) = \mathbf{A}\mathbf{x}$, and if we elect to use an orthonormal set of vectors $\mathbf{p}_1, \ldots, \mathbf{p}_n$ as a basis both for the domain space of A and the range space of A, then, as we saw in Sections 8.3 and 8.5, the transformation A is represented with respect to these bases by the matrix

$$\mathbf{A}' = \mathbf{P}^H\mathbf{A}\mathbf{P}, \quad \text{with} \quad \mathbf{P} = [\mathbf{p}_1, \ldots, \mathbf{p}_n],$$

so that $\mathbf{A}'$ is unitarily equivalent to $\mathbf{A}$. In this section we want to determine the simplest standard or *canonical* form $\mathbf{A}'$ to which we can transform a given matrix $\mathbf{A}$; since $\mathbf{A} = \mathbf{P}\mathbf{A}'\mathbf{P}^H$, we equivalently determine a "simple" decomposition of $\mathbf{A}$. The first important result here is that one "simple" form $\mathbf{T}$ possible for $\mathbf{A}'$ is *upper triangular*, so that $\mathbf{T} = [t_{ij}]$ has $t_{ij} = 0$ for $1 \leq j < i$ and $1 < i \leq n$.

THEOREM 9.1. Any square $n \times n$ matrix $\mathbf{A}$ can be reduced by a unitary transformation $\mathbf{P}^H\mathbf{A}\mathbf{P}$ to an upper triangular matrix $\mathbf{T}$ (so that $t_{ij} = 0$ for $1 \leq j < i$ and $1 < i \leq n$) with the eigenvalues of $\mathbf{A}$ on the diagonal of $\mathbf{T}$. $\mathbf{T}$ is called a Schur canonical form of $\mathbf{A}$ and the decomposition $\mathbf{A} = \mathbf{P}\mathbf{T}\mathbf{P}^H$ is called a Schur decomposition of $\mathbf{A}$. If $\mathbf{A}$ and its eigenvalues are real, then $\mathbf{P}$ may be taken to be real also.

Proof: Let λ_1 be an eigenvalue of $\mathbf{A}$ with an associated eigenvector $\mathbf{x}_1$ normalized so that $(\mathbf{x}_1, \mathbf{x}_1) = 1$. By Theorem 4.6 and **Key Theorem 4.12** we can choose vectors $\mathbf{w}_2, \ldots, \mathbf{w}_n$ so that the vectors $\mathbf{x}_1, \mathbf{w}_2, \ldots, \mathbf{w}_n$ form an orthonormal basis, that is, so that

$$\mathbf{Q} = [\mathbf{x}_1, \mathbf{w}_2, \ldots, \mathbf{w}_n] = [\mathbf{x}_1, \mathbf{W}]$$

is unitary. Since $(\mathbf{w}_i, \mathbf{x}_1) = 0$ for $i = 2, \ldots, n$ we have

$$\mathbf{W}^H\mathbf{x}_1 = \mathbf{0}.$$

Therefore, since $\mathbf{Ax}_1 = \lambda_1\mathbf{x}_1$,

$$\mathbf{Q}^H\mathbf{AQ} = \begin{bmatrix} \mathbf{x}_1^H \\ \mathbf{W}^H \end{bmatrix} \mathbf{A}[\mathbf{x}_1, \mathbf{W}] = \begin{bmatrix} \mathbf{x}_1^H \\ \mathbf{W}^H \end{bmatrix} [\lambda_1\mathbf{x}_1, \mathbf{AW}]$$
$$= \begin{bmatrix} \lambda_1 & \mathbf{x}_1^H\mathbf{AW} \\ \mathbf{0} & \mathbf{W}^H\mathbf{AW} \end{bmatrix} = \begin{bmatrix} \lambda_1 & \mathbf{b}^H \\ \mathbf{0} & \mathbf{C} \end{bmatrix}, \tag{9.1}$$

say. We proceed by induction. The theorem is true if $n = 2$ since (9.1) is already in the required form. Assume that $\mathbf{A}$ is $n \times n$ and the theorem is true for $n - 1$. Then $\mathbf{C} = \mathbf{W}^H\mathbf{AW}$ of (9.1) is of order $n - 1$, and a unitary matrix $\mathbf{V}$ of order $n - 1$ exists such that $\mathbf{V}^H\mathbf{CV}$ is upper triangular. The matrix

$$\mathbf{U} = \begin{bmatrix} 1 & \mathbf{0} \\ \mathbf{0} & \mathbf{V} \end{bmatrix}$$

is unitary and

$$\mathbf{U}^H(\mathbf{Q}^H\mathbf{AQ})\mathbf{U} = \begin{bmatrix} 1 & \mathbf{0} \\ \mathbf{0} & \mathbf{V}^H \end{bmatrix} \begin{bmatrix} \lambda_1 & \mathbf{b}^H \\ \mathbf{0} & \mathbf{C} \end{bmatrix} \begin{bmatrix} 1 & \mathbf{0} \\ \mathbf{0} & \mathbf{V} \end{bmatrix} = \begin{bmatrix} \lambda_1 & \mathbf{b}^H\mathbf{V} \\ \mathbf{0} & \mathbf{V}^H\mathbf{CV} \end{bmatrix}.$$

Hence

$$(\mathbf{QU})^H\mathbf{A}(\mathbf{QU})$$

is upper triangular. Since $\mathbf{P} = \mathbf{QU}$ is unitary, $\mathbf{A}$ has been reduced to upper triangular form $\mathbf{T}$ by a unitary transformation. The theorem follows by induction, so far as the triangularization is concerned. By Theorems 8.9(i) and 8.6(i), the eigenvalues of $\mathbf{A}$ and $\mathbf{T}$ are identical; since the eigenvalues of $\mathbf{T}$ are clearly its diagonal elements, the theorem is proved. The remark on the reality of $\mathbf{P}$ follows from **Key Theorem 8.2(i).**

In practice the Schur decomposition may be difficult to obtain because it requires knowledge of the eigensystem of $\mathbf{A}$, knowledge that is not usually readily available.

EXERCISE 9.1. Let

$$\mathbf{A} = \begin{bmatrix} 1 & 0 & 3 \\ 0 & 1 & 4 \\ 0 & 0 & 2 \end{bmatrix};$$

note that $\mathbf{A}$ is already a Schur canonical form. Show that $\mathbf{A}' = \mathbf{P}^H\mathbf{AP} \neq \mathbf{A}$ is also a Schur canonical form, where

$$\mathbf{P} = \begin{bmatrix} 0 & -1 & 0 \\ 1 & 0 & 0 \\ 0 & 0 & 1 \end{bmatrix},$$

and conclude that there is not a *unique* Schur form for a matrix (see Ex. 9.74).

EXERCISE 9.2. Consider the construction in Theorem 9.1 for the matrix

$$\mathbf{A} = \begin{bmatrix} 0.8 & 0.2 & 0.1 \\ 0.1 & 0.7 & 0.3 \\ 0.1 & 0.1 & 0.6 \end{bmatrix}$$

of (2.10) in Section 2.2, whose eigenvalues and eigenvectors we found at the start of Section 8.1. To the eigenvalue 0.6 there corresponds the eigenvector [see (8.1)]

$$\begin{bmatrix} 1 \\ -1 \\ 0 \end{bmatrix}$$

which can be normalized as

$$\frac{\sqrt{2}}{2}\begin{bmatrix} 1 \\ -1 \\ 0 \end{bmatrix}.$$

This can be used as the first column of a unitary matrix $\mathbf{Q}$, say

$$\mathbf{Q} = \begin{bmatrix} \frac{\sqrt{2}}{2} & \frac{\sqrt{2}}{2} & 0 \\ -\frac{\sqrt{2}}{2} & \frac{\sqrt{2}}{2} & 0 \\ 0 & 0 & 1 \end{bmatrix},$$

where we have chosen the last two columns of $\mathbf{Q}$ as simply as we could to make $\mathbf{Q}$ unitary; we then calculate

$$\mathbf{A}_1 = \mathbf{Q}^H\mathbf{A}\mathbf{Q} = \begin{bmatrix} 0.6 & 0.1 & -0.1\sqrt{2} \\ 0 & 0.9 & 0.2\sqrt{2} \\ 0 & 0.1\sqrt{2} & 0.6 \end{bmatrix}.$$

We next consider the 2×2 submatrix indicated above, namely

$$\begin{bmatrix} 0.9 & 0.2\sqrt{2} \\ 0.1\sqrt{2} & 0.6 \end{bmatrix}$$

which is easily found to have an eigenvalue $\lambda = 1$ with an associated normalized eigenvector

$$\begin{bmatrix} \frac{2\sqrt{2}}{3} \\ \frac{1}{3} \end{bmatrix}.$$

We add, as a second column, the simplest normalized orthogonal vector we can think of to get the unitary matrix

$$\begin{bmatrix} \frac{2\sqrt{2}}{3} & \frac{1}{3} \\ \frac{1}{3} & \frac{-2\sqrt{2}}{3} \end{bmatrix}$$

and then use the matrix

$$\begin{bmatrix} 1 & 0 & 0 \\ 0 & \frac{2\sqrt{2}}{3} & \frac{1}{3} \\ 0 & \frac{1}{3} & \frac{-2\sqrt{2}}{3} \end{bmatrix}$$

as the basis of a further unitary equivalence to be applied to $\mathbf{A}_1$, yielding for $\mathbf{A}$ the desired triangular form

$$\begin{bmatrix} 0.6 & \frac{0.1\sqrt{2}}{3} & \frac{1}{6} \\ 0 & 1 & -0.1\sqrt{2} \\ 0 & 0 & 0.5 \end{bmatrix}.$$

EXERCISE 9.3. Fill in the details in Ex. 9.2.

EXERCISE 9.4. Proceed as in Ex. 9.2 to reduce $\mathbf{A}^T$ to upper triangular form, where $\mathbf{A}$ is as in Ex. 9.2 (see Ex. 8.2).

Note that Theorem 9.1 can also be interpreted in terms of linear transformations and change of basis.

EXERCISE 9.5. Let the $n \times n$ matrix $\mathbf{A}$ define a linear transformation A from $\mathbb{C}^n$ to $\mathbb{C}^n$ by $A(\mathbf{x}) = \mathbf{Ax}$. Show by Theorem 9.1 that there is a basis of vectors $\mathbf{p}_1, \ldots, \mathbf{p}_n$ for $\mathbb{C}^n$ (both as the domain of A and as the range of A) with respect to which A is represented by an upper triangular matrix.

The result of Theorem 9.1 almost merits being called a **key** result, which may seem puzzling to the student; one reason for its importance stems from the startling result obtained by applying Theorem 9.1 to the special class of matrices known as *normal* matrices.

DEFINITION 9.1. A square matrix $\mathbf{A}$ which satisfies $\mathbf{AA}^H = \mathbf{A}^H\mathbf{A}$ is said to be a *normal* matrix.

Normal matrices can be viewed as a generalization of *hermitian* matrices (for which $\mathbf{A}^H = \mathbf{A}$; see Definition 1.8) which themselves include the *real symmetric* matrices (for which $\mathbf{A}^T = \mathbf{A}$ is real; see Definition 1.7); certainly if $\mathbf{A}^H = \mathbf{A}$ then

$$\mathbf{AA}^H = \mathbf{AA} = \mathbf{A}^H\mathbf{A},$$

so that hermitian matrices are normal. The class of normal matrices is important because the matrices that arise in real-world problems are very often normal (even hermitian or real symmetric), perhaps because of a basic symmetry inherent in nature.

EXERCISE 9.6. A *skew-hermitian* matrix $\mathbf{A}$ is one for which $\mathbf{A}^H = -\mathbf{A}$; show that such a matrix is normal.

SOLUTION: $\mathbf{A}^H\mathbf{A} = (-\mathbf{A})\mathbf{A} = -\mathbf{A}^2 = \mathbf{A}(-\mathbf{A}) = \mathbf{A}(\mathbf{A}^H)$.

EXERCISE 9.7. Show that every unitary matrix $\mathbf{A}$ is normal; see Definition 8.4.

We are now prepared to derive a **key** result by applying Theorem 9.1 to a normal matrix.

• *KEY THEOREM 9.2. An $n \times n$ matrix $\mathbf{A}$ is normal (for example: real symmetric hermitian, skew-hermitian, unitary) if and only if $\mathbf{A}$ can be reduced by a unitary transformation to a diagonal Schur canonical form $\mathbf{D} = \mathbf{P}^H\mathbf{A}\mathbf{P}$ where $\mathbf{P}$ is unitary and $\mathbf{D}$ is diagonal; the eigenvalues of $\mathbf{A}$ will be on the diagonal of $\mathbf{D}$. If $\mathbf{A}$ and its eigenvalues are real, then $\mathbf{P}$ can be taken to be real and hence orthogonal.*

Proof: Suppose that a unitary matrix $\mathbf{P}$ exists such that $\mathbf{P}^H\mathbf{A}\mathbf{P} = \mathbf{D}$, where $\mathbf{D}$ is a diagonal matrix. Then $\mathbf{A} = \mathbf{P}\mathbf{D}\mathbf{P}^H$ and

$$\mathbf{A}^H\mathbf{A} = \mathbf{P}\mathbf{D}^H\mathbf{D}\mathbf{P}^H, \qquad \mathbf{A}\mathbf{A}^H = \mathbf{P}\mathbf{D}\mathbf{D}^H\mathbf{P}^H.$$

But $\mathbf{D}^H\mathbf{D} = \mathbf{D}\mathbf{D}^H$ since diagonal matrices commute. Hence $\mathbf{A}\mathbf{A}^H = \mathbf{A}^H\mathbf{A}$ and $\mathbf{A}$ is normal. Conversely, suppose that $\mathbf{A}$ is normal. Theorem 9.1 shows that a unitary matrix $\mathbf{Q}$ exists such that $\mathbf{Q}^H\mathbf{A}\mathbf{Q} = \mathbf{T}$ is upper triangular. Since $\mathbf{A}$ is normal,

$$\mathbf{T}^H\mathbf{T} = \mathbf{Q}^H\mathbf{A}^H\mathbf{A}\mathbf{Q} = \mathbf{Q}^H\mathbf{A}\mathbf{A}^H\mathbf{Q} = \mathbf{T}\mathbf{T}^H,$$

which states that $\mathbf{T}$ is normal also. Since $t_{ij} = 0$ for $i > j$, the (i, i) element of $\mathbf{T}\mathbf{T}^H$ equals

$$\sum_{j=i}^{n} |t_{ij}|^2$$

while the (i, i) element of $\mathbf{T}^H\mathbf{T}$ equals

$$\sum_{j=1}^{i} |t_{ij}|^2.$$

Since $\mathbf{T}\mathbf{T}^H = \mathbf{T}^H\mathbf{T}$, these two expressions for the (i, i) elements are equal, which yields

$$\sum_{j=i+1}^{n} |t_{ij}|^2 = \sum_{j=1}^{i-1} |t_{ji}|^2 \tag{9.2}$$

after canceling the common term $|t_{ii}|^2$. If we let $i = 1$ in (9.2), we obtain

$$\sum_{j=2}^{n} |t_{1j}|^2 = 0$$

so that $t_{12} = t_{13} = \cdots = t_{1n} = 0$. If we next let $i = 2$ in (9.2), we obtain

$$\sum_{j=3}^{n} |t_{2j}|^2 = |t_{12}|^2 = 0$$

since we have just proved that $t_{12} = 0$. Therefore, $t_{23} = t_{24} = \cdots = t_{2n} = 0$. Continuing in this fashion we find $t_{ij} = 0$ for $j > i$, so that **T** is diagonal since we already had $t_{ij} = 0$ for $j < i$. The eigenvalues of **A** are on the diagonal of **T** by **Key Theorem 8.3.** Since the remark on the reality of **P** follows from Theorem 9.1, the theorem is proved.

We remark, as we did after Theorem 9.1, that in practice the diagonal canonical form is not easily obtained via **Key Theorem 9.2** because it requires knowledge of the eigensystem of **A**.

EXERCISE 9.8. Restate **Key Theorem 9.2** in the language of decompositions rather than in that of canonical forms.

EXERCISE 9.9. Restate **Key Theorem 9.2** in the language of linear transformations and changed bases rather than in that of canonical forms.

EXERCISE 9.10. The eigenvalues of the real symmetric (hence normal) matrix

$$\mathbf{A} = \begin{bmatrix} 5 & 4 & -4 \\ 4 & 5 & 4 \\ -4 & 4 & 5 \end{bmatrix}$$

are 9, 9, -3. Find associated eigenvectors and then find a unitary transformation reducing **A** to diagonal form.

The remarkable result of **Key Theorem 9.2** characterizing matrices diagonalizable by unitary transformations was given in the language of *canonical forms*, while the reader was asked to provide interpretations from the viewpoints of *decompositions* and of *representations of linear transformations* in Exs. 9.8 and 9.9; we have not yet spoken of the viewpoint involving the *structure of the eigensystem* of normal matrices because it is so important that we have saved it to be the main result of the next section.

9.3 EIGENSYSTEMS OF NORMAL MATRICES

We know from Section 8.3 and especially **Key Theorem 8.3** that decompositions such as described in **Key Theorem 9.2** are equivalent to the eigensystem having certain structures. The following **key** result follows immediately from

Key Theorems 8.3 and **9.2** since the columns (rows) of a unitary matrix form an *orthonormal* set (see Definition 4.14).

- *KEY THEOREM 9.3. An $n \times n$ matrix $\mathbf{A}$ is normal (for example: real symmetic, hermitian, skew-hermitian, unitary) if and only if $\mathbf{A}$ has a linearly independent set of n eigenvectors which may be chosen so as to form an orthonormal set. Moreover, for a normal matrix an eigenvalue of multiplicity s has associated with it an orthonormal set of s eigenvectors.*

  Proof: We merely sketch the proof. From **Key Theorem 9.2,** $\mathbf{A}$ is normal if and only if we can write $\mathbf{P}^H\mathbf{A}\mathbf{P} = \mathbf{D}$ for unitary $\mathbf{P} = [\mathbf{p}_1, \ldots, \mathbf{p}_n]$ and diagonal $\mathbf{D}$ having ith diagonal entry the eigenvalue λ_i of $\mathbf{A}$. From $\mathbf{P}^H\mathbf{A}\mathbf{P} = \mathbf{D}$ we get the equivalent $\mathbf{A}\mathbf{P} = \mathbf{P}\mathbf{D}$ and $\mathbf{A}\mathbf{p}_i = \lambda_i\mathbf{p}_i$ as in **Key Theorem 8.3,** so that the n eigenvectors can be taken to be the orthonormal vectors $\mathbf{p}_1, \ldots, \mathbf{p}_n$.

The above description of the eigensystem of a normal matrix followed primarily from **Key Theorem 9.2** and thence from Theorem 9.1; the construction used in Theorem 9.1 should not however be viewed as a way of computing the eigensystem of a normal matrix, because that construction itself already requires knowledge of the eigensystem, as in Ex. 9.10.

Some comment is necessary on the statement in **Key Theorem 9.3** that the eigenvectors "may be chosen" so as to be orthonormal. Since $c\mathbf{x}$ is an eigenvector for $c \neq 0$ if $\mathbf{x}$ is an eigenvector, we certainly can force the eigenvectors to have length unity. Eigenvectors associated with distinct eigenvalues are automatically orthogonal (see Ex. 9.24), but since according to **Key Theorem 9.3** an eigenvalue of multiplicity $s > 1$ has an invariant subspace of dimension $s > 1$ we have many different bases available for the invariant subspace and we must take care to choose an orthonormal basis to obtain the result in **Key Theorem 9.3**. The following example should clarify this point.

EXERCISE 9.11. The matrix

$$\mathbf{A} = \begin{bmatrix} 2 & 0 \\ 0 & 2 \end{bmatrix}$$

is normal and has the full set of eigenvectors

$$\begin{bmatrix} 1 \\ 0 \end{bmatrix}, \quad \begin{bmatrix} 1 \\ 1 \end{bmatrix}$$

which is *not* orthonormal. We may, however, instead choose, for example,

$$\begin{bmatrix} 1 \\ 0 \end{bmatrix}, \quad \begin{bmatrix} 0 \\ 1 \end{bmatrix}$$

as the eigenvectors if we wish to obtain an orthonormal set; other choices are also possible.

EXERCISE 9.12. Consider the real symmetric matrix

$$\mathbf{A} = \begin{bmatrix} 2 & -1 & 0 \\ -1 & 3 & -1 \\ 0 & -1 & 2 \end{bmatrix}$$

in the vibration problem of Section 8.1, whose eigenvectors were found in (8.7), (8.8), (8.9) to be

$$\mathbf{x}_1 = \begin{bmatrix} 1 \\ 2 \\ 1 \end{bmatrix}, \qquad \mathbf{x}_2 = \begin{bmatrix} 1 \\ 0 \\ -1 \end{bmatrix}, \qquad \mathbf{x}_3 = \begin{bmatrix} 1 \\ -1 \\ 1 \end{bmatrix}.$$

Show that $\{\mathbf{x}_1, \mathbf{x}_2, \mathbf{x}_3\}$ is an orthogonal set, while

$$\left\{\frac{\mathbf{x}_1}{\sqrt{6}}, \frac{\mathbf{x}_2}{\sqrt{2}}, \frac{\mathbf{x}_3}{\sqrt{3}}\right\}$$

is the orthonormal set of eigenvectors guaranteed by **Key Theorem 9.3.**

EXERCISE 9.13. Find an orthonormal set of eigenvectors for

$$\mathbf{A} = \begin{bmatrix} 7 & -16 & -8 \\ -16 & 7 & 8 \\ -8 & 8 & -5 \end{bmatrix}.$$

SOLUTION: The characteristic polynomial is found to be

$$\lambda^3 - 9\lambda^2 - 405\lambda - 2187 = 0,$$

with roots $\lambda_1 = 27$, $\lambda_2 = \lambda_3 = -9$. The equations $(\mathbf{A} - \lambda_1\mathbf{I})\mathbf{x} = \mathbf{0}$ readily yield precisely one eigenvector, as we should expect, since λ_1 is a simple root:

$$\mathbf{x}_1 = \alpha \begin{bmatrix} -2 \\ 2 \\ 1 \end{bmatrix},$$

where α is an arbitrary constant. The equations $(\mathbf{A} - \lambda_2\mathbf{I})\mathbf{x} = \mathbf{0}$ reduce to a single equation

$$2x_1 - 2x_2 - x_3 = 0. \tag{9.3}$$

This gives a two-dimensional subspace of solutions as we should expect, since λ_2 has multiplicity 2. For $\mathbf{x}_2$ we can choose any solution of this equation, say

$$\mathbf{x}_2 = \beta \begin{bmatrix} 1 \\ 1 \\ 0 \end{bmatrix},$$

where β is an arbitrary constant, since $\mathbf{x}_1$ and any solution $\mathbf{x}_2$ of (9.3) will automatically be orthogonal. The third eigenvector must satisfy the same equation (9.3) and it must also be orthogonal to $\mathbf{x}_2$, that is,

$$x_1 + x_2 = 0. \tag{9.4}$$

The solution of (9.3), (9.4) gives

$$\mathbf{x}_3 = \gamma \begin{bmatrix} 1 \\ -1 \\ 4 \end{bmatrix},$$

where γ is an arbitrary constant. Normalizing, we take

$$\mathbf{x}_1 = \begin{bmatrix} -\frac{2}{3} \\ \frac{2}{3} \\ \frac{1}{3} \end{bmatrix}, \quad \mathbf{x}_2 = \begin{bmatrix} \frac{1}{\sqrt{2}} \\ \frac{1}{\sqrt{2}} \\ 0 \end{bmatrix}, \quad \mathbf{x}_3 = \begin{bmatrix} \frac{\sqrt{2}}{6} \\ -\frac{\sqrt{2}}{6} \\ \frac{2\sqrt{2}}{3} \end{bmatrix}$$

as our orthonormal set.

EXERCISE 9.14. Prove that the eigenvalues of a normal matrix are all equal if and only if **A** is a scalar multiple of the identity matrix **I**.

EXERCISE 9.15. The eigenvalues of

$$\mathbf{A} = \begin{bmatrix} 5 & 4 & -4 \\ 4 & 5 & 4 \\ -4 & 4 & 5 \end{bmatrix}$$

are 9, 9, -3. Find an orthonormal set of three eigenvectors.

EXERCISE 9.16. Show that the matrix

$$\mathbf{A} = \begin{bmatrix} 1 & -4 \\ 1 & 1 \end{bmatrix}$$

is normal but is *not* symmetric, hermitian, skew-hermitian, or unitary. Find the eigenvalues and an orthonormal set of two eigenvectors of **A**.

To illustrate the power of the fundamental result of **Key Theorem 9.3** we give an application to the solution of systems of linear equations and then follow with a physical example; our theorem extends the Fredholm Alternative of **Key Theorem 3.8.**

THEOREM 9.4 (Fredholm Alternative). Consider the n linear equations in n unknowns described by $(\mathbf{A} - \lambda\mathbf{I})\mathbf{x} = \mathbf{b}$ *where* $\mathbf{A}$ *is an* $n \times n$ *normal matrix and* λ *is a given scalar. Precisely one of the following alternatives holds:*

(i) λ *is not one of the eigenvalues* λ_i *of* $\mathbf{A}$, *and the unique solution* $\mathbf{x}$ *can be written*

$$\mathbf{x} = \sum_{i=1}^{n} \frac{(\mathbf{x}_i, \mathbf{b})}{\lambda_i - \lambda} \mathbf{x}_i \tag{9.5}$$

where $\{\mathbf{x}_1, \ldots, \mathbf{x}_n\}$ *is an orthonormal basis consisting of eigenvectors of* $\mathbf{A}$.

(ii) $\lambda = \lambda_i$ *for some eigenvalue* λ_i *of* $\mathbf{A}$ *and there is a solution* $\mathbf{x}$ *if and only if* $\mathbf{b}$ *is orthogonal to all the eigenvectors* $\mathbf{x}_j$ *associated with the eigenvalue* λ_i. *If* $\mathbf{x}$ *is any solution then infinitely many solutions can be found by adding to* $\mathbf{x}$ *an arbitrary linear combination of the eigenvectors associated with* λ_i.

Proof: Since $\{\mathbf{x}_1, \ldots, \mathbf{x}_n\}$ forms an orthonormal basis we can write $\mathbf{b}$, according to Theorem 4.11(i), as $\mathbf{b} = (\mathbf{x}_1, \mathbf{b})\mathbf{x}_1 + \cdots + (\mathbf{x}_n, \mathbf{b})\mathbf{x}_n$ and we can write $\mathbf{x}$ as $\mathbf{x} = \alpha_1\mathbf{x}_1 + \cdots + \alpha_n\mathbf{x}_n$. Then the equation $(\mathbf{A} - \lambda\mathbf{I})\mathbf{x} = \mathbf{b}$ is equivalent to

$$\mathbf{b} = \sum_{j=1}^{n} (\mathbf{x}_j, \mathbf{b})\mathbf{x}_j = (\mathbf{A} - \lambda\mathbf{I}) \sum_{j=1}^{n} \alpha_j\mathbf{x}_j = \sum_{j=1}^{n} \alpha_j(\lambda_j - \lambda)\mathbf{x}_j$$

and hence a solution $\mathbf{x} = \alpha_1\mathbf{x}_1 + \cdots + \alpha_n\mathbf{x}_n$ exists if and only if

$$\alpha_j(\lambda_j - \lambda) = (\mathbf{x}_j, \mathbf{b})$$

for $j = 1, \ldots, n$. The theorem now follows easily.

Compare this theorem with the Fredholm Alternative of **Key Theorem 3.8.** That earlier theorem said that either $\mathbf{Ax} = \mathbf{b}$ is uniquely solvable for every $\mathbf{b}$ or there exists a nonzero $\mathbf{x}$ with $\mathbf{Ax} = \mathbf{0}$; the existence of a nonzero solution to $\mathbf{Ax} = \mathbf{0}$ of course is equivalent to $\mathbf{A}$ having $\lambda = 0$ as an eigenvalue. Thus, for normal matrices, Theorem 9.4 implies **Key Theorem 3.8** and can be viewed as a *stronger* version of the Fredholm Alternative since it also gives necessary and sufficient conditions on the solvability of $\mathbf{Ax} = \mathbf{b}$ when $\mathbf{A}$ is singular.

We apply the above theorem to forced vibration. Suppose that external transverse forces $F_r \cos(\omega t + \alpha_r)$, $r = 1, 2, 3$, are applied to the particles on the string discussed in Section 8.1. The equation of motion for the first mass becomes, for instance [cf. (8.2)],

$$m\frac{d^2X_1}{dt^2} + \frac{TX_1}{l} - \frac{T(X_2 - X_1)}{2l} = F_1 \cos(\omega t + \alpha_1) = \mathrm{Re}\,\{F_1 e^{j\omega t + j\alpha_1}\},$$

where Re stands for the real part of the expression in brackets and $j = \sqrt{-1}$. In the usual way we set

$$X_i = \mathrm{Re}\,\{x_i e^{j\omega t}\},\ \omega^2 = \frac{\lambda T}{2l},\ F_i e^{j\alpha_i} = \frac{1}{2}\left(\frac{T}{l}\right) f_i,$$

say, $i = 1, 2, 3$. Instead of the system (8.5) we now find

$$\begin{aligned}(3-\lambda)x_1 - \quad x_2 \quad\quad &= f_1\\ -x_1 + (2-\lambda)x_2 - \quad x_3 &= f_2\\ -x_2 + (3-\lambda)x_3 &= f_3.\end{aligned} \tag{9.6}$$

These are a set of three equations in three unknowns which we write in matrix form as

$$(\mathbf{A} - \lambda\mathbf{I})\mathbf{x} = \mathbf{f}.$$

The equations have a unique solution if $\det(\mathbf{A} - \lambda\mathbf{I}) \neq 0$. If $\det(\mathbf{A} - \lambda\mathbf{I}) = 0$, either there is no solution or there are infinitly many solutions.

Physically the situation can be interpreted in the following way, using Theorem 9.4. Remember that, from the analysis in Section 8.1, the eigenvalues of **A** correspond physically to the frequencies of free vibration of the system, and the corresponding eigenvectors represent the modes of vibration:

1. If $\det(\mathbf{A} - \lambda\mathbf{I}) \neq 0$, this means that λ is *not* one of the eigenvalues of **A**, that is, the frequency of the externally applied forces (the forcing frequency) does not coincide with a frequency of free vibration. In this case, the equations have a unique solution and the system will settle down to a unique mode of forced vibration.
2. If $\det(\mathbf{A} - \lambda\mathbf{I}) = 0$, that is, the forcing frequency coincides with a frequency of free vibration, say $\lambda = \lambda_i$, there are two possibilities:
   (a) If $(\mathbf{x}_i, \mathbf{f}) = 0$, that is, the vector **f** representing the applied force is orthogonal to the vector representing the mode of free vibration, then the equations can possess infinitely many solutions where the solution is determined to within an arbitrary multiple of the eigenvector corresponding to the mode of free vibration.
   (b) If $(\mathbf{x}_i, \mathbf{f}) \neq 0$, that is, the applied force is not orthogonal to the mode of free vibration, then the equations are inconsistent–i.e., no solution exists. Physically this corresponds to the case where resonance occurs and the amplitudes of vibration are infinite.

By means of Theorem 9.4(i), equation (9.5), we can write down the solution of the system (9.6) explicitly. The normalized eigenvectors $\mathbf{x}_i$ are given in Ex. 9.12. Substitution in (9.6) easily gives

$$\begin{aligned}
x_1 &= \tfrac{1}{6}(f_1 + 2f_2 + f_3)\frac{1}{1-\lambda} + \tfrac{1}{2}(f_1 - f_3)\frac{1}{3-\lambda} + \tfrac{1}{3}(f_1 - f_2 + f_3)\frac{1}{4-\lambda}\\
x_2 &= \tfrac{1}{3}(f_1 + 2f_2 + f_3)\frac{1}{1-\lambda} \qquad\qquad\qquad\qquad - \tfrac{1}{3}(f_1 - f_2 + f_3)\frac{1}{4-\lambda}\\
x_3 &= \tfrac{1}{6}(f_1 + 2f_2 + f_3)\frac{1}{1-\lambda} - \tfrac{1}{2}(f_1 - f_3)\frac{1}{3-\lambda} + \tfrac{1}{3}(f_1 - f_2 + f_3)\frac{1}{4-\lambda}.
\end{aligned} \tag{9.7}$$

If, for example, we are given $f_1 = 6, f_2 = -3, f_3 = -6$, then this formula gives

$$\begin{aligned} x_1 &= -\frac{1}{1-\lambda} + \frac{6}{3-\lambda} + \frac{1}{4-\lambda} \\ x_2 &= -\frac{2}{1-\lambda} \qquad\qquad - \frac{1}{4-\lambda} \\ x_3 &= -\frac{1}{1-\lambda} - \frac{6}{3-\lambda} + \frac{1}{4-\lambda}. \end{aligned} \tag{9.8}$$

The beauty of this result is that we see easily and directly the way in which the x_i vary with the forcing frequency, which is related to λ. In particular, the occurrence of resonance at $\lambda = 1, 3, 4$, is very clearly illustrated. A graph of x_1 against λ has been drawn in Figure 9.1. The exceptional behavior near $\lambda = 1, 3, 4$ is evident.

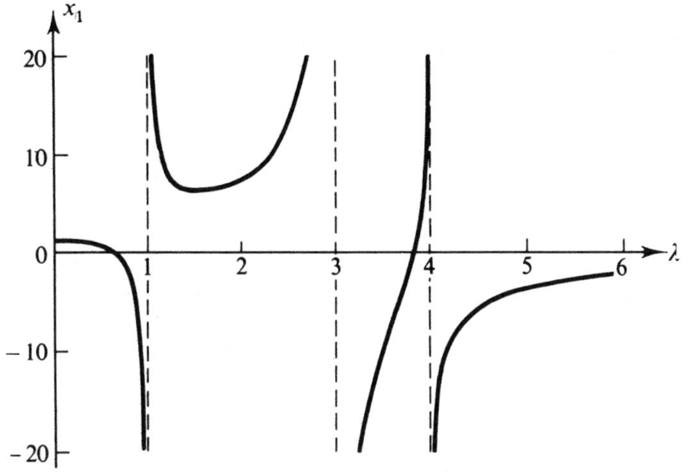

Figure 9.1. A graph of x_1 defined in (9.8) against λ.

Equations (9.7) contain a great deal of information in a clear and explicit form. It is difficult to extract this information from the original equations (9.6) and it would be difficult to obtain the solution (9.7) and a graph like that in Figure 9.1 from (9.6) without using the theory of eigenvalues and eigenvectors.

Further important applications of our results on eigensystems of normal matrices will be found in Sections 9.4, 9.5, 9.6, 9.7, 10.5, 10.6, and 10.7.

EXERCISE 9.17. Suppose that in (9.6) λ is one of the eigenvalues, say $\lambda = 3$; use (8.8) and Theorem 9.4 to write down necessary and sufficient conditions on f_1, f_2, f_3 in order that (9.6) be solvable.

EXERCISE 9.18. Find necessary and sufficient conditions on b_1 and b_2 using Theorem 9.4 in order that

$$\begin{bmatrix} 2-\lambda & 1 \\ 1 & 2-\lambda \end{bmatrix}\begin{bmatrix} x_1 \\ x_2 \end{bmatrix} = \begin{bmatrix} b_1 \\ b_2 \end{bmatrix}$$

be solvable when $\lambda = 1$ and again when $\lambda = 3$.

EXERCISE 9.19. Find necessary and sufficient conditions on b_1 and b_2 using Theorem 9.4 in order that

$$\begin{bmatrix} 3 & -9 \\ -9 & 27 \end{bmatrix}\begin{bmatrix} x_1 \\ x_2 \end{bmatrix} = \begin{bmatrix} b_1 \\ b_2 \end{bmatrix}$$

be solvable.

We can use **Key Theorem 9.2** to deduce still more about the structure of the eigensystem of a normal or hermitian matrix.

THEOREM 9.5. The eigenvalues of a hermitian matrix are real. If λ is an eigenvalue of the normal matrix **A** *corresponding to the eigenvector* **x** *of* **A** *then the complex conjugate $\bar{\lambda}$ is an eigenvalue of* $\mathbf{A}^H$ *corresponding to the eigenvector* **x** *of* $\mathbf{A}^H$.

Proof: We first suppose that **A** is hermitian; by **Key Theorem 9.2** we can write

$$\mathbf{A} = \mathbf{PDP}^H$$

for unitary **P** and diagonal **D**. Then from $\mathbf{A} = \mathbf{A}^H$ we get

$$\mathbf{PDP}^H = \mathbf{A} = \mathbf{A}^H = \mathbf{PD}^H\mathbf{P}^H$$

from which $\mathbf{D} = \mathbf{D}^H$ follows. Since $\mathbf{D}^H = (\bar{\mathbf{D}})^T$ and a diagonal matrix is its own transpose, we get $\mathbf{D} = \bar{\mathbf{D}}$ and the eigenvalues are real. For the second part, if **A** is normal we use **Key Theorem 9.2** to write

$$\mathbf{P}^H\mathbf{AP} = \mathbf{D} \tag{9.9}$$

as before. Taking the hermitian transpose of both sides of (9.9) gives

$$\mathbf{P}^H\mathbf{A}^H\mathbf{P} = \mathbf{D}^H = \bar{\mathbf{D}};$$

by **Key Theorem 9.2,** $\bar{\mathbf{D}}$ contains the eigenvalues of $\mathbf{A}^H$ and from $\mathbf{A}^H\mathbf{P} = \mathbf{PD}^H$ and $\mathbf{AP} = \mathbf{PD}$ it follows that the eigenvectors of **A** and $\mathbf{A}^H$ are the same, namely the columns of **P**, completing our proof.

EXERCISE 9.20. Verify the second part of Theorem 9.5 on the matrix **A** of Ex. 9.16.

EXERCISE 9.21. Prove that a normal matrix is hermitian if and only if all its eigenvalues are real.

EXERCISE 9.22. Suppose that $\mathbf{A}$ is skew-hermitian (see Ex. 9.6). Prove that $\sqrt{-1}\,\mathbf{A}$ is hermitian and that the eigenvalues of $\mathbf{A}$ are pure imaginary; conclude that $\mathbf{I} + z\mathbf{A}$ is nonsingular for all real numbers z.

EXERCISE 9.23. Fill in the details of the following argument from basic principles showing that every eigenvalue λ of a hermitian matrix $\mathbf{A}$ is real (see Theorem 9.5): If $\mathbf{Ax} = \lambda\mathbf{x}$ then $(\mathbf{x}, \mathbf{Ax}) = \lambda(\mathbf{x}, \mathbf{x})$, where $(\cdot, \cdot)$ denotes the inner product $(\mathbf{x}, \mathbf{y}) = \mathbf{x}^H\mathbf{y}$ from Section 4.7. Also

$$(\mathbf{x}, \mathbf{Ax}) = (\mathbf{A}^H\mathbf{x}, \mathbf{x}) = (\mathbf{Ax}, \mathbf{x}) = \bar{\lambda}(\mathbf{x}, \mathbf{x}).$$

Since $\lambda(\mathbf{x}, \mathbf{x}) = \bar{\lambda}(\mathbf{x}, \mathbf{x})$, $\lambda = \bar{\lambda}$ and λ must be real.

EXERCISE 9.24. Let $\lambda_1, \ldots, \lambda_s$ be a collection of *distinct* eigenvalues associated with eigenvectors $\mathbf{x}_1, \ldots, \mathbf{x}_s$ of a normal matrix $\mathbf{A}$; fill in the details of the following argument from basic principles showing that $\{\mathbf{x}_1, \ldots, \mathbf{x}_s\}$ is an orthogonal set: We have $\lambda_j(\mathbf{x}_i, \mathbf{x}_j) = (\mathbf{x}_i, \mathbf{Ax}_j) = (\mathbf{A}^H\mathbf{x}_i, \mathbf{x}_j)$, where $(\cdot, \cdot)$ is the inner product as in Ex. 9.23. Since $\mathbf{A}^H\mathbf{x}_i = \bar{\lambda}_i\mathbf{x}_i$, it follows that $(\mathbf{A}^H\mathbf{x}_i, \mathbf{x}_j) = \lambda_i(\mathbf{x}_i, \mathbf{x}_j)$ and hence $\lambda_j(\mathbf{x}_i, \mathbf{x}_j) = \lambda_i(\mathbf{x}_i, \mathbf{x}_j)$. We conclude that $(\mathbf{x}_i, \mathbf{x}_j) = 0$ if $\lambda_i \neq \lambda_j$.

9.4 THE QR DECOMPOSITION

Recall that **Key Theorem 9.2** told us not only that normal matrices *could* be diagonalized by unitary transformations but also that *only* normal matrices could be so diagonalized. In the equivalent language of linear transformations, since not all matrices are normal, we know that there are some linear transformations A from $\mathbb{C}^n$ to $\mathbb{C}^n$ which cannot be represented by a diagonal matrix simply by appropriately choosing a *single orthonormal basis for* $\mathbb{C}^n$ (*considered both as the domain of A and as the range of A*). This is unfortunate since linear transformations described by diagonal matrices are extremely simple to understand. If we wish to obtain diagonal representations, we must either *give up using orthonormal bases* or *give up using the same basis for both the domain and range.* We already described, in Section 8.4, the similarity transformations which result from giving up orthonormality; in Chapter 10 we will study what this lets us accomplish in the way of simple representations. In this section we study the alternative of using the *standard* orthonormal basis of unit vectors $\mathbf{e}_i$ for the *domain* of A but an *arbitrary* orthonormal basis for the *range* of A; we will see that this allows a simple (in fact, triangular) representation of A that can be computed *easily*, which is not the case for our earlier representations.

More precisely, suppose that the $m \times n$ matrix $\mathbf{A}$ defines the linear transformation A from $\mathbb{R}^n$ to $\mathbb{R}^m$ (or $\mathbb{C}^n$ to $\mathbb{C}^m$) by

$$A(\mathbf{x}) = \mathbf{Ax}$$

as usual; here we allow $m = n$ as a special case. Suppose that the standard orthonormal set of unit vectors $\mathbf{e}_1, \ldots, \mathbf{e}_n$ is used as a basis for the (domain) space $\mathbb{R}^n$ (or $\mathbb{C}^n$) while an arbitrary orthonormal set of vectors $\mathbf{q}_1, \ldots, \mathbf{q}_m$ is used as a basis for the (range) space $\mathbb{R}^m$ (or $\mathbb{C}^m$). If we define the $m \times m$ unitary matrix $\mathbf{Q}$ by

$$\mathbf{Q} = [\mathbf{q}_1, \ldots, \mathbf{q}_m]$$

then just as we found in (8.31) and (8.32) of Section 8.3 the new coordinates $\mathbf{y}'$ (with respect to the basis vectors $\mathbf{q}_i$) of a vector $\mathbf{y}$ in the range space $\mathbb{R}^m$ (or $\mathbb{C}^m$) are given by

$$\mathbf{y}' = \mathbf{Q}^H\mathbf{y}, \qquad \mathbf{y} = \mathbf{Q}\mathbf{y}'.$$

If we take for $\mathbf{y}$ the transformed vector $\mathbf{y} = A(\mathbf{x}) = \mathbf{Ax}$ then $\mathbf{y}' = \mathbf{Q}^H\mathbf{y} = \mathbf{Q}^H\mathbf{Ax}$, much as in (8.33) of Section 8.3. In short, *the linear transformation A defined by $\mathbf{A}$ is represented with respect to the new orthonormal basis in the range space by the $m \times n$ matrix*

$$\mathbf{A}' = \mathbf{Q}^H\mathbf{A}. \tag{9.10}$$

Our goal in this section, stated in matrix language, is to find an easily computed unitary matrix $\mathbf{Q}$ such that the matrix $\mathbf{A}'$ computed from $\mathbf{A}$ via (9.10) will be as "simple" as possible. Alternatively, we want as "simple" as possible a factorization $\mathbf{A} = \mathbf{QA}'$.

By rewriting (9.10) as

$$\mathbf{A} = \mathbf{QA}'$$

so that $\mathbf{Ax} = \mathbf{Q}(\mathbf{A}'\mathbf{x})$ for any vector $\mathbf{x}$, it is clear that any element $\mathbf{y} = \mathbf{Ax}$ of the column space of $\mathbf{A}$ is also in the column space of $\mathbf{Q}$ since $\mathbf{y} = \mathbf{Q}(\mathbf{A}'\mathbf{x})$. Thus the needed orthonormal columns $\mathbf{q}_1, \ldots, \mathbf{q}_m$ of $\mathbf{Q}$ must span at least the same space as the given columns $\mathbf{a}_1, \ldots, \mathbf{a}_n$ of $\mathbf{A}$; having phrased our problem in this way, we are reminded of the Gram-Schmidt process of **Key Theorem 4.12** in Section 4.7 which generates orthonormal vectors of this nature. Our given vectors $\mathbf{a}_1, \ldots, \mathbf{a}_n$ correspond with $\mathbf{u}_1, \ldots, \mathbf{u}_s$ in **Key Theorem 4.12**, while our desired vectors $\mathbf{q}_1, \ldots, \mathbf{q}_m$ correspond with $\mathbf{x}_1, \ldots, \mathbf{x}_s$. Since $\mathbf{A} = \mathbf{QA}'$ represents the columns $\mathbf{a}_i$ as linear combinations of the columns $\mathbf{q}_j$ for $1 \leq j \leq m$, while **Key Theorem 4.12** states that $\mathbf{u}_i$ can be written as a linear combination of the smaller set of vectors $\mathbf{x}_j$ for $1 \leq j \leq i$, we expect this latter result to hold in $\mathbf{A} = \mathbf{QA}'$ as well. Thus what we in fact *hope* to be able to accomplish is a representation of the form $\mathbf{A} = \mathbf{QA}'$ with $\mathbf{A}'$ being "right (or upper) triangular"; to remind the reader of the "right triangular" shape of $\mathbf{A}'$, we replace it by the symbol $\mathbf{R}$, and we seek a unitary matrix $\mathbf{Q}$ for which

$$\left.\begin{array}{l} \mathbf{A} = \mathbf{QR} \\ \text{where} \\ \mathbf{A} = [\mathbf{a}_1, \ldots, \mathbf{a}_n] \text{ is } m \times n, \\ \mathbf{Q} = [\mathbf{q}_1, \ldots, \mathbf{q}_m] \text{ is } m \times m \text{ and unitary,} \\ \mathbf{R} \text{ is } m \times n \text{ and right triangular:} \\ \mathbf{R} = \begin{bmatrix} r_{11} & r_{12} & \cdots & r_{1n} \\ 0 & r_{22} & \cdots & r_{2n} \\ & & \cdots & \\ 0 & 0 & \cdots & r_{nn} \\ 0 & 0 & \cdots & 0 \\ & & \cdots & \\ 0 & 0 & \cdots & 0 \end{bmatrix}. \end{array}\right\} \quad (9.11)$$

The preceding representation of $\mathbf{R}$ obviously assumes $m \geq n$; this is in fact the case of most practical interest, but we assume $m \geq n$ merely for the convenience of not having to make two statements throughout our arguments depending on whether $m \leq n$ or $m \geq n$. For $m \leq n$, see Ex. 9.89. Before studying the feasibility of the *QR decomposition* (9.11) in general, we consider a numerical example.

EXERCISE 9.25. Consider the 4×3 matrix

$$\mathbf{A} = \begin{bmatrix} 1 & 2 & -1 \\ 1 & -1 & 2 \\ 1 & -1 & 2 \\ -1 & 1 & 1 \end{bmatrix}$$

whose columns are just the vectors $\mathbf{u}_1$, $\mathbf{u}_2$, $\mathbf{u}_3$ in Ex. 4.50. We want to test our hypothesis that the vectors generated by the Gram-Schmidt process are precisely what we need as the columns $\mathbf{q}_i$ in the QR decomposition (9.11). From Ex. 4.50 we know that these vectors generated by Gram-Schmidt will be

$$\mathbf{q}_1 = \frac{1}{2}\begin{bmatrix} 1 \\ 1 \\ 1 \\ -1 \end{bmatrix}, \quad \mathbf{q}_2 = \frac{1}{2\sqrt{3}}\begin{bmatrix} 3 \\ -1 \\ -1 \\ 1 \end{bmatrix}, \quad \mathbf{q}_3 = \frac{1}{\sqrt{6}}\begin{bmatrix} 0 \\ 1 \\ 1 \\ 2 \end{bmatrix}.$$

We see that we only have three vectors $\mathbf{q}_i$ while the unitary matrix $\mathbf{Q}$ requires four vectors in order to be 4×4. We take *any* other normalized vector orthogonal to $\mathbf{q}_1, \mathbf{q}_2, \mathbf{q}_3$ to serve as $\mathbf{q}_4$; in this case the three equations $(\mathbf{q}_1, \mathbf{q}_4) = (\mathbf{q}_2, \mathbf{q}_4) = (\mathbf{q}_3, \mathbf{q}_4) = 0$, where $(\cdot, \cdot)$ is the usual inner product, restrict $\mathbf{q}_4$ to a one-dimensional subspace, and we select the vector

$$\mathbf{q}_4 = \frac{1}{\sqrt{2}}\begin{bmatrix} 0 \\ 1 \\ -1 \\ 0 \end{bmatrix}.$$

Finally we compute $\mathbf{A}' = \mathbf{Q}^H\mathbf{A}$ so that $\mathbf{A} = \mathbf{Q}\mathbf{A}'$, and we find

$$\mathbf{A}' = \begin{bmatrix} \frac{1}{2} & \frac{1}{2} & \frac{1}{2} & -\frac{1}{2} \\ \frac{3}{2\sqrt{3}} & \frac{-1}{2\sqrt{3}} & \frac{-1}{2\sqrt{3}} & \frac{1}{2\sqrt{3}} \\ 0 & \frac{1}{\sqrt{6}} & \frac{1}{\sqrt{6}} & \frac{2}{\sqrt{6}} \\ 0 & \frac{1}{\sqrt{2}} & \frac{-1}{\sqrt{2}} & 0 \end{bmatrix}\begin{bmatrix} 1 & 2 & -1 \\ 1 & -1 & 2 \\ 1 & -1 & 2 \\ -1 & 1 & 1 \end{bmatrix}$$

$$= \begin{bmatrix} 2 & -\frac{1}{2} & 1 \\ 0 & \frac{9}{2\sqrt{3}} & \frac{-3}{\sqrt{3}} \\ 0 & 0 & \frac{6}{\sqrt{6}} \\ 0 & 0 & 0 \end{bmatrix} = \mathbf{R};$$

$\mathbf{A}'$ is indeed right triangular and deserves to be called $\mathbf{R}$. When we write $\mathbf{A} = \mathbf{Q}\mathbf{R}$, we notice that the last column $\mathbf{q}_4$ of $\mathbf{Q}$ is always multiplied by a zero from the last row of $\mathbf{R}$, and hence $\mathbf{q}_4$ does not enter into the calculation. We therefore not only have the representation

$$\mathbf{A} = \mathbf{Q}\mathbf{R}$$

as desired for (9.11), but we also have the more compact decomposition

$$\mathbf{A} = \mathbf{Q}_0\mathbf{R}_0,$$

where

$$\mathbf{Q}_0 = [\mathbf{q}_1, \mathbf{q}_2, \mathbf{q}_3], \qquad \mathbf{R}_0 = \begin{bmatrix} 2 & -\frac{1}{2} & 2 \\ 0 & \frac{9}{2\sqrt{3}} & \frac{-3}{\sqrt{3}} \\ 0 & 0 & \frac{6}{\sqrt{6}} \end{bmatrix}.$$

Having been reassured by Ex. 9.25 that the **QR** decomposition in (9.11) is at least plausible, we now set out to prove that it can be found in general. Our simple approach to this will be to compute **Q** and **R** one column at a time by means of the representation (9.11). *If* (9.11) is to be valid, then we must have

$$\mathbf{a}_i = r_{1i}\mathbf{q}_1 + \cdots + r_{ii}\mathbf{q}_i; \tag{9.12}$$

from this we see already that the vectors $\mathbf{q}_{n+1}, \ldots, \mathbf{q}_m$ play no role and may in fact be deleted, as in the representation $\mathbf{A} = \mathbf{Q}_0\mathbf{R}_0$ in Ex. 9.25. Since the vectors $\mathbf{q}_1, \ldots, \mathbf{q}_m$ form an orthonormal set, it follows from (9.12) and Theorem 4.11(ii) that

$$r_{ji} = (\mathbf{q}_j, \mathbf{a}_i) \qquad (\text{for } 1 \leq j \leq i). \tag{9.13}$$

Since we want

$$\mathbf{a}_1 = r_{11}\mathbf{q}_1,$$

we clearly must start with

$$r_{11} = \|\mathbf{a}_1\|_2, \qquad \mathbf{q}_1 = \mathbf{a}_1/r_{11}; \tag{9.14}$$

if $r_{11} = 0$, we choose $\mathbf{q}_1$ as an arbitrary normalized vector (or decline to generate a **q** vector at all, as in Ex. 9.89). Once we have computed $\mathbf{q}_1, \ldots, \mathbf{q}_{i-1}$, then we can compute r_{ji} for $1 \leq j \leq i-1$ from (9.13) immediately; from (9.12) we then must have

$$r_{ii}\mathbf{q}_i = \mathbf{v}_i \equiv \mathbf{a}_i - r_{1i}\mathbf{q}_1 - \cdots - r_{i-1,i}\mathbf{q}_{i-1} \tag{9.15}$$

and hence we compute $r_{ii} = \|\mathbf{v}_i\|_2$. If $r_{ii} \neq 0$ we let $\mathbf{q}_i = \mathbf{v}_i/r_{ii}$; if $r_{ii} = 0$ then we choose $\mathbf{q}_i$ as any arbitrary normalized vector orthogonal to $\mathbf{q}_1, \ldots, \mathbf{q}_{i-1}$ (or decline to generate another **q** vector at all, as in Ex. 9.89). In this way we proceed to compute all the numbers r_{ji} and vectors $\mathbf{q}_i$ in (9.11), based on the *assumption* that the representation (9.11) is valid; it remains to show that the quantities we have computed actually do satisfy (9.11).

- *KEY THEOREM 9.6. Let* **A** *be an* $m \times n$ *matrix with* $m \geq n$ *(see* Ex. 9.89). *Then there exists an* $m \times m$ *unitary matrix* $\mathbf{Q} = [\mathbf{q}_1, \ldots, \mathbf{q}_m]$ *and an* $m \times n$ *right triangular matrix* **R** *as in* (9.11) *such that* $\mathbf{A} = \mathbf{QR}$. *Moreover, if we let* $\mathbf{Q}_0 = [\mathbf{q}_1, \ldots, \mathbf{q}_n]$ *denote the* $m \times n$ *matrix formed from the first n columns of* **Q** *and* $\mathbf{R}_0$ *denote the* $n \times m$ *right triangular matrix formed from the first n rows of* **R**, *then* $\mathbf{A} = \mathbf{Q}_0\mathbf{R}_0$. *The matrices* $\mathbf{Q}_0$ *and* $\mathbf{R}_0$ *can be computed as described in the preceding paragraph. These representations are called QR decompositions of* **A**.

Proof: We have already described how to compute **Q** and **R**; since the bottom $m - n$ rows of **R** are zero, it is clear that $\mathbf{A} = \mathbf{QR}$ if and only if $\mathbf{A} = \mathbf{Q}_0\mathbf{R}_0$, which in turn holds if and only if (9.12) is satisfied for $1 \leq i \leq n$. By construction in (9.14), with the described modification in case $r_{11} = 0$, it is clear that

(9.12) holds for $i = 1$. We use induction and assume that (9.12) holds for $1 \le i \le p$. We can then compute $r_{p+1,\,j}$ for $1 \le j \le p$ from (9.13), and the construction of $\mathbf{q}_{p+1}$ from (9.15) then guarantees that (9.12) is valid also for $i = p + 1$. By the principle of mathematical induction, (9.12) is valid for all i, $1 \le i \le n$; it only remains to prove that $\mathbf{q}_{p+1}$ is orthogonal to $\mathbf{q}_j$ for $1 \le j \le p$. This is true by construction if $r_{p+1,\,p+1} = 0$; if $r_{p+1,\,p+1} \ne 0$, then

$$\begin{aligned}(\mathbf{q}_j, \mathbf{q}_{p+1}) &= \frac{(\mathbf{q}_j, \mathbf{v}_{p+1})}{r_{p+1,\,p+1}} \\ &= \frac{1}{r_{p+1,\,p+1}}[(\mathbf{q}_j, \mathbf{a}_{p+1}) - r_{1,\,p+1}(\mathbf{q}_j, \mathbf{q}_1) - \cdots - r_{p,\,p+1}(\mathbf{q}_j, \mathbf{q}_p)].\end{aligned}$$

By the inductive hypothesis, $\{\mathbf{q}_1, \ldots, \mathbf{q}_p\}$ is already an orthonormal set; therefore this expression for $(\mathbf{q}_j, \mathbf{q}_{p+1})$ simplifies to

$$\frac{1}{r_{p+1,\,p+1}}[(\mathbf{q}_j, \mathbf{a}_{p+1}) - r_{j,\,p+1}] = 0,$$

by the definition of $r_{j,\,p+1}$ from (9.13). The theorem is thus proved.

We note that it is also possible to compute the QR decompositions by other methods, for example using the Householder transformations of Section 8.5 or the modified Gram-Schmidt process of Section 4.7; we do not pursue these details here, although in practical computational work these other methods are preferable (see Exs. 4.71, 4.72, Stewart [65]).

EXERCISE 9.26. Find the QR decompositions of

$$\mathbf{A} = \begin{bmatrix} 1 & 2 \\ 0 & 1 \\ 1 & 4 \end{bmatrix}.$$

SOLUTION: We have $\mathbf{a}_1 = r_{11}\mathbf{q}_1$ where

$$\mathbf{a}_1 = \begin{bmatrix} 1 \\ 0 \\ 1 \end{bmatrix}.$$

Since $\|\mathbf{q}_1\|_2 = 1$, we take

$$\mathbf{q}_1 = \begin{bmatrix} \frac{1}{\sqrt{2}} \\ 0 \\ \frac{1}{\sqrt{2}} \end{bmatrix}, \qquad r_{11} = \sqrt{2}.$$

Next we have $\mathbf{a}_2 = r_{12}\mathbf{q}_1 + r_{22}\mathbf{q}_2$ where

$$\mathbf{a}_2 = \begin{bmatrix} 2 \\ 1 \\ 4 \end{bmatrix}$$

and from (9.13) we obtain

$$r_{12} = (\mathbf{q}_1, \mathbf{a}_2) = 3\sqrt{2}$$

and hence

$$r_{22}\mathbf{q}_2 = \mathbf{v}_2 \equiv \mathbf{a}_2 - r_{12}\mathbf{q}_1 = \begin{bmatrix} -1 \\ 1 \\ 1 \end{bmatrix}.$$

Therefore

$$\mathbf{q}_2 = \begin{bmatrix} -\frac{1}{\sqrt{3}} \\ \frac{1}{\sqrt{3}} \\ \frac{1}{\sqrt{3}} \end{bmatrix}, \qquad r_{22} = \sqrt{3}$$

and hence the QR decomposition $\mathbf{A} = \mathbf{Q}_0\mathbf{R}_0$ of **Key Theorem 9.6** is

$$\begin{bmatrix} 1 & 2 \\ 0 & 1 \\ 1 & 4 \end{bmatrix} = \begin{bmatrix} \frac{1}{\sqrt{2}} & -\frac{1}{\sqrt{3}} \\ 0 & \frac{1}{\sqrt{3}} \\ \frac{1}{\sqrt{2}} & \frac{1}{\sqrt{3}} \end{bmatrix} \begin{bmatrix} \sqrt{2} & 3\sqrt{2} \\ 0 & \sqrt{3} \end{bmatrix}.$$

To write $\mathbf{A} = \mathbf{QR}$ for a 3×3 $\mathbf{Q}$ and a 3×2 $\mathbf{R}$, we need only add a row of zeros to the bottom of $\mathbf{R}_0$ to get $\mathbf{R}$ and add a normalized column $\mathbf{q}_3$ orthogonal to $\mathbf{q}_1$ and $\mathbf{q}_2$ on the right of $\mathbf{Q}_0$ to get $\mathbf{Q}$.

EXERCISE 9.27. Find the QR decompositions of

$$\mathbf{A} = \begin{bmatrix} 1 & 2 & 4 \\ 0 & 1 & 1 \\ 1 & 4 & 6 \end{bmatrix}$$

SOLUTION: Since the first two columns are the same as in Ex. 9.26, we naturally find (why?)

$$\mathbf{q}_1 = \begin{bmatrix} \frac{1}{\sqrt{2}} \\ 0 \\ \frac{1}{\sqrt{2}} \end{bmatrix}, \qquad \mathbf{q}_2 = \begin{bmatrix} -\frac{1}{\sqrt{3}} \\ \frac{1}{\sqrt{3}} \\ \frac{1}{\sqrt{3}} \end{bmatrix},$$

$$r_{11} = \sqrt{2}, \qquad r_{12} = 3\sqrt{2}, \qquad r_{22} = \sqrt{3}.$$

To find $\mathbf{q}_3$ we use (9.13) with

$$\mathbf{a}_3 = \begin{bmatrix} 4 \\ 1 \\ 6 \end{bmatrix}$$

to obtain

$$r_{13} = 5\sqrt{2}, \qquad r_{23} = \sqrt{3}.$$

Thus

$$r_{33}\mathbf{q}_3 = \mathbf{v}_3 \equiv \mathbf{a}_3 - r_{13}\mathbf{q}_1 - r_{23}\mathbf{q}_2 = \begin{bmatrix} 0 \\ 0 \\ 0 \end{bmatrix},$$

and we set $r_{33} = 0$ and choose $\mathbf{q}_3$ as *any* normalized vector orthogonal to $\mathbf{q}_1$ and $\mathbf{q}_2$, say

$$\mathbf{q}_3 = \begin{bmatrix} \dfrac{1}{\sqrt{6}} \\ \dfrac{2}{\sqrt{6}} \\ -\dfrac{1}{\sqrt{6}} \end{bmatrix},$$

so that the QR decomposition $\mathbf{A} = \mathbf{Q}_0\mathbf{R}_0$ of **Key Theorem 9.6** is

$$\begin{bmatrix} 1 & 2 & 4 \\ 0 & 1 & 1 \\ 1 & 4 & 6 \end{bmatrix} = \begin{bmatrix} \dfrac{1}{\sqrt{2}} & -\dfrac{1}{\sqrt{3}} & \dfrac{1}{\sqrt{6}} \\ 0 & \dfrac{1}{\sqrt{3}} & \dfrac{2}{\sqrt{6}} \\ \dfrac{1}{\sqrt{2}} & \dfrac{1}{\sqrt{3}} & -\dfrac{1}{\sqrt{6}} \end{bmatrix} \begin{bmatrix} \sqrt{2} & 3\sqrt{2} & 5\sqrt{2} \\ 0 & \sqrt{3} & \sqrt{3} \\ 0 & 0 & 0 \end{bmatrix}.$$

Since $m = n$ (=3), in this case we have $\mathbf{Q} = \mathbf{Q}_0$ and $\mathbf{R} = \mathbf{R}_0$. It is interesting to note however that the last row of $\mathbf{R} = \mathbf{R}_0$ is zero, so that the last column of $\mathbf{Q} = \mathbf{Q}_0$ is irrelevant to the decomposition. We could just as easily write

$$\begin{bmatrix} 1 & 2 & 4 \\ 0 & 1 & 1 \\ 1 & 4 & 6 \end{bmatrix} = \begin{bmatrix} \dfrac{1}{\sqrt{2}} & -\dfrac{1}{\sqrt{3}} \\ 0 & \dfrac{1}{\sqrt{3}} \\ \dfrac{1}{\sqrt{2}} & \dfrac{1}{\sqrt{3}} \end{bmatrix} \begin{bmatrix} \sqrt{2} & 3\sqrt{2} & 5\sqrt{2} \\ 0 & \sqrt{3} & \sqrt{3} \end{bmatrix}.$$

This in fact involves writing $\mathbf{A} = \mathbf{Q}_1\mathbf{R}_1$ where $\mathbf{Q}_1$ is $m \times k$ and has orthonormal columns and $\mathbf{R}_1$ is $k \times n$ and is right triangular, and where k is the *rank* of $\mathbf{A}$. See Ex. 9.89.

EXERCISE 9.28. Find the QR decompositions of

$$\mathbf{A} = \begin{bmatrix} 1 & 2 & 3 \\ 0 & 1 & 1 \\ 1 & 4 & 6 \end{bmatrix}.$$

EXERCISE 9.29. Find the QR decompositions of

$$\mathbf{A} = \begin{bmatrix} -3 & -3 & 3 \\ 3 & 2 & -4 \\ 3 & 1 & -5 \end{bmatrix}.$$

EXERCISE 9.30. Suppose that $\mathbf{A} = \mathbf{QR} = \mathbf{Q}_0\mathbf{R}_0$ are the QR decompositions of $\mathbf{A}$. Prove that the ranks of $\mathbf{A}$, of $\mathbf{R}$, and of $\mathbf{R}_0$ are equal.

EXERCISE 9.31. Let $\mathbf{A}$ be an $m \times n$ matrix of rank n. Show that the Gram-Schmidt process of **Key Theorem 4.12** actually computes $\mathbf{Q}_0$ explicitly as $\mathbf{A}\mathbf{R}_0^{-1}$. Use the Gram-Schmidt procedure to compute the matrix $\mathbf{Q}_0$ of Ex. 9.26, and compare the computation with that performed in Ex. 9.26.

EXERCISE 9.32. If $\mathbf{A} = \mathbf{Q}_0\mathbf{R}_0$ is the QR decomposition of $\mathbf{A}$ as in **Key Theorem 9.6,** show that $\mathbf{A}^H\mathbf{A} = \mathbf{R}_0^H\mathbf{R}_0$.

EXERCISE 9.33. Suppose that $\mathbf{A}$ is an $n \times n$ matrix having QR decomposition $\mathbf{A} = \mathbf{QR}$; show that $\mathbf{A}' = \mathbf{RQ}$ equals $\mathbf{Q}^H\mathbf{AQ}$. Deduce that $\mathbf{A}'$ and $\mathbf{A}$ have the same eigenvalues. This transformation is the basis of the famed *QR method* for computing the eigenvalues and eigenvectors of a matrix $\mathbf{A}$; we let $\mathbf{A}_0 = \mathbf{A}$, $\mathbf{A}_i = \mathbf{Q}_i\mathbf{R}_i$ and $\mathbf{A}_{i+1} \equiv \mathbf{R}_i\mathbf{Q}_i$, and often find that $\{\mathbf{A}_i\}_{i=0}^{\infty}$ converges to a matrix whose eigenvalues are easily calculated.

In this section we have considered the QR decomposition $\mathbf{A} = \mathbf{QR}$ only as a way of finding a simple matrix representation (namely $\mathbf{R}$) for the linear transformation A defined by $\mathbf{A}$ when we introduce the vectors $\mathbf{q}_1, \ldots, \mathbf{q}_m$ as a basis for the range space. The decomposition, in fact, is of great practical importance in numerical computations; we will see an indication of this later in Section 9.6.

9.5 THE SINGULAR-VALUE DECOMPOSITION

In the preceding section we found that we could easily compute a simplified representation for the linear transformation A defined by the $m \times n$ matrix

A by the simple device of introducing an arbitrary orthonormal basis only for the *range* space of A, while keeping the standard basis $\mathbf{e}_1, \ldots, \mathbf{e}_n$ for the domain space. In this section we take the next logical step and introduce new orthonormal bases for *both* the range space *and* the domain space of A. We will find that a very simple representation (namely, diagonal) can be found for A, at the cost of finding a great deal of information about some eigensystems related to **A**. The usefulness of the simplification (or decomposition) will outweigh the cost of finding it, however.

As usual, we let the linear transformation A be defined from the $m \times n$ matrix **A** via $A(\mathbf{x}) = \mathbf{Ax}$. We let the arbitrary orthonormal set of vectors $\mathbf{v}_1, \ldots, \mathbf{v}_n$ be used as a basis for the domain space $\mathbb{R}^n$ (or $\mathbb{C}^n$) while the arbitrary orthonormal set of vectors $\mathbf{u}_1, \ldots, \mathbf{u}_m$ is used as a basis for the range space $\mathbb{R}^m$ (or $\mathbb{C}^m$). If we define $n \times n$ and $m \times m$ unitary matrices **V** and **U**, respectively, by

$$\mathbf{V} = [\mathbf{v}_1, \ldots, \mathbf{v}_n], \qquad \mathbf{U} = [\mathbf{u}_1, \ldots, \mathbf{u}_m],$$

then, just as we found in (8.31) and (8.32) of Section 8.3, the new coordinates $\mathbf{x}'$ (with respect to the new basis vectors $\mathbf{v}_1, \ldots, \mathbf{v}_n$) of a vector **x** in the domain space $\mathbb{R}^n$ (or $\mathbb{C}^n$) are given by

$$\mathbf{x}' = \mathbf{V}^H\mathbf{x}, \qquad \mathbf{x} = \mathbf{V}\mathbf{x}',$$

while the relations in the range space are

$$\mathbf{y}' = \mathbf{U}^H\mathbf{y}, \qquad \mathbf{y} = \mathbf{U}\mathbf{y}'.$$

Taking $\mathbf{y} = A(\mathbf{x}) = \mathbf{Ax} = \mathbf{AVx}'$, we have $\mathbf{y}' = \mathbf{U}^H\mathbf{y} = \mathbf{U}^H\mathbf{AVx}'$. In short, *the linear transformation A defined by* **A** *is represented with respect to the two new orthonormal bases by the $m \times n$ matrix*

$$\mathbf{A}' = \mathbf{U}^H\mathbf{AV}. \tag{9.16}$$

Our goal then in this section, stated in matrix language, is to find two unitary matrices **U** and **V** such that the matrix $\mathbf{A}'$ computed from (9.16) will be as simple as possible.

When $m = n$ so that **A** is square, the Schur form (see Theorem 9.1) gives a reduction to the form (9.16) with $\mathbf{U} = \mathbf{V}$ and $\mathbf{A}'$ upper triangular. Without restricting ourselves to $m = n$ the QR decomposition gives a much more simple computational reduction to the form (9.16) with $\mathbf{A}'$ right-triangular, and $\mathbf{V} = \mathbf{I}_n$. We now consider what would be required of **U** and **V** in order that (9.16), in its full generality, produce the dramatic simplification to a *diagonal* form $\mathbf{A}'$.

To be precise, we change notation slightly and *assume* that

$$\left.\begin{aligned}&\mathbf{U}^H\mathbf{A}\mathbf{V} = \mathbf{\Sigma}\\ &\text{where}\\ &\quad\mathbf{A} \text{ is } m \times n,\\ &\quad\mathbf{U} \text{ and } \mathbf{V} \text{ are } m \times m \text{ and } n \times n \text{ unitary matrices, and}\\ &\quad\mathbf{\Sigma} = \begin{bmatrix} \sigma_1 & 0 & \dots & 0 \\ 0 & \sigma_2 & \dots & 0 \\ & & \dots & \\ 0 & 0 & \dots & \sigma_n \\ 0 & 0 & \dots & 0 \\ & & \dots & \\ 0 & 0 & \dots & 0 \end{bmatrix} \text{ is } m \times n.\end{aligned}\right\} \tag{9.17}$$

In (9.17) an assumption that $m > n$ is not intended, although we have displayed $\mathbf{\Sigma}$ in that fashion for convenience. The precise assumption on $\mathbf{\Sigma}$ is that it is zero except on the "main diagonal", that is,

$$\mathbf{\Sigma} = (\sigma_{ij}) \text{ satisfies } \sigma_{ij} = 0 \quad \text{if} \quad i \neq j. \tag{9.18}$$

If (9.17) holds, then we of course have the decomposition

$$\mathbf{A} = \mathbf{U}\mathbf{\Sigma}\mathbf{V}^H. \tag{9.19}$$

From (9.19) it follows that

$$\mathbf{A}^H\mathbf{A} = \mathbf{V}\mathbf{\Sigma}^H\mathbf{U}^H\mathbf{U}\mathbf{\Sigma}\mathbf{V}^H = \mathbf{V}(\mathbf{\Sigma}^H\mathbf{\Sigma})\mathbf{V}^H$$

where $\mathbf{\Sigma}^H\mathbf{\Sigma}$ is a *diagonal* $n \times n$ matrix because $\sigma_{ij} = 0$ for $i \neq j$. Since $\mathbf{V}$ is unitary, we know from Section 9.3 that the values σ_i^2, the diagonal elements of $\mathbf{\Sigma}^H\mathbf{\Sigma}$, are just the eigenvalues of $\mathbf{A}^H\mathbf{A}$, while the columns of $\mathbf{V}$ give the associated eigenvectors. Similarly, from

$$\mathbf{A}\mathbf{A}^H = \mathbf{U}(\mathbf{\Sigma}\mathbf{\Sigma}^H)\mathbf{U}^H,$$

where $\mathbf{\Sigma}\mathbf{\Sigma}^H$ is diagonal, we see that the numbers σ_i^2 are also the eigenvalues of $\mathbf{A}\mathbf{A}^H$ with associated eigenvectors given by the columns of $\mathbf{U}$. Conversely, we could *define* $\mathbf{U}$, $\mathbf{V}$, and $\mathbf{\Sigma}$ from these eigenvalue-eigenvector relationships and then hope to prove that (9.17) and (9.19) hold for some class of matrices $\mathbf{A}$; it is perhaps surprising that this decomposition is indeed possible for *all* $\mathbf{A}$. The nonzero diagonal entries σ_i of $\mathbf{\Sigma}$ are usually called the *singular values* of $\mathbf{A}$.

DEFINITION 9.2. The strictly positive square roots σ_i of the nonzero eigenvalues of $\mathbf{A}^H\mathbf{A}$ (and of $\mathbf{A}\mathbf{A}^H$, equivalently) are called the *singular values* of $\mathbf{A}$.

Before studying the possibility of generating a *singular-value decomposition* (9.19) in general, we consider a numerical example.

EXERCISE 9.34. Consider the 3 × 2 matrix

$$\mathbf{A} = \begin{bmatrix} 1 & 1 \\ 2 & 2 \\ 2 & 2 \end{bmatrix}.$$

Since

$$\mathbf{A}^H\mathbf{A} = \begin{bmatrix} 9 & 9 \\ 9 & 9 \end{bmatrix}$$

has eigenvalues 18 and 0, the singular values of **A** consist just of

$$\sigma_1 = \sqrt{18} = 3\sqrt{2}.$$

The eigenvectors of $\mathbf{A}^H\mathbf{A}$ can be taken to be

$$\mathbf{v}_1 = \frac{1}{\sqrt{2}}\begin{bmatrix} 1 \\ 1 \end{bmatrix}, \quad \mathbf{v}_2 = \frac{1}{\sqrt{2}}\begin{bmatrix} 1 \\ -1 \end{bmatrix},$$

as is easily verified; our work preceding this exercise leads us to try $\mathbf{V} = [\mathbf{v}_1, \mathbf{v}_2]$ in (9.17). Similarly,

$$\mathbf{A}\mathbf{A}^H = \begin{bmatrix} 2 & 4 & 4 \\ 4 & 8 & 8 \\ 4 & 8 & 8 \end{bmatrix}$$

has eigenvalues 18, 0, and 0 corresponding to the eigenvectors

$$\mathbf{u}_1 = \frac{1}{3}\begin{bmatrix} 1 \\ 2 \\ 2 \end{bmatrix}, \quad \mathbf{u}_2 = \frac{1}{\sqrt{5}}\begin{bmatrix} -2 \\ 1 \\ 0 \end{bmatrix}, \quad \mathbf{u}_3 = \frac{\sqrt{5}}{15}\begin{bmatrix} 2 \\ 4 \\ -5 \end{bmatrix},$$

as is easily verified; our work preceding this exercise leads us to try $\mathbf{U} = [\mathbf{u}_1, \mathbf{u}_2, \mathbf{u}_3]$ in (9.17). Evaluating (9.17) in this case we find

$$\mathbf{A}' = \mathbf{U}^H\mathbf{A}\mathbf{V} = \begin{bmatrix} \frac{1}{3} & \frac{2}{3} & \frac{2}{3} \\ \frac{-2}{\sqrt{5}} & \frac{1}{\sqrt{5}} & 0 \\ \frac{2\sqrt{5}}{15} & \frac{4\sqrt{5}}{15} & \frac{-5\sqrt{5}}{15} \end{bmatrix} \begin{bmatrix} 1 & 1 \\ 2 & 2 \\ 2 & 2 \end{bmatrix} \begin{bmatrix} \frac{1}{\sqrt{2}} & \frac{1}{\sqrt{2}} \\ \frac{1}{\sqrt{2}} & -\frac{1}{\sqrt{2}} \end{bmatrix}$$

$$= \begin{bmatrix} 3\sqrt{2} & 0 \\ 0 & 0 \\ 0 & 0 \end{bmatrix} = \begin{bmatrix} \sigma_1 & 0 \\ 0 & 0 \\ 0 & 0 \end{bmatrix} = \mathbf{\Sigma}.$$

Thus, at least in this case, creating **V** and **U** from the eigenvectors of $\mathbf{A}^H\mathbf{A}$ and $\mathbf{A}\mathbf{A}^H$ does lead to the reduction in (9.17).

We are now ready to present a statement to prove the **key** result that the reduction (9.17) and the singular-value decomposition (9.19) are *always* possible.

- *KEY THEOREM 9.7* (*Singular-Value Decomposition*). *Let the* $m \times n$ *matrix* **A** *have rank* k. *Then there exist numbers* $\sigma_1 \geq \sigma_2 \geq \cdots \geq \sigma_k > 0$, *the singular values of* **A** *from Definition 9.2, an* $m \times m$ *unitary matrix* $\mathbf{U} = [\mathbf{u}_1, \ldots, \mathbf{u}_m]$, *and an* $n \times n$ *unitary matrix* $\mathbf{V} = [\mathbf{v}_1, \ldots, \mathbf{v}_n]$, *such that* (9.17) *and* (9.19) *are valid, that is,* $\mathbf{\Sigma} = \mathbf{U}^H\mathbf{A}\mathbf{V}$ *and* $\mathbf{A} = \mathbf{U}\mathbf{\Sigma}\mathbf{V}^H$ *with the* $m \times n$ *matrix*

$$\mathbf{\Sigma} = \begin{bmatrix} \mathbf{D} & \mathbf{0} \\ \mathbf{0} & \mathbf{0} \end{bmatrix},$$

where **D** *is a* $k \times k$ *diagonal matrix with ith diagonal element* $d_{ii} = \sigma_i > 0$ *for* $1 \leq i \leq k$. *Moreover, for* $1 \leq i \leq k$, $\mathbf{u}_i = \sigma_i^{-1}\mathbf{A}\mathbf{v}_i$ *and* $\mathbf{v}_i = \sigma_i^{-1}\mathbf{A}^H\mathbf{u}_i$ *are eigenvectors of* $\mathbf{A}\mathbf{A}^H$ *and of* $\mathbf{A}^H\mathbf{A}$, *respectively, both associated with the eigenvalue* $\sigma_i^2 > 0$; *the vectors* $\mathbf{u}_i$ *for* $k + 1 \leq i \leq m$ *and* $\mathbf{v}_i$ *for* $k + 1 \leq i \leq n$ *are eigenvectors of* $\mathbf{A}\mathbf{A}^H$ *and* $\mathbf{A}^H\mathbf{A}$ *respectively, all associated with the eigenvalue zero. If* **A** *is real then* **U** *and* **V** *can be taken to be real (and hence orthogonal) matrices.*

Proof: We observe that if $\mathbf{A}^H\mathbf{A}\mathbf{z} = \mathbf{0}$, then $\|\mathbf{A}\mathbf{z}\|_2^2 = (\mathbf{A}\mathbf{z}, \mathbf{A}\mathbf{z}) = (\mathbf{z}, \mathbf{A}^H\mathbf{A}\mathbf{z}) = 0$, and hence $\mathbf{A}\mathbf{z} = \mathbf{0}$ also; it is therefore clear that $\mathbf{A}^H\mathbf{A}\mathbf{z} = \mathbf{0}$ if and only if $\mathbf{A}\mathbf{z} = \mathbf{0}$, that is, **z** is in the null space of **A**. By Theorem 5.2 of Section 5.3 this null space has dimension $n - k$; hence the hermitian matrix $\mathbf{A}^H\mathbf{A}$ has precisely $n - k$ eigenvectors associated with the eigenvalue zero, and the analogous argument gives the analogous result for $\mathbf{A}\mathbf{A}^H$ since $\mathbf{A}^H$ also has rank k. The remaining k eigenvalues are positive, since $\mathbf{A}^H\mathbf{A}\mathbf{z} = \lambda\mathbf{z}$ implies

$$\lambda \|\mathbf{z}\|_2^2 = \lambda(\mathbf{z}, \mathbf{z}) = (\mathbf{z}, \mathbf{A}^H\mathbf{A}\mathbf{z}) = (\mathbf{A}\mathbf{z}, \mathbf{A}\mathbf{z}) > 0;$$

therefore the k singular values $\sigma_1, \ldots, \sigma_k$ are well defined.

Let $\mathbf{v}_1, \ldots, \mathbf{v}_k$ be an orthonormal set of eigenvectors associated with the eigenvalues $\sigma_1^2, \ldots, \sigma_k^2$, respectively, of $\mathbf{A}^H\mathbf{A}$, and let us *define* $\mathbf{u}_1, \ldots, \mathbf{u}_k$ by $\mathbf{u}_i = \sigma_i^{-1}\mathbf{A}\mathbf{v}_i$ so that $\mathbf{A}\mathbf{v}_i = \sigma_i\mathbf{u}_i$; we will show that these vectors $\mathbf{u}_i$ form an orthonormal set of eigenvectors of $\mathbf{A}\mathbf{A}^H$ associated with the eigenvalues σ_i^2. This is straightforward since

$$\mathbf{A}\mathbf{A}^H\mathbf{u}_i = \mathbf{A}\mathbf{A}^H(\sigma_i^{-1}\mathbf{A}\mathbf{v}_i) = \sigma_i^{-1}\mathbf{A}(\mathbf{A}^H\mathbf{A}\mathbf{v}_i) = \sigma_i^{-1}\mathbf{A}(\sigma_i^2\mathbf{v}_i) = \sigma_i^2(\sigma_i^{-1}\mathbf{A}\mathbf{v}_i) = \sigma_i^2\mathbf{u}_i$$

as claimed, while for $i \neq j$ we have

$$(\mathbf{u}_i, \mathbf{u}_j) = (\sigma_i^{-1}\mathbf{A}\mathbf{v}_i, \sigma_j^{-1}\mathbf{A}\mathbf{v}_j) = \sigma_i^{-1}\sigma_j^{-1}(\mathbf{v}_i, \mathbf{A}^H\mathbf{A}\mathbf{v}_j) = \sigma_i^{-1}\sigma_j(\mathbf{v}_i, \mathbf{v}_j) = 0.$$

Since $\mathbf{A}^H\mathbf{A}$ and $\mathbf{A}\mathbf{A}^H$ have n and m eigenvalues, respectively, of which we showed earlier that precisely $n - k$ and $m - k$ are zero, and since we have displayed k eigenvalues $\sigma_i^2 > 0$ for each matrix, these eigenvalues σ_i^2 are the *only* positive ones and we can therefore extend the sets $\{\mathbf{u}_1, \ldots, \mathbf{u}_k\}$ and $\{\mathbf{v}_1, \ldots, \mathbf{v}_k\}$ to orthonormal sets $\{\mathbf{u}_1, \ldots, \mathbf{u}_m\}$ and $\{\mathbf{v}_1, \ldots, \mathbf{v}_n\}$ with $\mathbf{A}^H\mathbf{A}\mathbf{v}_i = \mathbf{0}$ for $k+1 \leq i \leq n$ and $\mathbf{A}\mathbf{A}^H\mathbf{u}_i = \mathbf{0}$ for $k+1 \leq i \leq m$. As we showed at the start of the proof, we therefore have $\mathbf{A}\mathbf{v}_i = \mathbf{0}$ for $k+1 \leq i \leq n$ as well. If we now define $\mathbf{U}$, $\mathbf{V}$, $\mathbf{\Sigma}$ as in the statement of the theorem, then, because of what we have already shown,

$$\mathbf{U}^H\mathbf{A}\mathbf{V} = \begin{bmatrix} \mathbf{u}_1^H \\ \cdot \\ \cdot \\ \cdot \\ \mathbf{u}_m^H \end{bmatrix} \mathbf{A}[\mathbf{v}_1, \ldots, \mathbf{v}_n] = \begin{bmatrix} \mathbf{u}_1^H \\ \cdot \\ \cdot \\ \cdot \\ \mathbf{u}_m^H \end{bmatrix} [\sigma_1\mathbf{u}_1, \ldots, \sigma_k\mathbf{u}_k, \mathbf{0}, \ldots, \mathbf{0}] = \mathbf{\Sigma},$$

and the theorem is proved.

We remark that $\mathbf{U}$ and $\mathbf{V}$ in the singular-value decomposition are not necessarily unique since they are composed of eigenvectors that are not uniquely determined. Also note that the main practical importance of **Key Theorem 9.7** is the existence of the decomposition via eigenvectors $\mathbf{u}_i$ and $\mathbf{v}_j$, not the relationships between $\mathbf{u}_i$ and $\mathbf{v}_i$.

EXERCISE 9.35. Find the singular-value decomposition of

$$\mathbf{A} = \begin{bmatrix} 1 & -1 & 2 \\ -1 & 1 & -2 \end{bmatrix}.$$

Note that only two vectors $\mathbf{q}_1$, $\mathbf{q}_2$ should be generated.

EXERCISE 9.36. In the specific case of Ex. 9.34 and Ex. 9.35, verify the relationship between $\mathbf{u}_i$ and $\mathbf{v}_i$ asserted in **Key Theorem 9.7.**

EXERCISE 9.37. Create an example of a matrix for which $\mathbf{U}$ and $\mathbf{V}$ in the singular-value decomposition are not uniquely determined.

As we shall see in the next section, one of the most important applications of the singular-value decomposition is in the solution of least-squares problems (see Section 2.6 also); for the moment we content ourselves with showing some other uses of the decomposition which are important in the statistical analysis of data.

EXERCISE 9.38. Define the *Frobenius norm* $\|\cdot\|_F$ of $m \times n$ matrices via

$$\|\mathbf{A}\|_F^2 = \sum_{i=1}^{m} \sum_{j=1}^{n} |a_{ij}|^2.$$

Note that the Frobenius norm of matrix $\mathbf{A}$ is just the Euclidean norm $\|\cdot\|_2$ of $\mathbf{A}$ considered as an element of $\mathbb{C}^{mn}$. Show that $\|\cdot\|_F$ is a norm in the sense that

$\|\mathbf{A}\|_F > 0$ unless $\mathbf{A} = \mathbf{0}$, when $\|\mathbf{0}\|_F = 0$, that $\|\alpha\mathbf{A}\|_F = |\alpha|\,\|\mathbf{A}\|_F$, and $\|\mathbf{A} + \mathbf{B}\|_F \leq \|\mathbf{A}\|_F + \|\mathbf{B}\|_F$. Compute $\|\mathbf{I}\|_F$ and deduce that the Frobenius norm is not an induced operator norm (See Definition 5.3.) when $m = n$ unless $m = n = 1$.

EXERCISE 9.39. Let $\mathbf{A}$ be an $m \times n$ matrix, and let $\mathbf{U}$ and $\mathbf{V}$ be $m \times m$ and $n \times n$ unitary matrices, respectively. Use Theorem 8.8(iv) to prove that $\|\mathbf{A}\|_F = \|\mathbf{UA}\|_F = \|\mathbf{AV}\|_F = \|\mathbf{UAV}\|_F$.

EXERCISE 9.40. Suppose that an $n \times n$ matrix $\mathbf{Q}_0$ is known for *theoretical* reasons to be unitary, but because of measurement errors $\mathbf{Q}_0$ is not *actually* unitary. Show that the unitary matrix $\mathbf{Q}_1$ which minimizes $\|\mathbf{Q} - \mathbf{Q}_0\|_F$ over all unitary matrices $\mathbf{Q}$ is just $\mathbf{Q}_1 = \mathbf{UV}^H$ where $\mathbf{U}$ and $\mathbf{V}$ are defined by the singular-value decomposition $\mathbf{Q}_0 = \mathbf{U\Sigma V}^H$.

SOLUTION: For any unitary $\mathbf{Q}$,

$$\|\mathbf{Q} - \mathbf{Q}_0\|_F = \|\mathbf{Q} - \mathbf{U\Sigma V}^H\|_F = \|\mathbf{U}^H\mathbf{QV} - \mathbf{\Sigma}\|_F$$

by Ex. 9.39, and hence the unitary matrix $\mathbf{P}_1 = \mathbf{U}^H\mathbf{Q}_1\mathbf{V}$ must be the closest unitary matrix to $\mathbf{\Sigma}$; we show that $\mathbf{P}_1$ must equal $\mathbf{I}$. For any unitary $\mathbf{P}$, we write

$$\begin{aligned}\|\mathbf{P} - \mathbf{\Sigma}\|_F^2 - \|\mathbf{I} - \mathbf{\Sigma}\|_F^2 &= \sum_{i \neq j} |p_{ij}|^2 + \sum_{i=1}^{n} |p_{ii} - \sigma_i|^2 - \sum_{i=1}^{n} |1 - \sigma_i|^2 \\ &= \sum_{i \neq j} |p_{ij}|^2 + \sum_{i=1}^{n} [|p_{ii}|^2 - \sigma_i(p_{ii} + \bar{p}_{ii}) + \sigma_i^2] \\ &\quad - \sum_{i=1}^{n}(1 - 2\sigma_i + \sigma_i^2) \qquad (9.20) \\ &= \sum_{i, j} |p_{ij}|^2 + \sum_{i=1}^{n} \sigma_i[2 - (p_{ii} + \bar{p}_{ii})] - n \\ &= \sum_{i=1}^{n} \sigma_i[2 - (p_{ii} + \bar{p}_{ii})]\end{aligned}$$

since $\mathbf{P}$ is unitary and hence has $\|\mathbf{P}\|_F^2 = n$. Since $\sum_{j=1}^{n} |p_{ij}|^2 = 1$, we of course have $|p_{ii}| \leq 1$, which implies that the real part $\frac{1}{2}(p_{ii} + \bar{p}_{ii})$ of p_{ii} also has absolute value not exceeding unity; therefore, since $\sigma_i \geq 0$, we have $\sigma_i[2 - (p_{ii} + \bar{p}_{ii})] \geq 0$ and, from (9.20),

$$\|\mathbf{P} - \mathbf{\Sigma}\|_F^2 - \|\mathbf{I} - \mathbf{\Sigma}\|_F^2 \geq 0.$$

Therefore $\mathbf{I}$ is the closest unitary matrix to $\mathbf{\Sigma}$, and we should therefore choose $\mathbf{Q}$ so that $\mathbf{P} = \mathbf{U}^H\mathbf{QV}$ is just $\mathbf{I}$, that is, $\mathbf{Q}_1 = \mathbf{UV}^H$. In short, $\mathbf{Q}_1 = \mathbf{UV}^H$ minimizes $\|\mathbf{Q} - \mathbf{Q}_0\|_F$ over all unitary matrices $\mathbf{Q}$ if $\mathbf{Q}_0 = \mathbf{U\Sigma V}^H$.

EXERCISE 9.41. Find the closest (in the sense of $\|\cdot\|_F$) unitary matrix $\mathbf{Q}_1$ to

$$\mathbf{Q}_0 = \begin{bmatrix} 1 & -0.1 \\ 0.1 & 1 \end{bmatrix}$$

and also the closest unitary matrix to

$$\mathbf{Q}_0 = \begin{bmatrix} \frac{3}{5} & 1 \\ \frac{4}{5} & 0 \end{bmatrix}.$$

EXERCISE 9.42. Let $\mathbf{A}_0$ be an $m \times n$ matrix of rank k, and let s be an integer with $0 < s < k$. Show that the matrix $\mathbf{A}_1$ which minimizes $\|\mathbf{A} - \mathbf{A}_0\|_F$ over all $m \times n$ matrices $\mathbf{A}$ of rank less than or equal to s is obtained from the singular-value decomposition $\mathbf{A}_0 = \mathbf{U\Sigma V}^H$ of $\mathbf{A}_0$ by defining $\mathbf{A}_1 = \mathbf{U\Sigma}_1\mathbf{V}^H$ where $\mathbf{\Sigma}_1$ is identical with $\mathbf{\Sigma}$ except that the diagonal elements $\sigma_{s+1}, \ldots, \sigma_k$ of $\mathbf{\Sigma}$ are replaced by zeros in $\mathbf{\Sigma}_1$. This result is also true for $\|\cdot\|_2$, but more difficult.

SOLUTION: The argument is much as for Ex. 9.40, but much too tedious for this book.

EXERCISE 9.43. As a special case of Ex. 9.42, suppose that $\mathbf{A}_0$ is a nonsingular square $n \times n$ matrix. In the sense of $\|\cdot\|_F$, find the closest singular matrix $\mathbf{A}_1$ to $\mathbf{A}_0$. Show that $\|\mathbf{A}_1 - \mathbf{A}_0\|_F = \sigma_n$, the smallest singular value of $\mathbf{A}_0$.

EXERCISE 9.44. Suppose that the $m \times n$ matrix $\mathbf{A}_0$ arises from some experiment and therefore contains measurement errors such that any other matrix $\mathbf{A}$ with $\|\mathbf{A} - \mathbf{A}_0\|_F \le \epsilon$ for some given estimate ϵ of the size of the measurement errors would be just as meaningful a result as $\mathbf{A}_0$. While $\mathbf{A}_0$ might by chance be of full rank, other matrices $\mathbf{A}$ with $\|\mathbf{A} - \mathbf{A}_0\|_F \le \epsilon$ may not be; we ask what is the least possible rank for all matrices $\mathbf{A}$ with $\|\mathbf{A} - \mathbf{A}_0\|_F \le \epsilon$. Solve this problem by using the result of Ex. 9.42.

9.6 LEAST SQUARES AND THE GENERALIZED INVERSE

In Section 2.6 we considered in some detail the problem of determining the parameters in a model so as to obtain the best fit in a certain sense to prescribed data on the actual behavior of the system being modeled. Specifically in (2.45) we modeled the position y of a body at time t via the equation

$$y = \alpha + vt$$

and then attempted to select the parameters α and v so as to minimize the sum of the squares of the distances between the observed positions y_i at times t_i and the predicted positions $\alpha + vt_i$. This produced the equations (2.49), (2.50). For clarity, the student should reread Section 2.6. With the tools we have developed in this chapter we can now study such problems in detail.

More generally we envision modeling a process in which the input of some system is described by some model parameters $\mathbf{x}$ and the output is roughly described as the result **Ax** of a linear transformation of **x**. If the vector **b** describes the observed actual behavior of the system, then the problem is to choose the parameters **x** so as to minimize some measure of the difference between the observed behavior **b** and the behavior **Ax** predicted by the model. If we measure the size of this difference via the norm $\|\cdot\|_2$, then the problem:

$$\text{find} \quad \mathbf{x} \quad \text{minimizing} \quad \|\mathbf{Ax} - \mathbf{b}\|_2, \tag{9.21}$$

where the $m \times n$ real matrix **A** and $m \times 1$ matrix **b** are given is called a *least-squares problem.* We consider another example drawn from Chapter 2.

EXERCISE 9.45. In Section 2.3 we discussed some simple models of population growth. In particular, (2.25) represented the behavior of a population of foxes and chickens in competition, where the parameters 0.6, 0.5, $-k$, and 1.2 in that model were selected randomly as an illustration. In practice, how might we determine such coefficients?

SOLUTION: We assume a model of the form

$$\begin{bmatrix} F_{i+1} \\ C_{i+1} \end{bmatrix} = \begin{bmatrix} x_1 & x_2 \\ x_3 & x_4 \end{bmatrix} \begin{bmatrix} F_i \\ C_i \end{bmatrix}$$

where we wish to estimate the model parameters described by the vector **x**. If we observe in practice the actual behavior of the population, such as in Table 2.2, then we could obtain say $p + 1$ measurements (of F_i and C_i for $i = 1, \ldots, p + 1$) and choose **x** to minimize the error between the predicted and observed populations. More precisely, we could choose **x** to solve (9.21) where

$$\mathbf{x} = \begin{bmatrix} x_1 \\ x_2 \\ x_3 \\ x_4 \end{bmatrix},$$

$$\mathbf{A} = \begin{bmatrix} F_1 & C_1 & 0 & 0 \\ 0 & 0 & F_1 & C_1 \\ F_2 & C_2 & 0 & 0 \\ 0 & 0 & F_2 & C_2 \\ \cdot & \cdot & \cdot & \cdot \\ \cdot & \cdot & \cdot & \cdot \\ \cdot & \cdot & \cdot & \cdot \\ F_p & C_p & 0 & 0 \\ 0 & 0 & F_p & C_p \end{bmatrix}, \qquad \mathbf{b} = \begin{bmatrix} F_2 \\ C_2 \\ F_3 \\ C_3 \\ \cdot \\ \cdot \\ \cdot \\ F_{p+1} \\ C_{p+1} \end{bmatrix}.$$

Note that this problem splits into two independent minimization problems, one involving x_1, x_2 only, the other involving x_3, x_4 only.

In Section 2.6 we deduced that the solution of the least-squares problem (9.21) was given by the solution to the system of equations

$$\mathbf{A}^T\mathbf{A}\mathbf{x} = \mathbf{A}^T\mathbf{b} \tag{9.22}$$

whose matrix $\mathbf{A}^T\mathbf{A}$ is an $n \times n$ real symmetric matrix. Since a real symmetric matrix is normal, we can use the Fredholm Alternative of Theorem 9.4 to discover whether or not (9.22) has a solution (see also **Key Theorem 3.8**). There are two alternatives, depending on whether or not $\mathbf{A}^T\mathbf{A}$ is nonsingular. If $\mathbf{A}^T\mathbf{A}$ is nonsingular, we of course have a unique solution for (9.22), namely $\mathbf{x} = (\mathbf{A}^T\mathbf{A})^{-1}\mathbf{A}^T\mathbf{b}$. The proof of **Key Theorem 9.7** showed that $\mathbf{A}^T\mathbf{A}$ is nonsingular if and only if the rank of $\mathbf{A}$ equals n. To see this another way, we let

$$\mathbf{A} = \mathbf{U}\mathbf{\Sigma}\mathbf{V}^H \tag{9.23}$$

be the singular-value decomposition of $\mathbf{A}$ as described by **Key Theorem 9.7**, and we let $\sigma_1 \geq \sigma_2 \geq \cdots \geq \sigma_k > 0$ be the singular values, where k is the rank of $\mathbf{A}$; since $\mathbf{A}$ is real, we may assume that $\mathbf{U}$ and $\mathbf{V}$ are real as well. Then $\mathbf{A}^T\mathbf{A} = \mathbf{V}\mathbf{\Sigma}^T\mathbf{\Sigma}\mathbf{V}^T$ and since $\mathbf{\Sigma}^T\mathbf{\Sigma}$ is an $n \times n$ diagonal matrix with diagonal elements $\sigma_1^2, \ldots, \sigma_k^2$, $\mathbf{A}^T\mathbf{A}$ is nonsingular if and only if the rank k of $\mathbf{A}$ equals n since this is what makes $\mathbf{\Sigma}^T\mathbf{\Sigma}$ nonsingular. Therefore (9.22) has a unique solution if and only if the rank of $\mathbf{A}$ equals n.

On the other hand, if $\mathbf{A}^T\mathbf{A}$ is singular, then Theorem 9.4 tells us that (9.22) is solvable if and only if $\mathbf{A}^T\mathbf{b}$ is orthogonal to all eigenvectors of $\mathbf{A}^T\mathbf{A}$ associated with the eigenvalue zero. Since according to **Key Theorem 9.7** these eigenvectors are precisely the columns $\mathbf{v}_{k+1}, \ldots, \mathbf{v}_n$, (9.22) is solvable if and only if $\mathbf{v}_i^T(\mathbf{A}^T\mathbf{b}) = 0$ for $i = k + 1, \ldots, n$. Since $\mathbf{A}^T = \mathbf{V}\mathbf{\Sigma}^T\mathbf{U}^T$ and $\mathbf{V}$ is orthogonal,

$$\mathbf{v}_i^T\mathbf{A}^T\mathbf{b} = \mathbf{v}_i^T\mathbf{V}\mathbf{\Sigma}^T\mathbf{U}^T\mathbf{b} = \mathbf{e}_i^T\mathbf{\Sigma}^T\mathbf{U}^T\mathbf{b} = \mathbf{0}^T\mathbf{U}^T\mathbf{b} = 0$$

for $i > k$. Therefore, (9.22) has a solution even when $\mathbf{A}^T\mathbf{A}$ is singular. We restate this **key** result on the solvability of least-squares problems.

- KEY THEOREM 9.8. *The normal equations* (9.22) *for the least-squares problem* (9.21) *always have a solution; the solution is unique if and only if the rank of the $m \times n$ matrix* $\mathbf{A}$ *equals n.*

At this point it would be natural to assume that the normal equations (9.22) provide the way to *compute* a solution to the least-squares problem; this assumption is correct, but one should avoid actually *computing* $\mathbf{A}^T\mathbf{A}$ as the following example shows.

EXERCISE 9.46. Consider the least-squares problem (9.21) with

$$\mathbf{A} = \begin{bmatrix} 1 & 1 \\ 1 & 1 \\ 1 & 1+\epsilon \end{bmatrix}, \qquad \mathbf{b} = \begin{bmatrix} 2 \\ 3 \\ 2 \end{bmatrix}$$

where ϵ is small enough that $1 + \epsilon^2$ is evaluated on a particular computer to equal 1, while ϵ is large enough that $1 + \epsilon$ is evaluated accurately; for example, on a decimal computer carrying 8 decimal figures $\epsilon = 10^{-4}$ will suffice. Discuss the normal equations (9.22).

SOLUTION: In perfect arithmetic we would compute

$$\mathbf{A}^T\mathbf{A} = \begin{bmatrix} 3 & 3+\epsilon \\ 3+\epsilon & 3+2\epsilon+\epsilon^2 \end{bmatrix}, \qquad \mathbf{A}^T\mathbf{b} = \begin{bmatrix} 7 \\ 7+2\epsilon \end{bmatrix}$$

so that the unique solution to the ideal normal equations (9.22) would be

$$\mathbf{x} = \frac{1}{\epsilon}\begin{bmatrix} \frac{1}{2} + \frac{5\epsilon}{2} \\ -\frac{1}{2} \end{bmatrix}. \tag{9.24}$$

On the computer, however, we would compute $\mathbf{A}^T\mathbf{A}$ and $\mathbf{A}^T\mathbf{b}$ to be

$$\begin{bmatrix} 3 & 3+\epsilon \\ 3+\epsilon & 3+2\epsilon \end{bmatrix}, \qquad \begin{bmatrix} 7 \\ 7+2\epsilon \end{bmatrix},$$

respectively; the exact solution of these *computed* normal equations is

$$\frac{1}{\epsilon}\begin{bmatrix} 1-2\epsilon \\ -1 \end{bmatrix}$$

which differs from the true solution in (9.24) by about a factor of two. Moreover, if the *computed* normal equations are solved by Gauss elimination on our mythical computer, the *computed* matrix $\mathbf{A}^T\mathbf{A}$ will usually be determined to be singular (why?). See also Ex. 9.47.

The point of Ex. 9.46 is that the actual computation of $\mathbf{A}^T\mathbf{A}$ may cause difficulties not inherent in the original least-squares problem itself. One good way around the difficulty illustrated in Ex. 9.46 is to use the QR decomposition of Section 9.4. If $\mathbf{A} = \mathbf{Q}_0\mathbf{R}_0$ is the decomposition in **Key Theorem 9.7**, so that $\mathbf{Q}_0$ is $m \times n$ and has real orthonormal columns while $\mathbf{R}_0$ is $n \times n$ and right triangular, then the normal equations $\mathbf{A}^T\mathbf{A}\mathbf{x} = \mathbf{A}^T\mathbf{b}$ from (9.22) can be written

$$(\mathbf{Q}_0\mathbf{R}_0)^T(\mathbf{Q}_0\mathbf{R}_0)\mathbf{x} = (\mathbf{Q}_0\mathbf{R}_0)^T\mathbf{b}$$

and become

$$\mathbf{R}_0^T\mathbf{R}_0\mathbf{x} = \mathbf{R}_0^T\mathbf{Q}_0^T\mathbf{b} \tag{9.25}$$

since $\mathbf{Q}_0^T\mathbf{Q}_0 = \mathbf{I}$. By **Key Theorem 9.8**, we know that (9.25) always has a solution. If the rank of $\mathbf{A}$ equals n then the rank of $\mathbf{R}_0$ equals n and hence $\mathbf{R}_0$ and $\mathbf{R}_0^T$ are nonsingular; in this case we conclude from (9.25) that

$$\mathbf{R}_0\mathbf{x} = \mathbf{Q}_0^T\mathbf{b}, \tag{9.26}$$

which is simply a *triangular* system of equations to be solved for the least squares solution $\mathbf{x}$.

EXERCISE 9.47. Use the QR decomposition as described in (9.26) to solve the least-squares problem in Ex. 9.46.

SOLUTION: On our mythical computer described in Ex. 9.46, the process of generating the QR decomposition of the matrix $\mathbf{A}$ in Ex. 9.46 yields $\mathbf{A} = \mathbf{Q}_0\mathbf{R}_0$ with

$$\mathbf{Q}_0 = \begin{bmatrix} \frac{\sqrt{3}}{3} & -\frac{\sqrt{6}}{6} \\ \frac{\sqrt{3}}{3} & -\frac{\sqrt{6}}{6} \\ \frac{\sqrt{3}}{3} & \frac{\sqrt{6}}{3} \end{bmatrix}, \qquad \mathbf{R}_0 = \begin{bmatrix} \sqrt{3} & \sqrt{3} + \frac{\sqrt{3}}{3}\epsilon \\ 0 & \frac{\sqrt{6}}{3}\epsilon \end{bmatrix},$$

as we easily verify directly. We then compute

$$\mathbf{Q}_0^T\mathbf{b} = \begin{bmatrix} \frac{7\sqrt{3}}{3} \\ -\frac{\sqrt{6}}{6} \end{bmatrix}$$

and solve (9.26):

$$\begin{bmatrix} \sqrt{3} & \sqrt{3} + \frac{\sqrt{3}}{3}\epsilon \\ 0 & \frac{\sqrt{3}}{3}\epsilon \end{bmatrix}\mathbf{x} = \begin{bmatrix} \frac{7\sqrt{3}}{3} \\ -\frac{\sqrt{6}}{6} \end{bmatrix}.$$

We obtain the solution

$$\mathbf{x} = \frac{1}{\epsilon}\begin{bmatrix} \frac{1+5\epsilon}{2} \\ -\frac{1}{2} \end{bmatrix}$$

which is the exact solution. On a computer, of course, we would not exactly evaluate $\sqrt{3}$ and hence would not get the exact answer. Using an eight-figure decimal computer and $\epsilon = 10^{-4}$ we actually approximately obtained (by using the QR decomposition) the solution

$$\begin{bmatrix} 4967 \\ -4965 \end{bmatrix}$$

as opposed to the true solution

$$\begin{bmatrix} 5002.5 \\ -5000 \end{bmatrix}$$

and the *much* worse approximate solution

$$\begin{bmatrix} 333 \\ -331 \end{bmatrix}$$

obtained by using the same computer directly on the normal equations as in Ex. 9.46.

EXERCISE 9.48. In Ex. 9.45, let $p = 5$ and use the data from Table 2.2 of Section 2.3 to estimate the parameters x_1, x_2, x_3, x_4 by the method of least squares.

EXERCISE 9.49. Use the method of (9.25), (9.26) to solve the example least-squares problem discussed throughout Section 2.6.

The method of (9.25), (9.26) is useful when $\mathbf{A}^T\mathbf{A}$ and $\mathbf{R}$ are nonsingular, that is, when $\mathbf{A}$ has full rank. Another approach, which works even when $\mathbf{A}^T\mathbf{A}$ is singular, uses the singular-value decomposition of $\mathbf{A}$ described in **Key Theorem 9.7.** By Theorem 8.8(iv), we have for all $\mathbf{x}$ that

$$\|\mathbf{Ax} - \mathbf{b}\|_2 = \|\mathbf{U\Sigma V}^T\mathbf{x} - \mathbf{b}\|_2 = \|\mathbf{\Sigma y} - \mathbf{U}^T\mathbf{b}'\|_2, \quad \text{where} \quad \mathbf{y} = \mathbf{V}^T\mathbf{x}.$$

Therefore $\mathbf{x}$ solves the least-squares problem (9.21) if and only if $\mathbf{y} = \mathbf{V}^T\mathbf{x}$ solves:

$$\text{minimize } \|\mathbf{\Sigma y} - \mathbf{b}'\|_2, \quad \text{where} \quad \mathbf{b}' = \mathbf{U}^T\mathbf{b}.$$

But since

$$\|\mathbf{\Sigma y} - \mathbf{b}'\|_2 = [(\sigma_1 y_1 - b_1')^2 + \cdots + (\sigma_k y_k - b_k')^2 + b_{k+1}'^2 + \cdots + b_m'^2]^{1/2},$$

this latter problem is solved by letting $y_i = b_i'/\sigma_i$ for $i = 1, \ldots, k$ and letting y_i for $i = k + 1, \ldots, m$ be arbitrary. Since $\|\mathbf{x}\|_2 = \|\mathbf{y}\|_2$ is minimized over all such solutions $\mathbf{y}$ by setting $y_i = 0$ for $i = k + 1, \ldots, m$, we deduce the following.

THEOREM 9.9. The solutions $\mathbf{x}$ *that minimize* $\|\mathbf{Ax} - \mathbf{b}\|_2$ *when the* $m \times n$ *matrix* $\mathbf{A}$ *has rank* k *are precisely those vectors* $\mathbf{x}$ *of the form* $\mathbf{x} = \mathbf{Vy}$ *where* $y_i = b'_i/\sigma_i$ *for* $i = 1, \ldots, k$ *and where* $y_{k+1}, \ldots, y_m$ *are arbitrary, where* $\mathbf{b}' = \mathbf{U}^T\mathbf{b}$, *and where* $\mathbf{A} = \mathbf{U\Sigma V}^T$ *is the singular-value decomposition of* $\mathbf{A}$ *as described by **Key Theorem 9.7.** Moreover, the solution* $\mathbf{x}$ *of least* $\|\cdot\|_2$ *norm is given by setting* $y_{k+1} = \cdots = y_m = 0$.

EXERCISE 9.50. Use the singular-value decomposition to solve the least-squares problem (9.21) when

$$\mathbf{A} = \begin{bmatrix} 1 & 1 \\ 1 & 1 \end{bmatrix}, \qquad \mathbf{b} = \begin{bmatrix} 2 \\ 4 \end{bmatrix}.$$

SOLUTION: Since

$$\mathbf{A}^T\mathbf{A} = \begin{bmatrix} 2 & 2 \\ 2 & 2 \end{bmatrix} = \mathbf{AA}^T$$

the singular-value decomposition is $\mathbf{A} = \mathbf{U\Sigma V}^T$ with

$$\mathbf{U} = \mathbf{V} = \begin{bmatrix} \frac{\sqrt{2}}{2} & \frac{\sqrt{2}}{2} \\ \frac{\sqrt{2}}{2} & -\frac{\sqrt{2}}{2} \end{bmatrix}, \qquad \mathbf{\Sigma} = \begin{bmatrix} 2 & 0 \\ 0 & 0 \end{bmatrix}.$$

Hence

$$\mathbf{b}' = \mathbf{U}^T\mathbf{b} = \begin{bmatrix} 3\sqrt{2} \\ -\sqrt{2} \end{bmatrix}, \qquad y_1 = \frac{3\sqrt{2}}{2}, \qquad y_2 = \text{arbitrary},$$

$$\mathbf{x} = \mathbf{Vy} = \begin{bmatrix} \frac{3}{2} \\ \frac{3}{2} \end{bmatrix} + y_2 \begin{bmatrix} \frac{\sqrt{2}}{2} \\ \frac{-\sqrt{2}}{2} \end{bmatrix}.$$

Any such $\mathbf{x}$ solves (9.21), and the solution $\mathbf{x}$ of least $\|\cdot\|_2$-norm is

$$\mathbf{x} = \begin{bmatrix} \frac{3}{2} \\ \frac{3}{2} \end{bmatrix}.$$

EXERCISE 9.51. Use the method of Theorem 9.9 to solve the example least-squares problem discussed throughout Section 2.6.

The important result of Theorem 9.9 is often expressed in a different but equivalent fashion by using the concept of a *generalized inverse* of a matrix, to which we now turn our attention. We consider in more detail the mapping from $\mathbf{b}$ to the least-squares solution $\mathbf{x}$ given in Theorem 9.9. If from the $m \times n$ matrix $\mathbf{\Sigma}$ we define $\mathbf{\Sigma}^+$ as the $n \times m$ matrix

$$\mathbf{\Sigma}^+ = \begin{bmatrix} \mathbf{E} & \mathbf{0} \\ \mathbf{0} & \mathbf{0} \end{bmatrix}, \tag{9.27}$$

where $\mathbf{E}$ *is the* $k \times k$ *diagonal matrix whose ith diagonal element is* $e_{ii} = \sigma_i^{-1}$ *for* $1 \leq i \leq k$, then writing the result of Theorem 9.9 in matrix notation, we have $\mathbf{y} = \mathbf{\Sigma}^+\mathbf{b}'$, $\mathbf{b}' = \mathbf{U}^T\mathbf{b}$, and $\mathbf{x} = \mathbf{V}\mathbf{y}$, so that

$$\mathbf{x} = \mathbf{V}\mathbf{\Sigma}^+\mathbf{U}^T\mathbf{b}.$$

Thus the least-squares solution $\mathbf{x}$ of (9.21) having least $\|\cdot\|_2$ norm can be written as a linear transformation applied to $\mathbf{b}$; by analogy with the case in which $\mathbf{A}$ is square and nonsingular so that $\mathbf{x} = \mathbf{A}^{-1}\mathbf{b}$, we call this transformation matrix a *generalized inverse* of $\mathbf{A}$.

DEFINITION 9.3. Let the $m \times n$ matrix $\mathbf{A}$ of rank k have the singular-value decomposition $\mathbf{A} = \mathbf{U}\mathbf{\Sigma}\mathbf{V}^H$ as in **Key Theorem 9.7** with $\sigma_1 \geq \sigma_2 \geq \cdots \geq \sigma_k > 0$. Then the (Moore-Penrose) *generalized inverse* (*pseudoinverse*) $\mathbf{A}^+$ of $\mathbf{A}$ is the $n \times m$ matrix $\mathbf{A}^+ = \mathbf{V}\mathbf{\Sigma}^+\mathbf{U}^H$, where $\mathbf{\Sigma}^+$ is given by (9.27).

We remark that $\mathbf{A}^+$ can be shown to be unique although $\mathbf{U}$ and $\mathbf{V}$ may not be.

EXERCISE 9.52. Find the generalized inverse $\mathbf{A}^+$ of the matrix $\mathbf{A}$ in Ex. 9.36.

SOLUTION: In Ex. 9.36 we found the singular-value decomposition

$$\mathbf{A} = \begin{bmatrix} \frac{1}{3} & \frac{-2\sqrt{5}}{5} & \frac{2\sqrt{5}}{15} \\ \frac{2}{3} & \frac{\sqrt{5}}{5} & \frac{4\sqrt{5}}{15} \\ \frac{2}{3} & 0 & \frac{-\sqrt{5}}{15} \end{bmatrix} \begin{bmatrix} 3\sqrt{2} & 0 \\ 0 & 0 \\ 0 & 0 \end{bmatrix} \begin{bmatrix} \frac{\sqrt{2}}{2} & \frac{\sqrt{2}}{2} \\ \frac{\sqrt{2}}{2} & \frac{-\sqrt{2}}{2} \end{bmatrix}$$

so we must have

$$\mathbf{A}^+ = \begin{bmatrix} \frac{\sqrt{2}}{2} & \frac{\sqrt{2}}{2} \\ \frac{\sqrt{2}}{2} & -\frac{\sqrt{2}}{2} \end{bmatrix} \begin{bmatrix} \frac{1}{3\sqrt{2}} & 0 & 0 \\ 0 & 0 & 0 \end{bmatrix} \begin{bmatrix} \frac{1}{3} & \frac{2}{3} & \frac{2}{3} \\ -\frac{2\sqrt{5}}{5} & \frac{\sqrt{5}}{5} & 0 \\ \frac{2\sqrt{5}}{15} & \frac{4\sqrt{5}}{15} & \frac{-\sqrt{5}}{15} \end{bmatrix}$$

$$= \begin{bmatrix} \frac{1}{18} & \frac{1}{9} & \frac{1}{9} \\ \frac{1}{18} & \frac{1}{9} & \frac{1}{9} \end{bmatrix}.$$

EXERCISE 9.53. Show that *any* unitary $\mathbf{U} = \mathbf{V}$ may be used in the singular-value decomposition of $\mathbf{I}$, but that the generalized inverse does not change.

EXERCISE 9.54. If $\mathbf{A}$ is square and nonsingular, prove that $\mathbf{A}^+ = \mathbf{A}^{-1}$.

With this terminology, we can restate the key result of Theorem 9.9.

- *KEY COROLLARY 9.1. The solution* $\mathbf{x}$ *of least* $\|\cdot\|_2$*-norm to the least-squares problem of minimizing* $\|\mathbf{Ax} - \mathbf{b}\|_2$ *is* $\mathbf{x} = \mathbf{A}^+\mathbf{b}$, *where* $\mathbf{A}^+$ *is the (Moore-Penrose) generalized inverse of* $\mathbf{A}$ *in Definition 9.3.*

Using **Key Corollary 9.1** and Ex. 9.54, we see that the generalized inverse $\mathbf{A}^+$ truly generalizes our earlier notion of the usual inverse $\mathbf{A}^{-1}$. It is natural to wonder what properties of the true inverse carry over to the generalized inverse. We know, of course, that we cannot expect $\mathbf{A}^+\mathbf{A}$ to equal the unit matrix $\mathbf{I}$; however, we do have $\mathbf{A}^+\mathbf{A} = \mathbf{V\Sigma}^+\mathbf{U}^H\mathbf{U\Sigma V}^H = \mathbf{V\Sigma}^+\mathbf{\Sigma V}^H$, and $\mathbf{\Sigma}^+\mathbf{\Sigma}$ is $n \times n$ and

$$\mathbf{\Sigma}^+\mathbf{\Sigma} = \begin{bmatrix} \mathbf{I}_k & \mathbf{0} \\ \mathbf{0} & \mathbf{0} \end{bmatrix}.$$

EXERCISE 9.55. Use the decomposition $\mathbf{A} = \mathbf{U\Sigma V}^H$ to show that the set of the last $n - k$ columns of $\mathbf{V}$ spans the null space of $\mathbf{A}$. (See also the proof of **Key Theorem 9.7.**)

EXERCISE 9.56. Use Ex. 9.55 and the above representation of $\mathbf{A}^+\mathbf{A}$ to deduce that for every vector $\mathbf{b}$ the vector $\mathbf{b}' = (\mathbf{I}_n - \mathbf{A}^+\mathbf{A})\mathbf{b}$ is the *orthogonal projection* of $\mathbf{b}$ onto the null space of $\mathbf{A}$, that is, $\mathbf{b}'$ is in the null space of $\mathbf{A}$ and $\mathbf{b} - \mathbf{b}'$ is orthogonal to every vector in the null space of $\mathbf{A}$. Similarly, we know that we cannot expect $\mathbf{AA}^+$ to equal the unit matrix $\mathbf{I}_m$; we do have

$$\mathbf{AA}^+ = \mathbf{U\Sigma V}^H\mathbf{V\Sigma}^+\mathbf{U}^H = \mathbf{U\Sigma\Sigma}^+\mathbf{U}^H,$$

where $\mathbf{\Sigma\Sigma}^+$ is $m \times m$ and

$$\mathbf{\Sigma\Sigma}^+ = \begin{bmatrix} \mathbf{I}_k & \mathbf{0} \\ \mathbf{0} & \mathbf{0} \end{bmatrix}.$$

EXERCISE 9.57. Use the decomposition $\mathbf{A} = \mathbf{U\Sigma V}^H$ to show that the set of the first k columns of $\mathbf{U}$ spans the exact range space (column space) of $\mathbf{A}$.

EXERCISE 9.58. Use Ex. 9.57 and the above representation of $\mathbf{AA}^+$ to deduce that for every vector $\mathbf{b}$ the vector $\mathbf{b}' = \mathbf{AA}^+\mathbf{b}$ is the orthogonal projection of $\mathbf{b}$ onto the exact range space of $\mathbf{A}$, that is, $\mathbf{b}'$ is in the exact range space of $\mathbf{A}$ and $\mathbf{b} - \mathbf{b}'$ is orthogonal to every vector in the exact range space of $\mathbf{A}$.

EXERCISE 9.59. Interpret the conclusions of Exs. 9.55–9.58 for the case in which $\mathbf{A}$ is $n \times n$ and nonsingular.

In the preceding four problems we have given some of the properties of $\mathbf{A}^+$ with respect to $\mathbf{A}$ which extend the relationship between $\mathbf{A}^{-1}$ and $\mathbf{A}$ for nonsingular matrices $\mathbf{A}$. In the next theorem we give some further properties which in fact *uniquely characterize* the (Moore-Penrose) generalized inverse.

THEOREM 9.10. Let $\mathbf{A}$ *and* $\mathbf{A}^+$ *be as in Definition 9.3. Then an* $n \times m$ *matrix* $\mathbf{A}_0$ *is the (Moore-Penrose) generalized inverse* $\mathbf{A}^+$ *of* $\mathbf{A}$ *if and only if the following three statements are true:*

(i) $\mathbf{AA_0A} = \mathbf{A}$;
(ii) $\mathbf{A_0AA_0} = \mathbf{A_0}$;
(iii) $\mathbf{AA_0}$ *and* $\mathbf{A_0A}$ *are hermitian.*

Moreover, the generalized inverse $\mathbf{A}^+$ *is unique.*

Proof: First we show that $\mathbf{A_0} = \mathbf{A}^+$ satisfies (i)–(iii). For (i),

$$\mathbf{AA^+A} = \mathbf{U\Sigma V}^H\mathbf{V\Sigma^+U}^H\mathbf{U\Sigma V}^H = \mathbf{U\Sigma\Sigma^+\Sigma V}^H = \mathbf{U\Sigma V}^H = \mathbf{A}$$

as asserted; the arguments for (ii) and (iii) are equally simple. The proof that (i)–(iii) imply $\mathbf{A_0} = \mathbf{A}^+$ and that $\mathbf{A}^+$ is unique is complicated and not particularly instructive, so we omit it.

We emphasize that $\mathbf{A}^+$ is uniquely determined by $\mathbf{A}$, even though $\mathbf{U}$ and $\mathbf{V}$ in the singular-value decomposition used to define $\mathbf{A}^+$ are not generally unique.

EXERCISE 9.60. Verify that (i)–(iii) of Theorem 9.10 hold for the $\mathbf{A}$ and $\mathbf{A}^+$ of Ex. 9.52.

While the details of the proof of Theorem 9.10 are not important for us, the characterization of $\mathbf{A}^+$ by (i)–(iii) of Theorem 9.10 has quite some significance. One minor point is that we can now explain our regular parenthetical use of "Moore-Penrose" in reference to $\mathbf{A}^+$. While the early work of Moore and Penrose related to matrices $\mathbf{A_0}$ satisfying (i)–(iii), later authors have discussed still more general inverses for which not all of (i)–(iii) are assumed to hold. More important, perhaps, the characterization of $\mathbf{A}^+$ through Theorem 9.10 allows us to *compute* the generalized inverse without using the singular-value decomposition and hence without requiring us to find eigenvalues and eigenvectors.

THEOREM 9.11. Suppose that the $m \times n$ *matrix* $\mathbf{A}$ *is of the form*

$$\mathbf{A} = \mathbf{BC}$$

where $\mathbf{B}$ *is* $m \times k$ *and* $\mathbf{C}$ *is* $k \times n$ *and where all three matrices* $\mathbf{A}, \mathbf{B},$ *and* $\mathbf{C}$ *have rank* k*. Then the generalized inverse* $\mathbf{A}^+$ *of* $\mathbf{A}$ *is given by*

$$\mathbf{A}^+ = \mathbf{C}^T(\mathbf{CC}^T)^{-1}(\mathbf{B}^T\mathbf{B})^{-1}\mathbf{B}^T. \tag{9.28}$$

Proof: We simply prove that the matrix $\mathbf{A_0}$ defined as in (9.28) satisfies (i)–(iii) of Theorem 9.10. For (i), we write

$$\mathbf{AA_0A} = \mathbf{BCC}^T(\mathbf{CC}^T)^{-1}(\mathbf{B}^T\mathbf{B})^{-1}\mathbf{B}^T\mathbf{BC} = \mathbf{BC} = \mathbf{A}$$

as required. Both (ii) and (iii) follow similarly, proving the theorem.

In order to evaluate the generalized inverse of **A** from (9.28) we must decompose **A** in the form **BC**. This can be done by means of the following theorem (see also Ex. 9.63).

THEOREM 9.12. If **A** *is m × n and of rank k, and we can partition* **A** *in the form*

$$\mathbf{A} = \begin{bmatrix} \mathbf{A}_{11} & \mathbf{A}_{12} \\ \mathbf{A}_{21} & \mathbf{A}_{22} \end{bmatrix} \tag{9.29}$$

where $\mathbf{A}_{11}$ *is a nonsingular k × k matrix, then*

$$\mathbf{A} = \begin{bmatrix} \mathbf{I} \\ \mathbf{P} \end{bmatrix}[\mathbf{A}_{11} \quad \mathbf{A}_{12}] = \begin{bmatrix} \mathbf{I} \\ \mathbf{P} \end{bmatrix}\mathbf{A}_{11}[\mathbf{I} \quad \mathbf{Q}] = \begin{bmatrix} \mathbf{A}_{11} \\ \mathbf{A}_{12} \end{bmatrix}[\mathbf{I} \quad \mathbf{Q}], \tag{9.30}$$

where

$$\mathbf{P} = \mathbf{A}_{21}\mathbf{A}_{11}^{-1}, \qquad \mathbf{Q} = \mathbf{A}_{11}^{-1}\mathbf{A}_{12}. \tag{9.31}$$

Proof: Since the rank of **A** is k, the last $m - k$ rows of (9.29) are linear combinations of the first k rows, so that a matrix **P** exists such that

$$\mathbf{A}_{21} = \mathbf{P}\mathbf{A}_{11}, \qquad \mathbf{A}_{22} = \mathbf{P}\mathbf{A}_{12}.$$

Similarly, the last $n - k$ columns of (9.29) are linear combinations of the first k, so that a matrix **Q** exists such that

$$\mathbf{A}_{12} = \mathbf{A}_{11}\mathbf{Q}, \qquad \mathbf{A}_{22} = \mathbf{A}_{21}\mathbf{Q}.$$

Since $\mathbf{A}_{11}$ is nonsingular, these equations give the expressions for **P** and **Q** in (9.31) and the theorem then follows easily.

EXERCISE 9.61. Find the generalized inverse of

$$\mathbf{A} = \begin{bmatrix} -1 & 0 & 1 & 2 \\ -1 & 1 & 0 & -1 \\ 0 & -1 & 1 & 3 \\ 0 & 1 & -1 & -3 \\ 1 & -1 & 0 & 1 \\ 1 & 0 & -1 & -2 \end{bmatrix}.$$

SOLUTION: On reducing this matrix to row-echelon form, it is found that its rank is 2. The 2 × 2 matrix in the upper left of **A** is nonsingular so that we can choose

$$\mathbf{A}_{11} = \begin{bmatrix} -1 & 0 \\ -1 & 1 \end{bmatrix}, \qquad \mathbf{A}_{11}^{-1} = \begin{bmatrix} -1 & 0 \\ -1 & 1 \end{bmatrix}.$$

Working in terms of the third formula in (9.30), we compute

$$\mathbf{Q} = \mathbf{A}_{11}^{-1}\mathbf{A}_{12} = \begin{bmatrix} -1 & -2 \\ -1 & -3 \end{bmatrix}.$$

In the terminology of Theorem 9.11, we now set

$$\mathbf{B} = \begin{bmatrix} \mathbf{A}_{11} \\ \mathbf{A}_{21} \end{bmatrix} = \begin{bmatrix} -1 & 0 \\ -1 & 1 \\ 0 & -1 \\ 0 & 1 \\ 1 & -1 \\ 1 & 0 \end{bmatrix}, \qquad \mathbf{C} = [\mathbf{I} \quad \mathbf{Q}] = \begin{bmatrix} 1 & 0 & -1 & -2 \\ 0 & 1 & -1 & -3 \end{bmatrix}.$$

Substitution in (9.28) yields

$$\mathbf{A}^+ = \frac{1}{102}\begin{bmatrix} -15 & -18 & 3 & -3 & 18 & 15 \\ 8 & 13 & -5 & 5 & -13 & -8 \\ 7 & 5 & 2 & -2 & -5 & -7 \\ 6 & -3 & 9 & -9 & 3 & -6 \end{bmatrix}.$$

EXERCISE 9.62. Find the same generalized inverse as found in Ex. 9.52 but by the method of Ex. 9.61.

EXERCISE 9.63. A *permutation matrix* is a matrix obtained by interchanging rows or columns of the unit matrix. Show that for any matrix $\mathbf{B}$ of rank k, permutation matrices $\mathbf{P}, \mathbf{Q}$ can be found so that $\mathbf{PBQ}$ can be partitioned in the form (9.29), where the matrix $\mathbf{A}_{11}$ of order k is nonsingular. Show that $\mathbf{B}^+ = \mathbf{QA}^+\mathbf{P}$. Generalize Theorem 9.11 to show that *any* $m \times n$ matrix of rank k can be written in the form $\mathbf{BC}$ where $\mathbf{B}$ is $m \times k$, $\mathbf{C}$ is $k \times n$, and each is of rank k.

EXERCISE 9.64. Prove that

(a) The rank of $\mathbf{A}^+$ is the same as the rank of $\mathbf{A}$.
(b) If $\mathbf{A}$ is symmetric, then $\mathbf{A}^+$ is symmetric.
(c) $(c\mathbf{A})^+ = (1/c)\mathbf{A}^+$ for $c \neq 0$.
(d) $(\mathbf{A}^+)^T = (\mathbf{A}^T)^+$
(e) $(\mathbf{A}^+)^+ = \mathbf{A}$.
(f) Show by a counterexample that in general $(\mathbf{AB})^+ \neq \mathbf{B}^+\mathbf{A}^+$.
(g) If $\mathbf{A}$ is $m \times r$, $\mathbf{B}$ is $r \times n$, and both matrices are of rank r, then $(\mathbf{AB})^+ = \mathbf{B}^+\mathbf{A}^+$.

EXERCISE 9.65. Prove that:

(a) The generalized inverse of the null matrix is the null matrix.
(b) If $\mathbf{u}$ is a nonzero column vector, then $\mathbf{u}^+ = (\mathbf{u}^T\mathbf{u})^{-1}\mathbf{u}^T$.

(c) If $\mathbf{A} = \mathbf{u}\mathbf{v}^T$ then

$$\mathbf{A}^+ = \frac{\mathbf{A}^T}{(\mathbf{v}^T\mathbf{v})(\mathbf{u}^T\mathbf{u})}.$$

(d) The generalized inverse of a 2×2 matrix $\mathbf{A}$ of rank one is given by

$$\frac{1}{a_{11}^2 + a_{12}^2 + a_{21}^2 + a_{22}^2}\mathbf{A}^T.$$

(e) If all the elements of an $m \times n$ matrix $\mathbf{A}$ are unity, then $\mathbf{A}^+ = (1/mn)\mathbf{A}^T$.

9.7 ANALYZING INFLUENTIAL FACTORS IN EXPERIMENTS

One problem we have not discussed at all is how mathematical models are built; instead we have restricted ourselves to showing various ways linear algebra may be used in the *analysis* of mathematical models. Linear algebra methods can be useful in the *design* of models as well, however; we now present an application of linear algebra in statistics which is useful in attempting to determine how many independent factors are really at work in some system.

Suppose that we are able to observe and measure some n aspects $x_1, \ldots, x_n$ of a complicated system of interest (where in technical statistical language we assume the x_i to be random variables with nonsingular multinormal distributions and zero means). We suspect that the behavior of these n aspects is essentially determined by some m independent factors $y_1, \ldots, y_m$ with $m < n$, and we wish to test this suspicion. More precisely we want to test the conjecture that there are some variables $y_1, \ldots, y_m$ and $e_1, \ldots, e_n$ and constants b_{ij} for $i = 1, \ldots, n$ and $j = 1, \ldots, m$ for which

$$x_i = b_{i1}y_1 + \cdots + b_{im}y_m + e_i$$

(where in technical statistical language we assume the variables y_i to be independently and normally distributed with zero means and unit variances while the e_i are independently and normally distributed variables with zero means and unknown variances d_i). Under these hypotheses the variance of x_i would be $b_{i1}^2 + \cdots + b_{im}^2 + d_i$ while the covariance of x_i and x_j would be $b_{i1}b_{j1} + \cdots + b_{im}b_{jm}$. If we define the $n \times m$ matrix $\mathbf{B}$ from the b_{ij}, and the $n \times n$ *diagonal* matrix $\mathbf{D}$ with $d_{ii} = d_i$ and $d_{ij} = 0$ if $i \neq j$, then the covariance matrix of the x_i under these assumptions should be just $\mathbf{B}\mathbf{B}^T + \mathbf{D}$. Since we can estimate the actual $n \times n$ symmetric covariance matrix $\mathbf{A}$ of the x_i from experimental data, the problem becomes one of trying to represent $\mathbf{A}$ in the form $\mathbf{A} = \mathbf{B}\mathbf{B}^T + \mathbf{D}$.

Since the reader lacking a background in statistics may have found the preceding motivation incomprehensible, we restate our problem as purely one of linear algebra. In many situations in which one wants to discover whether or not the behavior of some system described by n components can in fact be adequately represented by some m influential factors with $m < n$, the following linear algebra problem results: given an $n \times n$ real symmetric matrix $\mathbf{A}$ can we find matrices $\mathbf{D}$ and $\mathbf{B}$ such that $\mathbf{D}$ is an $n \times n$ diagonal matrix with positive diagonal elements and $\mathbf{B}$ is an $n \times m$ real matrix of rank m and such that

$$\mathbf{A} = \mathbf{B}\mathbf{B}^T + \mathbf{D}? \tag{9.32}$$

EXERCISE 9.66. Suppose that $n = 2$, $m = 1$, and

$$\mathbf{A} = \begin{bmatrix} 5 & 0 \\ 0 & 2 \end{bmatrix}.$$

Then there are obviously many solutions to (9.32), for example,

$$\mathbf{A} = \begin{bmatrix} 2 \\ 0 \end{bmatrix}[2 \quad 0] + \begin{bmatrix} 1 & 0 \\ 0 & 2 \end{bmatrix} = \begin{bmatrix} 0 \\ 1 \end{bmatrix}[0 \quad 1] + \begin{bmatrix} 5 & 0 \\ 0 & 1 \end{bmatrix}.$$

If $\mathbf{P}$ is any $m \times m$ orthogonal matrix so that $\mathbf{P}\mathbf{P}^T = \mathbf{I}$, then clearly if $\mathbf{B}$ and $\mathbf{D}$ satisfy (9.32), then so do $\mathbf{BP}$ and $\mathbf{D}$ since

$$(\mathbf{BP})(\mathbf{BP})^T + \mathbf{D} = \mathbf{B}\mathbf{P}\mathbf{P}^T\mathbf{B}^T + \mathbf{D} = \mathbf{B}\mathbf{B}^T + \mathbf{D} = \mathbf{A}.$$

Thus there is no hope of finding a unique $\mathbf{B}$, but we might hope to identify some particular $\mathbf{B}$ in terms of $\mathbf{A}$ and $\mathbf{D}$. More precisely, suppose that (9.32) is satisfied, so that

$$\mathbf{A} - \mathbf{D} = \mathbf{B}\mathbf{B}^T$$

has rank m [since $\mathbf{B}$ has rank m]. Since the real symmetric matrix $\mathbf{A} - \mathbf{D}$ is normal, by **Key Theorem 9.2** there is an $n \times n$ orthogonal matrix $\mathbf{P}$ such that

$$\mathbf{P}^H(\mathbf{A} - \mathbf{D})\mathbf{P} = \mathbf{\Lambda}, \tag{9.33}$$

where the diagonal matrix $\mathbf{\Lambda}$ contains the eigenvalues of $\mathbf{A} - \mathbf{D}$ on its main diagonal; since $\mathbf{A} - \mathbf{D}$ has rank m, so does $\mathbf{\Lambda}$, so that we may assume

$$\mathbf{\Lambda} = \begin{bmatrix} \lambda_1 & 0 & \cdots & 0 & \cdots & 0 \\ 0 & \lambda_2 & \cdots & 0 & \cdots & 0 \\ & & \cdots & & & \\ 0 & 0 & \cdots & \lambda_m & \cdots & 0 \\ 0 & 0 & \cdots & 0 & \cdots & 0 \\ & & \cdots & & & \\ 0 & 0 & \cdots & 0 & \cdots & 0 \end{bmatrix} \tag{9.34}$$

with $\lambda_i \neq 0$ for $i = 1, \ldots, m$. If $\mathbf{p}_1, \ldots, \mathbf{p}_n$ denote the columns of $\mathbf{P}$, so that

$$\mathbf{P} = [\mathbf{p}_1, \ldots, \mathbf{p}_n],$$

then from (9.33) and (9.34) we have

$$\mathbf{A} - \mathbf{D} = [\mathbf{p}_1, \ldots, \mathbf{p}_m] \begin{bmatrix} \lambda_1 & 0 & \ldots & 0 \\ 0 & \lambda_2 & \ldots & 0 \\ & & \ldots & \\ 0 & 0 & \ldots & \lambda_m \end{bmatrix} \begin{bmatrix} \mathbf{p}_1^T \\ \cdot \\ \cdot \\ \cdot \\ \mathbf{p}_m^T \end{bmatrix},$$

so that one solution $\mathbf{B}$ to (9.32) is to let the ith column of $\mathbf{B}$, say $\mathbf{b}_i$, be given by

$$\mathbf{b}_i = \sqrt{\lambda_i}\, \mathbf{p}_i \qquad (\text{for } i = 1, \ldots, m)$$

if $\lambda_i > 0$. In sentence form, the columns of $\mathbf{B}$ in (9.32) can be chosen to be the eigenvectors of $\mathbf{A} - \mathbf{D}$ corresponding to the m nonzero eigenvalues $\lambda_1, \ldots, \lambda_m$ of $\mathbf{A} - \mathbf{D}$, with each such eigenvector normalized to have length $\sqrt{\lambda_i}$. This realization suggests a process for computing an approximate solution to (9.32): given $\mathbf{D}$, find the eigenvalues and eigenvectors of $\mathbf{A} - \mathbf{D}$ and see if $\lambda_1 > 0, \ldots, \lambda_m > 0, \lambda_{m+1} = \cdots = \lambda_n = 0$ so that we can construct $\mathbf{B}$; if we cannot construct $\mathbf{B}$, then modify $\mathbf{D}$ and try again. For example, we might start with

$$\mathbf{D}_0 = \mathbf{0}, \qquad \mathbf{A}_0 = \mathbf{A} - \mathbf{D}_0,$$

and find the eigenvalues and eigenvectors of $\mathbf{A}_0$, so that

$$\mathbf{A}_0 = \mathbf{P}_0 \mathbf{\Lambda}_0 \mathbf{P}_0^T.$$

If $\mathbf{A}_0$ does not have m positive eigenvalues and $n - m$ zero eigenvalues we define $\tilde{\mathbf{\Lambda}}_0$ as $\mathbf{\Lambda}_0$ with the $n - m$ smallest eigenvalues of $\mathbf{A}_0$ replaced by zeros and we then let

$$\mathbf{D}_1 = \text{diagonal of } (\mathbf{A} - \mathbf{P}_0 \tilde{\mathbf{\Lambda}}_0 \mathbf{P}_0^T), \qquad \mathbf{A}_1 = \mathbf{A} - \mathbf{D}_1$$

and continue the iteration to obtain $\mathbf{P}_1, \mathbf{D}_2, \mathbf{A}_2, \ldots$.

EXERCISE 9.67. Use the iteration described above to solve (9.32) with $m = 1$ when

$$\mathbf{A} = \begin{bmatrix} 3 & 1 \\ 1 & 3 \end{bmatrix}.$$

SOLUTION: We write $\mathbf{A}_0 = \mathbf{A} = \mathbf{P}_0 \boldsymbol{\Lambda}_0 \mathbf{P}_0^T$, where

$$\mathbf{P}_0 = \begin{bmatrix} \frac{\sqrt{2}}{2} & \frac{\sqrt{2}}{3} \\ \frac{\sqrt{2}}{2} & \frac{-\sqrt{2}}{2} \end{bmatrix}, \qquad \boldsymbol{\Lambda}_0 = \begin{bmatrix} 4 & 0 \\ 0 & 2 \end{bmatrix}$$

and then define $\mathbf{D}_1$ as the diagonal of $\mathbf{A}_0 - \mathbf{P}_0 \tilde{\boldsymbol{\Lambda}}_0 \mathbf{P}_0^T$ where

$$\tilde{\boldsymbol{\Lambda}}_0 = \begin{bmatrix} 4 & 0 \\ 0 & 0 \end{bmatrix}$$

so that $\mathbf{D}_1$ and $\mathbf{A}_1 = \mathbf{A} - \mathbf{D}_1$ are

$$\mathbf{D}_1 = \begin{bmatrix} 1 & 0 \\ 0 & 1 \end{bmatrix}, \qquad \mathbf{A}_1 = \begin{bmatrix} 2 & 1 \\ 1 & 2 \end{bmatrix}.$$

Continuing in this fashion, we obtain

$$\mathbf{D}_2 = \begin{bmatrix} -\frac{1}{2} & 0 \\ 0 & -\frac{1}{2} \end{bmatrix}, \qquad \mathbf{A}_2 = \begin{bmatrix} \frac{7}{2} & 1 \\ 1 & \frac{7}{2} \end{bmatrix},$$

$$\mathbf{D}_3 = \begin{bmatrix} \frac{3}{4} & 0 \\ 0 & \frac{3}{4} \end{bmatrix}, \qquad \mathbf{A}_3 = \begin{bmatrix} \frac{9}{4} & 1 \\ 1 & \frac{9}{4} \end{bmatrix},$$

$$\mathbf{D}_5 = \begin{bmatrix} \frac{27}{16} & 0 \\ 0 & \frac{27}{16} \end{bmatrix}, \qquad \mathbf{A}_5 = \begin{bmatrix} \frac{21}{16} & 1 \\ 1 & \frac{21}{16} \end{bmatrix},$$

and we eventually see that $\mathbf{D}_i$ is converging to $2\mathbf{I}$, so that the solution to (9.32) is

$$\mathbf{A} = \mathbf{B}\mathbf{B}^T + \mathbf{D}$$

with

$$\mathbf{B} = \begin{bmatrix} 1 \\ 1 \end{bmatrix}, \qquad \mathbf{D} = \begin{bmatrix} 2 & 0 \\ 0 & 2 \end{bmatrix}.$$

EXERCISE 9.68. Use the iteration described above to solve (9.32) with $m = 1$ when

$$\mathbf{A} = \begin{bmatrix} 6 & 2 \\ 2 & 9 \end{bmatrix}.$$

EXERCISE 9.69. Suppose that $\mathbf{A} = \mathbf{B}\mathbf{B}^T + \mathbf{D}$ with $\mathbf{D}$ a diagonal matrix with positive diagonal elements. If $\mathbf{A}\mathbf{x} = \lambda\mathbf{x}$ with $\mathbf{x} \neq \mathbf{0}$, show that

$$\lambda = (\mathbf{A}\mathbf{x}, \mathbf{x})/(\mathbf{x}, \mathbf{x}) > 0,$$

so that when such a $\mathbf{D}$ exists (9.32) can be solved only if all eigenvalues of $\mathbf{A}$ are positive.

EXERCISE 9.70. By explicitly finding **B** and **D**, show that every real symmetric 2×2 matrix **A** having strictly positive eigenvalues can be written as $\mathbf{A} = \mathbf{B}\mathbf{B}^T + \mathbf{D}$, where **D** is a diagonal matrix with positive diagonal elements and **B** is 2×1. Show that there are infinitely many solutions.

EXERCISE 9.71. Show that the eigenvalues of

$$\mathbf{A} = \begin{bmatrix} 2 & -1 & 0 \\ -1 & 2 & -1 \\ 0 & -1 & 2 \end{bmatrix}$$

are $\lambda_1 = 2 + \sqrt{2}$, $\lambda_2 = 2$, $\lambda_3 = 2 - \sqrt{2}$, so that the eigenvalues of **A** are positive as in Ex. 9.69, but that **A** cannot be decomposed as in Ex. 9.69 with **B** being 3×1.

EXERCISE 9.72. Apply the algorithm described in this section to try to decompose the matrix **A** of Ex. 9.71 in the fashion that Ex. 9.71 has shown to be impossible; what happens with the sequence of approximate factorizations?

MISCELLANEOUS EXERCISES 9

EXERCISE 9.73. Rigorously deduce **Key Theorem 3.8** for normal matrices from Theorem 9.4.

EXERCISE 9.74. Suppose that $\mathbf{A} = \mathbf{P}\mathbf{T}\mathbf{P}^H$ and $\mathbf{A} = \mathbf{P}'\mathbf{T}'\mathbf{P}'^H$ are two Schur decompositions of the same matrix **A**. Show that $\mathbf{T}' = \mathbf{Q}\mathbf{T}\mathbf{Q}^H$ for some unitary matrix **Q**. Discuss to what extent **P** and **T** in a Schur decomposition are uniquely determined by **A**. (See Ex. 9.1.)

EXERCISE 9.75. In addition to the assumptions of Ex. 9.74, suppose that **A** is normal and **T** and **T**′ are diagonal. Discuss to what extent **P** and the diagonal **T** in a diagonal Schur decomposition are uniquely determined by **A**.

EXERCISE 9.76. Use the Householder transformations of Section 8.5 to solve Ex. 9.26.

SOLUTION: First we need a Householder transformation $\mathbf{P}_1$ to transform $\mathbf{u}_1 = [1 \quad 0 \quad 1]^T$ into $\mathbf{v}_1 = \mathbf{P}_1\mathbf{u}_1 = \mathbf{e}_1\sqrt{2}$. This is given by $\mathbf{P}_1 = \mathbf{I} - 2\mathbf{w}_1\mathbf{w}_1^H$ where $\mathbf{w}_1 = (\mathbf{v} - \mathbf{u})/\|\mathbf{v} - \mathbf{u}\|_2$. We find

$$\mathbf{P}_1 = \begin{bmatrix} \frac{\sqrt{2}}{2} & 0 & \frac{\sqrt{2}}{2} \\ 0 & 1 & 0 \\ \frac{\sqrt{2}}{2} & 0 & \frac{-\sqrt{2}}{2} \end{bmatrix}, \qquad \mathbf{P}_1\mathbf{A} = \begin{bmatrix} \sqrt{2} & 3\sqrt{2} \\ 0 & 1 \\ 0 & -\sqrt{2} \end{bmatrix}.$$

To operate on the matrix $\mathbf{P}_1\mathbf{A}$ we take $\mathbf{P}_2 = \mathbf{I} - 2\mathbf{w}_2\mathbf{w}_2^H$ with

$$\mathbf{w}_2 = (\mathbf{v}_2 - \mathbf{u}_2)/\|\mathbf{v}_2 - \mathbf{u}_2\|_2$$

where

$$\mathbf{u}_2 = \begin{bmatrix} 3\sqrt{2} \\ 1 \\ -\sqrt{2} \end{bmatrix}, \qquad \mathbf{v}_2 = \begin{bmatrix} 3\sqrt{2} \\ \sqrt{3} \\ 0 \end{bmatrix}.$$

This gives

$$\mathbf{P}_2 = \begin{bmatrix} 1 & 0 & 0 \\ 0 & \frac{\sqrt{3}}{3} & -\frac{\sqrt{6}}{3} \\ 0 & -\frac{\sqrt{6}}{3} & \frac{\sqrt{3}}{3} \end{bmatrix}, \qquad \mathbf{P}_2\mathbf{P}_1\mathbf{A} = \begin{bmatrix} \sqrt{2} & 3\sqrt{2} \\ 0 & \sqrt{3} \\ 0 & 0 \end{bmatrix} = \tilde{\mathbf{R}}.$$

Since

$$\tilde{\mathbf{Q}} = \mathbf{P}_1^H\mathbf{P}_2^H = \begin{bmatrix} \frac{\sqrt{2}}{2} & \frac{-\sqrt{3}}{3} & \frac{-\sqrt{6}}{6} \\ 0 & \frac{\sqrt{3}}{3} & \frac{-\sqrt{6}}{3} \\ \frac{\sqrt{2}}{2} & \frac{\sqrt{3}}{3} & \frac{\sqrt{6}}{6} \end{bmatrix},$$

we obtain the same $\mathbf{Q}$ and $\mathbf{R}$ as in Ex. 9.26.

EXERCISE 9.77. Use Householder transformations to solve Ex. 9.27.

EXERCISE 9.78. Use Householder transformations to solve Ex. 9.28.

EXERCISE 9.79. If $\mathbf{x} = [\xi, \mathbf{y}]$, where $\|\mathbf{x}\| = 1$, ξ is a real scalar, and $\mathbf{y}$ is a $1 \times (n - 1)$ row vector, show that

$$\mathbf{Q} = \begin{bmatrix} \xi & \mathbf{y} \\ \mathbf{y}^H & -\mathbf{I} + (1 + \xi)^{-1}\mathbf{y}^H\mathbf{y} \end{bmatrix}$$

is both hermitian and unitary.

EXERCISE 9.80. Let $\mathbf{A}$ be a real symmetric matrix and $\mathbf{u}_1$ an arbitrary real vector. Set

$$\mathbf{v}_1 = \mathbf{A}\mathbf{u}_1, \qquad \mathbf{u}_2 = \mathbf{v}_1 - \alpha_1\mathbf{u}_1,$$

where α_1 is determined so as to make $\mathbf{u}_2$ and $\mathbf{u}_1$ orthogonal, which gives $\alpha_1 = (\mathbf{u}_1, \mathbf{v}_1)/(\mathbf{u}_1, \mathbf{u}_1)$. Next, form

$$\mathbf{v}_2 = \mathbf{A}\mathbf{u}_2, \qquad \mathbf{u}_3 = \mathbf{v}_2 - \alpha_2\mathbf{u}_2 - \beta_1\mathbf{u}_1,$$

where α_2, β_1 are determined so that $\mathbf{u}_3$ is orthogonal to $\mathbf{u}_2$ and $\mathbf{u}_1$. This gives

$\alpha_2 = (\mathbf{u}_2, \mathbf{v}_2)/(\mathbf{u}_2, \mathbf{u}_2)$, $\beta_1 = (\mathbf{u}_1, \mathbf{v}_2)/(\mathbf{u}_1, \mathbf{u}_1)$. Next, form

$$\mathbf{v}_3 = \mathbf{A}\mathbf{u}_3, \qquad \mathbf{u}_4 = \mathbf{v}_3 - \alpha_3\mathbf{u}_3 - \beta_2\mathbf{u}_2 - \gamma_1\mathbf{u}_1,$$

where α_3, β_2, γ_1 are determined so that $\mathbf{u}_4$ is orthogonal to $\mathbf{u}_3$, $\mathbf{u}_2$, $\mathbf{u}_1$. Prove that this gives $\gamma_1 = 0$. Show that at the general step we have

$$\mathbf{v}_r = \mathbf{A}\mathbf{u}_r, \qquad \mathbf{u}_{r+1} = \mathbf{v}_r - \alpha_r\mathbf{u}_r - \beta_{r-1}\mathbf{u}_{r-1},$$

where $\mathbf{u}_{r+1}$ is orthogonal to $\mathbf{u}_1, \ldots, \mathbf{u}_r$. Prove also that $\mathbf{u}_{n+1}$ must be identically zero. Let $\mathbf{x}$ be an eigenvector of $\mathbf{A}$ and set

$$\mathbf{x} = c_1\mathbf{u}_1 + \cdots + c_n\mathbf{u}_n.$$

Form $\mathbf{Ax}$ and express $\mathbf{A}\mathbf{u}_r = \mathbf{v}_r$ in terms of the $\mathbf{u}_s$ by the above formulas. Deduce that the eigenvalues of $\mathbf{A}$ coincide with those of the tridiagonal matrix

$$\begin{bmatrix} \alpha_1 & \beta_1 & 0 & \ldots & 0 \\ 1 & \alpha_2 & \beta_2 & \ldots & 0 \\ 0 & 1 & \alpha_3 & \ldots & 0 \\ & & \ldots & & \\ 0 & 0 & 0 & \ldots & \alpha_n \end{bmatrix}.$$

(The orthogonality conditions used to determine the α_i, β_i also minimize the $\|\mathbf{u}_i\|_2$. This is why the method, due to C. Lanczos, is also known as the *method of minimized iterations*.)

EXERCISE 9.81. Prove that Householder-transformation matrices are hermitian.

EXERCISE 9.82. Let $\mathbf{A}$ be an $n \times n$ matrix, and let $\mathbf{D}$ be a diagonal matrix with $d_{ii} = \epsilon^i$ for some $\epsilon \neq 0$. Describe $\mathbf{D}^{-1}\mathbf{AD}$ (see Ex. 9.83).

EXERCISE 9.83. Let $\mathbf{T} = \mathbf{P}^H\mathbf{AP}$ be a Schur canonical form for $\mathbf{A}$. Use Ex. 9.82 to show that there exists a nonsingular matrix $\mathbf{P}_0$ making the off-diagonal elements of the upper triangular matrix $\mathbf{P}_0^{-1}\mathbf{AP}_0$ as small as we choose.

EXERCISE 9.84. If the vector norm $\|\cdot\|$ induces the operator norm $\|\cdot\|$ on matrices and if $\mathbf{P}_0$ is nonsingular, show that the vector norm $|||\cdot|||$ defined by $|||\mathbf{x}||| = \|\mathbf{P}_0^{-1}\mathbf{x}\|$ induces the operator norm $|||\cdot|||$ with $|||\mathbf{A}||| = \|\mathbf{P}_0^{-1}\mathbf{AP}_0\|$. Use Ex. 9.83 to show that there exists some induced operator norm $|||\cdot|||$ for a matrix $\mathbf{A}$ for which $|||\mathbf{A}|||$ is arbitrarily close to $\rho(\mathbf{A}) = \max\{|\lambda_1|, |\lambda_2|, \ldots, |\lambda_n|\}$ where the λ_i are the eigenvalues of $\mathbf{A}$.

EXERCISE 9.85. Use the singular-value decomposition of $\mathbf{A}$ to show that $\|\mathbf{A}\|_2 = \sigma_1$, the largest singular value of $\mathbf{A}$.

EXERCISE 9.86. Let $\mathbf{A} = \mathbf{Q}_1\mathbf{R}_1 = \mathbf{Q}_2\mathbf{R}_2$ be two QR decompositions of an $m \times m$ matrix $\mathbf{A}$ of rank $n \leq m$. Deduce that $\mathbf{R}_1 = \mathbf{Q}\mathbf{R}_2$ with $\mathbf{Q} = \mathbf{Q}_1^H\mathbf{Q}_2$ and thus determine to what extent the QR decomposition is unique under these hypotheses.

EXERCISE 9.87. Use the QR decomposition to prove that every $n \times m$ matrix $\mathbf{A}$ with $n \leq m$ can be written as $\mathbf{A} = \mathbf{L}\mathbf{P}$ where $\mathbf{L}$ is lower-triangular and $\mathbf{P}$ has orthonormal rows.

EXERCISE 9.88. Explain why in **Key Theorem 9.6** as stated we must assume $n \leq m$; see Ex. 9.89.

EXERCISE 9.89. Provide the details in the following argument extending **Key Theorem 9.6** so as to write any $m \times n$ matrix $\mathbf{A}$ of rank k (with no restrictions on the relative size of m and n) in the form $\mathbf{A} = \mathbf{Q}\mathbf{R}$ where $\mathbf{Q}$ is $m \times k$ and has orthonormal columns while $\mathbf{R}$ is $k \times n$ and is "upper-triangular" in the sense that the (i, j) element r_{ij} of $\mathbf{R}$ equals zero if $i > j$. Our argument uses the construction beginning with and immediately following (9.17). Whenever $\mathbf{v}_i = \mathbf{a}_i - r_{1i}\mathbf{q}_1 - \cdots - r_{i-1,i}\mathbf{q}_{i-1} = \mathbf{0}$, we do not introduce a new vector $\mathbf{q}_i$ since we already are able to write $\mathbf{a}_1, \ldots, \mathbf{a}_i$ all as linear combinations of $\mathbf{q}_1, \ldots, \mathbf{q}_{i-1}$. By this device we guarantee that $r_{ii} \neq 0$ for all i and that any given sequence $\mathbf{a}_1, \ldots, \mathbf{a}_i$ of columns of $\mathbf{A}$ spans precisely the same space as does the set of columns of $\mathbf{Q}$ computed from them. Thus, since the dimension of the column space of $\mathbf{A}$ is k, the process must stop with k columns in $\mathbf{Q}$ as asserted.

EXERCISE 9.90. Find the QR decomposition in the sense of Ex. 9.89 for the matrix $\mathbf{A}$ in Ex. 9.27.

SOLUTION: As shown in Ex. 9.27, we find $r_{11} = \sqrt{2}$ and $r_{22} = \sqrt{2}$ since both the computed $\mathbf{v}_1 \neq \mathbf{0}$ and $\mathbf{v}_2 \neq \mathbf{0}$. Since we find $\mathbf{v}_3 = \mathbf{0}$, we do not add a vector $\mathbf{q}_3$. We thus obtain the decomposition at the very end of Ex. 9.27.

EXERCISE 9.91. Find the QR decomposition of

$$\mathbf{A} = \begin{bmatrix} 1 & 1 \\ 1 & 1 \\ 1 & 1 \end{bmatrix}$$

both in the sense of **Key Theorem 9.6** and in the sense of Ex. 9.89.

EXERCISE 9.92. Find the QR decomposition of

$$\mathbf{A} = \begin{bmatrix} 1 & 1 & 1 \\ 1 & 1 & 1 \end{bmatrix}$$

in the sense of Ex. 9.89.

EXERCISE 9.93. Find the QR decomposition of

$$\mathbf{A} = \begin{bmatrix} 1 & 0 & 1 \\ 2 & 1 & 4 \end{bmatrix}$$

in the sense of Ex. 9.89.

EXERCISE 9.94. In using the QR decomposition to solve least squares problems via (9.25), (9.26), we asserted that $\mathbf{R}$ is nonsingular if $\mathbf{A}^T\mathbf{A}$ is nonsingular. Prove that $\mathbf{R}$ is nonsingular if and only if $\mathbf{A}^T\mathbf{A}$ is nonsingular.

CHAPTER TEN

SIMILARITY TRANSFORMATIONS, EIGENSYSTEMS, AND APPLICATIONS

This chapter details precisely what can be accomplished in the simplification of matrices (or linear transformations) by means of the similarity transformations introduced in Section 8.3. **Key Theorem 10.2**, in particular, describes the Jordan canonical form to which every square matrix **A** can be reduced, while **Key Theorem 10.3** interprets this result in terms of the structure of the eigensystem of **A**. The remainder of the chapter presents some applications of matrices which can best be treated—in the light of the simplifications possible—by similarity (or unitary) transformations. **Key Theorems 10.6, 10.7,** and **10.8** analyze the behavior of powers of matrices (important, for example, in discrete modeling), while the behavior of matrix exponentials (arising, for example, in continuous modeling) is analyzed in **Key Theorem 10.15**.

10.1 INTRODUCTION

The important question of the structure of the eigensystem of a matrix **A** is equivalent to questions concerning decompositions of **A** and changes of basis that simplify the representation of the standard linear transformation defined by **A** (see Chapter 8 and Figure 8.4). In particular, Section 8.3 shows that changing to the basis of vectors $\mathbf{p}_1, \ldots, \mathbf{p}_n$ changes the representation from **A** to

$$\mathbf{A}' = \mathbf{P}^{-1}\mathbf{A}\mathbf{P}, \tag{10.1}$$

where $\mathbf{P} = [\mathbf{p}_1, \ldots, \mathbf{p}_n]$ defines the *similarity transformation* of (10.1), whose general properties were studied in Section 8.4. Chapter 9 studied similarity transformations based on the special class of unitary matrices **P** and showed that only normal matrices could be tremendously simplified (diagonalized) by such transformations unless we were prepared to use different bases for the domain space and range space, as we did in Section 9.5. Figure 10.1 shows all of this in terms of matrix transformations. Note that unitary transformations are special cases of generalized unitary transformations and also of similarity transformations.

Name	Transformation	Assumptions	Place Analyzed
Unitary transformation	$\mathbf{P}^H\mathbf{A}\mathbf{P}$	**A**: $n \times n$ **P**: $n \times n$ unitary	Sections 9.2, 9.3
Generalized unitary transformation	$\mathbf{U}^H\mathbf{A}\mathbf{V}$	**A**: $m \times n$ **U**: $m \times m$ unitary **V**: $n \times n$ unitary	Section 9.5
Similarity transformation	$\mathbf{P}^{-1}\mathbf{A}\mathbf{P}$	**A**: $n \times n$ **P**: $n \times n$ nonsingular	Not yet done (see Section 10.3).

Figure 10.1. Transformations to simplify matrices.

Alternatively, we can consider the linear transformation A from a vector space V ($\mathbb{R}^n$ or $\mathbb{C}^n$) to a vector space W ($\mathbb{R}^m$ or $\mathbb{C}^m$) and its matrix representation when we use bases B_V and B_W for V and W, respectively. From this viewpoint we can extend the picture in Figure 10.1 to that of Figure 10.2.

We need not consider two different general (not orthonormal) bases B_V and B_W (or, equivalently, matrix transformation $\mathbf{P}^{-1}\mathbf{A}\mathbf{Q}$) since we can

Name of Matrix Transformation	Vector Space Assumption	Basis Assumption
Unitary transformation	$V = W$	$B_V = B_W$ orthonormal
Generalized unitary transformation	V and W independent	B_V and B_W both orthonormal
Similarity transformation	$V = W$	$B_V = B_W$ general

Figure 10.2. Basis changes and equivalent matrix transformations.

already diagonalize **A** by a specialization of this type of transformation in which **P** and **Q** are unitary, that is, via the generalized unitary transformations of Section 9.5; let us, therefore, consider to what extent a matrix **A** can be simplified by a similarity transformation (10.1). We emphasized in Section 8.3 that this is, of course, equivalent to studying the structure of the eigensystem of the matrix **A**.

10.2 DEFECTIVE MATRICES: EXAMPLES AND THEIR MEANINGS

In Section 8.3 we gave one **key** result on the simplification of matrices by similarity transformations; in this section, for completeness, we restate **Key Theorem 8.3.**

- *REVIEW OF **KEY THEOREM 8.3.** An $n \times n$ matrix* **A** *can be reduced to a diagonal matrix* $\mathbf{\Lambda}$ *by a similarity transformation* $\mathbf{\Lambda} = \mathbf{P}^{-1}\mathbf{A}\mathbf{P}$ *if and only if* **A** *has a linearly independent set of n eigenvectors; these eigenvectors are the columns of* **P** *and their corresponding eigenvalues are the diagonal entries in* $\mathbf{\Lambda}$.

Thus the matrices left for us to study are precisely those which *fail* to have a linearly independent set of n eigenvectors.

DEFINITION 10.1. An $n \times n$ matrix **A** which fails to have a linearly independent set of n eigenvectors is said to be *defective*.

The study of defective matrices reduces to the study of the dimension of the subspace of eigenvectors associated with an eigenvalue λ_i of multiplicity m_i (see Definition 8.2). The matrix

$$\mathbf{A} = \begin{bmatrix} 2 & 1 \\ 0 & 2 \end{bmatrix},$$

in Ex. 8.7, shows that the dimension of this subspace can definitely be less than the multiplicity m_i of λ_i, although the dimension is at least unity because of **Key Theorem 8.2.** We now give some examples to show that this dimension can be *any* integer between unity and the multiplicity m_i; by Ex. 8.81, it cannot exceed m_i.

To illustrate what can happen, we will consider the matrix

$$\mathbf{A} = \begin{bmatrix} 2 & a & 0 & 0 \\ 0 & 2 & b & 0 \\ 0 & 0 & 2 & 0 \\ 0 & 0 & 0 & 3 \end{bmatrix} \tag{10.2}$$

for various values of a and b. Independent of the values of a and b,

$$\det(\mathbf{A} - \lambda\mathbf{I}) = (2 - \lambda)^3(3 - \lambda)$$

so that $\lambda = 2$ is an eigenvalue of multiplicity three $(=m_1)$ while $\lambda = 3$ is a simple eigenvalue. Independent of a and b, the unit column vector $\mathbf{e}_4$ spans the subspace of eigenvectors associated with $\lambda = 3$, as we see by solving $(\mathbf{A} - 3\mathbf{I})\mathbf{x} = \mathbf{0}$.

EXERCISE 10.1. Suppose that $a = b = 0$ in (10.2). Discuss the eigensystem.

SOLUTION: The unit column vectors $\mathbf{e}_1$, $\mathbf{e}_2$, and $\mathbf{e}_3$ are all eigenvectors associated with $\lambda = 2$, the subspace of eigenvectors associated with $\lambda = 2$ has dimension three $(= m_1)$, and the set of all four eigenvectors $\mathbf{e}_1, \mathbf{e}_2, \mathbf{e}_3, \mathbf{e}_4$ is linearly independent.

EXERCISE 10.2. Suppose that $a = 0$ and $b = 1$ in (10.2). Discuss the eigensystem.

SOLUTION: We want to find the eigenvectors associated with $\lambda = 2$; we try to solve $\mathbf{Ax} - 2\mathbf{x} = \mathbf{0}$ via the augmented matrix

$$\begin{bmatrix} 0 & 0 & 0 & 0 & 0 \\ 0 & 0 & 1 & 0 & 0 \\ 0 & 0 & 0 & 0 & 0 \\ 0 & 0 & 0 & 1 & 0 \end{bmatrix}$$

which gives $x_3 = x_4 = 0$ while x_1 and x_2 are arbitrary. Thus the set of vectors $\mathbf{e}_1$, $\mathbf{e}_2$ spans the subspace of eigenvectors associated with $\lambda = 2$ and we cannot find the desired third associated eigenvector since the subspace of eigenvectors has only dimension two $(= m_1 - 1)$.

EXERCISE 10.3. Suppose, finally, that $a = b = 1$ in (10.2). Describe the eigensystem.

SOLUTION: Again, to find eigenvectors associated with $\lambda_1 = 2$ we solve $(\mathbf{A} - 2\mathbf{I})\mathbf{x} = \mathbf{0}$ and examine the augmented matrix

$$\begin{bmatrix} 0 & 1 & 0 & 0 & 0 \\ 0 & 0 & 1 & 0 & 0 \\ 0 & 0 & 0 & 0 & 0 \\ 0 & 0 & 0 & 1 & 0 \end{bmatrix}$$

which yields $x_2 = x_3 = x_4 = 0$ while x_1 is arbitrary. Thus the set formed by the single vector $\mathbf{e}_1$ spans the subspace of eigenvectors associated with $\lambda_1 = 2$, so that this subspace has only dimension unity $(= m_1 - 2)$.

The three preceding examples show that the dimension of the subspace of eigenvectors associated with an eigenvalue λ_i of multiplicity m_i can take any integer value from unity to m_i. The question is how to conveniently describe this sort of information and the equivalent information on simplification by similarity transformations. The key to this description will be the notion of an *invariant subspace* (see Definition 8.2).

Recall from the introduction in Section 8.1 that the discovery of low-dimensional invariant subspaces V_0 of a linear transformation A defined on a vector space V seemed a helpful tool for simplifying the study of a highly complex application described by A. The best result occurred when V_0 was one-dimensional, so that V_0 was spanned by a single eigenvector. Ideally we should like to break up the entire space V into a set of essentially disjoint (nonoverlapping) invariant subspaces each of as low dimension as possible. From this viewpoint **Key Theorem 8.3**, reviewed above, describes precisely when the n-dimensional space V can be broken up into n invariant subspaces each of dimension unity. We now reexamine our two examples of Exs. 10.2 and 10.3 to describe the decomposition of $\mathbb{R}^4$ into small invariant subspaces.

EXERCISE 10.4. Suppose that $a = 0$ and $b = 1$ in (10.2); see Ex. 10.2. Describe the invariant subspaces of **A**.

SOLUTION: While $\mathbf{e}_4$ is an eigenvector for $\lambda = 3$ and $\mathbf{e}_1$ and $\mathbf{e}_2$ are both eigenvectors for $\lambda = 2$, the vector $\mathbf{e}_3$ is not an eigenvector at all. We note that

$$\mathbf{A}\mathbf{e}_3 = \begin{bmatrix} 0 \\ 1 \\ 2 \\ 0 \end{bmatrix} = 2\mathbf{e}_3 + \mathbf{e}_2;$$

since also $\mathbf{A}\mathbf{e}_2 = 2\mathbf{e}_2$, we conclude that

$$A(\alpha\mathbf{e}_2 + \beta\mathbf{e}_3) = (2\alpha + 1)\mathbf{e}_2 + (2\beta)\mathbf{e}_3$$

which says that the space spanned by $\{\mathbf{e}_2, \mathbf{e}_3\}$ is an invariant subspace. Thus we have $\mathbb{R}^4$ broken into the three invariant subspaces spanned, respectively, by $\{\mathbf{e}_1\}$, $\{\mathbf{e}_2, \mathbf{e}_3\}$ and $\{\mathbf{e}_4\}$. Moreover, our basis $\{\mathbf{e}_2, \mathbf{e}_3\}$ for the second subspace is very simple in that it consists of an eigenvector $\mathbf{e}_2$ satisfying $\mathbf{A}\mathbf{e}_2 = 2\mathbf{e}_2$ and some sort of generalization $\mathbf{e}_3$ of an eigenvector since $\mathbf{e}_3$ satisfies $\mathbf{A}\mathbf{e}_3 = 2\mathbf{e}_3 + \mathbf{e}_2$.

EXERCISE 10.5. Suppose that $a = b = 1$ in (10.2); see Ex. 10.3. Describe the invariant subspaces of **A**.

SOLUTION: While $\mathbf{e}_4$ is an eigenvector for $\lambda = 3$ and $\mathbf{e}_1$ is an eigenvector for $\lambda = 2$, neither $\mathbf{e}_2$ nor $\mathbf{e}_3$ is an eigenvector at all. We note that

$$\mathbf{A}\mathbf{e}_2 = \begin{bmatrix} 1 \\ 2 \\ 0 \\ 0 \end{bmatrix} = 2\mathbf{e}_2 + \mathbf{e}_1$$

so that $\{\mathbf{e}_1, \mathbf{e}_2\}$ spans an invariant subspace, as of course does $\{\mathbf{e}_4\}$. Unfortunately the remaining vector $\mathbf{e}_3$ satisfies

$$\mathbf{A}\mathbf{e}_3 = \begin{bmatrix} 0 \\ 1 \\ 2 \\ 0 \end{bmatrix} = 2\mathbf{e}_3 + \mathbf{e}_2$$

which involves $\mathbf{e}_2$ again so that $\{\mathbf{e}_3\}$ does not yield us a basis for another invariant subspace. Of course we do have

$$A(\alpha\mathbf{e}_1 + \beta\mathbf{e}_2 + \gamma\mathbf{e}_3) = (2\alpha + \beta)\mathbf{e}_1 + (2\beta + \gamma)\mathbf{e}_2 + (2\gamma)\mathbf{e}_3$$

which says that the space spanned by $\{\mathbf{e}_1, \mathbf{e}_2, \mathbf{e}_3\}$ *is* an invariant subspace. Thus again we have $\mathbb{R}^4$ broken up into the invariant subspaces spanned, respectively, by $\{\mathbf{e}_1, \mathbf{e}_2, \mathbf{e}_3\}$ and $\{\mathbf{e}_4\}$. Moreover, our basis $\{\mathbf{e}_1, \mathbf{e}_2, \mathbf{e}_3\}$ is very simple in that it consists of an eigenvector $\mathbf{e}_1$ satisfying $\mathbf{A}\mathbf{e}_1 = 2\mathbf{e}_1$ and two generalized eigenvectors $\mathbf{e}_2$ and $\mathbf{e}_3$ satisfying $\mathbf{A}\mathbf{e}_2 = 2\mathbf{e}_2 + \mathbf{e}_1$ and $\mathbf{A}\mathbf{e}_3 = 2\mathbf{e}_3 + \mathbf{e}_2$.

These examples show that for the multiple eigenvalue $\lambda = 2$ of multiplicity $m_1 = 3$ we always find three $(= m_1)$ vectors, at least one of which is an eigenvector, while the others are eigenvectors or generalized eigenvectors in some sense, which we can combine somehow into bases for invariant subspaces of as low dimension as possible. More generally, of course, *we cannot always expect the relevant vectors to be as simple as* $\mathbf{e}_1$, $\mathbf{e}_2$, $\mathbf{e}_3$, and $\mathbf{e}_4$; these special vectors arose because of the very simple form of (10.2), which we chose in order to make the calculations simple in the examples. In other cases we will require general vectors $\mathbf{p}_1, \ldots, \mathbf{p}_n$ to replace $\mathbf{e}_1$, $\mathbf{e}_2$, $\mathbf{e}_3$, and $\mathbf{e}_4$ and we will find that general matrices $\mathbf{A}$ can be transformed only into such simple forms as (10.2) by appropriate similarity transformations, as we now see.

Consider the matrix

$$\mathbf{A}' = \begin{bmatrix} 2 - a - b & -\frac{1}{2} - a - \frac{3b}{2} & a + b & \frac{1}{2} + a + \frac{b}{2} \\ b & \frac{5}{2} + \frac{3b}{2} & -b & -\frac{1}{2} - \frac{b}{2} \\ -a & \frac{1}{2} - a & 2 + a & -\frac{1}{2} + a \\ b & -\frac{1}{2} + \frac{3b}{2} & -b & \frac{5}{2} - \frac{b}{2} \end{bmatrix} \tag{10.3}$$

for various values of a and b. If we define the 4×4 nonsingular matrix $\mathbf{P} = [\mathbf{p}_1, \mathbf{p}_2, \mathbf{p}_3, \mathbf{p}_4]$ with

$$\mathbf{p}_1 = \begin{bmatrix} 1 \\ 0 \\ 1 \\ 0 \end{bmatrix}, \quad \mathbf{p}_2 = \begin{bmatrix} -1 \\ 1 \\ 0 \\ 1 \end{bmatrix}, \quad \mathbf{p}_3 = \begin{bmatrix} 1 \\ 1 \\ 1 \\ 1 \end{bmatrix}. \quad \mathbf{p}_4 = \begin{bmatrix} -1 \\ 1 \\ 1 \\ -1 \end{bmatrix},$$

we find

$$\mathbf{A} = \mathbf{P}^{-1}\mathbf{A}'\mathbf{P} = \begin{bmatrix} 2 & a & 0 & 0 \\ 0 & 2 & b & 0 \\ 0 & 0 & 2 & 0 \\ 0 & 0 & 0 & 3 \end{bmatrix}; \tag{10.4}$$

note that this is precisely the matrix $\mathbf{A}$ of (10.2).

EXERCISE 10.6. Suppose $a = b = 0$ in (10.3). Describe the eigensystem and invariant subspaces of $\mathbf{A}'$.

SOLUTION: By (10.4) $\mathbf{A}'$ is similar to a diagonal matrix. The eigenvalues of $\mathbf{A}'$ are $\lambda_1 = 2$ with a multiplicity $m_1 = 3$ and $\lambda_2 = 3$ with a multiplicity $m_2 = 1$. Moreover, there is a linearly independent set of three ($= m_1$) eigenvectors $\mathbf{p}_1, \mathbf{p}_2, \mathbf{p}_3$ associated with λ_1, while there is one ($= m_2$) eigenvector for λ_2. We can break up $\mathbb{R}^4$ into invariant subspaces spanned, respectively, by $\{\mathbf{p}_1\}, \{\mathbf{p}_2\}, \{\mathbf{p}_3\}, \{\mathbf{p}_4\}$.

EXERCISE 10.7. Suppose that $a = 0$ and $b = 1$ in (10.3). Describe the eigensystem and invariant subspaces of $\mathbf{A}'$.

SOLUTION: By (10.4), $\mathbf{A}'$ is similar to the matrix $\mathbf{A}$ in Exs. 10.2 and 10.4. The eigenvalues of $\mathbf{A}'$ are $\lambda_1 = 2$ with a multiplicity $m_1 = 3$ and $\lambda_2 = 3$ with a multiplicity $m_2 = 1$. Those two earlier exercises show a linearly independent set of two eigenvectors $\{\mathbf{p}_1, \mathbf{p}_2\}$ associated with $\lambda_1 = 2$ and one eigenvector $\mathbf{p}_4$ associated with $\lambda_2 = 3$. We can break up $\mathbb{R}^4$ into invariant subspaces spanned, respectively, by $\{\mathbf{p}_1\}, \{\mathbf{p}_2, \mathbf{p}_3\}$, and $\{\mathbf{p}_4\}$. Regarding $\{\mathbf{p}_2, \mathbf{p}_3\}$, we have $\mathbf{A}\mathbf{p}_2 = 2\mathbf{p}_2$, $\mathbf{A}\mathbf{p}_3 = 2\mathbf{p}_3 + \mathbf{p}_2$.

EXERCISE 10.8. Suppose that $a = b = 1$ in (10.3); see Ex. 10.5. Describe the eigensystem and invariant subspaces of $\mathbf{A}'$.

EXERCISE 10.9. Suppose that $a = 1$ and $b = 0$ in (10.3). Describe the eigensystem and invariant subspaces of $\mathbf{A}'$.

EXERCISE 10.10. Describe the eigensystem and invariant subspaces of the matrix

$$\mathbf{A} = \begin{bmatrix} 7 & 1 & 2 \\ -1 & 7 & 0 \\ 1 & -1 & 6 \end{bmatrix}.$$

The simplest form to which $\mathbf{A}'$ in (10.3) can be reduced has been described in (10.4) and clarified in Exs. 10.6 and 10.7; we have in addition interpreted this simplification in terms of the structure of the eigensystem and invariant subspaces. We follow this same pattern in the next two sections, first stating how we can simplify a matrix (Section 10.3) and then interpreting this in the language of eigensystems and invariant subspaces (Section 10.4).

10.3 JORDAN FORM OF AN $n \times n$ MATRIX

In this section we give the barest outline of the reduction, using similarity transformations, of a general $n \times n$ square matrix $\mathbf{A}$ to an especially simple form, generalizing (10.2). The first step in this process is to reduce $\mathbf{A}$ to an upper-triangular form as in the Schur form (see Section 9.2); since we are no longer restricting ourselves to unitary matrices we may use any similarity transformation to accomplish this step.

EXERCISE 10.11. Reduce the matrix

$$\mathbf{A} = \begin{bmatrix} 5 & 4 & 3 \\ -1 & 0 & -3 \\ 1 & -2 & 1 \end{bmatrix}$$

to upper-triangular form by a similarity transformation.

SOLUTION: The eigenvalues are found to be $\lambda_1 = -2$, $\lambda_2 = \lambda_3 = 4$, with only two eigenvectors. We choose any eigenvector as the first column of a matrix, say $\mathbf{x}_1 = [1 \quad -1 \quad -1]^T$, corresponding to $\lambda_1 = -2$. We choose the second and third columns so that the resulting matrix $\mathbf{Q}$ is nonsingular. We can always choose these to be unit vectors, and we find, for instance,

$$\mathbf{Q} = \begin{bmatrix} 1 & 0 & 0 \\ -1 & 1 & 0 \\ -1 & 0 & 1 \end{bmatrix}, \qquad \mathbf{Q}^{-1} = \begin{bmatrix} 1 & 0 & 0 \\ 1 & 1 & 0 \\ 1 & 0 & 1 \end{bmatrix}, \qquad \mathbf{Q}^{-1}\mathbf{A}\mathbf{Q} = \begin{bmatrix} -2 & 4 & 3 \\ 0 & 4 & 0 \\ 0 & 2 & 4 \end{bmatrix}.$$

The submatrix

$$\begin{bmatrix} 4 & 0 \\ 2 & 4 \end{bmatrix}$$

has a repeated eigenvalue of 4, as it must have. There is only one corresponding eigenvector $\mathbf{z}_1 = [0 \quad 1]^T$ which we use as the first column of a 2×2 matrix $\mathbf{R}$, picking any suitable second column that makes the matrix nonsingular. This is then used to form

$$\mathbf{R} = \begin{bmatrix} 0 & 1 \\ 1 & 0 \end{bmatrix}, \quad \mathbf{S} = \begin{bmatrix} 1 & 0 & 0 \\ 0 & 0 & 1 \\ 0 & 1 & 0 \end{bmatrix}, \quad \mathbf{S}^{-1} = \begin{bmatrix} 1 & 0 & 0 \\ 0 & 0 & 1 \\ 0 & 1 & 0 \end{bmatrix},$$

$$\mathbf{S}^{-1}(\mathbf{Q}^{-1}\mathbf{A}\mathbf{Q})\mathbf{S} = \begin{bmatrix} -2 & 3 & 4 \\ 0 & 4 & 2 \\ 0 & 0 & 4 \end{bmatrix}.$$

This is upper triangular, as required.

The next step on the road to the Jordan form, and the only step we will rigorously justify, is to reduce the Schur upper-triangular form to a *block-diagonal upper-triangular* form.

THEOREM 10.1. Suppose that an upper-triangular matrix $\mathbf{U}$ *has the following structure:*

$$\mathbf{U} = \begin{bmatrix} \mathbf{U}_{11} & \mathbf{U}_{12} & \dots & \mathbf{U}_{1s} \\ 0 & \mathbf{U}_{22} & \dots & \mathbf{U}_{2s} \\ & & \dots & \\ 0 & 0 & \dots & \mathbf{U}_{ss} \end{bmatrix} \tag{10.5}$$

where each $\mathbf{U}_{ii}$ *is an upper-triangular matrix, all of whose diagonal elements are equal to* λ_i. *Also,* $\lambda_1, \dots, \lambda_s$ *are distinct. Then there exists a nonsingular matrix* $\mathbf{R}$ *such that*

$$\mathbf{R}^{-1}\mathbf{U}\mathbf{R} = \begin{bmatrix} \mathbf{V}_1 & 0 & \dots & 0 \\ 0 & \mathbf{V}_2 & \dots & 0 \\ & & \dots & \\ 0 & 0 & \dots & \mathbf{V}_s \end{bmatrix} \tag{10.6}$$

where $\mathbf{V}_i$ *is an upper-triangular matrix of the same size as* $\mathbf{U}_{ii}$ *and all of whose diagonal elements are equal to* λ_i.

Proof: Let $\mathbf{Q} = \mathbf{I} + \mathbf{K}$, where $\mathbf{K}$ is a matrix all of whose elements are zero except the (p, q) element which equals k, where $p < q$. It is easily verified that $\mathbf{Q}^{-1} = \mathbf{I} - \mathbf{K}$. The similarity transformation $\mathbf{Q}^{-1}\mathbf{U}\mathbf{Q}$, where $\mathbf{U}$ is upper triangular, transforms u_{pq} $(p < q)$ into

$$u_{pq} - k(u_{qq} - u_{pp}), \tag{10.7}$$

and otherwise modifies only elements in $\mathbf{U}$ in the pth row to the right of u_{pq}, and in the qth column above u_{pq}. If $u_{pp} \neq u_{qq}$, we can choose $k = u_{pq}/(u_{qq} - u_{pp})$ and then (10.7) shows that the (p, q) element in $\mathbf{Q}^{-1}\mathbf{U}\mathbf{Q}$ is zero. Returning to the notation (10.5), the elements u_{pq} for which the corresponding diagonal elements u_{pp}, u_{qq} are distinct lie in rectangular blocks. A sequence of similarity transformations of the above type can be carried out to reduce these u_{pq} to zero, row by row, starting at the left of the bottom row of the lowest rectangular block. This will give a matrix of the form (10.6).

EXERCISE 10.12. Reduce the upper-triangular matrix derived in Ex. 10.11 to the form (10.6) by means of a similarity transformation.

SOLUTION: The proof of the theorem tells us that we first reduce the (1, 2) element to zero. Equation (10.7) gives $k = \frac{1}{2}$ and we find that $\mathbf{Q}^{-1}\mathbf{U}\mathbf{Q}$ equals

$$\begin{bmatrix} 1 & -\frac{1}{2} & 0 \\ 0 & 1 & 0 \\ 0 & 0 & 1 \end{bmatrix} \begin{bmatrix} -2 & 3 & 4 \\ 0 & 4 & 2 \\ 0 & 0 & 4 \end{bmatrix} \begin{bmatrix} 1 & \frac{1}{2} & 0 \\ 0 & 1 & 0 \\ 0 & 0 & 1 \end{bmatrix} = \begin{bmatrix} -2 & 0 & 3 \\ 0 & 4 & 2 \\ 0 & 0 & 4 \end{bmatrix}.$$

The (1, 3) element is reduced to zero in a similar way. However, the (2, 3) element *cannot* be reduced to zero since the (2, 2) and (3, 3) diagonal elements are equal.

We note that the form (10.6) is really very instructive. If we start with a matrix $\mathbf{A}$, reduce it to Schur upper-triangular form $\mathbf{T} = \mathbf{P}_0^{-1}\mathbf{A}\mathbf{P}_0$, and then reduce $\mathbf{T}$ to block-diagonal, upper-triangular form $\mathbf{D} = \mathbf{R}^{-1}\mathbf{T}\mathbf{R}$, overall we have $\mathbf{D} = (\mathbf{P}_0\mathbf{R})^{-1}\mathbf{A}(\mathbf{P}_0\mathbf{R})$ so that, with respect to the basis formed from the columns of $\mathbf{P}_0\mathbf{R}$, the linear transformation A defined by $A(\mathbf{x}) = \mathbf{A}\mathbf{x}$ is represented by the simplified matrix $\mathbf{D}$. The block-diagonal nature of $\mathbf{D}$ is revealing since, if we partition vectors $\mathbf{y}$ as

$$\mathbf{y} = \begin{bmatrix} \mathbf{y}_1 \\ \cdot \\ \cdot \\ \cdot \\ \mathbf{y}_s \end{bmatrix}$$

compatibly with (10.6), then $\mathbf{D}$ (or A) transforms any vector of the form

$$\begin{bmatrix} \mathbf{y}_1 \\ \mathbf{0} \\ \cdot \\ \cdot \\ \cdot \\ \mathbf{0} \end{bmatrix}$$

into one of exactly the same form, and similarly, in general, for

$$\begin{bmatrix} \mathbf{0} \\ \cdot \\ \cdot \\ \cdot \\ \mathbf{0} \\ \mathbf{y}_i \\ \mathbf{0} \\ \cdot \\ \cdot \\ \cdot \\ \mathbf{0} \end{bmatrix}.$$

Thus the transformation A we needed to study in our application has been decomposed into s different "small transformations" which can be studied more easily. This is the real meaning of (10.6). Actually, as we shall see, each "small transformation" described by $\mathbf{V}_i$ in (10.6) also has form special enough for us to learn more about it and, in fact, to simplify its structure still further by another similarity transformation.

EXERCISE 10.13. Describe the form (10.6) derived in Ex. 10.12 for the matrix **A** of Ex. 10.11 in terms of splitting up a linear transformation.

Each matrix $\mathbf{V}_i$ in (10.6) is of the form

$$\mathbf{V} = \begin{bmatrix} \lambda & v_{12} & \dots & v_{1q} \\ 0 & \lambda & \dots & v_{2q} \\ & & \dots & \\ 0 & 0 & \dots & \lambda \end{bmatrix}$$

which is clearly quite special. By some technically rather complex arguments which we shall omit, it is possible to show that any such matrix **V** can be transformed by a similarity transformation into the form

$$\mathbf{S}^{-1}\mathbf{V}\mathbf{S} = \begin{bmatrix} \mathbf{J}_1 & \mathbf{0} & \dots & \mathbf{0} \\ \mathbf{0} & \mathbf{J}_2 & \dots & \mathbf{0} \\ & & \dots & \\ \mathbf{0} & \mathbf{0} & \dots & \mathbf{J}_q \end{bmatrix}$$

whose diagonal elements all equal λ and where each $\mathbf{J}_i$ is a *Jordan block* in the following sense.

DEFINITION 10.2. A *Jordan block* is a square matrix whose elements are zero except for those on the principal diagonal, which are all equal, and those on the first superdiagonal, which are all equal to unity; thus,

$$\mathbf{J} = \begin{bmatrix} \lambda & 1 & 0 & \dots & 0 \\ 0 & \lambda & 1 & \dots & 0 \\ & & \dots & & \\ 0 & 0 & 0 & \dots & \lambda \end{bmatrix}.$$

By applying such transformations to the form (10.6), we finally obtain our **key** result.

- *KEY THEOREM 10.2 (Jordan Canonical Form). If* $\mathbf{A}$ *is a general square* $n \times n$ *matrix then a nonsingular matrix* $\mathbf{Q}$ *exists such that*

$$\mathbf{Q}^{-1}\mathbf{A}\mathbf{Q} = \begin{bmatrix} \mathbf{J}_1 & \mathbf{0} & \dots & \mathbf{0} \\ \mathbf{0} & \mathbf{J}_2 & \dots & \mathbf{0} \\ & \dots & & \\ \mathbf{0} & \mathbf{0} & \dots & \mathbf{J}_k \end{bmatrix} = \mathbf{J}, \tag{10.8}$$

where the $\mathbf{J}_i$ *are* $n_i \times n_i$ *Jordan blocks. The same eigenvalues may occur in different blocks, but the number of distinct blocks corresponding to a given eigenvalue is equal to the number of eigenvectors corresponding to that eigenvalue and forming an independent set. The number* k *and the set of numbers* $n_1, \dots, n_k$ *are uniquely determined by* $\mathbf{A}$.

Although we have not *proved* the existence of the form (10.8) or shown how to find it, the mere knowledge of its existence in fact makes it easy to see how to set about finding it. Multiplying both sides of (10.8) by $\mathbf{Q}$, we obtain

$$\mathbf{AQ} = \mathbf{QJ}.$$

This replaces the equation $\mathbf{AP} = \mathbf{P\Lambda}$, which is the equation for diagonalization that we have met so often before. If the columns of $\mathbf{Q}$ are denoted by $\mathbf{q}_i$,

$$\mathbf{Q} = [\mathbf{q}_1, \dots, \mathbf{q}_n],$$

the form of $\mathbf{J}$ tells us that the equation $\mathbf{AQ} = \mathbf{QJ}$ separates into equations of the form

$$\mathbf{A}\mathbf{q}_i = \lambda_i \mathbf{q}_i + \nu_i \mathbf{q}_{i-1}$$

where ν_i may be either 0 or 1, depending on $\mathbf{J}$. More precisely, let us recall that the Jordan block $\mathbf{J}_i$ is $n_i \times n_i$; then the columns of $\mathbf{Q}$ which are affected by $\mathbf{J}_i$ in the equation $\mathbf{AQ} = \mathbf{QJ}$ are just those numbered $n_1 + n_2 + \cdots + n_{i-1} + 1$ to $n_1 + n_2 + \cdots + n_i$. For convenience we denote these columns of $\mathbf{Q}$ by $\mathbf{v}_{i,1}, \ldots, \mathbf{v}_{i,n_i}$, so that

$$\mathbf{v}_{i,j} = \mathbf{q}_{n_1 + \cdots + n_{i-1} + j}.$$

It then follows from $\mathbf{AQ} = \mathbf{QJ}$ that

$$\mathbf{Av}_{i,1} = \lambda_i \mathbf{v}_{i,1}, \ \mathbf{Av}_{i,j+1} = \lambda_i \mathbf{v}_{i,j+1} + \mathbf{v}_{i,j} \quad (\text{for } j = 1, 2, \ldots, n_i - 1), \tag{10.9}$$

an equation to which we will return in Section 10.4. For the moment, (10.9) can serve as the basis for actually finding the matrix $\mathbf{Q}$ required in **Key Theorem 10.2**; we first find the eigenvectors $\mathbf{v}_{i,1}$ and then the remaining vectors $\mathbf{v}_{i,j}$ for $j > 1$.

To illustrate the point (and the previous theory) by a specific example, consider

$$\mathbf{Q}^{-1}\mathbf{AQ} = \begin{bmatrix} \mathbf{J}_1 & \mathbf{0} & \mathbf{0} \\ \mathbf{0} & \mathbf{J}_2 & \mathbf{0} \\ \mathbf{0} & \mathbf{0} & \mathbf{J}_3 \end{bmatrix} = \mathbf{J},$$

where

$$\mathbf{J}_1 = \begin{bmatrix} \alpha & 1 & 0 \\ 0 & \alpha & 1 \\ 0 & 0 & \alpha \end{bmatrix}, \qquad \mathbf{J}_2 = [\alpha], \qquad \mathbf{J}_3 = \begin{bmatrix} \beta & 1 \\ 0 & \beta \end{bmatrix}.$$

We first of all find the eigenvectors of $\mathbf{J}$. Suppose that $\mathbf{y}$ is an eigenvector corresponding to the eigenvalue α. We assume $\alpha \neq \beta$. Then

$$(\mathbf{J} - \alpha\mathbf{I})\mathbf{y} = \begin{bmatrix} 0 & 1 & 0 & 0 & 0 & 0 \\ 0 & 0 & 1 & 0 & 0 & 0 \\ 0 & 0 & 0 & 0 & 0 & 0 \\ 0 & 0 & 0 & 0 & 0 & 0 \\ 0 & 0 & 0 & 0 & \beta - \alpha & 1 \\ 0 & 0 & 0 & 0 & 0 & \beta - \alpha \end{bmatrix} \begin{bmatrix} y_1 \\ y_2 \\ y_3 \\ y_4 \\ y_5 \\ y_6 \end{bmatrix} = \mathbf{0}.$$

These equations have an independent set of precisely two solutions which can be taken to be the unit vectors $\mathbf{e}_1$ and $\mathbf{e}_4$. In a similar way, there is only one eigenvector corresponding to the second eigenvalue β, namely $\mathbf{e}_5$. Hence, $\mathbf{J}$ (and therefore $\mathbf{A}$) has an independent set of precisely three eigenvectors.

The equations (10.9) for this example are

$$\begin{aligned} (\mathbf{A} - \alpha\mathbf{I})\mathbf{q}_1 &= \mathbf{0} \\ (\mathbf{A} - \alpha\mathbf{I})\mathbf{q}_2 &= \mathbf{q}_1, \\ (\mathbf{A} - \alpha\mathbf{I})\mathbf{q}_3 &= \mathbf{q}_2, \end{aligned} \qquad (\mathbf{A} - \alpha\mathbf{I})\mathbf{q}_4 = \mathbf{0}, \qquad \begin{aligned} (\mathbf{A} - \beta\mathbf{I})\mathbf{q}_5 &= \mathbf{0}, \\ (\mathbf{A} - \beta\mathbf{I})\mathbf{q}_6 &= \mathbf{q}_5. \end{aligned}$$

If $\mathbf{u}$ is an eigenvector of $\mathbf{J}$ corresponding to an eigenvalue λ, so that $\mathbf{Ju} = \lambda\mathbf{u}$, this means that

$$\mathbf{Q}^{-1}\mathbf{AQu} = \lambda\mathbf{u} \quad \text{or} \quad \mathbf{A}(\mathbf{Qu}) = \lambda(\mathbf{Qu}).$$

Since $\mathbf{e}_1, \mathbf{e}_4, \mathbf{e}_5$ are eigenvectors of $\mathbf{J}$, this means that $\mathbf{q}_1, \mathbf{q}_4, \mathbf{q}_5$ are eigenvectors of $\mathbf{A}$. The other vectors, namely $\mathbf{q}_2, \mathbf{q}_3, \mathbf{q}_6$, are known as *generalized* eigenvectors.

Suppose now that we are given the matrix $\mathbf{A}$ in this example, and we know nothing beforehand about its eigenvalues and eigenvectors. If we compute its eigenvalues, we will find α of multiplicity 4 and β of multiplicity 2. If we compute the eigenvectors, we will find two eigenvectors $\mathbf{x}_1, \mathbf{x}_2$ corresponding to α, and one eigenvector $\mathbf{x}_3$ corresponding to β. If we try to solve

$$(\mathbf{A} - \beta\mathbf{I})\mathbf{x} = \mathbf{x}_3$$

we will find a solution, say $\mathbf{x}_4$, which is a generalized eigenvector. The multiplicity of β is 2, and we have found an eigenvector $\mathbf{x}_3$ and a generalized eigenvector $\mathbf{x}_4$, so we need not consider the eigenvalue β further. (The vectors $\mathbf{x}_3, \mathbf{x}_4$ correspond to $\mathbf{q}_5, \mathbf{q}_6$ in our previous notation.) To find the two generalized eigenvectors $\mathbf{q}_2$ and $\mathbf{q}_3$ corresponding to α, we first seek $\mathbf{q}_2$ by trying to solve

$$(\mathbf{A} - \alpha\mathbf{I})\mathbf{x} = \mathbf{q}_1 \equiv a\mathbf{x}_1 + b\mathbf{x}_2,$$

where a and b are arbitrary initially, but we immediately find that a solution exists for only one ratio of a to b. The corresponding vector $a\mathbf{x}_1 + b\mathbf{x}_2$ is $\mathbf{q}_1$ in our previous notation, while the solution $\mathbf{x}$ is $\mathbf{q}_2$. The remaining generalized eigenvector $\mathbf{q}_3$ can only come from $(\mathbf{A} - \alpha\mathbf{I})\mathbf{x} = \mathbf{q}_2$. The eigenvector $\mathbf{q}_4$ is any combination of $\mathbf{x}_1$ and $\mathbf{x}_2$ independent of $\mathbf{q}_1$. The multiplicity of α is 4 and we have found two eigenvectors $\mathbf{q}_1$ and $\mathbf{q}_4$ and two generalized eigenvectors $\mathbf{q}_2$ and $\mathbf{q}_3$.

The following examples should clarify the procedure. The point is that in order to carry out the reduction to the Jordan canonical form in practice we do not follow the sequence of steps outlined in the course of the theorems, since it is much simpler to operate directly on the original matrix. The theorems tell us the kind of thing we are looking for, and we go after this directly.

EXERCISE 10.14. Reduce the following matrix to Jordan canonical form:

$$\mathbf{A} = \begin{bmatrix} 5 & 4 & 3 \\ -1 & 0 & -3 \\ 1 & -2 & 1 \end{bmatrix}.$$

SOLUTION: A straightforward calculation gives eigenvalues $\lambda_1 = -2$, $\lambda_2 = \lambda_3 = 4$, with only two eigenvectors,

$$\mathbf{x}_1 = \begin{bmatrix} 1 \\ -1 \\ -1 \end{bmatrix} \quad (\text{for } \lambda_1) \qquad \mathbf{x}_2 = \begin{bmatrix} 1 \\ -1 \\ 1 \end{bmatrix} \quad (\text{for } \lambda_2).$$

We obtain a third vector by solving $(\mathbf{A} - 4\mathbf{I})\mathbf{x}_3 = \mathbf{x}_2$. The solution of this equation is, of course, arbitrary to within a multiple of $\mathbf{x}_2$, and we obtain

$$\mathbf{x}_3 = \begin{bmatrix} 0 \\ 1 \\ -1 \end{bmatrix} + k\mathbf{x}_2,$$

where k is an arbitrary constant, the choice of which is not important for present purposes. If we take $k = 0$ and therefore set

$$\mathbf{Q} = \begin{bmatrix} 1 & 1 & 0 \\ -1 & -1 & 1 \\ -1 & 1 & -1 \end{bmatrix}, \qquad \mathbf{Q}^{-1} = \frac{1}{2}\begin{bmatrix} 0 & -1 & -1 \\ 2 & 1 & 1 \\ 2 & 2 & 0 \end{bmatrix},$$

we find

$$\mathbf{Q}^{-1}\mathbf{A}\mathbf{Q} = \begin{bmatrix} -2 & 0 & 0 \\ 0 & 4 & 1 \\ 0 & 0 & 4 \end{bmatrix} = \mathbf{J}.$$

It is clear that the matrix $\mathbf{Q}$ is not unique.

EXERCISE 10.15. Reduce the following matrix to Jordan canonical form.

$$\mathbf{A} = \begin{bmatrix} 2 & 2 & -1 \\ -1 & -1 & 1 \\ -1 & -2 & 2 \end{bmatrix}.$$

SOLUTION: We find $\lambda_1 = \lambda_2 = \lambda_3 = 1$, with two eigenvectors, say

$$\mathbf{x}'_1 = \begin{bmatrix} 1 \\ 0 \\ 1 \end{bmatrix}, \qquad \mathbf{x}'_2 = \begin{bmatrix} 0 \\ 1 \\ 2 \end{bmatrix}.$$

It is clear from the general theory that, since there are two eigenvectors, there are two Jordan canonical boxes $\mathbf{J}_1$ and $\mathbf{J}_2$. The third vector $\mathbf{q}_2$ required to form $\mathbf{Q}$ is obtained by solving

$$(\mathbf{A} - \mathbf{I})\mathbf{x}'_3 = \mathbf{q}_1 \equiv \alpha\mathbf{x}'_1 + \beta\mathbf{x}'_2$$

where α, β are constants chosen so that these equations have a nonzero solution. Trying to solve the equations we find that we must set $\beta = -\alpha$. A suitable solution $\mathbf{q}_2 = \mathbf{x}'_3$ then has $x_1 = \alpha$, $x_2 = x_3 = 0$. We must take one of the columns

$\mathbf{q}_1$ of $\mathbf{Q}$ to be $\alpha\mathbf{x}_1' + \beta\mathbf{x}_2'$ and the other column $\mathbf{q}_3$ to be an eigenvector independent of $\mathbf{q}_1$. The value of α is arbitrary. If we take $\alpha = 1$, we might choose

$$\mathbf{Q} = [\mathbf{x}_1' - \mathbf{x}_2', \mathbf{x}_3', \mathbf{x}_1'] = \begin{bmatrix} 1 & 1 & 1 \\ -1 & 0 & 0 \\ -1 & 0 & 1 \end{bmatrix}$$

which gives

$$\mathbf{Q}^{-1}\mathbf{A}\mathbf{Q} = \begin{bmatrix} 1 & 1 & 0 \\ 0 & 1 & 0 \\ 0 & 0 & 1 \end{bmatrix}.$$

In Sections 10.5, 10.6, and 10.7 we will examine some applications which use the specific Jordan form (10.8), although even in these cases the less specific block-diagonal form (10.6) would suffice. The reader need not be concerned over not having seen the detailed derivation of the Jordan form, since it is more important to understand what it *means* (see also Section 10.4) and how it can be *applied* than how it is derived.

EXERCISE 10.16. Reduce the matrix to upper-triangular form:

$$\mathbf{A} = \begin{bmatrix} 0 & 1 & 0 \\ 0 & 0 & 1 \\ 6 & -1 & -4 \end{bmatrix}.$$

EXERCISE 10.17. Reduce the matrix to Jordan canonical form:

$$\mathbf{A} = \begin{bmatrix} 0 & 1 & 0 \\ 0 & 0 & 1 \\ 6 & -1 & -4 \end{bmatrix}$$

EXERCISE 10.18. Reduce the upper-triangular matrix

$$\begin{bmatrix} 1 & -3 & 2 \\ 0 & 1 & -1 \\ 0 & 0 & 2 \end{bmatrix}$$

to block-diagonal upper-triangular form (10.6).

EXERCISE 10.19. Reduce the upper-triangular matrix

$$\mathbf{A} = \begin{bmatrix} 1 & -2 & 3 & -4 \\ 0 & 1 & -1 & -2 \\ 0 & 0 & 1 & 4 \\ 0 & 0 & 0 & -3 \end{bmatrix}$$

to block-diagonal upper-triangular form (10.6).

EXERCISE 10.20. Reduce the matrix **A** in Ex. 9.5 to Jordan canonical form.

EXERCISE 10.21. If **J** is a 2×2 Jordan block with zero on the diagonal, show that $\mathbf{J}^2 = \mathbf{0}$. If **J** is a 3×3 Jordan block with zero on the diagonal, show that $\mathbf{J}^3 = \mathbf{0}$.

EXERCISE 10.22. If **J** is an $n \times n$ Jordan block with zero on the diagonal, show that $\mathbf{J}^n = \mathbf{0}$.

EXERCISE 10.23. Use Ex. 10.22 to show that if **J** is a Jordan block with an arbitrary number λ_0 on the diagonal and f is the characteristic polynomial of **J**, then $f(\mathbf{J}) = \mathbf{0}$.

EXERCISE 10.24. Use Exs. 10.22 and 10.23 to show that if **A** is a square matrix and f is its characteristic polynomial, then $f(\mathbf{A}) = \mathbf{0}$.

EXERCISE 10.25. Much as in the material in Section 2.2, suppose that four businesses share a market with the fraction held by the jth business in the ith month being the jth component of the vector $\mathbf{x}_i$ in $\mathbb{R}^4$. Suppose that the development of the market is described by $\mathbf{x}_{i+1} = \mathbf{A}\mathbf{x}_i$, where

$$\mathbf{A} = \begin{bmatrix} 1.0 & 0.4 & 0.2 & 0.1 \\ 0 & 0.6 & 0.3 & 0.2 \\ 0 & 0 & 0.5 & 0.2 \\ 0 & 0 & 0 & 0.5 \end{bmatrix}.$$

Reduce **A** to block-diagonal upper-triangular form **B** as in (10.6) and explain what the representation **B** reveals about the market process.

EXERCISE 10.26. Prove that two matrices that can be diagonalized are similar to each other if and only if they have the same characteristic equation.

EXERCISE 10.27. If **A** is a real 2×2 matrix with complex conjugate eigenvalues $\alpha \pm i\beta$, prove that a *real* nonsingular matrix **T** exists such that

$$\mathbf{T}^{-1}\mathbf{A}\mathbf{T} = \begin{bmatrix} \alpha & \beta \\ -\beta & \alpha \end{bmatrix}.$$

10.4 JORDAN FORM AND EIGENSYSTEMS OF GENERAL MATRICES

In Exs. 10.6 and 10.7, we interpreted the simplification of the matrix $\mathbf{A}'$ in (10.3) to the Jordan form in (10.4) by describing the structure of the eigensystem and invariant subspaces of $\mathbf{A}'$. We can interpret the **key** result in **Key**

Theorem 10.2 on the Jordan form in similar terms. This gives the following **key** result.

- KEY THEOREM 10.3. *Let* $\mathbf{A}$ *be a general square* $n \times n$ *matrix and let* k *be the maximum possible number of eigenvectors of* $\mathbf{A}$ *in a linearly independent set. Then there exist positive integers* $n_1, n_2, \ldots, n_k$ *and a linearly independent set of* n *vectors* $\mathbf{v}_{1,1}, \mathbf{v}_{1,2}, \ldots, \mathbf{v}_{1,n_1}, \mathbf{v}_{2,1}, \ldots, \mathbf{v}_{2,n_2}, \ldots, \mathbf{v}_{k,1}, \mathbf{v}_{k,2}, \ldots, \mathbf{v}_{k,n_k}$ *such that:*

  (i) $\{\mathbf{v}_{1,1}, \mathbf{v}_{2,1}, \ldots, \mathbf{v}_{k,1}\}$ *is a linearly independent set of (the maximum number* k *of) eigenvectors of* $\mathbf{A}$;

  (ii) *for each* i *with* $1 \le i \le k$ *the set of* n_i *vectors* $\mathbf{v}_{i,1}, \mathbf{v}_{i,2}, \ldots, \mathbf{v}_{i,n_i}$ *spans an invariant subspace* V_i *of* $\mathbf{A}$ *which cannot be further decomposed into two invariant subspaces having in common only the vector* $\mathbf{0}$;

  (iii) *if the eigenvalue associated with the eigenvector* $\mathbf{v}_{i,1}$ *is denoted by* λ_i, *then the generalized eigenvectors* $\mathbf{v}_{i,2}, \ldots, \mathbf{v}_{i,n_i}$ *satisfy* $\mathbf{A}\mathbf{v}_{i,j+1} = \lambda_i \mathbf{v}_{i,j+1} + \mathbf{v}_{i,j}$ *for* $1 \le j \le n_i - 1$.

Proof: If we let $\mathbf{v}_{i,j}$ be defined as the column numbered $n_1 + n_2 + \cdots + n_{i-1} + j$ in the matrix $\mathbf{Q}$ of **Key Theorem 10.2,** then as we saw before it follows that (10.9) holds, which immediately proves (i) and (iii). The fact from (10.9) that $\mathbf{A}\mathbf{v}_{i,j}$ only involves the vectors $\mathbf{v}_{i,r}$ for some values of r says that any linear combination of $\mathbf{v}_{i,1}, \ldots, \mathbf{v}_{i,n_i}$ will be mapped by $\mathbf{A}$ into another such linear combination, meaning that we have an invariant subspace. The impossibility of further decomposing this invariant subspace follows from the uniqueness of k and of the set of numbers $n_1, \ldots, n_k$. This completes the proof of (ii) and hence of the theorem.

Because the problem is complicated, the result on the structure of eigensystems is complicated. Figure 10.3, depicting the functions of the $\mathbf{v}_{i,j}$, may help clarify the problem. We also reconsider our earlier examples of Exs. 10.1–10.5 in this notation.

	Bases for Invariant Subspaces $V_1, \ldots, V_k$			
	For subspace V_1 ↓	For subspace V_2 ↓	. . .	For subspace V_k ↓
Eigensystem →	$\mathbf{v}_{1,1}$ (for λ_1)	$\mathbf{v}_{2,1}$ (for λ_2)	. . .	$\mathbf{v}_{k,1}$ (for λ_k)
	$\mathbf{v}_{1,2}$	$\mathbf{v}_{2,2}$	. . .	$\mathbf{v}_{k,2}$
	⋮	⋮	. . .	⋮
	$\mathbf{v}_{1,n_1}$	$\mathbf{v}_{2,n_2}$		$\mathbf{v}_{k,n_k}$

Figure 10.3. Meaning of $\mathbf{v}_{i,j}$ in **Key Theorem 10.3.**

Consider the 4×4 matrix of Ex. 10.1. We have $\lambda_1 = 2$, $m_1 = 3$, $\lambda_2 = 3$, $m_2 = 1$, $k = 4$, and $n_1 = n_2 = n_3 = n_4 = 1$ so that there are four eigenvectors and four invariant subspaces, namely those spanned by each of the eigenvectors $\mathbf{v}_{1,1} = \mathbf{e}_1$, $\mathbf{v}_{2,1} = \mathbf{e}_2$, $\mathbf{v}_{3,1} = \mathbf{e}_3$, and $\mathbf{v}_{4,1} = \mathbf{e}_4$.

Consider the 4×4 matrix of Exs. 10.2 and 10.4. We have $\lambda_1 = 2$, $m_1 = 3$, $\lambda_2 = 3$, $m_2 = 1$, $k = 3$, and $n_1 = 1$, $n_2 = 2$, $n_3 = 1$, so that there are three eigenvectors $\mathbf{v}_{1,1} = \mathbf{e}_1$, $\mathbf{v}_{2,1} = \mathbf{e}_2$, and $\mathbf{v}_{3,1} = \mathbf{e}_4$, while the three invariant subspaces are those spanned by $\{\mathbf{v}_{1,1} = \mathbf{e}_1\}$, $\{\mathbf{v}_{2,1} = \mathbf{e}_2, \mathbf{v}_{2,2} = \mathbf{e}_3\}$, and $\{\mathbf{v}_{3,1} = \mathbf{e}_4\}$.

EXERCISE 10.28. Interpret Exs. 10.3 and 10.5 in the notation of **Key Theorem 10.3.**

EXERCISE 10.29. In the notation of **Key Theorem 10.3** describe the eigenvectors and invariant subspaces of

$$\mathbf{A} = \begin{bmatrix} 2 & 2 & -2 \\ 0 & 2 & -3 \\ 0 & 0 & -1 \end{bmatrix}.$$

EXERCISE 10.30. Interpret the results of Exs. 10.6, 10.7, and 10.8 in the notation of **Key Theorem 10.3.**

It is possible to interpret the Jordan form of **Key Theorem 10.2** in a slightly different fashion, one that is often useful; we will do this in Theorem 10.5, but first we must develop some tools. In Ex. 8.2 we used the fact that, when $\mathbf{A} - \lambda\mathbf{I}$ is singular, not only is there a nonzero *column* vector (eigenvector) $\mathbf{x}$ with $(\mathbf{A} - \lambda\mathbf{I})\mathbf{x} = \mathbf{0}$, but there is also a nonzero *row* vector $\mathbf{y}^H$ for which $\mathbf{y}^H(\mathbf{A} - \lambda\mathbf{I}) = \mathbf{0}$. Such a vector $\mathbf{y}$ is called a *left eigenvector* of $\mathbf{A}$. We might be tempted to call λ a left eigenvalue also, but since the left eigenvector $\mathbf{y}^H$ exists if and only if $\mathbf{A} - \lambda\mathbf{I}$ is singular, that is, if and only if λ is an eigenvalue of $\mathbf{A}$, we see that this concept would coincide with our usual one. From $\mathbf{y}^H(\mathbf{A} - \lambda\mathbf{I}) = \mathbf{0}$ it follows by taking transposes that $(\mathbf{A}^T - \lambda\mathbf{I})\bar{\mathbf{y}} = \mathbf{0}$, where $\bar{\mathbf{y}}$ is the complex conjugate of $\mathbf{y}$, so that eigenvalues of $\mathbf{A}$ are eigenvalues of $\mathbf{A}^T$ while left eigenvectors of $\mathbf{A}$ are complex-conjugates of eigenvectors of $\mathbf{A}^T$. We restate all of what we have demonstrated more formally.

DEFINITION 10.3. Let the $n \times n$ matrix $\mathbf{A}$ have as its distinct eigenvalues the numbers $\lambda_1, \ldots, \lambda_s$. A nonzero vector $\mathbf{y}$ for which $\mathbf{y}^H\mathbf{A} = \lambda_i\mathbf{y}^H$ for some λ_i is said to be a *left eigenvector* of $\mathbf{A}$ associated with the eigenvalue λ_i; if required for clarity, we call the usual eigenvectors of Definition 8.1 *right eigenvectors.*

THEOREM 10.4. The eigenvalues of $\mathbf{A}$ *and of* $\mathbf{A}^T$ *are the same, while the eigenvalues of* $\mathbf{A}^H$ *are the complex-conjugates of those of* $\mathbf{A}$*. The left eigenvectors of* $\mathbf{A}$ *associated with an eigenvalue* λ *of* $\mathbf{A}$ *are precisely the complex conjugates of the right eigenvectors of* $\mathbf{A}^T$ *associated with* λ.

EXERCISE 10.31. Verify Theorem 10.4 when

$$\mathbf{A} = \begin{bmatrix} 0.8 & 0.1 & 0.1 \\ 0.2 & 0.7 & 0.1 \\ 0.1 & 0.3 & 0.6 \end{bmatrix}.$$

SOLUTION: $\mathbf{A}^T = \mathbf{A}^H$ is just the matrix of (2.10) whose eigenvalues $\lambda_1, \lambda_2, \lambda_3$ and right eigenvectors $\mathbf{v}_1, \mathbf{v}_2, \mathbf{v}_3$ were found in Section 8.1 to be $\lambda_1 = 0.5$, $\lambda_2 = 0.6$, $\lambda_3 = 1.0$,

$$\mathbf{v}_1 = \begin{bmatrix} 1 \\ -2 \\ 1 \end{bmatrix}, \quad \mathbf{v}_2 = \begin{bmatrix} 1 \\ -1 \\ 0 \end{bmatrix}, \quad \mathbf{v}_3 = \begin{bmatrix} 0.45 \\ 0.35 \\ 0.20 \end{bmatrix}.$$

To verify that $\bar{\mathbf{v}}_i = \mathbf{v}_i$ is a left eigenvector of $\mathbf{A}$ associated with λ_i, we merely compute

$$\mathbf{v}_1^H\mathbf{A} = \mathbf{v}_1^T\mathbf{A} = [0.5 \quad -1.0 \quad 0.5] = 0.5[1 \quad -2 \quad 1] = \lambda_1\mathbf{v}_1^H$$

as required, and similarly $\mathbf{v}_2^H\mathbf{A} = \lambda_2\mathbf{v}_2^H$, $\mathbf{v}_3^H\mathbf{A} = \lambda_3\mathbf{v}_3^H$.

EXERCISE 10.32. Verify Theorem 10.4 when the matrix $\mathbf{A}$ of Ex. 10.31 is replaced by its transpose (see Ex. 8.2).

EXERCISE 10.33. Suppose that the vector $\mathbf{x}_i$ represents the state of some system at time i and that the system evolves according to

$$\mathbf{x}_{i+1} = \mathbf{A}\mathbf{x}_i$$

for a given square matrix $\mathbf{A}$. Suppose that $\mathbf{y}$ is a left eigenvector of $\mathbf{A}$ associated with the eigenvalue λ. Prove that

$$\mathbf{y}^H\mathbf{x}_{i+1} = \lambda\mathbf{y}^H\mathbf{x}_i = \lambda^{i+1}\mathbf{y}^H\mathbf{x}_0.$$

We have seen in Exs. 8.2 and 10.33 that left eigenvectors can be useful in understanding simple aspects of the behavior of a complicated system. They are equally important, however, because of a fundamental orthogonality relation between the left eigenvectors and the right eigenvectors of $\mathbf{A}$; since by **Key Theorem 9.3** we cannot generally expect the right eigenvectors themselves to form an orthogonal set for a general matrix, this *bi-orthogonality* will have to replace the orthogonality we could exploit for normal matrices. To develop these results, we consider again the Jordan form

$$\mathbf{Q}^{-1}\mathbf{A}\mathbf{Q} = \mathbf{J}. \qquad (10.10)$$

When we rewrote this as

$$\mathbf{AQ} = \mathbf{QJ}$$

and considered what this said about the columns of **Q**, we obtained (10.9) and our information in **Key Theorem 10.3** on the eigensystem of **A**. This time we rewrite $\mathbf{AQ} = \mathbf{QJ}$ as

$$\mathbf{Q}^{-1}\mathbf{A} = \mathbf{J}\mathbf{Q}^{-1} \tag{10.11}$$

and see what this says about the *rows* of $\mathbf{Q}^{-1}$.

EXERCISE 10.34. Interpret (10.11) for the matrices in Ex. 10.14.

SOLUTION: In this case (10.11) is

$$\begin{bmatrix} 0 & -0.5 & -0.5 \\ 1 & 0.5 & 0.5 \\ 1 & 1 & 0 \end{bmatrix} \mathbf{A} = \begin{bmatrix} -2 & 0 & 0 \\ 0 & 4 & 1 \\ 0 & 0 & 4 \end{bmatrix} \begin{bmatrix} 0 & -0.5 & -0.5 \\ 1 & 0.5 & 0.5 \\ 1 & 1 & 0 \end{bmatrix}.$$

If we simply write

$$\mathbf{y}_1^H = [0 \quad -0.5 \quad -0.5],\ \mathbf{y}_2^H = [1 \quad 0.5 \quad 0.5],\ \mathbf{y}_3^H = [1 \quad 1 \quad 0]$$

for the rows of $\mathbf{Q}^{-1}$, then we obtain

$$\mathbf{y}_1^H\mathbf{A} = -2\mathbf{y}_1^H = \lambda_1\mathbf{y}_1^H,$$
$$\mathbf{y}_2^H\mathbf{A} = 4\mathbf{y}_2^H + \mathbf{y}_3^H = \lambda_2\mathbf{y}_2^H + \mathbf{y}_3^H,$$
$$\mathbf{y}_3^H\mathbf{A} = 4\mathbf{y}_3^H = \lambda_2\mathbf{y}_3^H.$$

Thus $\mathbf{y}_1$ and $\mathbf{y}_3$ are left eigenvectors while $\mathbf{y}_2$ is a *generalized left eigenvector.*

EXERCISE 10.35. As in Ex. 10.34, interpret (10.11) for the matrices in Ex. 10.15.

It should now be clear that (10.11) can be interpreted as a statement about left eigenvectors of **A**, and in fact that there is a left eigenvector of **A** for each Jordan block $\mathbf{J}_i$ in the Jordan form; more precisely, in the notation of **Key Theorem 10.2** there are k left eigenvectors $\mathbf{y}_1, \ldots, \mathbf{y}_k$, where, $\mathbf{y}_i^H$, for $1 \leq i \leq k$, is just that row of $\mathbf{Q}^{-1}$ numbered $n_1 + \cdots + n_i$. Since columns of **Q** are right eigenvectors or generalized right eigenvectors of **A** and since the relation $\mathbf{Q}^{-1}\mathbf{Q} = \mathbf{I}$ states that the ith row of $\mathbf{Q}^{-1}$ is orthogonal to the jth column of **Q** if $i \neq j$, the following result is immediate.

THEOREM 10.5. Let **A**, k, $n_1, \ldots, n_k$, *and the vectors* $\mathbf{v}_{i,j}$ *be as in Theorem 10.3. Then the maximum number of left eigenvectors of* **A** *in a linearly independent set is* k, *the same as for right eigenvectors, and there exists a linearly independent set* *of left eigenvectors* $\mathbf{y}_1, \ldots, \mathbf{y}_k$ *such that the left eigenvector* $\mathbf{y}_i$ *and the right eigenvector* $\mathbf{v}_{i,1}$ *correspond to the same eigenvalue, and such that*

$$\begin{aligned} (\mathbf{y}_i, \mathbf{v}_{i,n_i}) &= 1 \qquad (\textit{for } i = 1, \ldots, k), \\ (\mathbf{y}_i, \mathbf{v}_{i,j}) &= 0 \qquad (\textit{for } i = 1, \ldots, k \quad \textit{and} \quad j = 1, \ldots, n_i - 1), \\ (\mathbf{y}_i, \mathbf{v}_{j,l}) &= 0 \qquad (\textit{for } j \neq i \quad \textit{and} \quad l = 1, \ldots, n_j). \end{aligned}$$

In particular, $(\mathbf{y}_i, \mathbf{v}_{i,1}) = 0$ *if* $n_i > 1$ *and* $(\mathbf{y}_i, \mathbf{v}_{i,1}) = 1$ *if* $n_i = 1$.

EXERCISE 10.36. Verify Theorem 10.5 for the 3×3 matrix $\mathbf{A}$ of (2.10).

SOLUTION: In Section 8.1 we found the eigenvalues of $\mathbf{A}$ to be $\lambda_1 = 0.5$, $\lambda_2 = 0.6$, and $\lambda_3 = 1.0$, with eigenvectors

$$\mathbf{v}_{1,1} = \begin{bmatrix} 1 \\ -2 \\ 1 \end{bmatrix}, \qquad \mathbf{v}_{2,1} = \begin{bmatrix} 1 \\ -1 \\ 0 \end{bmatrix}, \qquad \mathbf{v}_{3,1} = \begin{bmatrix} 0.45 \\ 0.35 \\ 0.20 \end{bmatrix}$$

so that $n_1 = n_2 = n_3 = 1$ and $k = 3$. The corresponding left eigenvectors were found in Ex. 8.2 to be

$$\mathbf{y}_1' = \begin{bmatrix} 1 \\ 1 \\ -4 \end{bmatrix}, \qquad \mathbf{y}_2' = \begin{bmatrix} 3 \\ -1 \\ -5 \end{bmatrix}, \qquad \mathbf{y}_3' = \begin{bmatrix} 1 \\ 1 \\ 1 \end{bmatrix}.$$

We readily compute $(\mathbf{y}_1', \mathbf{v}_{2,1}) = 1(1) + 1(-1) + (-4)(0) = 0$ as asserted, and similarly $(\mathbf{y}_1', \mathbf{v}_{3,1}) = (\mathbf{y}_2', \mathbf{v}_{1,1}) = (\mathbf{y}_2', \mathbf{v}_{3,1}) = (\mathbf{y}_3', \mathbf{v}_{1,1}) = (\mathbf{y}_3', \mathbf{v}_{2,1}) = 0$. We also compute $(\mathbf{y}_1', \mathbf{v}_{1,1}) = 1(1) + 1(-2) + (-4)(1) = -5$, so that if we replace $\mathbf{y}_1'$ by $\mathbf{y}_1 = -\frac{1}{5}\mathbf{y}_1'$ and similarly $\mathbf{y}_2'$ by $\mathbf{y}_2 = \frac{1}{4}\mathbf{y}_2'$ and set $\mathbf{y}_3 = \mathbf{y}_3'$ then the theorem is verified for this case.

EXERCISE 10.37. Verify Theorem 10.5 for the 4×4 matrix of Ex. 10.1.

EXERCISE 10.38. Verify Theorem 10.5 for the 4×4 matrix of Ex. 10.2.

EXERCISE 10.39. Verify Theorem 10.5 for the 4×4 matrix of Ex. 10.3.

EXERCISE 10.40. As we did in Theorem 9.4 for normal matrices, prove the following strong version of the Fredholm Alternative of **Key Theorem 3.8:** For the n linear equations $(\mathbf{A} - \lambda\mathbf{I})\mathbf{x} = \mathbf{b}$ in n unknowns, precisely one of the following alternatives holds:

(a) λ is not an eigenvalue of $\mathbf{A}$ and the unique solution $\mathbf{x}$ can be written

$$\mathbf{x} = \sum_{i=1}^{k} \sum_{j=1}^{n_i} \alpha_{ij}\mathbf{v}_{i,j}$$

with the $\mathbf{v}_{i,j}$ as in **Key Theorem 10.3,** and where

$$\alpha_{i,n_i} = \beta_{i,n_i}/(\lambda_i - \lambda) \qquad (\text{for } 1 \leq i \leq k),$$
$$\alpha_{i,j} = (\beta_{ij} - \alpha_{i,j+1})/(\lambda_i - \lambda) \qquad (\text{for } 1 \leq i \leq k \quad \text{and} \quad n_i - 1 \geq j \geq 1),$$
$$\mathbf{b} = \sum_{i=1}^{k} \sum_{j=1}^{n_i} \beta_{ij}\mathbf{v}_{i,j}$$

(b) λ is an eigenvalue λ_i of **A** and there is a solution **x** if and only if **b** is orthogonal to all the left eigenvectors of **A** associated with the eigenvalue λ_i. If **x** is any such solution then infinitely many solutions can be obtained by adding to **x** any linear combination of the right eigenvectors of **A** associated with λ_i.

By setting $\lambda = 0$, we see that a restatement of this theorem is: *either* $\mathbf{Ax} = \mathbf{b}$ *is uniquely solvable for every* **b** *or it is solvable for a particular* **b** *if and only if* **b** *is orthogonal to all vectors* **y** *solving* $\mathbf{y}^H\mathbf{A} = \mathbf{0}$.

EXERCISE 10.41. Find necessary and sufficient conditions on b_1 and b_2 in order that

$$6x_1 + 2x_2 = b_1$$
$$3x_1 + x_2 = b_2$$

be solvable.

SOLUTION: Clearly the relevant matrix

$$\mathbf{A} = \begin{bmatrix} 6 & 2 \\ 3 & 1 \end{bmatrix}$$

is singular, and the only vectors **y** for which $\mathbf{y}^H\mathbf{A} = \mathbf{0}$ are multiples of

$$\mathbf{y} = \begin{bmatrix} 1 \\ -2 \end{bmatrix}.$$

Thus, by Ex. 10.40, the condition is $b_1 - 2b_2 = 0$.

EXERCISE 10.42. Find necessary and sufficient conditions on b_1, b_2, b_3, and b_4 in order that

$$\begin{aligned} 2x_1 + 2x_2 + 4x_3 + x_4 &= b_1 \\ -3x_1 - 3x_2 - 6x_3 + 2x_4 &= b_2 \\ -6x_1 - 6x_2 - 12x_3 + 3x_4 &= b_3 \\ x_1 + x_2 + 2x_3 + x_4 &= b_4 \end{aligned}$$

be solvable; see Ex. 10.40.

10.5 DISCRETE SYSTEM EVOLUTION AND POWERS OF MATRICES

In Section 2.2 we considered how the shares of a market controlled by three dairies evolved from month to month; in (2.7) we were able to describe this evolution via

$$\mathbf{x}_{r+1} = \mathbf{A}\mathbf{x}_r$$

where the three components of $\mathbf{x}_r$ represent the fractions of the market held by the dairies at the end of the rth month, and where $\mathbf{A}$, the transition matrix, describes the transformation of the market shares during one month. Similarly, in Section 2.3 we considered how the population of chickens and foxes evolved with time; in (2.26) we were able to describe this evolution via

$$\mathbf{x}_{i+1} = \mathbf{A}\mathbf{x}_i$$

where the two components of $\mathbf{x}_i$ represent the numbers of foxes and chickens at the ith time instant and where $\mathbf{A}$ describes the transformation of these populations from one time period to the next.

More generally, it is very common in applied problems for a vector $\mathbf{x}_i$ in $\mathbb{R}^n$, say, to represent the state of some complex system at the ith point in time and for the $n \times n$ matrix $\mathbf{A}$ to represent the transformation of the state during one time period, so that again

$$\mathbf{x}_{i+1} = \mathbf{A}\mathbf{x}_i.$$

The examples of Sections 2.2 and 2.3 were of interest in learning how the system behaves as time passes: Does the market tend towards an equilibrium distribution? Do the populations explode? Since, of course,

$$\mathbf{x}_i = \mathbf{A}^i\mathbf{x}_0, \tag{10.12}$$

we are really asking for the behavior of the powers of a fixed matrix $\mathbf{A}$. Fortunately, it is easy to understand this via the Jordan canonical form

$$\mathbf{A} = \mathbf{Q}\mathbf{J}\mathbf{Q}^{-1}$$

of (10.8) in **Key Theorem 10.2.** Since

$$\mathbf{A}^i = \mathbf{Q}\mathbf{J}^i\mathbf{Q}^{-1} \tag{10.13}$$

(see Theorem 8.7), we need to study the behavior of $\mathbf{J}^i$; since

$$\mathbf{J} = \begin{bmatrix} \mathbf{J}_1 & \mathbf{0} & \dots & \mathbf{0} \\ \mathbf{0} & \mathbf{J}_2 & \dots & \mathbf{0} \\ & & \dots & \\ \mathbf{0} & \mathbf{0} & \dots & \mathbf{J}_k \end{bmatrix}, \tag{10.14}$$

we have

$$\mathbf{J}^i = \begin{bmatrix} \mathbf{J}_1^i & \mathbf{0} & \dots & \mathbf{0} \\ \mathbf{0} & \mathbf{J}_2^i & \dots & \mathbf{0} \\ & & \dots & \\ \mathbf{0} & \mathbf{0} & \dots & \mathbf{J}_k^i \end{bmatrix}, \tag{10.15}$$

and hence we need only study the behavior of $\mathbf{J}_p^i$ where $\mathbf{J}_p$ is a Jordan block as described in Definition 10.2. For simplicity we first consider the case in which the Jordan form $\mathbf{J}$ in (10.13) is strictly diagonal; by **Key Theorems 10.2** and **8.3** we know that $\mathbf{J}$ is diagonal precisely when $\mathbf{A}$ is nondefective, that is, when $\mathbf{A}$ has a linearly independent set of n eigenvectors. *In this special case of nondefective* $\mathbf{A}$ we therefore replace $\mathbf{J}$ in (10.13) by

$$\mathbf{J} = \mathbf{\Lambda} = \begin{bmatrix} \lambda_1 & 0 & \dots & 0 \\ 0 & \lambda_2 & \dots & 0 \\ & & \dots & \\ 0 & 0 & \dots & \lambda_n \end{bmatrix},$$

hence

$$\mathbf{\Lambda}^i = \begin{bmatrix} \lambda_1^i & 0 & \dots & 0 \\ 0 & \lambda_2^i & \dots & 0 \\ & & \dots & \\ 0 & 0 & \dots & \lambda_n^i \end{bmatrix}. \tag{10.16}$$

When λ is just a number, it is clear that, as i tends to infinity,

1. λ^i tends to zero precisely when $|\lambda| < 1$,
2. $|\lambda^i|$ tends to infinity precisely when $|\lambda| > 1$,
3. λ^i is bounded (that is, for some constant c we have $|\lambda| \leq c$ for all i) precisely when $|\lambda| \leq 1$.

(10.17)

We want to extend (10.17) so as to apply to $\mathbf{\Lambda}^i$ in (10.16), rather than just to λ^i, because we could then analyze the behavior of $\mathbf{A}^i = \mathbf{Q}\mathbf{\Lambda}^i\mathbf{Q}^{-1}$. First we need to understand the notions of "converges," "tends to," and "is bounded" for sequences of matrices, a matter we discussed somewhat in the paragraph preceding Ex. 5.16 on page 167.

DEFINITION 10.4. A sequence $\{\mathbf{A}_i\}$ of $m \times n$ matrices is said to *converge to* or *tend to* the matrix $\mathbf{B}$ if and only if the sequence of (j, k) elements of $\mathbf{A}_i$ converges to the (j, k) element of $\mathbf{B}$ for every j and k with $1 \leq j \leq m$ and $1 \leq k \leq n$. The sequence $\{\mathbf{A}_i\}$ is said to be *bounded* if and only if the sequence of (j, k) elements of $\mathbf{A}_i$ is bounded for every j and k with $1 \leq j \leq m$ and $1 \leq k \leq n$.

EXERCISE 10.43. Show that the sequence of matrices

$$\mathbf{A}_i = \begin{bmatrix} 1 + \frac{1}{i} & 2 - \left(\frac{1}{2}\right)^i & 3 \\ -1 + \frac{1}{i^2} & -4 & \left(\frac{1}{3}\right)^i \end{bmatrix}$$

converges to

$$\mathbf{B} = \begin{bmatrix} 1 & 2 & 3 \\ -1 & -4 & 0 \end{bmatrix}.$$

EXERCISE 10.44. Show that the sequence of matrices

$$\mathbf{A}_i = \begin{bmatrix} \frac{1}{i} & (-1)^i \\ 10 & 2 - (-1)^i \end{bmatrix}$$

is bounded but does not converge to any matrix $\mathbf{B}$.

EXERCISE 10.45. Show that the sequence of matrices

$$\mathbf{A}_i = \begin{bmatrix} -\frac{1}{i} & 3 \\ (-1)^i & 1 - i \\ 2^i & \left(\frac{1}{2}\right)^i \end{bmatrix}$$

is not bounded.

Our definitions of convergence and boundedness are essentially "element-wise" definitions. These can be shown to be equivalent to conditions that deal more with the matrices as independent entities by using our old concept of a *norm* of a matrix (see Definitions 5.5 and 5.6 in Section 5.3); we state the following **key** characterization of Definition 10.4 without proving it in general.

- *KEY THEOREM 10.6. The sequence* $\{\mathbf{A}_i\}$ *of* $m \times n$ *matrices converges to* $\mathbf{B}$ *if and only if the sequence of real numbers* $\|\mathbf{A}_i - \mathbf{B}\|$ *converges to zero. The sequence* $\{\mathbf{A}_i\}$ *is bounded if and only if the sequence of real numbers* $\|\mathbf{A}_i\|$ *is bounded. Here the conditions on* $\|\cdot\|$ *need only be checked for some one single norm* $\|\cdot\|$ *entirely of our own choosing.*

Proof of Special Case: As we noted following (5.6) in Section 5.3, we deduce from (5.6) that the sequences of norms $\|\mathbf{A}_i\|_1$, $\|\mathbf{A}_i\|_2$, $\|\mathbf{A}_i\|_\infty$ must all behave the same, so that, for example, if one sequence is bounded then all three are. Since for any $n \times m$ matrix, $\|\mathbf{A}\|_\infty \leq n \max_{i,j} |a_{ij}| \leq n \|\mathbf{A}\|_\infty$, the theorem follows immediately from Definition 10.4 if $\|\cdot\|$ denotes any of the three special norms $\|\cdot\|_1$, $\|\cdot\|_2$, $\|\cdot\|_\infty$. The proof of the result for general norms proceeds similarly, although the central inequalities are harder to obtain.

EXERCISE 10.46. Solve Ex. 10.43 using **Key Theorem 10.6.**

SOLUTION: We choose to use the norm $\|\cdot\|_1$. Then

$$\|\mathbf{A}_i - \mathbf{B}\|_1 = \left\| \begin{bmatrix} \frac{1}{i} & -\left(\frac{1}{2}\right)^i & 0 \\ \frac{1}{i^2} & 0 & \left(\frac{1}{3}\right)^i \end{bmatrix} \right\|_1$$

$$= \max \left\{ \frac{1}{i} + \frac{1}{i^2}, \left(\frac{1}{2}\right)^i, \left(\frac{1}{3}\right)^i \right\} = \frac{1}{i} + \frac{1}{i^2}$$

which converges to zero; therefore $\mathbf{A}_i$ converges to $\mathbf{B}$.

EXERCISE 10.47. Solve Ex. 10.44 using **Key Theorem 10.6.**

SOLUTION: We choose to use the norm $\|\cdot\|_\infty$. Then

$$\|\mathbf{A}_i\|_\infty = \max \left\{ \frac{1}{i} + 1, 10 + |2 - (-1)^i| \right\} \leq 13 \text{ for all } i;$$

therefore the sequence $\{\mathbf{A}_i\}$ is bounded.

EXERCISE 10.48. Solve Ex. 10.45 using **Key Theorem 10.6.**

Now we return to the problem of analyzing the behavior of $\mathbf{A}^i$ in (10.16).

EXERCISE 10.49. Determine the behavior of $\mathbf{A}^i$ when

$$\mathbf{A} = \begin{bmatrix} \lambda_1 & 0 \\ 0 & \lambda_2 \end{bmatrix}.$$

SOLUTION: We have

$$\mathbf{A}^i = \begin{bmatrix} \lambda_1^i & 0 \\ 0 & \lambda_2^i \end{bmatrix}$$

and hence

$$\|\mathbf{A}^i\|_\infty = \max \{|\lambda_1|^i, |\lambda_2|^i\} = \{\max [|\lambda_1|, |\lambda_2|]\}^i.$$

Clearly then $\mathbf{A}^i$ converges to zero if and only if $\max [|\lambda_1|, |\lambda_2|] < 1$, while $\mathbf{A}^i$ is bounded if and only if $\max [|\lambda_1|, |\lambda_2|] \leq 1$.

The same argument used in Ex. 10.49 clearly applies when $\mathbf{\Lambda}$ is $n \times n$ instead of just 2×2, so that $\mathbf{\Lambda}^i$ converges to zero if and only if

$$\max[|\lambda_1|, \ldots, |\lambda_n|] < 1,$$

while $\mathbf{\Lambda}^i$ is bounded if and only if $\max[|\lambda_1|, \ldots, |\lambda_n|] \leq 1$. Since

$$||\mathbf{A}^i||_\infty = ||\mathbf{Q}\mathbf{\Lambda}^i\mathbf{Q}^{-1}||_\infty \leq ||\mathbf{Q}||_\infty ||\mathbf{\Lambda}^i||_\infty ||\mathbf{Q}^{-1}||_\infty,$$

the characterization of the behavior of $\mathbf{\Lambda}^i$ carries over directly to characterize the behavior of $\mathbf{A}^i$. Since

$$||\mathbf{A}^i\mathbf{x}_0||_\infty \leq ||\mathbf{A}^i||_\infty ||\mathbf{x}_0||$$

while

$$||\mathbf{A}^i\mathbf{x}||_\infty = ||\lambda^i\mathbf{x}||_\infty = |\lambda|^i ||\mathbf{x}||_\infty$$

if $\mathbf{x}$ is an eigenvector of $\mathbf{A}$ associated with the eigenvalue λ, our characterization of the behavior of $\mathbf{A}^i$ also provides one for $\mathbf{A}^i\mathbf{x}_0$. We summarize the situation for nondefective matrices; see also **Key Theorem 10.8.**

- KEY THEOREM 10.7. *Let* $\mathbf{A}$ *be a nondefective* $n \times n$ *matrix (for example,* $\mathbf{A}$ *might be normal). Then, as i tends to plus infinity:*
  - (i) $\mathbf{A}^i$ *converges to zero if and only if* $\mathbf{A}^i\mathbf{x}_0$ *converges to zero for every fixed* $\mathbf{x}_0$, *which in turn holds if and only if* $|\lambda| < 1$ *for all eigenvalues* λ *of* $\mathbf{A}$;
  - (ii) $\mathbf{A}^i$ *is bounded if and only if* $\mathbf{A}^i\mathbf{x}_0$ *is bounded for every fixed* $\mathbf{x}_0$, *which in turn holds if and only if* $|\lambda| \leq 1$ *for all eigenvalues* λ *of* $\mathbf{A}$.
  - (iii) $||\mathbf{A}^i||$ *tends to infinity for every norm* $||\cdot||$ *if and only if* $||\mathbf{A}^i\mathbf{x}_0||$ *tends to infinity for every norm* $||\cdot||$ *and for some fixed* $\mathbf{x}_0$, *which in turn holds if and only if* $|\lambda| > 1$ *for some eigenvalue* λ *of* $\mathbf{A}$.

We remark that in **Key Theorem 10.7**(iii) if *some* eigenvalue λ satisfies $|\lambda| < 1$ then $\mathbf{A}^i\mathbf{x}_0$ tends to zero for *some* $\mathbf{x}_0$, namely an eigenvector associated with λ; similarly if *some* eigenvalue λ satisfies $|\lambda| \leq 1$ then $\mathbf{A}^i\mathbf{x}_0$ is bounded for *some* $\mathbf{x}_0$. We can only guarantee that $\mathbf{A}^i\mathbf{x}_0$ tends to infinity for *every* $\mathbf{x}_0$ when *every* eigenvalue λ of $\mathbf{A}$ satisfies $|\lambda| > 1$.

We now show how to use **Key Theorem 10.7**; we apply it to our examples of discrete evolution in Sections 2.2 and 2.3.

EXERCISE 10.50. Analyze the behavior of the shares $\mathbf{x}_r$ of the market situation in Section 2.2 as r tends to infinity.

SOLUTION: According to (2.10) we have

$$\mathbf{A} = \begin{bmatrix} 0.8 & 0.2 & 0.1 \\ 0.1 & 0.7 & 0.3 \\ 0.1 & 0.1 & 0.6 \end{bmatrix}$$

whose eigenvalues were found in Section 8.1 to be $\lambda_1 = 0.5$, $\lambda_2 = 0.6$, $\lambda_3 = 1.0$ with eigenvectors $\mathbf{v}_1$, $\mathbf{v}_2$, $\mathbf{v}_3$ as given there also; therefore $\mathbf{A}$ is nondefective. According to **Key Theorem 10.7**(ii), therefore, $\mathbf{A}^i\mathbf{x}_0$ is bounded for every $\mathbf{x}_0$. Moreover, since $\{\mathbf{v}_1, \mathbf{v}_2, \mathbf{v}_3\}$ is linearly independent, it spans $\mathbb{R}^3$ and hence we can write $\mathbf{x}_0 = a_1\mathbf{v}_1 + a_2\mathbf{v}_2 + a_3\mathbf{v}_3$ so that $\mathbf{A}^i\mathbf{x}_0 = a_1\lambda_1^i\mathbf{v}_2 + a_2\lambda_2^i\mathbf{v}_2 + a_3\mathbf{v}_3$ which converges to $a_3\mathbf{v}_3$, a multiple of

$$\mathbf{v}_3 = \begin{bmatrix} 0.45 \\ 0.35 \\ 0.20 \end{bmatrix}.$$

Since we showed in Section 2.2 that the components of $\mathbf{A}^i\mathbf{x}_0$ are nonnegative and sum to one, the same holds for the limit $a_3\mathbf{v}_3$ and hence $a_3 = 1$ and

$$\lim_{r\to\infty} \mathbf{x}_r = \mathbf{v}_3$$

as asserted in Section 2.2 (see Theorem 10.9).

EXERCISE 10.51. Analyze theoretically the behavior of the chicken and fox populations described experimentally in Table 2.2 in Section 2.3.

SOLUTION: In this case we have

$$\mathbf{A} = \begin{bmatrix} 0.6 & 0.5 \\ -0.1 & 1.2 \end{bmatrix}$$

whose eigenvalues are easily found to be $\lambda_1 = 0.7$ and $\lambda_2 = 1.1$ with associated eigenvectors

$$\mathbf{v}_1 = \begin{bmatrix} 5 \\ 1 \end{bmatrix}, \qquad \mathbf{v}_2 = \begin{bmatrix} 1 \\ 1 \end{bmatrix}.$$

Since $\mathbf{A}$ is nondefective, we can apply **Key Theorem 10.7**(iii) to conclude that $\mathbf{A}^i\mathbf{x}_0$ tends to infinity for some vectors $\mathbf{x}_0$. If we write $\mathbf{x}_0 = a_1\mathbf{v}_1 + a_2\mathbf{v}_2$ then $\mathbf{A}^i\mathbf{x}_0 = a_1(0.7)^i\mathbf{v}_1 + a_2(1.1)^i\mathbf{v}_2$ so that $\mathbf{A}^i\mathbf{x}_0$ tends to infinity with $\mathbf{x}_0 = a_1\mathbf{v}_1 + a_2\mathbf{v}_2$ if and only if $a_2 \neq 0$. In the case of Table 2.2 we have

$$\mathbf{x}_0 = \begin{bmatrix} 100 \\ 1000 \end{bmatrix} = -225\mathbf{v}_1 + 1225\mathbf{v}_2$$

and hence we have proved that $\mathbf{x}_i$ behaves like $(1.1)^i\mathbf{v}_2$, that is, both F_i and C_i tend to infinity while F_i/C_i tends to unity, as verified experimentally in Table 2.2.

EXERCISE 10.52. Much as in Ex. 10.51, analyze theoretically the behavior of the chicken and fox populations described experimentally in Table 2.3 in Section 2.3.

EXERCISE 10.53. When x is a real number with $|x| < 1$, we show that $(1 - x)^{-1} = 1 + x + x^2 + \cdots$ by *first* showing that the series converges and *second* writing $(1 - x)(1 + x + \cdots + x^n) = 1 - x^{n+1}$ and taking the limit as n tends to infinity. Use a similar argument to show that if $\mathbf{A}$ is an $n \times n$ (nondefective) matrix all of whose eigenvalues λ satisfy $|\lambda| < 1$ then $\mathbf{I} - \mathbf{A}$ is nonsingular and its inverse is given by the convergent series

$$(\mathbf{I} - \mathbf{A})^{-1} = \mathbf{I} + \mathbf{A} + \mathbf{A}^2 + \cdots.$$

EXERCISE 10.54. Determine the behavior of $\mathbf{A}^i$ and of $\mathbf{A}^i\mathbf{x}_0$ for various $\mathbf{x}_0$ when

$$\mathbf{A} = \begin{bmatrix} \frac{1}{6} & \frac{\sqrt{6}}{3} & \frac{-\sqrt{2}}{6} \\ \frac{\sqrt{6}}{3} & 0 & \frac{\sqrt{3}}{3} \\ \frac{-\sqrt{2}}{6} & \frac{\sqrt{3}}{3} & \frac{1}{3} \end{bmatrix}.$$

EXERCISE 10.55. Find the value of the kill rate k of chickens by foxes in Section 2.3 in order that both populations F_i and C_i converge to nonzero stable totals F_∞ and C_∞ as i tends to infinity. If the initial populations are 100 and 1000, respectively, find the stable populations F_∞ and C_∞.

Our assumption in **Key Theorem 10.7** that $\mathbf{A}$ is nondefective was made merely to simplify the analysis of the behavior of $\mathbf{A}^i = \mathbf{J}^i$ in (10.12); when $\mathbf{A}$ is defective so that $\mathbf{J}$ is not diagonal, the situation is more complex.

EXERCISE 10.56. Analyze the behavior of $\mathbf{J}^i$, when

$$\mathbf{J} = \begin{bmatrix} \lambda & 1 \\ 0 & \lambda \end{bmatrix}.$$

SOLUTION: We easily find

$$\mathbf{J}^2 = \begin{bmatrix} \lambda^2 & 2\lambda \\ 0 & \lambda^2 \end{bmatrix}, \quad \mathbf{J}^3 = \begin{bmatrix} \lambda^3 & 3\lambda^2 \\ 0 & \lambda^3 \end{bmatrix}, \quad \mathbf{J}^4 = \begin{bmatrix} \lambda^4 & 4\lambda^3 \\ 0 & \lambda^4 \end{bmatrix},$$

and more generally,

$$\mathbf{J}^i = \begin{bmatrix} \lambda^i & i\lambda^{i-1} \\ 0 & \lambda^i \end{bmatrix} = \lambda^i \begin{bmatrix} 1 & \frac{i}{\lambda} \\ 0 & 1 \end{bmatrix}.$$

Since $i\lambda^{i-1}$ tends to zero if and only if $|\lambda| < 1$, we find that $\mathbf{J}^i$ tends to zero if and only if $|\lambda| < 1$, and $\|\mathbf{J}^i\|$ tends to infinity if $|\lambda| > 1$ as before for nondefective matrices. In this case, however, if $|\lambda| = 1$ then

$$\|\mathbf{J}^i\|_\infty = \left\| \lambda^i \begin{bmatrix} 1 & \frac{i}{\lambda} \\ 0 & 1 \end{bmatrix} \right\|_\infty = |\lambda^i| \left\| \begin{bmatrix} 1 & \frac{i}{\lambda} \\ 0 & 1 \end{bmatrix} \right\|_\infty$$
$$= 1 + i,$$

which tends to infinity. We *cannot* conclude that $\mathbf{J}^i$ is bounded when $|\lambda| = 1$ in this case.

The situation illustrated in Ex. 10.56 is typical; we state the following generalization of **Key Theorem 10.7** without proof. A proof can be constructed along the lines of that for **Key Theorem 10.7** but modified so as to allow for the situation in Ex. 10.56.

- *KEY THEOREM 10.8. Let* **A** *be an* $n \times n$ *matrix. Then as* i *tends to infinity:*

  (i) $\mathbf{A}^i$ *converges to zero if and only if* $\mathbf{A}^i\mathbf{x}_0$ *converges to zero for every* $\mathbf{x}_0$*, which in turn holds if and only if* $|\lambda| < 1$ *for all eigenvalues* λ *of* **A**.

  (ii) $\mathbf{A}^i$ *is bounded if and only if* $\mathbf{A}^i\mathbf{x}_0$ *is bounded for every* $\mathbf{x}_0$*, which in turn holds if and only if* $|\lambda| \leq 1$ *for all eigenvalues* λ *of* **A** *and, for those eigenvalues* λ *with* $|\lambda| = 1$*, the number of associated eigenvectors in an independent set equals the multiplicity of the eigenvalue* λ *as a root of the characteristic polynomial of* **A**.

  (iii) $\|\mathbf{A}^i\|$ *tends to infinity for each norm* $\|\cdot\|$ *if and only if* $\|\mathbf{A}^i\mathbf{x}_0\|$ *tends to infinity for each norm* $\|\cdot\|$ *for some vector* $\mathbf{x}_0$*, which in turn holds if and only if either some eigenvalue* λ *of* **A** *has* $|\lambda| > 1$ *or some eigenvalue* λ *of* **A** *has* $|\lambda| = 1$ *and the number of associated eigenvectors in an independent set is strictly less than the multiplicity of* λ.

EXERCISE 10.57. Extend Ex. 10.53 to apply to defective matrices as well.

The fact that we could show in Ex. 10.50 that the states $\mathbf{x}_r$ in the Markov chain (describing the division of a market among milk producers) converge to a limit distribution is no coincidence; this can be proved in general and is important in applications. In such cases the $n \times n$ matrix **A** is nonnegative in the sense that $a_{ij} \geq 0$ for all i and j and moreover

$$\sum_{i=1}^{n} a_{ij} = 1 \qquad (\text{for } 1 \leq j \leq n). \tag{10.18}$$

This tells us that $\|\mathbf{A}\|_1 = 1$ and therefore that

$$\begin{aligned} \|\mathbf{A}^i\|_1 &\leq \|\mathbf{A}\|_1^i \leq 1, \\ \|\mathbf{A}^i\mathbf{x}_0\|_1 &\leq \|\mathbf{A}^i\|_1 \|\mathbf{x}_0\|_1 \leq \|\mathbf{x}_0\|_1. \end{aligned} \tag{10.19}$$

Thus

$$\mathbf{x}_i = \mathbf{A}^i\mathbf{x}_0$$

is bounded as i tends to infinity. We want to prove, however, that $\mathbf{x}_i$, in addition, converges to some limit $\mathbf{v}$. From (10.19) and **Key Theorem 10.8**(ii) we know that all eigenvalues λ of **A** satisfy $|\lambda| \leq 1$ and that those with $|\lambda| = 1$ must have a full set of associated eigenvectors. Because of (10.18) we know that

$$\mathbf{e}^T\mathbf{A} = \mathbf{e}^T, \qquad \mathbf{e}^T = [1, 1, \ldots, 1],$$

and hence that $\lambda = 1$ is an eigenvalue of $\mathbf{A}$. If we had $|\lambda| = 1$ but $\lambda \neq 1$ for some other eigenvalue λ of $\mathbf{A}$ associated with an eigenvector $\mathbf{x}_0$, then

$$\mathbf{A}^i\mathbf{x}_0 = \lambda^i\mathbf{x}_0$$

would not converge. For us to prove, therefore, that $\mathbf{A}^i\mathbf{x}_0$ converges for every $\mathbf{x}_0$ we must somehow rule out the possibility of having an eigenvalue λ of $\mathbf{A}$ with $|\lambda| = 1$ but $\lambda \neq 1$.

EXERCISE 10.58. Let

$$\mathbf{A} = \begin{bmatrix} 0 & 1 \\ 1 & 0 \end{bmatrix}.$$

Show that $a_{ij} \geq 0$, that (10.18) holds, but that $\lambda = -1$ is an eigenvalue.

The point of Ex. 10.58 is that some additional conditions are needed to guarantee that $\lambda = 1$ is the only eigenvalue of $\mathbf{A}$ with $|\lambda| = 1$. If we make the stronger assumption, for example, that $\mathbf{A}$ is *strictly positive* (or in fact that $\mathbf{A}^p$ is strictly positive for some integer $p \geq 1$), that is, that

$$a_{ij} > 0 \quad \text{for} \quad 1 \leq i \leq n, \qquad 1 \leq j \leq n,$$

then it can be shown (see Ex. 10.61) that $\lambda = 1$ is the only eigenvalue of $\mathbf{A}$ with $|\lambda| = 1$ and that λ has multiplicity unity. We can now prove our desired result.

THEOREM 10.9. Let $\mathbf{A}$ be a strictly positive $n \times n$ Markov (or stochastic) matrix, so that

$$a_{ij} > 0 \quad \text{for} \quad 1 \leq i \leq n, \qquad 1 \leq j \leq n, \qquad \sum_{i=1}^{n} a_{ij} = 1.$$

Then there is a fixed vector $\mathbf{v} \neq \mathbf{0}$ (actually the eigenvector of $\mathbf{A}$ associated with the eigenvalue $\lambda = 1$) such that, for every $\mathbf{x}_0$, the Markov-chain sequence defined by $\mathbf{x}_i = \mathbf{A}\mathbf{x}_{i-1} = \mathbf{A}^i\mathbf{x}_0$ converges to a scalar multiple of $\mathbf{v}$.

Proof: We write $\mathbf{A}$ via its Jordan form $\mathbf{J}$ as

$$\mathbf{A} = \mathbf{Q}\mathbf{J}\mathbf{Q}^{-1}$$

where $\mathbf{v}$ is the first column of $\mathbf{Q}$, and $\mathbf{J}$ is as in (10.14); by the remark immediately preceding this theorem and by Ex. 10.61, we can take $\mathbf{J}_1$ in (10.14) to be 1×1 with $\mathbf{J}_1 = [1]$, while each other Jordan block $\mathbf{J}_i$ is associated with an eigenvalue λ with $|\lambda| < 1$. Hence $\mathbf{J}^i$ converges to the $n \times n$ matrix $\mathbf{Z}$ with $z_{11} = 1$ and $z_{ij} = 0$ unless $i = j = 1$. This in turn states that

$$\mathbf{A}^i\mathbf{x}_0 = \mathbf{Q}\mathbf{J}^i\mathbf{Q}^{-1}\mathbf{x}_0$$

converges to $\mathbf{QZQ}^{-1}\mathbf{x}_0$. Since the first column of $\mathbf{QZ}$ equals the first column of $\mathbf{Q}$ (namely $\mathbf{v}$) while the other columns equal the zero vector, $(\mathbf{QZ})(\mathbf{Q}^{-1}\mathbf{x}_0)$ is just a scalar multiple of $\mathbf{v}$, as required by the theorem.

EXERCISE 10.59. Generalize Theorem 10.9 to hold when $\mathbf{A}^p$ is strictly positive for some integer $p \geq 1$.

EXERCISE 10.60. Let $\mathbf{A}$ be as in Theorem 10.9 or its generalization in Ex. 10.59. By taking $\mathbf{x}_0 = \mathbf{e}$, prove that the eigenvector $\mathbf{v}$ mentioned in Theorem 10.9 can be taken to have strictly positive components. This result, with the remark (see Ex. 10.61) preceding Theorem 10.9, is essentially the famed *Perron-Frobenius Theorem.*

EXERCISE 10.61. Provide the details in the following outline of the proof that an $n \times n$ matrix $\mathbf{A}$, as in Theorem 10.9, has only the simple eigenvalue $\lambda = 1$ satisfying $|\lambda| = 1$. The outline is as follows. If $|\lambda| = 1$ and λ is an eigenvalue of $\mathbf{A}$, then let $\mathbf{x} \neq \mathbf{0}$ satisfy $\mathbf{x}^T\mathbf{A} = \lambda\mathbf{x}^T$ and $\max\{|x_1|, |x_2|, \ldots, |x_n|\} = |x_{i_0}| = 1$ for some i_0. From the i_0th equation in $\mathbf{x}^T\mathbf{A} = \lambda\mathbf{x}^T$ deduce that

$$|x_{i_0}| \leq \sum_{j=1}^{n} a_{ji_0}|x_j| \leq |x_{i_0}| + \sum_{j=1}^{n} a_{ji_0}[|x_j| - |x_{i_0}|],$$

that therefore $\sum_{j=1}^{n} a_{ji_0}[|x_j| - |x_{i_0}|] = 0$, and hence that $|x_j| = |x_{i_0}|$ for all j. From $\lambda x_1 = \sum_{j=1}^{n} a_{j1}x_j$ and $|\lambda x_1| = |x_1| = \cdots = |x_n| = 1$ deduce that $x_1 = \cdots = x_n$ so that $\mathbf{x} = c\mathbf{e}$, $\mathbf{e}^T = [1, \ldots, 1]$. From

$$\lambda c\mathbf{e}^T = \lambda\mathbf{x}^T = \mathbf{x}^T\mathbf{A} = c\mathbf{e}^T\mathbf{A} = c\mathbf{e}^T,$$

deduce that $\lambda = 1$. Conclude that $\lambda = 1$ is a simple eigenvalue since $\|\mathbf{A}^i\|_1 \leq 1$. This proves the result.

EXERCISE 10.62. Let $\mathbf{J}$ denote a $q \times q$ Jordan block; see Definition 10.2. Show that the (i, j) element of $\mathbf{J}^r$ equals

$$\lambda^{r+i-j}\frac{r!}{(j-i)!(r-j+i)!} \qquad \text{if } i \leq j \leq \max\{r+i, q\}$$

and equals zero for all other i, j.

EXERCISE 10.63. Use Ex. 10.62 to prove **Key Theorem 10.8.**

10.6 ITERATIVE SOLUTION OF LINEAR EQUATIONS

In dealing with mathematical models of very large and complex physical or social systems, we often have to solve systems of equations involving thousands of variables. Even on very fast, modern computers the storage for representing such equations and the time needed to solve them can often be prohibitively expensive or even unavailable. Fortunately, however, it is often the case that while there are thousands of equations, in as many variables, each specific equation may involve very few variables. This arises from the fact that even in very large and complex models each parameter typically interacts directly with only a small number of the other parameters; in a model of an economic system, for example, each individual probably buys from, and sells to, only a small number of the remaining individuals. Thus the matrix describing a system of equations in such circumstances may well be what is called *sparse*: it contains a relatively low percentage of nonzero elements. This certainly saves space when we store the matrix, since we can ignore zeros, but it may not help much when we use Gauss elimination because the elimination process tends to destroy zeros.

EXERCISE 10.64. Suppose that we have a 10×10 matrix $\mathbf{A}$ with the following structure, where "x" represents a nonzero element in the original $\mathbf{A}$, "0" represents a zero in the original $\mathbf{A}$ whose presence can be used to advantage during Gauss elimination, and "$\otimes$" represents a zero in the original $\mathbf{A}$ which is replaced by a nonzero number during elimination before elimination can proceed far enough to take advantage of the zero.

$$\mathbf{A} = \begin{bmatrix} x & x & 0 & 0 & 0 & 0 & 0 & 0 & 0 & x \\ x & x & x & 0 & 0 & 0 & 0 & 0 & 0 & x \\ x & x & x & x & 0 & 0 & 0 & 0 & 0 & x \\ x & \otimes & x & x & x & 0 & 0 & 0 & 0 & x \\ x & \otimes & \otimes & x & x & x & 0 & 0 & 0 & x \\ x & \otimes & \otimes & \otimes & x & x & x & 0 & 0 & x \\ x & \otimes & \otimes & \otimes & \otimes & x & x & x & 0 & x \\ x & \otimes & \otimes & \otimes & \otimes & \otimes & x & x & x & x \\ x & \otimes & \otimes & \otimes & \otimes & \otimes & \otimes & x & x & x \\ x & \otimes & \otimes & \otimes & \otimes & \otimes & \otimes & \otimes & x & x \end{bmatrix}.$$

Although only 44 of the 100 elements of $\mathbf{A}$ are nonzero, fully half of the 56 zero elements have to be treated as nonzero during elimination, so that we can take advantage only of 28 zeros.

In such circumstances it is important to be able to solve systems of equations without so modifying the matrix as to destroy or significantly reduce its sparseness. For this purpose *iterative* methods are very important.

We start by describing two special classical iterative procedures. Suppose that the equations are $\mathbf{Ax} = \mathbf{b}$.

1. In the *Jacobi* method (or method of *simultaneous corrections*), if we know the approximations $x_i^{(r)}$ to x_i at the rth stage, we obtain the esti-

mates at the $(r+1)$ stage from

$$a_{ii}x_i^{(r+1)} = -a_{i,1}x_1^{(r)} - \cdots - a_{i,i-1}x_{i-1}^{(r)} - a_{i,i+1}x_{i+1}^{(r)} - \cdots - a_{i,n}x_n^{(r)} + b_i. \tag{10.20}$$

2. In the *Gauss-Seidel* method (or method of *successive corrections*), in order to compute $x_i^{(r+1)}$ we make use of the estimates at the $(r+1)$ stage that are available at that point in the calculation:

$$a_{ii}x_i^{(r+1)} = -a_{i,1}x_1^{(r+1)} - \cdots - a_{i,i-1}x_{i-1}^{(r+1)} - a_{i,i+1}x_{i+1}^{(r)} - \cdots - a_{i,n}x_n^{(r)} + b_i. \tag{10.21}$$

EXERCISE 10.65. Describe the Jacobi iteration for solving

$$2u + v = 4$$
$$u + 2v = 5.$$

SOLUTION: The new estimates u_r, v_r are obtained from the old ones u_{r-1}, v_{r-1} according to (10.20) by

$$u_r = 2 - \tfrac{1}{2}v_{r-1}$$
$$v_r = \frac{5}{2} - \tfrac{1}{2}u_{r-1}.$$

EXERCISE 10.66. Describe the Gauss-Seidel iteration for the equations in Ex. 10.65.

SOLUTION: This time, according to (10.21), we use u_r to compute v_r, so that

$$u_r = 2 - \tfrac{1}{2}v_{r-1}$$
$$v_r = \frac{5}{2} - \tfrac{1}{2}u_r.$$

EXERCISE 10.67. In the approximate solution of certain partial differential equations we must often solve systems of equations described by an $N^2 \times N^2$ matrix $\mathbf{A}$ satisfying $a_{ii} = 4$ for $i = 1, \ldots, N^2$, $a_{i,i+1} = a_{i+1,i} = -1$ for $i = 1, \ldots, N^2 - 1$, $a_{i,i+N} = a_{i+N,i} = -1$ for $i = 1, \ldots, N^2 - N$, and all other $a_{ij} = 0$. Thus only $5N^2 - 2N - 2$ of the N^4 elements of $\mathbf{A}$ are nonzero; when $\mathrm{N} = 100$, for example, this represents only about 0.04% of the matrix elements. The Jacobi iteration (10.20) in this case for $\mathbf{Ax} = \mathbf{b}$ becomes simply

$$x_i^{(r+1)} = \tfrac{1}{4}b_i + \tfrac{1}{4}(x_{i+1}^{(r)} + x_{i-1}^{(r)} + x_{i+N}^{(r)} + x_{i-N}^{(r)})$$

for $i = N, \ldots, N^2 - N$, with small modifications for large and small i. The analogous equation for the Gauss-Seidel iteration is

$$x_i^{(r+1)} = \tfrac{1}{4}b_i + \tfrac{1}{4}(x_{i+1}^{(r)} + x_{i-1}^{(r+1)} + x_{i+N}^{(r)} + x_{i-N}^{(r+1)}).$$

A general class of iterative procedures that includes these as special cases can be discussed in the following way. We *split* the matrix $\mathbf{A}$ in the form $\mathbf{A} = \mathbf{E} - \mathbf{F}$ and rearrange $\mathbf{Ax} = \mathbf{b}$ in the form

$$\mathbf{Ex} = \mathbf{Fx} + \mathbf{b}. \tag{10.22}$$

The iterative procedure we wish to discuss is obtained by writing

$$\mathbf{Ex}^{(r+1)} = \mathbf{Fx}^{(r)} + \mathbf{b}, \tag{10.23}$$

where $\mathbf{x}^{(r)}$ is the approximation to $\mathbf{x}$ at the rth stage, and we start by choosing $\mathbf{x}^{(0)}$ to be an arbitrary column vector. The matrix $\mathbf{E}$ must obviously be chosen to be nonsingular, and for practical reasons it is clear that $\mathbf{E}$ should be chosen so that it is easy to solve any system of the form $\mathbf{Ex} = \mathbf{g}$. In (1) above, $\mathbf{E}$ equals the diagonal of $\mathbf{A}$ and $\mathbf{F}$ has zeros on its diagonal. In (2) above, $\mathbf{E}$ is lower triangular and $\mathbf{F}$ is upper triangular with zeros on the diagonal.

Instead of working in terms of the $\mathbf{x}^{(r)}$ it is often more convenient to work in terms of the corrections

$$\mathbf{c}^{(r+1)} = \mathbf{x}^{(r+1)} - \mathbf{x}^{(r)}.$$

The advantage of this is that if we subtract from (10.23) the same equation with $r - 1$ in place of r, we find

$$\mathbf{Ec}^{(r+1)} = \mathbf{Fc}^{(r)}$$

$$\mathbf{x}^{(r+1)} = \mathbf{x}^{(0)} + \mathbf{c}^{(1)} + \cdots + \mathbf{c}^{(r+1)}.$$

If we subtract (10.22) from (10.23) and introduce

$$\boldsymbol{\delta}^{(r)} = \mathbf{x}^{(r)} - \mathbf{x} \tag{10.24}$$

this gives

$$\mathbf{E}\boldsymbol{\delta}^{(r+1)} = \mathbf{F}\boldsymbol{\delta}^{(r)} \quad \text{or} \quad \boldsymbol{\delta}^{(r+1)} = \mathbf{H}\boldsymbol{\delta}^{(r)}, \tag{10.25}$$

where we have introduced

$$\mathbf{H} = \mathbf{E}^{-1}\mathbf{F}. \tag{10.26}$$

We emphasize there that computationally we probably do not *compute* $\mathbf{E}^{-1}$ or $\mathbf{H}$; (10.25) and (10.26) are merely convenient tools for analysis which, in practice, are discarded in favor of simple formulas such as those in Ex. 10.67.

EXERCISE 10.68. Verify that (10.25), (10.26) hold in Ex. 10.65.

SOLUTION: The solution of this system is $u = 1$, $v = 2$. Since

$$u_r - 1 = 1 - \tfrac{1}{2}v_{r-1} = -\tfrac{1}{2}(v_{r-1} - 2)$$
$$v_r - 2 = \tfrac{1}{2} - \tfrac{1}{2}u_{r-1} = -\tfrac{1}{2}(u_{r-1} - 1),$$

we have

$$\begin{bmatrix} u_r \\ v_r \end{bmatrix} - \begin{bmatrix} 1 \\ 2 \end{bmatrix} = \begin{bmatrix} 0 & -\frac{1}{2} \\ -\frac{1}{2} & 0 \end{bmatrix} \left\{ \begin{bmatrix} u_{r-1} \\ v_{r-1} \end{bmatrix} - \begin{bmatrix} 1 \\ 2 \end{bmatrix} \right\}. \tag{10.27}$$

In this case $\mathbf{A} = \mathbf{E} - \mathbf{F}$ with

$$\mathbf{E} = \begin{bmatrix} 2 & 0 \\ 0 & 2 \end{bmatrix}, \qquad \mathbf{F} = \begin{bmatrix} 0 & -1 \\ -1 & 0 \end{bmatrix},$$

so that by (10.26)

$$\mathbf{H} = \mathbf{E}^{-1}\mathbf{F} = \begin{bmatrix} \frac{1}{2} & 0 \\ 0 & \frac{1}{2} \end{bmatrix} \begin{bmatrix} 0 & -1 \\ -1 & 0 \end{bmatrix} = \begin{bmatrix} 0 & -\frac{1}{2} \\ -\frac{1}{2} & 0 \end{bmatrix}$$

which, with (10.27), verifies (10.25) and (10.26).

EXERCISE 10.69. Verify that (10.25), (10.26) hold in Ex. 10.66, and that

$$\mathbf{H} = \begin{bmatrix} 0 & -\frac{1}{2} \\ 0 & \frac{1}{4} \end{bmatrix}.$$

EXERCISE 10.70. Describe some splittings $\mathbf{A} = \mathbf{E} - \mathbf{F}$ which might be convenient for a matrix $\mathbf{A}$ structured as in Ex. 10.64.

From (10.25) we know that the error $\boldsymbol{\delta}^{(r)}$ at the rth step is given by

$$\boldsymbol{\delta}^{(r)} = \mathbf{H}^r\boldsymbol{\delta}^{(0)}; \tag{10.28}$$

since $\mathbf{x}^{(r)}$ converges to $\mathbf{x}$ if and only if $\boldsymbol{\delta}^{(r)}$ converges to zero, from (10.28) and **Key Theorem 10.8**(i) we know that $\mathbf{x}^{(r)}$ converges to the solution $\mathbf{x}$ for every initial estimate $\mathbf{x}^{(0)}$ of $\mathbf{x}$ if and only if all eigenvalues λ of $\mathbf{H}$ satisfy $|\lambda| < 1$. From our analysis preceding **Key Theorem 10.7**, especially (10.16), we see that the smaller the eigenvalues of $\mathbf{H}$, the more rapidly we can expect the sequence $\mathbf{x}^{(r)}$ to converge to $\mathbf{x}$. Thus the Jordan canonical form has given us some aid in deciding which methods (or splittings $\mathbf{A} = \mathbf{E} - \mathbf{F}$) yield the most rapid convergence; we study this question very briefly.

Let us introduce the notation

$$\mathbf{A} = \mathbf{L} + \mathbf{D} + \mathbf{U},$$

where $\mathbf{L}$ is a lower triangular matrix with zeros on the diagonal, $\mathbf{D}$ is a diagonal matrix, and $\mathbf{U}$ is an upper triangular matrix with zeros on the diagonal. In this notation the methods of simultaneous and successive corrections, (10.20), (10.21), read

$$\mathbf{D}\mathbf{x}^{(r+1)} = -(\mathbf{L} + \mathbf{U})\mathbf{x}^{(r)} + \mathbf{b}, \tag{10.29}$$

$$(\mathbf{L} + \mathbf{D})\mathbf{x}^{(r+1)} = -\mathbf{U}\mathbf{x}^{(r)} + \mathbf{b}. \tag{10.30}$$

We examine a generalization of (10.30) known as the *successive over-relaxation* (*S.O.R*) (or *accelerated Gauss-Seidel*) method:

$$(\mathbf{L} + p\mathbf{D})\mathbf{x}^{(r+1)} = -\{(1 - p)\mathbf{D} + \mathbf{U}\}\mathbf{x}^{(r)} + \mathbf{b}, \qquad (10.31)$$

where p is a constant that will be determined so as to yield the fastest rate of convergence. If we rearrange (10.31) in the form

$$\mathbf{x}^{(r+1)} = \mathbf{x}^{(r)} + w\mathbf{D}^{-1}\{-\mathbf{L}\mathbf{x}^{(r+1)} - (\mathbf{D} + \mathbf{U})\mathbf{x}^{(r)} + \mathbf{b}\}, \qquad (10.32)$$

where $w = 1/p$ is known as the *over-relaxation factor*, we see that $\mathbf{x}^{(r+1)}$ is obtained by adding to $\mathbf{x}^{(r)}$ a multiple of a scaled residual or error at the given point in the calculation. As we shall see, it often turns out that the optimum value of w for most rapid convergence is greater than unity, and this is the reason for the name *over*relaxation. The Gauss-Seidel method is the special case $w = 1$ of (10.32). The method (10.31) corresponds to that of (10.22) with $\mathbf{E} = \mathbf{L} + p\mathbf{D}$ and $\mathbf{F} = -(1 - p)\mathbf{D} - \mathbf{U}$.

EXERCISE 10.71. Describe the method (10.31) for the equations of Ex. 10.65.

SOLUTION: We have

$$\mathbf{D} = \begin{bmatrix} 2 & 0 \\ 0 & 2 \end{bmatrix}, \quad \mathbf{L} = \begin{bmatrix} 0 & 0 \\ 1 & 0 \end{bmatrix}, \quad \mathbf{U} = \begin{bmatrix} 0 & 1 \\ 0 & 0 \end{bmatrix},$$

so that by (10.31) we have

$$\begin{bmatrix} 2p & 0 \\ 1 & 2p \end{bmatrix}\begin{bmatrix} u_r \\ v_r \end{bmatrix} = \begin{bmatrix} 4 \\ 5 \end{bmatrix} - \left\{\begin{bmatrix} 2(1-p) & 1 \\ 0 & 2(1-p) \end{bmatrix}\begin{bmatrix} u_{r-1} \\ v_{r-1} \end{bmatrix}\right\}.$$

Therefore,

$$u_r = \frac{2}{p} - \frac{1-p}{p}u_{r-1} - \frac{1}{2p}v_{r-1}$$

$$v_r = \frac{5p-2}{2p^2} + \frac{1-p}{2p^2}u_{r-1} + \frac{1-4p+4p^2}{4p^2}v_{r-1}.$$

In the notation of (10.25), (10.26), we have

$$\mathbf{H} = -(\mathbf{L} + p\mathbf{D})^{-1}[(1-p)\mathbf{D} + \mathbf{U}]$$

$$= \begin{bmatrix} \dfrac{p-1}{p} & \dfrac{-1}{2p} \\[2ex] \dfrac{1-p}{2p^2} & \dfrac{1-4p+4p^2}{4p^2} \end{bmatrix}.$$

We want to compare the rates of convergence expected from the methods (10.29)–(10.31); we saw that to do so we must choose that method whose largest eigenvalue is smallest.

DEFINITION 10.5. The *spectral radius* $\rho(\mathbf{A})$ of an $n \times n$ square matrix **A** is the value of the modulus of the eigenvalue of **A** with largest modulus, that is, $\rho(\mathbf{A}) = \max \{|\lambda_i| \mid i = 1, \ldots, n\}$.

Using this terminology, we want to choose that method from among (10.29)–(10.31) whose **H** matrix has the smallest spectral radius.

EXERCISE 10.72. Compare the spectral radii of the **H** matrices for (10.29), (10.30), and (10.31) with $p = 0.9$ for the equations in Ex. 10.65.

SOLUTION: In Exs. 10.68, 10.69, and 10.71 we have found the **H** matrices for (10.29)–(10.31) to be, respectively,

$$\mathbf{H}_1 = \begin{bmatrix} 0 & -\frac{1}{2} \\ -\frac{1}{2} & 0 \end{bmatrix},$$

$$\mathbf{H}_2 = \begin{bmatrix} 0 & -\frac{1}{2} \\ 0 & \frac{1}{4} \end{bmatrix},$$

$$\mathbf{H}_3 = \begin{bmatrix} \dfrac{p-1}{p} & \dfrac{-1}{2p} \\ \dfrac{1-p}{2p^2} & \dfrac{1-4p+4p^2}{4p^2} \end{bmatrix}.$$

A simple calculation shows the spectral radii to be

$$\rho(\mathbf{H}_1) = \frac{1}{2}, \qquad \rho(\mathbf{H}_2) = \frac{1}{4}, \qquad \rho(\mathbf{H}_3) = \frac{1}{9} \qquad (\text{for } p = 0.9)$$

so that the S.O.R. method should converge fastest.

EXERCISE 10.73. Compare the actual speed of convergence of the methods of Jacobi, Gauss-Seidel, and S.O.R. with $p = 0.9$ on the equations in Ex. 10.65, starting with $u_0 = v_0 = 0$.

SOLUTION: By actual computation with the Jacobi method, we find the errors $\boldsymbol{\delta}_i$ satisfy

$$\|\boldsymbol{\delta}_0\|_\infty = 2, \qquad \|\boldsymbol{\delta}_6\|_\infty = \frac{1}{32}, \qquad \|\boldsymbol{\delta}_{11}\|_\infty = \frac{1}{1024}, \qquad \|\boldsymbol{\delta}_i\|_\infty = \frac{1}{2^{i-1}}.$$

For the Gauss-Seidel method, we find

$$\|\boldsymbol{\delta}_0\|_\infty = 2,$$

$$\|\boldsymbol{\delta}_6\|_\infty = \frac{1}{1024} = 0.00098,$$

$$\|\boldsymbol{\delta}_{11}\|_\infty = \frac{1}{1,048,576} = 0.96 \times 10^{-6},$$

$$\|\boldsymbol{\delta}_i\|_\infty = \frac{1}{4^{i-1}}.$$

For the S.O.R. method with $p = 0.9$, we find

$$\|\boldsymbol{\delta}_0\|_\infty = 2, \qquad \|\boldsymbol{\delta}_6\|_\infty = 0.000012, \qquad \|\boldsymbol{\delta}_{11}\| = 10^{-10}.$$

Clearly the comparisons among the actual convergence rates are similar to those among the theoretical convergence rates in Ex. 10.72.

An instructive elementary discussion of rates of convergence can be given for the special case of a tridiagonal matrix, the elements of which are zero except on the principal diagonal and the immediately adjacent diagonals. This means that in the decomposition

$$\mathbf{A} = \mathbf{L} + \mathbf{D} + \mathbf{U},$$

L and **U** have the forms

$$\mathbf{L} = \begin{bmatrix} 0 & 0 & \dots & 0 & 0 \\ l_1 & 0 & \dots & 0 & 0 \\ & & \dots & & \\ 0 & 0 & \dots & 0 & 0 \\ 0 & 0 & \dots & l_{n-1} & 0 \end{bmatrix}, \qquad \mathbf{U} = \begin{bmatrix} 0 & u_1 & \dots & 0 & 0 \\ 0 & 0 & \dots & 0 & 0 \\ & & \dots & & \\ 0 & 0 & \dots & 0 & u_{n-1} \\ 0 & 0 & \dots & 0 & 0 \end{bmatrix} \tag{10.33}$$

We will compare the rates of convergence of (10.29) and (10.31); this means that we wish to compare the spectral radii of

$$-\mathbf{D}^{-1}(\mathbf{L} + \mathbf{U}) \quad \text{and} \quad -(\mathbf{L} + p\mathbf{D})^{-1}\{(1 - p)\mathbf{D} + \mathbf{U}\}.$$

In other words, we wish to compare the eigenvalues λ and μ satisfying

$$[(\mathbf{L}_0 + \mathbf{U}_0) + \lambda\mathbf{I}]\mathbf{u} = \mathbf{0}, \tag{10.34}$$

$$\{[(1 - p)\mathbf{I} + \mathbf{U}_0] + \mu(\mathbf{L}_0 + p\mathbf{I})\}\mathbf{v} = \mathbf{0}, \tag{10.35}$$

$$\mathbf{L}_0 = \mathbf{D}^{-1}\mathbf{L}, \qquad \mathbf{U}_0 = \mathbf{D}^{-1}\mathbf{U}.$$

THEOREM 10.10. The values of the eigenvalues λ, μ defined by (10.34), (10.35) *are related by:*

$$\{(1-p)+\mu p\}^2 = \mu\lambda^2. \tag{10.36}$$

Proof: Let **E**, **F** be diagonal matrices with nonzero diagonal elements. The following eigenvalue problem is completely equivalent to (10.35):

$$\mathbf{E}[\{(1-p)\mathbf{I}+\mathbf{U}_0\}+\mu(\mathbf{L}_0+p\mathbf{I})]\mathbf{F}(\mathbf{F}^{-1}\mathbf{v}) = \mathbf{0}. \tag{10.37}$$

We are going to show that **E** and **F** can be chosen so that the matrix multiplying $\mathbf{F}^{-1}\mathbf{v}$ in (10.37) is identical with the matrix multiplying **u** in (10.34)—i.e.,

$$\mathbf{E}[\{(1-p)\mathbf{I}+\mathbf{U}_0\}+\mu(\mathbf{L}_0+p\mathbf{I})]\mathbf{F} = k\{(\mathbf{L}_0+\mathbf{U}_0)+\lambda\mathbf{I}\}$$

for some constant k. Because of the nature of $\mathbf{L}_0$, **I**, $\mathbf{U}_0$ this implies that we must have

$$\mathbf{E}\mathbf{U}_0\mathbf{F} = k\mathbf{U}_0, \qquad \mu\mathbf{E}\mathbf{L}_0\mathbf{F} = k\mathbf{L}_0,$$
$$\{(1-p)+\mu p\}\mathbf{E}\mathbf{F} = k\lambda\mathbf{I}.$$

It is easy to check that these equations are satisfied if

$$e_i f_{i+1} = k, \qquad \mu e_{i+1} f_i = k, \qquad (i = 1, \ldots, n-1),$$
$$e_i f_i = q, \qquad q = \frac{k\lambda}{\{(1-p)+\mu p\}} \qquad (i = 1, \ldots, n),$$

where e_i, f_i are the diagonal elements of **E**, **F**. These equations are consistent if (10.36) is true, which proves the theorem.

We can now compare our methods.

THEOREM 10.11. If the methods of Jacobi and Gauss-Seidel are applied to a tridiagonal system, either both methods converge or both diverge. If both converge, Gauss-Seidel converges faster than Jacobi.

Proof: The Gauss-Seidel method corresponds to taking $p = 1$ in (10.31). Hence the relation between the eigenvalues λ in the Jacobi method and μ in the Gauss-Seidel method, for the matrices that determine the rate of convergence, is given by $p = 1$ in (10.36)—i.e., $\mu = \lambda^2$. The results stated in the theorem follow immediately.

THEOREM 10.12. If the Jacobi method converges when applied to a tridiagonal system with real eigenvalues and if the **H** *matrix for the Jacobi method has largest eigenvalue λ_1, then the optimum value of p in the S.O.R. method is given by*

$$p = p_1 = \tfrac{1}{2}\{1+(1-\lambda_1^2)^{1/2}\}, \tag{10.38}$$

and the corresponding eigenvalue μ_1 that determines the rate of convergence of *the accelerated method is given by*

$$\mu_1 = \frac{1-(1-\lambda_1^2)^{1/2}}{1+(1-\lambda_1^2)^{1/2}}. \tag{10.39}$$

Proof: We have to find the value of p in (10.36) that gives the smallest μ for a given λ, $0 \leq |\lambda| < 1$. Equation (10.36) is a quadratic in μ. Of the two roots we must choose the one such that $\mu = \lambda^2$ when $p = 1$ since, if $p = 1$, (10.36) gives $\mu = 0$ or λ^2 but the eigenvalue μ of largest modulus is certainly not zero. This gives

$$\mu = \tfrac{1}{2}\lambda^2 w^2 - w + 1 + \tfrac{1}{2}|\lambda w|\{\lambda^2 w^2 - 4w + 4\}^{1/2},$$

where $w = 1/p$. The value of μ decreases as w increases until μ becomes complex, which occurs for $w_1 < w < w_2$ where w_1 and w_2 are the zeros of the quadratic under the square root. We find

$$w_1 = \frac{2}{1+(1-\lambda^2)^{1/2}}, \qquad w_2 = \frac{2}{1-(1-\lambda^2)^{1/2}}. \tag{10.40}$$

Verify that $|\mu| = w - 1$ for $w_1 < w < w_2$ so that $|\mu|$ is then increasing with w, and the increase continues for $w > w_2$. Hence the minimum $|\mu|$ occurs for $w = w_1$ and it is given by $|\mu| = w_1 - 1$. Each eigenvalue of the tridiagonal system gives rise to a μ. We must choose the largest, which corresponds to taking $\lambda = \lambda_1$ in (10.40). The results of the theorem follow.

The value of the S.O.R. method can be illustrated by supposing that $\lambda_1 = 1 - \epsilon$, where ϵ is very small so that the ordinary Gauss Seidel method converges very slowly. Then (10.39) gives

$$\mu_1 \approx \{1 - \sqrt{(2\epsilon)}\}\{1 + \sqrt{(2\epsilon)}\}^{-1} \approx 1 - 2\sqrt{(2\epsilon)}.$$

If $\lambda_1 = 0.995$, $\epsilon = 0.005$, then $\mu_1 \simeq 0.8$ and a remarkable increase in the rate of convergence is achieved. Our two theorems tell us more generally that if the Jacobi method converges, then so do the other two, with the fastest of all being S.O.R. with the relaxation factor p picked by (10.38). This still leaves the problem of when even the Jacobi method will converge. In many cases of practical importance the matrix $\mathbf{A}$ not only is sparse but also is *strictly diagonally dominant*, that is,

$$\sum_{\substack{j=1 \\ j \neq i}}^{n} |a_{ij}| < a_{ii};$$

in this case we can easily prove that the Jacobi method converges and hence so do the others.

EXERCISE 10.74. Suppose that $\mathbf{A} = \mathbf{I} + \mathbf{L} + \mathbf{U}$ and that $\mathbf{A}$ is strictly diagonally dominant. Deduce that the Jacobi iteration matrix $\mathbf{H} = -(\mathbf{L} + \mathbf{U})$ has $\|\mathbf{H}\|_\infty < 1$ and then conclude that the Jacobi method is convergent. Extend this to $\mathbf{A} = \mathbf{D} + \mathbf{L} + \mathbf{U}$ for general $\mathbf{D}$ and strictly diagonally dominant $\mathbf{A}$.

EXERCISE 10.75. Show that the matrix $\mathbf{A}$ in Ex. 10.67 is not strictly diagonally dominant. In this case, however, the spectral radius of the Jacobi iteration matrix is $\cos [\pi/(N + 1)]$; conclude that the methods are convergent.

EXERCISE 10.76. Find the optimum value of p for the S.O.R. method for the equations in Ex. 10.65. Show that the spectral radius of the $\mathbf{H}$ matrix in this case equals approximately 0.0718 as compared with 0.5 and 0.25 for the Jacobi and Gauss-Seidel methods, respectively, and with 0.11 for S.O.R. with $p = 0.9$.

EXERCISE 10.77. Assuming (10.36) to hold more generally for the matrix $\mathbf{A}$ of Ex. 10.67, find and compare for large N the spectral radii for the iteration matrices arising from Jacobi iteration, Gauss-Seidel iteration, and optimal successive over-relaxation.

10.7 CONTINUOUS SYSTEM EVOLUTION AND MATRIX EXPONENTIALS

In Section 10.5 we generalized our examples from Sections 2.2 and 2.3 to allow us to consider the evolution of systems whose state at a specified point in time is given as a linear transformation of the state at a specified preceding point in time. If the time interval between these two points is extremely small, or, perhaps, if the change in the state during the time period is small relative to the size of the state itself, it is sometimes just as good a model to assume that the state $\mathbf{x}$ is defined for all time t by the function $\mathbf{x}(t)$ and that information on how the state changes is described by means of the *derivative* $\dot{\mathbf{x}}$, where $\dot{\mathbf{x}}$ denotes the vector function with

$$\dot{\mathbf{x}}(t) = \begin{bmatrix} \frac{dx_1}{dt}(t) \\ \frac{dx_2}{dt}(t) \\ \cdot \\ \cdot \\ \cdot \\ \frac{dx_n}{dt}(t) \end{bmatrix}$$

if $\mathbf{x}$ is an $n \times 1$ column vector. In other instances, particularly in engineering and the physical sciences, time is most naturally considered to be a continuously sampled variable, so that the natural descriptions of how the state **x** of a system changes with time involve derivatives.

EXERCISE 10.78. The relation (2.23) of Section 2.2 arose from assuming that the population p_i at the ith instant evolved according to

$$p_{i+1} - p_i = (b - d)p_i;$$

that is, the change is proportional to the total population. If the time interval is extremely small, then the rate of change is given by the derivative of p with respect to time, yielding the model

$$\dot{p}(t) = (\beta - \delta)p(t).$$

Analogously, the model (2.25) of the fox-chicken population might be replaced by

$$\begin{aligned} \dot{F}(t) &= -0.4F(t) + 0.5C(t), \qquad & F(t_0) &= 100 \\ \dot{C}(t) &= -kF(t) + 0.2C(t), \qquad & C(t_0) &= 1000 \end{aligned}$$

or, in matrix notation,

$$\dot{\mathbf{x}} = \mathbf{A}\mathbf{x}, \qquad \mathbf{x}(t_0) = \mathbf{x}_0$$

where

$$\mathbf{x} = \begin{bmatrix} F \\ C \end{bmatrix}, \qquad \mathbf{A} = \begin{bmatrix} -0.4 & 0.5 \\ -k & 0.2 \end{bmatrix}, \qquad \mathbf{x}_0 = \begin{bmatrix} 100 \\ 1000 \end{bmatrix}.$$

EXERCISE 10.79. Suppose that $v(t)$ denotes the total income at time t of a given business enterprise (or perhaps the GNP of an economy), and suppose that, if $i(t)$ of this income is reinvested, the rate of increase of v will be proportional to i, so that

$$\dot{v}(t) = gi(t)$$

for some growth rate g. We propose to reinvest a fixed fraction r of income, so that $i(t) = rv(t)$, and we suppose that the remaining income, $(1 - r)v(t)$, is to be distributed as profits to the shareholders after taxes and other expenses are deducted, so that the rate of increase of profit p is proportional to $(1 - r)v$, that is,

$$\dot{p} = s(1 - r)v.$$

Altogether then, we have the model

$$\begin{aligned} \dot{v} &= grv, \\ \dot{p} &= s(1 - r)v, \end{aligned}$$

or, in matrix notation,

$$\dot{\mathbf{x}} = \mathbf{A}\mathbf{x},$$

where

$$\mathbf{x} = \begin{bmatrix} v \\ p \end{bmatrix}, \qquad \mathbf{A} = \begin{bmatrix} gr & 0 \\ s(1-r) & 0 \end{bmatrix}.$$

EXERCISE 10.80. Consider the vibration of the three beads depicted in Figure 8.1 of Section 8.1, but suppose, in addition, that external forces, such as the force due to gravity, are applied to each bead. If X_1, X_2, and X_3 denote, as before, the displacements of the beads, show, for example, that

$$m\ddot{X}_1 = -\frac{TX_1}{l} + \frac{T(X_2 - X_1)}{2l} + F_1(t) \tag{10.41}$$

where $F_1(t)$ denotes the external force. We introduce the 6×1 column vector $\mathbf{x}$ with $x_1 = X_1$, $x_2 = \dot{X}_1$, $x_3 = X_2$, $x_4 = \dot{X}_2$, $x_5 = X_3$, and $x_6 = \dot{X}_3$; show that by this device (10.41), for example, is replaced by the *pair* of equations

$$\dot{x}_1 = x_2$$

$$\dot{x}_2 = -\frac{T}{ml}x_1 + \frac{T}{2ml}(x_3 - x_1) + \frac{1}{m}F_1(t).$$

Show that replacing each of the three equations following Figure 8.1 with such a pair leads finally to

$$\dot{\mathbf{x}} = \mathbf{A}\mathbf{x} + \mathbf{f},$$

where

$$\mathbf{A} = \begin{bmatrix} 0 & 1 & 0 & 0 & 0 & 0 \\ \frac{-3T}{2ml} & 0 & \frac{T}{2ml} & 0 & 0 & 0 \\ 0 & 0 & 0 & 1 & 0 & 0 \\ \frac{T}{2ml} & 0 & \frac{-T}{ml} & 0 & \frac{T}{2ml} & 0 \\ 0 & 0 & 0 & 0 & 0 & 1 \\ 0 & 0 & \frac{T}{2ml} & 0 & \frac{-3T}{2ml} & 0 \end{bmatrix}, \qquad \mathbf{f} = \begin{bmatrix} 0 \\ \frac{1}{m}F_1 \\ 0 \\ \frac{1}{m}F_2 \\ 0 \\ \frac{1}{m}F_3 \end{bmatrix}.$$

In each of the three preceding examples we were led to the differential equation

$$\dot{\mathbf{x}} = \mathbf{A}\mathbf{x} + \mathbf{f} \tag{10.42}$$

where $\mathbf{A}$ is a given fixed $n \times n$ matrix, $\mathbf{f}$ is a given vector-valued function, and the vector-valued function $\mathbf{x}$ is to be found so as to satisfy (10.42) identically. Typically, in models of evolutionary processes, the state $\mathbf{x}$ of the system is given at some initial time t_0, so that

$$\mathbf{x}(t_0) = \mathbf{x}_0,$$

say, where $\mathbf{x}_0$ is given. For simplicity, we first deal with the case where $t_0 = 0$ and $\mathbf{f}$ in (10.42) is zero, so that we have to solve

$$\dot{\mathbf{x}} = \mathbf{A}\mathbf{x}. \tag{10.43}$$

If $\mathbf{A}$ can be reduced to diagonal form by a similarity transformation, then the solution of this equation is straightforward. Suppose more precisely that $\mathbf{A}$ is nondefective; we know then that $\mathbf{P}$ exists such that $\mathbf{P}^{-1}\mathbf{A}\mathbf{P} = \mathbf{\Lambda}$ is a diagonal matrix, with diagonal elements equal to the eigenvalues of $\mathbf{A}$. We multiply (10.43) by $\mathbf{P}^{-1}$ and rearrange in the form

$$\frac{d}{dt}(\mathbf{P}^{-1}\mathbf{x}) = (\mathbf{P}^{-1}\mathbf{A}\mathbf{P})\mathbf{P}^{-1}\mathbf{x}, \tag{10.44}$$

where, on the left-hand side, $\mathbf{P}^{-1}$ can be shifted through the differentiation sign because it is a matrix of constants. (If you doubt this, write out everything in detail.) For simplicity, introduce

$$\mathbf{y} = \mathbf{P}^{-1}\mathbf{x}. \tag{10.45}$$

Since $\mathbf{P}^{-1}\mathbf{A}\mathbf{P} = \mathbf{\Lambda}$ is diagonal, (10.44) can be written as the separate equations

$$\frac{dy_r}{dt} = \lambda_r y_r \qquad (r = 1, \ldots, n) \tag{10.46}$$

which can be integrated immediately to give

$$y_r = (y_r)_0 e^{\lambda_r t}, \tag{10.47}$$

where $(y_r)_0$ denotes the value of y_r at $t = 0$. If we denote the column vector consisting of these initial values by $\mathbf{y}_0$, equation (10.47) can be written

$$\mathbf{y} = \mathbf{L}\mathbf{y}_0 \tag{10.48}$$

where

$$\mathbf{L}(t) = \begin{bmatrix} e^{\lambda_1 t} & 0 & \ldots & 0 \\ 0 & e^{\lambda_2 t} & \ldots & 0 \\ & & \ldots & \\ 0 & 0 & \ldots & e^{\lambda_n t} \end{bmatrix}. \tag{10.49}$$

From (10.45),

$$\mathbf{x} = \mathbf{P}\mathbf{y}, \qquad \mathbf{y}_0 = \mathbf{P}^{-1}\mathbf{x}_0.$$

Equation (10.48) gives the final solution

$$\mathbf{x} = \mathbf{P}\mathbf{L}\mathbf{y}_0 = \mathbf{P}\mathbf{L}\mathbf{P}^{-1}\mathbf{x}_0. \tag{10.50}$$

It is instructive to write this solution in the form

$$\mathbf{x} = a_1\mathbf{p}_1e^{\lambda_1 t} + a_2\mathbf{p}_2e^{\lambda_2 t} + \cdots + a_n\mathbf{p}_ne^{\lambda_n t},$$

where $a_i = (y_i)_0$, and $\mathbf{p}_i$ is the ith column of $\mathbf{P}$. The $\mathbf{p}_i$ are the n eigenvectors of $\mathbf{A}$ and the a_i are arbitrary constants that are determined from the initial conditions. We summarize.

THEOREM 10.13. Let $\mathbf{A}$ *be a nondefective* $n \times n$ *matrix, with* $\mathbf{P}^{-1}\mathbf{A}\mathbf{P} = \mathbf{\Lambda}$ *diagonal and with the eigenvalues* $\lambda_1, \ldots, \lambda_n$ *of* $\mathbf{A}$ *on the diagonal of* $\mathbf{\Lambda}$. *Then the solution* $\mathbf{x}(t)$ *to the equation* $\dot{\mathbf{x}} = \mathbf{A}\mathbf{x}$ *of* (10.43), *subject to the given initial condition* $\mathbf{x}(0) = \mathbf{x}_0$, *is* $\mathbf{x}(t) = \mathbf{P}\mathbf{L}(t)\mathbf{P}^{-1}\mathbf{x}_0$, *where* $\mathbf{L}(t)$ *is the diagonal matrix in* (10.49).

EXERCISE 10.81. Solve the equations $\dot{\mathbf{x}} = \mathbf{A}\mathbf{x}$ of Ex. 10.78 with $k = 0.16$, $t_0 = 0$.

SOLUTION: In this case

$$\mathbf{A} = \begin{bmatrix} -0.4 & 0.5 \\ -0.16 & 0.2 \end{bmatrix}$$

has eigenvalues $\lambda_1 = 0$, $\lambda_2 = -0.2$ and is reduced to diagonal Jordan form by

$$\mathbf{P} = \begin{bmatrix} 5 & 5 \\ 4 & 2 \end{bmatrix}, \qquad \mathbf{P}^{-1} = \begin{bmatrix} -0.2 & 0.5 \\ 0.4 & -0.5 \end{bmatrix}$$

From Theorem 10.13 we have $\mathbf{x} = \mathbf{P}\mathbf{L}\mathbf{P}^{-1}\mathbf{x}_0$ which equals

$$\mathbf{x}(t) = \begin{bmatrix} 2400 - 2300e^{-0.2t} \\ 1920 - 920e^{-0.2t} \end{bmatrix}.$$

EXERCISE 10.82. Solve the equations of Ex. 10.78 with $k = 0.18$, $t_0 = 0$.

EXERCISE 10.83. Solve $\dot{\mathbf{x}} = \mathbf{A}\mathbf{x}$, $\mathbf{x}(0) = \mathbf{x}_0$, where

$$\mathbf{A} = \begin{bmatrix} 0 & 1 & 0 \\ 0 & 0 & 1 \\ 6 & -1 & -4 \end{bmatrix}, \qquad \mathbf{x}_0 = \begin{bmatrix} 12 \\ -12 \\ 12 \end{bmatrix}.$$

The situation is almost as simple when the matrix cannot be reduced to diagonal form and we have to use the Jordan canonical form. If $\mathbf{Q}^{-1}\mathbf{A}\mathbf{Q} = \mathbf{J}$, where $\mathbf{J}$ is the Jordan canonical form, the equation becomes, on introducing $\mathbf{Q}^{-1}\mathbf{x} = \mathbf{y}$ [compare (10.44)]:

$$\dot{\mathbf{y}} = \mathbf{J}\mathbf{y}. \tag{10.51}$$

Suppose first that $\mathbf{J}$ is a single Jordan canonical block of the form given in Definition 10.2. On writing out the separate equations in (10.51) we obtain

$$\begin{aligned}
\frac{dy_1}{dt} &= \lambda y_1 + y_2, &\quad \text{that is,}\quad \frac{d(y_1 e^{-\lambda t})}{dt} &= y_2 e^{-\lambda t} \\
&\cdots & &\cdots \\
\frac{dy_{n-1}}{dt} &= \lambda y_{n-1} + y_n, &\quad \text{that is,}\quad \frac{d(y_{n-1} e^{-\lambda t})}{dt} &= y_n e^{-\lambda t} \\
\frac{dy_n}{dt} &= \lambda y_n, &\quad \text{that is,}\quad \frac{d(y_n e^{-\lambda t})}{dt} &= 0.
\end{aligned} \tag{10.52}$$

These can be solved in turn, starting from the last equation. It is left to the reader to show that $y_r e^{-\lambda t}$ can be expressed as a polynomial of degree $n - r$ in t, and that the final results can be written in the form

$$\mathbf{y} = \mathbf{K}\mathbf{y}_0, \tag{10.53}$$

where the rth element of $\mathbf{y}_0$ is the value of y_r at $t = 0$, and

$$\mathbf{K}(t) = e^{\lambda t}\begin{bmatrix} 1 & t & \frac{1}{2}t^2 & \cdots & \dfrac{t^{n-1}}{(n-1)!} \\ 0 & 1 & t & \cdots & \dfrac{t^{n-2}}{(n-2)!} \\ & & \cdots & & \\ 0 & 0 & 0 & \cdots & 1 \end{bmatrix}. \tag{10.54}$$

In the general case, when $\mathbf{J}$ has the structure (10.8) we see that (10.51) and its solution involve

$$\mathbf{J} = \begin{bmatrix} \mathbf{J}_1 & \cdots & \mathbf{0} \\ & \cdots & \\ \mathbf{0} & \cdots & \mathbf{J}_k \end{bmatrix}, \qquad \mathbf{K} = \begin{bmatrix} \mathbf{K}_1 & \cdots & \mathbf{0} \\ & \cdots & \\ \mathbf{0} & \cdots & \mathbf{K}_k \end{bmatrix}, \tag{10.55}$$

where $\mathbf{K}_i$ is of the form (10.54) with λ_i in place of λ and n_i in place of n. To sum up the situation in this case: the solution of $\dot{\mathbf{x}} = \mathbf{A}\mathbf{x}$ is given by

$$\mathbf{x} = \mathbf{Q}\mathbf{y}, \quad \text{where} \quad \mathbf{y} = \mathbf{K}\mathbf{y}_0, \quad \text{and} \quad \mathbf{y}_0 = \mathbf{Q}^{-1}\mathbf{x}_0. \tag{10.56}$$

We summarize

THEOREM 10.14. Let the $n \times n$ matrix $\mathbf{A}$ have the Jordan form $\mathbf{J}$, with $\mathbf{Q}^{-1}\mathbf{A}\mathbf{Q} = \mathbf{J}$ as in (10.49) and (10.55). Then the solution $\mathbf{x}(t)$ to the equation

$\dot{\mathbf{x}} = \mathbf{A}\mathbf{x}$ *of* (10.43), *subject to the given initial condition* $\mathbf{x}(0) = \mathbf{x}_0$, *is* $\mathbf{x}(t) = \mathbf{Q}\mathbf{K}(t)\mathbf{Q}^{-1}\mathbf{x}_0$, *where* $\mathbf{K}(t)$ *is as in (10.55) and each* $\mathbf{K}_i$ *is as in (10.54) with* λ *replaced by the eigenvalue* λ_i *appearing in the Jordan block* $\mathbf{J}_i$ *and* n *replaced by the order* n_i *of that same Jordan block* $\mathbf{J}_i$.

EXERCISE 10.84. Solve $\dot{\mathbf{x}} = \mathbf{A}\mathbf{x}$ where $\mathbf{A}$ is given in Ex. 10.14 and

$$\mathbf{x}(0) = \begin{bmatrix} 2 \\ -2 \\ 2 \end{bmatrix}.$$

SOLUTION: We found in Ex. 10.14 that

$$\mathbf{Q} = \begin{bmatrix} 1 & 1 & 0 \\ -1 & -1 & 1 \\ -1 & 1 & -1 \end{bmatrix}, \qquad \mathbf{Q}^{-1} = \begin{bmatrix} 0 & -0.5 & -0.5 \\ 1 & 0.5 & 0.5 \\ 1 & 1 & 0 \end{bmatrix}$$

reduce $\mathbf{A}$ to the Jordan form

$$\mathbf{J} = \begin{bmatrix} -2 & 0 & 0 \\ 0 & 4 & 1 \\ 0 & 0 & 4 \end{bmatrix}.$$

From Theorem 10.14 we then find

$$\mathbf{x}(t) = \begin{bmatrix} 4e^{4t}(1+t) - 2e^{-2t} \\ -4e^{4t}(1+t) + 2e^{-2t} \\ 4te^{4t} + 2e^{-2t} \end{bmatrix}.$$

EXERCISE 10.85. Solve

$$\begin{aligned} \dot{x}_1 &= 2x_1 + x_2, & x_1(0) &= 1 \\ \dot{x}_2 &= 2x_2, & x_2(0) &= 1. \end{aligned}$$

So far we have confined our attention to the equation $\dot{\mathbf{x}} = \mathbf{A}\mathbf{x}$ of (10.43). A slight extension of the method will take care of the more general equation (10.42). Consider only the case where $\mathbf{A}$ can be reduced to diagonal form. If $\mathbf{P}^{-1}\mathbf{A}\mathbf{P} = \mathbf{\Lambda}$ where $\mathbf{\Lambda}$ is diagonal, equation (10.42) can be written [compare (10.44)]

$$\frac{d}{dt}(\mathbf{P}^{-1}\mathbf{x}) = (\mathbf{P}^{-1}\mathbf{A}\mathbf{P})\mathbf{P}^{-1}\mathbf{x} + \mathbf{P}^{-1}\mathbf{f},$$

or

$$\dot{\mathbf{y}} = \mathbf{\Lambda}\mathbf{y} + \mathbf{g}, \tag{10.57}$$

where we have introduced the notation $\mathbf{y} = \mathbf{P}^{-1}\mathbf{x}$, $\mathbf{g}(t) = \mathbf{P}^{-1}\mathbf{f}(t)$. The rth equation in (10.57) is

$$\frac{dy_r}{dt} = \lambda_r y_r + g_r(t), \tag{10.58}$$

or

$$\frac{d}{dt}(y_r e^{-\lambda_r t}) = g_r(t)e^{-\lambda_r t}$$

which then gives

$$y_r = e^{\lambda_r t}\left[(y_r)_0 + \int_0^t g_r(\tau)e^{-\lambda_r \tau}\,d\tau\right]. \tag{10.59}$$

This equation generalizes (10.47). Instead of (10.48), we obtain

$$\mathbf{y} = \mathbf{L}\{\mathbf{y}_0 + \mathbf{w}(t)\},$$

where $\mathbf{L}$, $\mathbf{y}_0$ were defined in connection with (10.49) and $\mathbf{w}(t)$ is a vector whose rth element is given by the integral in (10.59). The solution is given by

$$\mathbf{x} = \mathbf{P}\mathbf{y} = \mathbf{P}\mathbf{L}\{\mathbf{y}_0 + \mathbf{w}(t)\}.$$

The case where $\mathbf{A}$ cannot be reduced to diagonal form can be solved in the same way, but instead of solving (10.58) we have to solve (10.52) with additional nonhomogeneous terms. Nothing new is introduced in principle, and we do not pursue the matter.

EXERCISE 10.86. Solve $\dot{\mathbf{x}} = \mathbf{A}\mathbf{x} + \mathbf{f}$ with $\mathbf{A}$ and $\mathbf{x}_0$ as in Ex. 10.81 but with

$$\mathbf{f}(t) = \begin{bmatrix} 1 \\ 1 \end{bmatrix}.$$

EXERCISE 10.87. Solve $\dot{\mathbf{x}} = \mathbf{A}\mathbf{x} + \mathbf{f}$ with $\mathbf{A}$ and $\mathbf{x}_0$ as in Ex. 10.84 but with

$$\mathbf{f}(t) = \begin{bmatrix} 1 \\ 0 \\ t \end{bmatrix}.$$

It should now be clear from (10.47), (10.49), (10.54), and (10.59) that exponential functions e^{at} play a fundamental role in the solution of $\dot{\mathbf{x}} = \mathbf{A}\mathbf{x}$ and $\dot{\mathbf{x}} = \mathbf{A}\mathbf{x} + \mathbf{f}$; of course if x and A are scalar quantities, then the solution to $\dot{x} = Ax$, $x(0) = x_0$ is precisely the exponential function $\exp(At) = e^{At}$, so the involvement of exponentials more generally should be no surprise. For the matrix case, we want to assign a meaning to $\exp(\mathbf{A}t)$ which will extend $\exp(At)$ and will simplify the solution of $\dot{\mathbf{x}} = \mathbf{A}\mathbf{x}$, for example. We first consider how to define $\exp(\mathbf{B})$ for $n \times n$ matrices $\mathbf{B}$. If we replace $\mathbf{B}$ by the scalar B, then exp (B) is defined by its convergent power series

$$\exp(B) = 1 + B + \frac{B^2}{2} + \cdots = \sum_{i=0}^{\infty} \frac{B^i}{i!}, \tag{10.60}$$

and we might hope to use such a definition for exp $(\mathbf{B})$. If

$$\mathbf{Q}^{-1}\mathbf{B}\mathbf{Q} = \mathbf{J}$$

is the Jordan form of $\mathbf{B}$, then $\mathbf{B}^i = \mathbf{Q}\mathbf{J}^i\mathbf{Q}^{-1}$ and hence

$$\sum_{i=0}^{N} \frac{\mathbf{B}^i}{i!} = \mathbf{Q}\left[\sum_{i=0}^{N} \frac{\mathbf{J}^i}{i!}\right]\mathbf{Q}^{-1} \tag{10.61}$$

which leads us to expect

$$\exp(\mathbf{B}) = \mathbf{Q}[\exp(\mathbf{J})]\mathbf{Q}^{-1}, \tag{10.62}$$

leaving us only the problem of defining exp $(\mathbf{J})$. By explicitly summing the series (10.60) with B replaced by $\mathbf{J}$ it is easy to see that if $\mathbf{J}$ has the usual structure of (10.8) and (10.55) then

$$\exp(\mathbf{J}) = \begin{bmatrix} \exp(\mathbf{J}_1) & \cdots & \mathbf{0} \\ & \cdots & \\ \mathbf{0} & \cdots & \exp(\mathbf{J}_k) \end{bmatrix}, \tag{10.63}$$

where

$$\exp(\mathbf{J}_i) = e^{\lambda_i} \begin{bmatrix} 1 & 1 & \frac{1}{2} & \cdots & \frac{1}{(n_i - 1)!} \\ 0 & 1 & 1 & \cdots & \frac{1}{(n_i - 2)!} \\ & & & \cdots & \\ 0 & 0 & 0 & \cdots & 1 \end{bmatrix}. \tag{10.64}$$

Thus, the representations of $\mathbf{x}(t)$ in Theorems 10.13 and 10.14 are the same as saying

$$\mathbf{x} = \exp(t\mathbf{A})\mathbf{x}_0.$$

EXERCISE 10.88. Find exp $(\mathbf{A})$ when $\mathbf{A}$ is the matrix of Ex. 10.81.

SOLUTION: We found $\mathbf{P}$ and $\mathbf{P}^{-1}$ in Ex. 10.81 so that $\mathbf{P}^{-1}\mathbf{A}\mathbf{P} = \mathbf{J}$, where

$$\mathbf{J} = \begin{bmatrix} 0 & 0 \\ 0 & -0.2 \end{bmatrix}.$$

Then by (10.63), (10.64)

$$\exp(\mathbf{J}) = \begin{bmatrix} 1 & 0 \\ 0 & e^{-0.2} \end{bmatrix}$$

and hence $\exp(\mathbf{A}) = \mathbf{P} \exp(\mathbf{J}) \mathbf{P}^{-1}$ is given by

$$\exp(\mathbf{A}) = \begin{bmatrix} -1 + 2e^{-0.2} & 2.5(1 - e^{-0.2}) \\ -0.08 - 0.2e^{-0.2} & 0.2 + 0.25e^{-0.2} \end{bmatrix}.$$

EXERCISE 10.89. Find exp (**A**) with **A** as in Ex. 10.82.

EXERCISE 10.90. Find exp (**A**) with **A** as in Ex. 10.84.

Just as in Section 10.5 where the Jordan form of $\mathbf{A}^i$ allowed us to describe in **Key Theorem 10.6** the behavior of $\mathbf{A}^i$ as i tended to infinity, our Jordan form

$$\exp(t\mathbf{A}) = \mathbf{Q} \exp(t\mathbf{J})\mathbf{Q}^{-1}$$

allows us to describe the behavior of exp (t**A**) and hence of solutions

$$\mathbf{x}(t) = \exp(t\mathbf{A})\mathbf{x}_0 \tag{10.65}$$

of $\dot{\mathbf{x}} = \mathbf{A}\mathbf{x}$, $\mathbf{x}(0) = \mathbf{x}_0$, as t tends to plus infinity; in fact it follows from (10.65), and (10.62)–(10.64) that the following **key** facts hold (compare **Key Theorem 10.6**).

● *KEY THEOREM 10.15. Let* **A** *be an* $n \times n$ *matrix. Then, as* t *tends to plus infinity:*

(i) *exp* (t**A**) *tends to zero if and only if all the eigenvalues of* **A** *have strictly negative real parts;*

(ii) *exp* (t**A**) *is bounded (in the sense that* $\|\exp(t\mathbf{A})\| \le c$ *for all* t *and some fixed constant* c*) if and only if all eigenvalues of* **A** *have nonpositive real parts and, for those pure imaginary eigenvalues* λ*, the number of eigenvectors associated with* λ *in an independent set equals the multiplicity of* λ *as a root of the characteristic polynomial of* **A***;*

(iii) $\|\exp(t\mathbf{A})\|$ *tends to infinity if and only if either some eigenvalue of* **A** *has positive real part or some purely imaginary eigenvalue* λ *has its maximum number of associated eigenvectors in an independent set being strictly less than the multiplicity of* λ.

EXERCISE 10.91. Determine the behavior of the populations modeled in our Ex. 10.78 for the values $k = 0.18$ and $k = 0.10$.

SOLUTION: For $k = 0.18$, the eigenvalues of **A** are $\lambda_1 = \lambda_2 = -0.1$; therefore, by **Key Theorem 10.15,** exp(t**A**) tends to zero and hence by (10.65) so do all solutions $\mathbf{x}(t)$. For $k = 0.10$, the eigenvalues of **A** are $\lambda_1 = 0.1$, $\lambda_2 = -0.3$; therefore, by **Key Theorem 10.15,** $\|\exp(t\mathbf{A})\|$ tends to infinity and hence from (10.65) we see that *some* solutions **x** do also. It is left to the reader to show that $\|\mathbf{x}\|$ tends to infinity for the initial conditions $\mathbf{x}_0$ in Ex. 10.78.

EXERCISE 10.92. Find the value of k in Ex. 10.78 so that all solutions remain bounded but not all solutions tend to zero.

EXERCISE 10.93. If $\mathbf{AB} = \mathbf{BA}$ where $\mathbf{A}$ and $\mathbf{B}$ are square, prove that $\exp(\mathbf{A} + \mathbf{B}) = \exp(\mathbf{A}) \exp(\mathbf{B})$; show by a 2×2 example that if $\mathbf{AB} \neq \mathbf{BA}$ then we may have $\exp(\mathbf{A} + \mathbf{B}) \neq \exp(\mathbf{A}) \exp(\mathbf{B})$.

EXERCISE 10.94. Prove that $\exp(\mathbf{A})$ is nonsingular for every square matrix $\mathbf{A}$ and that in fact $[\exp(\mathbf{A})]^{-1} = \exp(-\mathbf{A})$.

MISCELLANEOUS EXERCISES 10

EXERCISE 10.95. Show that the class of nondefective matrices with multiple eigenvalues is strictly larger than the class of normal matrices by considering

$$\mathbf{A} = \begin{bmatrix} 2 & 0 & 0 \\ 0 & 3 & 1 \\ 0 & 0 & 2 \end{bmatrix}.$$

EXERCISE 10.96. Give a detailed proof of Theorem 10.5.

EXERCISE 10.97. Find the eigenvalues and the vectors $\mathbf{v}_{i,j}$ of **Key Theorem 10.3** for

$$\mathbf{A} = \begin{bmatrix} 5 & 4 & 3 \\ -1 & 0 & -3 \\ 1 & -2 & 1 \end{bmatrix}.$$

EXERCISE 10.98. Find the eigenvalues and the vectors $\mathbf{v}_{i,j}$ of **Key Theorem 10.3** for

$$\mathbf{A} = \begin{bmatrix} 2 & 2 & -1 \\ -1 & -1 & 1 \\ -1 & -2 & 4 \end{bmatrix}.$$

EXERCISE 10.99. Some insight into the Jordan canonical form can be obtained from the following argument, which we present in a nonrigorous form. Suppose that

$$\mathbf{A}\mathbf{x}_1 = \lambda_1 \mathbf{x}_1, \qquad \mathbf{A}\mathbf{x}_2 = \lambda_2 \mathbf{x}_2, \qquad \lambda_1 \neq \lambda_2$$

and that $\mathbf{A} = \mathbf{A}_0 + \epsilon \mathbf{B}$, where $\mathbf{A}_0$ has a repeated eigenvalue λ_1. Suppose that, for small ϵ, $\lambda_2 \approx \lambda_1 + \epsilon$, $\mathbf{x}_2 = \mathbf{x}_1 + \epsilon \mathbf{z}$. Subtracting the two equations above, we find, neglecting second-order terms, that

$$\mathbf{A}_0 \mathbf{z} = \lambda_1 \mathbf{z} + \mathbf{x}_1.$$

Hence **z** corresponds to what we have called a generalized eigenvector. It appears when eigenvalues and eigenvectors coalesce.

EXERCISE 10.100. Suppose that the $n \times n$ matrices $\mathbf{A}$ and $\mathbf{B}$ commute, and each has a linearly independent set of n eigenvectors. If $\mathbf{P}^{-1}\mathbf{B}\mathbf{P} = \mathbf{\Lambda}$, where $\mathbf{\Lambda}$ is diagonal, the distinct eigenvalues of $\mathbf{B}$ being $\lambda_1, \ldots, \lambda_s$ with multiplicities $m_1, \ldots, m_s$, show that $\mathbf{P}^{-1}\mathbf{A}\mathbf{P}$ has a block-diagonal form,

$$\mathbf{P}^{-1}\mathbf{A}\mathbf{P} = \begin{bmatrix} \mathbf{C}_1 & \ldots & \mathbf{0} \\ & \ldots & \\ \mathbf{0} & \ldots & \mathbf{C}_s \end{bmatrix},$$

where $\mathbf{C}_i$ is $m_i \times m_i$.

EXERCISE 10.101. The result in the last exercise means that if we can find $\mathbf{B}$ such that $\mathbf{AB} = \mathbf{BA}$, where $\mathbf{B}$ has an independent set of n eigenvectors (the columns of $\mathbf{P}$) that can be found easily, then it is often easy to find the eigenvectors of $\mathbf{A}$ by deducing them from those for $\mathbf{P}^{-1}\mathbf{A}\mathbf{P}$. A suitable form for $\mathbf{B}$ is often indicated on physical grounds when symmetry is present. As a simple example, consider the system of resistors and capacitors in Figure 10.4. If q_i is the charge on the ith capacitor, remembering that the voltage across a capacitor C containing charge q is q/C, and the current through it is $i = dq/dt$, prove that the equations for the circuit are

$$RC\frac{d\mathbf{q}}{dt} = \mathbf{A}\mathbf{q}, \qquad \mathbf{A} = \begin{bmatrix} -2 & 1 & 1 \\ 1 & -2 & 1 \\ 1 & 1 & -2 \end{bmatrix}, \qquad \mathbf{q} = \begin{bmatrix} q_1 \\ q_2 \\ q_3 \end{bmatrix}.$$

Because of symmetry, the circuit is unchanged if capacitor 1 is called 2, 2 is called 3, and 3 is called 1. This suggests that we try the following permutation matrix for $\mathbf{B}$. The eigenvectors of $\mathbf{B}$, which give the columns of $\mathbf{P}$, are easily found $[\omega = \frac{1}{2}(-1 + i\sqrt{3}),\ \omega^3 = 1,\ \omega^2 + \omega + 1 = 0]$:

$$\mathbf{B} = \begin{bmatrix} 0 & 1 & 0 \\ 0 & 0 & 1 \\ 1 & 0 & 0 \end{bmatrix},$$

$$\mathbf{P} = \begin{bmatrix} 1 & 1 & 1 \\ 1 & \omega & \omega^2 \\ 1 & \omega^2 & \omega \end{bmatrix},$$

$$\mathbf{P}^{-1} = \frac{1}{3}\begin{bmatrix} 1 & 1 & 1 \\ 1 & \omega^2 & \omega \\ 1 & \omega & \omega^2 \end{bmatrix}.$$

We readily find **AB** = **BA** so that we can use the method in the last exercise. This gives

$$\mathbf{P}^{-1}\mathbf{A}\mathbf{P} = \begin{bmatrix} 0 & 0 & 0 \\ 0 & -3 & 0 \\ 0 & 0 & -3 \end{bmatrix}$$

so that the eigenvalues of **A** are 0, −3, −3 and the eigenvectors are the columns of **P**.

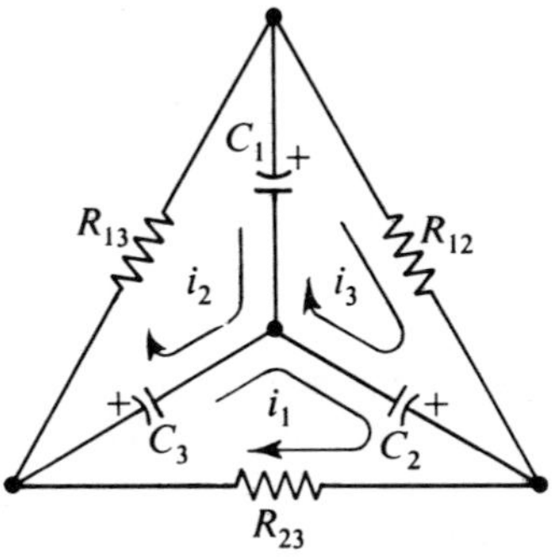

Figure 10.4. Resistor-capacitor network for Ex. 10.101.

EXERCISE 10.102. Show that the equations of motion for the ozone molecule in Figure 10.5 are given by

$$m\frac{d^2\mathbf{x}}{dt^2} = -k\mathbf{A}\mathbf{x}$$

where $\mathbf{x}$ is 6×1, m is the mass of each atom, k is the spring constant and, if $s = \sin(\pi/6)$, $c = \cos(\pi/6)$,

$$\mathbf{A} = \begin{bmatrix} \mathbf{Q} & \mathbf{R} \\ \mathbf{R}^T & \mathbf{S} \end{bmatrix},$$

where

$$\mathbf{Q} = c^2\begin{bmatrix} 2 & 1 & 1 \\ 1 & 2 & 1 \\ 1 & 1 & 2 \end{bmatrix},$$

$$\mathbf{R} = sc\begin{bmatrix} 0 & 1 & -1 \\ -1 & 0 & 1 \\ 1 & -1 & 0 \end{bmatrix},$$

$$\mathbf{S} = s^2\begin{bmatrix} 2 & -1 & -1 \\ -1 & 2 & -1 \\ -1 & -1 & 2 \end{bmatrix}.$$

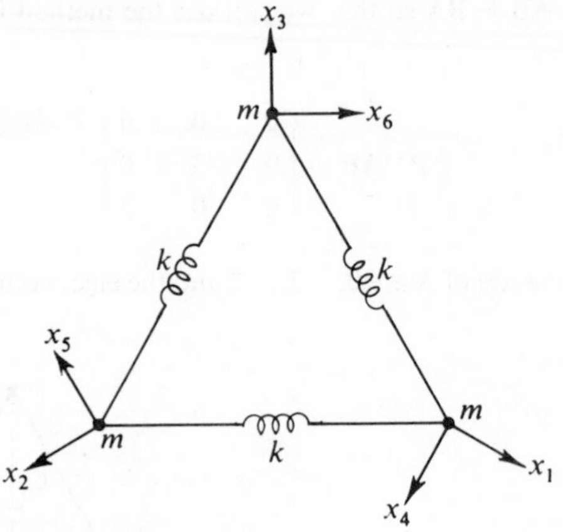

Figure 10.5. Ozone molecule for Ex. 10.102.

The symmetry in 1, 2, 3, and 4, 5, 6 suggests that we try to see if the following matrix **C** commutes with **A**:

$$\mathbf{C} = \begin{bmatrix} \mathbf{B} & \mathbf{0} \\ \mathbf{0} & \mathbf{B} \end{bmatrix},$$

where **B** was defined in Ex. 10.101. It is left to the reader to check that **AC** = **CA**. We therefore form, using **P** defined in Ex. 10.101,

$$\begin{bmatrix} \mathbf{P}^{-1} & \mathbf{0} \\ \mathbf{0} & \mathbf{P}^{-1} \end{bmatrix} \begin{bmatrix} \mathbf{Q} & \mathbf{R} \\ \mathbf{R}^T & \mathbf{S} \end{bmatrix} \begin{bmatrix} \mathbf{P} & \mathbf{0} \\ \mathbf{0} & \mathbf{P} \end{bmatrix} = \frac{3}{4} \begin{bmatrix} 4 & 0 & 0 & 0 & 0 & 0 \\ 0 & 1 & 0 & 0 & i & 0 \\ 0 & 0 & 1 & 0 & 0 & -i \\ 0 & 0 & 0 & 0 & 0 & 0 \\ 0 & -i & 0 & 0 & 1 & 0 \\ 0 & 0 & i & 0 & 0 & 1 \end{bmatrix}.$$

This can be reduced to a simpler block-diagonal form by permuting rows and columns. Whether this is done or not, it is a simple matter to deduce the eigenvalues, namely 0 with multiplicity 3, 1.5 with multiplicity 2, and 3 with multiplicity 1. The corresponding eigenvectors can be shown to be

$$\begin{bmatrix} 1 \\ -s \\ -s \\ 0 \\ -c \\ c \end{bmatrix}, \quad \begin{bmatrix} 0 \\ c \\ -c \\ 1 \\ -s \\ -s \end{bmatrix}, \quad \begin{bmatrix} 0 \\ 0 \\ 0 \\ 1 \\ 1 \\ 1 \end{bmatrix}, \quad \begin{bmatrix} 1 \\ -s \\ -s \\ 0 \\ c \\ -c \end{bmatrix}, \quad \begin{bmatrix} 0 \\ -c \\ c \\ 1 \\ -s \\ -s \end{bmatrix}, \quad \begin{bmatrix} 1 \\ 1 \\ 1 \\ 0 \\ 0 \\ 0 \end{bmatrix}.$$

The first three are rigid motions (two translations and a rotation). Only three nontrivial modes exist, of which the last is a dilation.

EXERCISE 10.103. Suppose that the $n \times n$ matrix $\mathbf{A}$ has one simple eigenvalue λ_1 with $|\lambda_1| = 1$ and that all other eigenvalues λ satisfy $|\lambda| < 1$. Describe how one might use the behavior of $\mathbf{A}^i$ or of $\mathbf{A}^i\mathbf{x}_0$ to compute an approximation to λ_1.

EXERCISE 10.104. Characterize those $\mathbf{x}_0$ for which the Markov-chain sequence $\mathbf{x}_i$ in Theorem 10.9 converges to zero.

EXERCISE 10.105. Verify that the "fill-in" described in Ex. 10.64 actually occurs.

EXERCISE 10.106. Describe the "fill-in" that would result from using Gauss elimination in Ex. 10.67.

CHAPTER ELEVEN

QUADRATIC FORMS AND VARIATIONAL PRINCIPLES

This chapter continues the study of eigenvalues and eigenvectors but uses a different (geometrical) approach from that used before. **Key Theorem 11.3** relates the new concept of quadratic forms to eigenvalues in a very simple fashion. **Key Theorems 11.7** (and its extension **11.8**) and **11.9** give much more precise and complex information on the relationships between these concepts.

11.1 INTRODUCTION

After the constant functions and the linear functions (which were important to us in Chapter 7 on linear programming), the quadratic functions are next in order of complexity. Such functions arise in diverse areas of application, but fortunately matrix techniques allow a unified study of their properties; conversely, it happens that certain quadratic functions are useful in studying matrices, especially their eigensystems. We will study the special functions evaluated as $(\mathbf{x}, \mathbf{Ax}) = \mathbf{x}^H\mathbf{Ax}$ (the so-called *quadratic forms*), where $\mathbf{A}$ is an $n \times n$ hermitian matrix, the variable $\mathbf{x}$ ranges over $\mathbb{R}^n$ or $\mathbb{C}^n$, and $(\cdot, \cdot)$ denotes the inner product of Section 4.7. We start off by describing some situations in which quadratic forms arise naturally:

1. In two dimensions, the so-called *conic sections* are the simplest curves and play a fundamental role in two-dimensional geometry. For example, the ellipse described by $2x_1^2 + x_2^2 = 1$ can equivalently be described by $\mathbf{x}^H\mathbf{A}\mathbf{x} = 1$ where $\mathbf{x}$ is in $\mathbb{R}^2$ and

$$\mathbf{A} = \begin{bmatrix} 2 & 0 \\ 0 & 1 \end{bmatrix};$$

thus an ellipse in two dimensions is a set of points on which a certain quadratic form is constant. In n dimensions, the equation $\mathbf{x}^H\mathbf{A}\mathbf{x} = 1$, where $\mathbf{A}$ is $n \times n$, similarly gives rise to the analogues of conic sections and plays a fundamental role in n-dimensional geometry.

2. In Chapter 7 on linear programming we saw indications of the wide range of applications that can be described mathematically as attempting to locate a point maximizing or minimizing some real-valued function $f(\mathbf{x})$ of n real variables $x_1, \ldots, x_n$ (written as the components of $\mathbf{x}$ in $\mathbb{R}^n$). Assuming for simplicity that f is extremized at the origin $\mathbf{x} = \mathbf{0}$, we can get an idea of the behavior of f near there by examining the Taylor series

$$f(x_1, \ldots, x_n) = (f)_0 + \sum_{i=1}^{n} (f_i)_0 x_i + \sum_{i=1}^{n} \sum_{j=1}^{n} (f_{ij})_0 x_i x_j + \cdots$$

where

$$f_i = \frac{\partial f}{\partial x_i}, \qquad f_{ij} = \frac{\partial^2 f}{\partial x_i \, \partial x_j},$$

and the notation $(\ )_0$ means that the quantities within the parentheses are evaluated at the origin $\mathbf{0}$. At an extremizing point the first derivatives f_i of course equal zero, so that the local behavior of f is approximately described by the quadratic form

$$\sum_{i=1}^{n} \sum_{j=1}^{n} (f_{ij})_0 x_i x_j.$$

This generalizes the theory for $n = 1$, in which case the nature of the extremum (minimum, maximum, or inflection point) depends on d^2f/dx^2.

3. In studying the dynamics of physical systems, important physical quantities such as kinetic energy and potential energy are often approximated by quadratic forms in the neighborhood of an equilibrium state of the system. Thus the theory of the behavior of small vibrations about equilibrium in many-coordinate systems naturally involves quadratic forms.

4. As in Section 9.7, in the statistical analysis of data described by random variables $x_1, \ldots, x_n$, it is often important to try to find some small set of factors which explains the experimental results. If the means of the varlables x_i are assumed to be zero and if $\mathbf{S}$ is the $n \times n$ so-called *covariance matrix* whose (i, j) element s_{ij} measures the mean of $x_i x_j$, a common approach is to seek a new variable y_1 as a linear combination

$$y_1 = a_1 x_1 + \cdots + a_n x_n$$

of the old variables so that the variance

$$\sum_{i=1}^{n} \sum_{j=1}^{n} a_i a_j s_{ij} \tag{11.1}$$

of y_1 is as large as possible (subject to the technical constraint that $a_1^2 + \cdots + a_n^2 = 1$). If we let $\mathbf{a}$ denote the column vector with components $a_1, \ldots, a_n$, then the mathematical statement of this statistics problem becomes:

$$\text{maximize} \quad (\mathbf{a}, \mathbf{Sa}) = \mathbf{a}^H\mathbf{Sa} \tag{11.2}$$

$$\text{subject to} \quad (\mathbf{a}, \mathbf{a}) = 1. \tag{11.3}$$

This general problem will be studied closely in Section 11.5.

5. In Sections 2.6 and 9.6 we studied the least-squares problem of minimizing with respect to $\mathbf{x}$ the expression

$$\|\mathbf{Ax} - \mathbf{b}\|_2^2 = (\mathbf{Ax} - \mathbf{b}, \mathbf{Ax} - \mathbf{b}) = (\mathbf{x}, \mathbf{A}^H\mathbf{Ax}) - 2(\mathbf{A}^H\mathbf{b}, \mathbf{x}) + (\mathbf{b}, \mathbf{b})$$

which again involves a quadratic form, namely $(\mathbf{x}, \mathbf{A}^H\mathbf{Ax})$.

Having seen how quadratic forms often arise in applications, we return to (1), the simplest of our examples, to begin a more detailed study of quadratic forms.

11.2 A GEOMETRIC EXAMPLE

We begin to develop the subject of quadratic forms by considering a simple geometrical example involving conic sections. We first solve the problem by elementary algebra, and then show that a matrix description leads naturally to eigenvalues and eigenvectors. The virtue of the eigenvector approach is that it provides a method for dealing with n-variable quadratic expressions that are generalizations of the equation for conics in two dimensions.

Consider the problem of discovering the nature of the curve represented by the equation

$$ax^2 + 2kxy + by^2 = c, \tag{11.4}$$

where a, k, b, c are given real numbers. The clue to the analysis lies in transforming the equation to a new set of perpendicular axes X, Y, rotated through an angle θ with respect to the old x-, y-axes, and such that the resulting equation has no cross-term in XY. In order to perform the rotation of axes algebraically, we set (see Figure 11.1)

$$\begin{aligned} x &= X\cos\theta - Y\sin\theta, \\ y &= X\sin\theta + Y\cos\theta \end{aligned} \tag{11.5}$$

where θ is an angle which is at our disposal. If we substitute (11.5) in (11.4) we obtain

$$\alpha X^2 + 2\kappa XY + \beta Y^2 = c, \tag{11.6}$$

where

$$\begin{aligned} \alpha &= a\cos^2\theta + 2k\sin\theta\cos\theta + b\sin^2\theta, \\ \kappa &= (b-a)\sin\theta\cos\theta + k(\cos^2\theta - \sin^2\theta), \\ \beta &= a\sin^2\theta - 2k\sin\theta\cos\theta + b\cos^2\theta, \end{aligned} \tag{11.7}$$

and we also find that

$$\kappa^2 - \alpha\beta = k^2 - ab \tag{11.8}$$

The quantity κ vanishes if

$$\tan 2\theta = \frac{2k}{(a-b)}; \tag{11.9}$$

we interpret (11.9) as including the case $a = b$, when we set $\theta = \pi/4$. We then have form (11.6) and (11.8) that

$$\alpha X^2 + \beta Y^2 = c, \qquad \alpha\beta = ab - k^2. \tag{11.10}$$

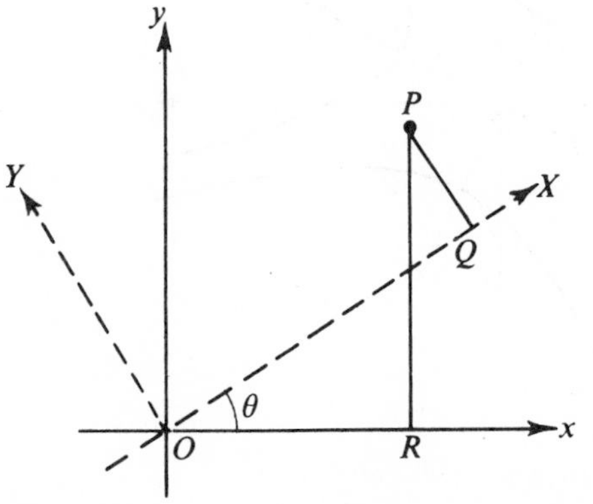

$$\begin{aligned} OR &= OQ\cos\theta - PQ\sin\theta \\ RP &= OQ\sin\theta + PQ\cos\theta \end{aligned}$$

Figure 11.1. Rotation of axes.

By inspection of (11.10), we see that there are three cases to consider:

1. $k^2 < ab$ (so that α and β have the same sign). If α, β, c have the same sign, then (11.4) represents an ellipse [Figure 11.2(a)]. If the sign of c is opposite to that of α and β, then there are no points x, y that satisfy (11.4).
2. $k^2 = ab$ (so that α or β is zero). Apart from the trivial case $\alpha = \beta = 0$, (11.4) represents two parallel straight lines [Figure 11.2(b)].
3. $k^2 > ab$ (so that α and β have opposite signs). If $c \neq 0$, then (11.4) represents a hyperbola [Figure 11.2(c)]. If $c = 0$, the equation represents two straight lines through the origin.

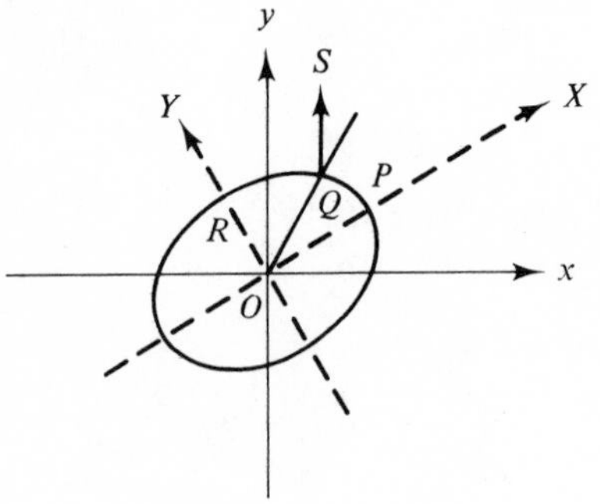

(a) Ellipse, $k^2 < ab$

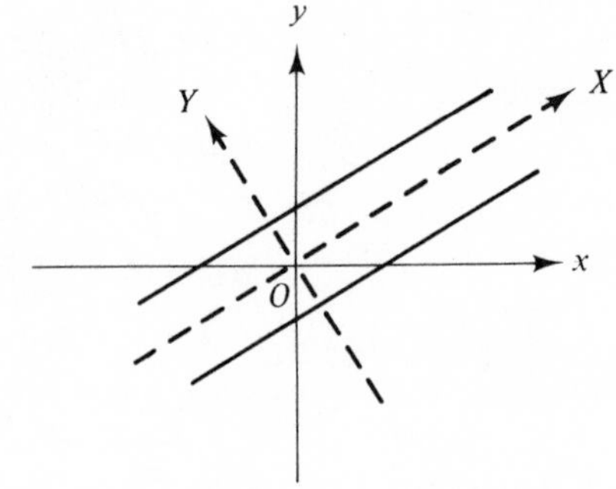

(b) Parallel straight lines, $k^2 = ab$.

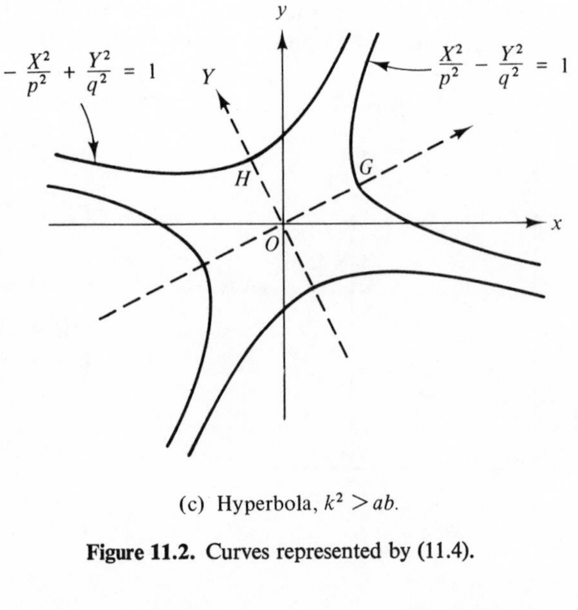

(c) Hyperbola, $k^2 > ab$.

Figure 11.2. Curves represented by (11.4).

In matrix notation, the analysis leading from (11.4) to (11.10) can be rewritten in the following way. We introduce the matrices

$$\mathbf{x} = \begin{bmatrix} x \\ y \end{bmatrix}, \qquad \mathbf{A} = \begin{bmatrix} a & k \\ k & b \end{bmatrix}, \qquad \mathbf{c} = [c], \tag{11.11}$$

$$\mathbf{X} = \begin{bmatrix} X \\ Y \end{bmatrix}, \qquad \mathbf{S} = \begin{bmatrix} \cos\theta & -\sin\theta \\ \sin\theta & \cos\theta \end{bmatrix}, \qquad \mathbf{D} = \begin{bmatrix} \alpha & 0 \\ 0 & \beta \end{bmatrix}. \tag{11.12}$$

It is easy to verify that (11.4), (11.5), (11.10) give us

$$\mathbf{x}^T\mathbf{A}\mathbf{x} = \mathbf{c}, \tag{11.13}$$

$$\mathbf{x} = \mathbf{S}\mathbf{X}, \tag{11.14}$$

$$\mathbf{X}^T\mathbf{D}\mathbf{X} = \mathbf{c}. \tag{11.15}$$

Substituting the expression (11.14) for $\mathbf{x}$ in (11.13), we obtain

$$\mathbf{X}^T(\mathbf{S}^T\mathbf{A}\mathbf{S})\mathbf{X} = \mathbf{c}.$$

A comparison of this equation with (11.15) shows that, since these equations must be true for all $\mathbf{X}$, we must have

$$\mathbf{S}^T\mathbf{A}\mathbf{S} = \mathbf{D}, \tag{11.16}$$

that is, $\mathbf{S}$ is a matrix that transforms $\mathbf{A}$ to diagonal form via (11.16).

In matrix notation, the problem of simplifying the original equation (11.4) can therefore be posed as follows: Find a matrix $\mathbf{S}$ representing a rotation such that $\mathbf{S}^T\mathbf{A}\mathbf{S}$ is a diagonal matrix. One of the key phrases here is "representing a rotation." As we learned in Section 8.5, $\mathbf{S}$ will be an *orthogonal* matrix—i.e., $\mathbf{S}$ has the property

$$\mathbf{S}\mathbf{S}^T = \mathbf{S}^T\mathbf{S} = 1. \tag{11.17}$$

Premultiplying both sides of (11.16) by $\mathbf{S}$ and using (11.17), we see that

$$\mathbf{A}\mathbf{S} = \mathbf{S}\mathbf{D}.$$

If we set $\mathbf{S} = [\mathbf{s}_1, \mathbf{s}_2]$, this means, on introducing $\mathbf{D}$ from (11.12), that

$$\mathbf{A}[\mathbf{s}_1, \mathbf{s}_2] = [\alpha\mathbf{s}_1, \beta\mathbf{s}_2],$$

that is,

$$\mathbf{A}\mathbf{s}_1 = \alpha\mathbf{s}_1, \qquad \mathbf{A}\mathbf{s}_2 = \beta\mathbf{s}_2.$$

Hence α, β are the *eigenvalues* of $\mathbf{A}$ and $\mathbf{s}_1$, $\mathbf{s}_2$ are the corresponding *eigenvectors*, as also follows from **Key Theorems 9.2** and **9.3**.

From this point on, we therefore replace equation (11.4), which represents the conic section in terms of the x, y-axes, by the entirely equivalent equation:

$$\lambda_1 X^2 + \lambda_2 Y^2 = 1 \tag{11.18}$$

in terms of the X, Y-axes, where we have also replaced c by 1. This choice of c means, of course, that we are starting with $c = 1$ in (11.4) as well. Provided $c \neq 0$ [that is, (11.4) does not represent two intersecting straight lines], there is no loss of generality in doing this, since it corresponds simply to dividing (11.4) throughout by a constant, which does not change the curves.

We can now give a geometrical interpretation of the eigenvalues by discussing the three cases listed above in connection with (11.10), but now using (11.18) as the basis for the discussion.

1′. If λ_1 and λ_2 have the same (positive) sign, we compare (11.18) with the canonical equation for an ellipse whose major and minor axes coincide with the X, Y-axes [Figure 11.2(a)],

$$\frac{X^2}{p^2} + \frac{Y^2}{q^2} = 1,$$

where $p = OP$ and $q = OR$ are the lengths of the axes. A comparison of this equation with (11.18) yields

$$p^2 = OP^2 = \frac{1}{\lambda_1}, \qquad q^2 = OR^2 = \frac{1}{\lambda_2}.$$

Hence the eigenvalues λ_1, λ_2 are inversely proportional to the squares of the lengths of the axes of the ellipse.

2′. If one of the eigenvalues is zero, (11.18) shows immediately that this corresponds to the case of two parallel straight lines, and then the nonzero (positive) eigenvalue is proportional to the inverse of the square of the distance from the origin to either of the lines.

3′. If λ_1 and λ_2 have opposite signs, we compare (11.18) with the canonical equation for a hyperbola

$$\pm\left\{\frac{X^2}{p^2} - \frac{Y^2}{q^2}\right\} = 1.$$

The situation is slightly different from that for an ellipse since we must include both signs on the left of this equation in order to obtain a geometrical interpretation for both the eigenvalues λ_1 and λ_2. Suppose that $\lambda_1 > 0$ (i.e., $\lambda_2 < 0$). We first take the plus sign on the left. In this case, there is no real Y corresponding to $X = 0$, that is, the curve does not cut the Y-axis. In Figure 11.2(c), the equation represents the curve that cuts the X-axis in G, and its reflection in the Y-axis. At G we have $Y = 0$, so that $OG^2 = p^2 = 1/\lambda_1$. If we take the minus sign on the left of the above equation, we find that we are dealing with the other hyperbola in Figure 11.2(c) which cuts the Y-axis in H, where $X = 0$ so that $OH^2 = q^2$. Remembering that we are still assuming $\lambda_1 > 0$, $\lambda_2 < 0$, the equation for this conic is not (11.18) but

$$-\lambda_1 X^2 - \lambda_2 Y^2 = 1.$$

Hence $OH^2 = q^2 = -\lambda_2$. Summing up, we have

$$OG^2 = p^2 = \frac{1}{\lambda_1}, \qquad OH^2 = q^2 = -\frac{1}{\lambda_2}.$$

This geometrical discussion has been a digression, so we return to the original problem which was to discover the nature of the conic (11.4) by means of a change of variable corresponding to a rotation of axes. In view of the analysis in (11.13)–(11.18), we can now solve this problem by a method which is quite different from the elementary geometrical argument given at the beginning of this section. The end result of the analysis will be precisely the same as before, namely (11.10).

We start by solving the eigenvalue problem—i.e., we find the constants λ for which the following matrix equation has nonzero solutions:

$$\mathbf{Ax} = \lambda\mathbf{x},$$

where $\mathbf{A}$ is the matrix given in (11.11). The corresponding solutions $\mathbf{x}$ are the eigenvectors. On writing out this equation in detail, we obtain

$$\begin{bmatrix} a & k \\ k & b \end{bmatrix}\begin{bmatrix} x \\ y \end{bmatrix} = \lambda\begin{bmatrix} x \\ y \end{bmatrix},$$

that is,

$$\begin{aligned} (a - \lambda)x + ky &= 0 \\ kx + (b - \lambda)y &= 0 \end{aligned} \tag{11.19}$$

This is a set of two homogeneous equations in two unknowns. From the theory of such systems we know that nonzero solutions exist only if the determinant of the coefficients is zero,

$$\begin{vmatrix} a - \lambda & k \\ k & b - \lambda \end{vmatrix} = 0,$$

or

$$\lambda^2 - (a + b)\lambda + (ab - k^2) = 0.$$

This is a quadratic in λ with two roots,

$$\tfrac{1}{2}(a+b) \pm \tfrac{1}{2}\{(a-b)^2 + 4k^2\}^{1/2}.$$

It is left to the reader to check that these are precisely the quantities α, β defined by (11.7), as the above theory suggests. We can also find the eigenvectors corresponding to the two values of λ and check that these, when normalized, are the columns of **S** defined in (11.12).

Rather than carry through these algebraic verifications in detail, it is perhaps equally instructive to consider a numerical example.

EXERCISE 11.1. Investigate the nature of the curve

$$x^2 + 4xy - 2y^2 = c.$$

SOLUTION: The matrix **A** is

$$\mathbf{A} = \begin{bmatrix} 1 & 2 \\ 2 & -2 \end{bmatrix}$$

and the eigenvalue problem is to solve (11.19),

$$\begin{aligned} (1-\lambda)x + 2y &= 0 \\ 2x - (2+\lambda)y &= 0 \end{aligned} \qquad (11.20)$$

The eigenvalues are the roots of the quadratic

$$\lambda^2 + \lambda - 6 = 0,$$

that is, $\lambda_1 = 2$, $\lambda_2 = -3$. Solution of the equations obtained by substituting these values in (11.20) gives the corresponding normalized eigenvectors,

$$x_1 = \frac{1}{\sqrt{5}}\begin{bmatrix} 2 \\ 1 \end{bmatrix}, \qquad x_2 = \frac{1}{\sqrt{5}}\begin{bmatrix} -1 \\ 2 \end{bmatrix}.$$

This gives

$$\mathbf{S} = \frac{1}{\sqrt{5}}\begin{bmatrix} 2 & -1 \\ 1 & 2 \end{bmatrix}, \qquad (11.21)$$

and the reader can readily check that

$$\mathbf{S}^T\mathbf{S} = \mathbf{I}, \qquad \mathbf{S}^T\mathbf{A}\mathbf{S} = \begin{bmatrix} 2 & 0 \\ 0 & -3 \end{bmatrix}.$$

Comparing **S** defined in (11.12) with (11.21), we see that $\cos\theta = 2/\sqrt{5}$, $\sin\theta = 1/\sqrt{5}$, and it is easy to deduce that $\tan 2\theta$ is then $\frac{4}{3}$ which is the value that can be found independently from (11.9). The end result is that if the coordinates (x, y) are transformed into coordinates (X, Y) by means of (11.14), where **S** is given by (11.21), the conic becomes

$$2X^2 - 3Y^2 = c,$$

so that the conic is a hyperbola.

Another way of looking at the original problem depicted in Figure 11.2 is to reflect that the eigenvalues in Figures 11.2(a), (c) are related to the distances from the origin to the points P, R and G, H, respectively. At all of these points, the normal to the curve is in the same direction as the radius vector. This is not true at the general point—e.g., Q, at which OQ and QS are not in the same direction. The slope of the curve, dy/dx, is given by implicit differentiation of (11.4) with respect to x,

$$2ax + 2k\left(y + x\frac{dy}{dx}\right) + 2by\frac{dy}{dx} = 0.$$

The slope of the normal at a point (x, y) is therefore given by

$$-\frac{1}{(dy/dx)} = \frac{kx + by}{ax + ky}.$$

We wish this to be identical with the slope of the radius vector—i.e., we require

$$\frac{y}{x} = \frac{kx + by}{ax + ky},$$

or

$$ax + ky = \lambda x$$
$$kx + by = \lambda y,$$

where λ is a factor of proportionality. These equations are identical with (11.19). Suppose that these have solutions $\lambda = \lambda_1, \lambda_2$ with corresponding eigenvectors $(x_i, y_i)^T$, $i = 1, 2$. If we substitute (x_i, y_i) for (x, y) in (11.4), we obtain

$$c = (ax_i + ky_i)x_i + (kx_i + by_i)y_i = \lambda_i(x_i^2 + y_i^2),$$

that is, for the special points in which we are interested, the distance from the origin to the conic is inversely proportional to the square root of the corresponding eigenvalue. We have already obtained and discussed this result, from a slightly different point of view, in 1′, 2′ above. In addition, the eigenvectors specify y_i/x_i which are the slopes of the lines being considered—i.e., the eigenvectors give the directions of the radius vectors specifying the axes of the ellipse or hyperbola.

EXERCISE 11.2. Check that the eigenvalues in (11.19) are the quantities α, β given in (11.7), as suggested in the text. Deduce the result $\alpha\beta = ab - k^2$ as an almost trivial consequence.

EXERCISE 11.3. Suppose that we translate coordinates by shifting the origin to x_0, y_0, so that the new coordinates ξ, η are related to the old by

$$x = \xi + x_0, \qquad y = \eta + y_0.$$

Show that the general quadratic curve

$$ax^2 + 2kxy + by^2 + 2cx + 2dy + e = 0$$

becomes

$$a\xi^2 + 2k\xi\eta + b\eta^2 + 2C\xi + 2D\eta + E = 0,$$

where

$$C = ax_0 + ky_0 + c, \qquad D = kx_0 + by_0 + d.$$

Show that if $ab - k^2 \neq 0$ or $ab - k^2 = 0$ and $ad = bc$, we can choose x_0, y_0 so that the equation for the curve becomes

$$a\xi^2 + 2k\xi\eta + b\eta^2 + E = 0,$$

which can be analyzed by the method used to deal with (11.4). However, if $ab = k^2$ and $ad \neq bc$, then there is no loss of generality in assuming $a \neq 0$ (since if $a = b = 0$, then $k = 0$, and we have no quadratic). We can then choose $y_0 = 0$, $x_0 = -c/a$, so that $C = 0$. The equation for the curve now becomes

$$a\xi^2 + 2k\xi\eta + b\eta^2 + 2D\eta + E = 0.$$

The rotation of axes to reduce the quadratic terms to the form (11.10) will give

$$\alpha X^2 + 2\gamma Y = c,$$

which is the equation for a parabola.

EXERCISE 11.4. Investigate the nature of the curves given by the following equations. Sketch the curves.

(a) $x^2 + 12xy + y^2 = 10.$
(b) $x^2 - 4xy + 4y^2 = 0.$
(c) $5x^2 + 8xy + 5y^2 = 10.$
(d) $5x^2 - 8xy + 5y^2 - 18x + 18y + 8 = 0.$
(e) $9x^2 - 12xy + 4y^2 - 42x - 2y + 7 = 0.$

11.3 GENERAL QUADRATIC FORMS

We generalize some of the ideas introduced in the last section. A *quadratic form* in n real variables $x_1, x_2, \ldots, x_n$ associated with a real matrix $\mathbf{A}$ is a scalar quantity consisting of a sum of multiples of the products and squares of the variables:

$$F = \sum_{i=1}^{n} \sum_{j=1}^{n} a_{ij} x_i x_j$$
$$= a_{11}x_1^2 + (a_{12} + a_{21})x_1 x_2 + a_{22}x_2^2 + \cdots. \tag{11.22}$$

If we introduce the matrices $\mathbf{x} = [x_i]$, $\mathbf{A} = [a_{ij}]$, the quadratic form can be written, using inner product notation, as

$$F = \sum_{i=1}^{n} x_i \sum_{j=1}^{n} a_{ij} x_j = \sum_{i=1}^{n} x_i (\mathbf{Ax})_i = (\mathbf{x}, \mathbf{Ax}). \tag{11.23}$$

According to the definition just given, we can associate a quadratic form with *any* matrix $\mathbf{A}$, but it is clear when the quadratic form is written out in detail, as in (11.22), that the terms involving a_{ij} and a_{ji} will give a contribution

$$(a_{ij} + a_{ji})x_i x_j, \tag{11.24}$$

that is, the contribution depends on $a_{ij} + a_{ji}$ and not on a_{ij}, a_{ji} separately. This means that if we start with a general matrix $\mathbf{A}$ and obtain from it a second matrix $\mathbf{A}' = \frac{1}{2}(\mathbf{A} + \mathbf{A}^T)$, whose (i, j) and (j, i) elements are both $\frac{1}{2}(a_{ij} + a_{ji})$, the quadratic forms $(\mathbf{x}, \mathbf{Ax})$ and $(\mathbf{x}, \mathbf{A}'\mathbf{x})$ are identical. Hence without loss of generality we can assume that the quadratic form is associated with a symmetric matrix when the a_{ij} and x_i are real.

When we are dealing with complex elements, recall that there is a natural way to generalize the real inner product that occurs in (11.23) (see Definition 4.12), and, then, hermitian matrices play the same role as symmetric matrices in the real case. This suggests that the natural generalization of (11.23) is still $(\mathbf{x}, \mathbf{Ax})$ but now $\mathbf{A}$ is hermitian and $\mathbf{x}$ may have complex elements; despite the fact that $\mathbf{A}$ and $\mathbf{x}$ are complex, the quadratic form $(\mathbf{x}, \mathbf{Ax})$ is real when $\mathbf{A}$ is hermitian since $(\mathbf{x}, \mathbf{Ax}) = \overline{(\mathbf{Ax}, \mathbf{x})} = \overline{(\mathbf{x}, \mathbf{A}^H\mathbf{x})} = \overline{(\mathbf{x}, \mathbf{Ax})}$. We formalize the preceding as follows.

DEFINITION 11.1. The *quadratic form* associated with the hermitian matrix $\mathbf{A}$ is the real quantity $(\mathbf{x}, \mathbf{Ax})$. When $\mathbf{A}$ is a real symmetric matrix, we refer to the *real quadratic form* $(\mathbf{x}, \mathbf{Ax})$.

Note that in many textbooks the quadratic form $(\mathbf{x}, \mathbf{Ax})$ is written as $\mathbf{x}^T\mathbf{Ax}$ in the real case, and $\bar{\mathbf{x}}^T\mathbf{Ax}$ in the complex case. Also, the theory for the real case is often worked out in detail, and the corresponding theory for the complex case is left as an exercise. However, we saw in Chapter 4 that the inner product notation provides a convenient tool for dealing with the real and complex cases together, and this is the policy we shall adopt here.

As illustrated in the last section, the key to the study of quadratic forms is the idea of a change of variable from coordinates $\mathbf{x}$ to coordinates $\mathbf{y}$ by means of a transformation

$$\mathbf{x} = \mathbf{S}\mathbf{y},$$

where $\mathbf{S}$ is a nonsingular matrix. We restrict ourselves to nonsingular $\mathbf{S}$ since it is only in this case that we can write $\mathbf{y} = \mathbf{S}^{-1}\mathbf{x}$, so that a given $\mathbf{x}$ determines a unique $\mathbf{y}$, and a given $\mathbf{y}$ determines a unique $\mathbf{x}$. As we noted at some length in Section 8.3, this corresponds to choosing a new basis (the set of columns of $\mathbf{S}$) for representing our vectors; if we are interested in *linear transformations* described by a matrix $\mathbf{A}$ this leads us to the similarity transformation $\mathbf{S}^{-1}\mathbf{A}\mathbf{S}$ and to the question of the extent to which a matrix can be simplified by a similarity transformation. At present, however, we are interested in *quadratic forms* rather than in linear transformations, and so we should ask how the quadratic form is changed by the change of variables $\mathbf{x} = \mathbf{S}\mathbf{y}$. Substitution of $\mathbf{x} = \mathbf{S}\mathbf{y}$ into $f = (\mathbf{x}, \mathbf{A}\mathbf{x})$ gives

$$F = (\mathbf{S}\mathbf{y}, \mathbf{A}\mathbf{S}\mathbf{y}) = (\mathbf{y}, \mathbf{S}^H\mathbf{A}\mathbf{S}\mathbf{y}) = (\mathbf{y}, \mathbf{B}\mathbf{y}),$$

where

$$\mathbf{B} = \mathbf{S}^H\mathbf{A}\mathbf{S}. \qquad (11.25)$$

If we can find $\mathbf{S}$ so that $\mathbf{B}$ is a diagonal matrix with diagonal elements β_i, then we have obtained a representation for F of the form

$$F = \beta_1\bar{y}_1y_1 + \cdots + \beta_n\bar{y}_ny_n,$$

and the form of F has been very considerably simplified. So far we have said nothing about $\mathbf{S}$ except that it must be nonsingular. In many applications of quadratic forms we wish to compare the sizes of various quantities that are involved, so that, if we change variables by a formula of the type $\mathbf{x} = \mathbf{S}\mathbf{y}$, it is important that lengths be preserved—i.e., that $(\mathbf{x}, \mathbf{x}) = (\mathbf{y}, \mathbf{y})$. Then $(\mathbf{x}, \mathbf{x}) = (\mathbf{S}\mathbf{y}, \mathbf{S}\mathbf{y}) = (\mathbf{y}, \mathbf{S}^H\mathbf{S}\mathbf{y}) = (\mathbf{y}, \mathbf{y})$ for all $\mathbf{x}$, so that $\mathbf{S}^H\mathbf{S} = \mathbf{I}$. This means that, in order to preserve distances, $\mathbf{S}$ must be unitary, and then (11.25) is a unitary transformation. (See also Exs. 8.39 and 8.40.)

Our basic result on the simplification of quadratic forms is a direct result of **Key Theorems 9.2** and **9.3**: if $\mathbf{A}$ is hermitian, there exists a unitary matrix $\mathbf{P}$ such that $\mathbf{P}^H\mathbf{A}\mathbf{P}$ is a diagonal matrix whose diagonal elements are the eigenvalues of $\mathbf{A}$.

THEOREM 11.1. *If* $\mathbf{A}$ *is a hermitian matrix of order* n *and* $\mathbf{P}$ *is a unitary matrix whose columns are the normalized eigenvectors of* $\mathbf{A}$, *then*

$$(\mathbf{x}, \mathbf{A}\mathbf{x}) = \lambda_1\bar{y}_1y_1 + \lambda_2\bar{y}_2y_2 + \cdots + \lambda_n\bar{y}_ny_n \qquad (11.26)$$

where $\mathbf{x} = \mathbf{Py}$ *and the* λ_i *are the eigenvalues of* $\mathbf{A}$. *If* $\mathbf{A}$ *is real symmetric and* $\mathbf{P}$ *is an orthogonal matrix whose columns are the normalized eigenvectors of* $\mathbf{A}$, *then, if* $\mathbf{x} = \mathbf{Py}$,

$$(\mathbf{x}, \mathbf{Ax}) = \lambda_1 y_1^2 + \lambda_2 y_2^2 + \cdots + \lambda_n y_n^2. \tag{11.27}$$

A quadratic form containing no cross products is said to be in *diagonal form*. We say that the transformation $\mathbf{x} = \mathbf{Py}$ in the above theorem has *diagonalized* the form. The theorem above shows that any quadratic form can be diagonalized by a unitary transformation.

EXERCISE 11.5. Diagonalize the following quadratic form as outlined below:

$$F = 3x_1^2 + 2x_2^2 + 3x_3^2 - 2x_1x_2 - 2x_2x_3.$$

Show that $F = (\mathbf{x}, \mathbf{Ax})$, where

$$A = \begin{bmatrix} 3 & -1 & 0 \\ -1 & 2 & -1 \\ 0 & -1 & 3 \end{bmatrix},$$

and that the eigenvalues of $\mathbf{A}$ are 1, 3, 4. Set $\mathbf{x} = \mathbf{Py}$ for an appropriate $\mathbf{P}$ and find

$$F = y_1^2 + 3y_2^2 + 4y_3^2. \tag{11.28}$$

EXERCISE 11.6. As an example of an application of diagonalization, consider the following optimization problem in chemical engineering. A chemical reaction involves the temperature T, the concentration c of one of the reactants, and the time t of the reaction. It is required to investigate how the yield η of the reaction varies when T, c, t are altered within certain limits. Preliminary experiments led to the levels $T = 167°\text{C}$, $c = 27.5\%$, $t = 6.5$ hr., and indicated that T could be varied by $\pm 5°\text{C}$, c by $\pm 2.5\%$, and t by ± 1.5 hr. We introduce the standardized variables

$$x_1 = \frac{(T - 167)}{5}, \qquad x_2 = \frac{(c - 27.5)}{2.5}, \qquad x_3 = \frac{(t - 6.5)}{1.5}, \tag{11.29}$$

where we are now interested in the range $-1 \leq x_i \leq +1$.

By measuring η for various values of the x_i around $x_i = 0$, it is possible to represent η approximately by a second-degree surface in the neighborhood of $x_i = 0$, $i = 1, 2, 3$. The technical details of how this is done are irrelevant here, but can be found, for instance, in the original paper of G. E. P. Box on which the present discussion is based; see *The Design and Analysis of Industrial Experiments* (Oliver and Boyd, 1956), edited by O. L. Davies. The numbers have been changed slightly so that the arithmetic can be carried out by pencil and paper, since in practice the calculations would be performed on a digital computer.

Suppose that the following equation is found, relating the yield η to the independent variables $\mathbf{x}$, near $\mathbf{x} = \mathbf{0}$:

$$\eta = 57.822 + 1.83x_1 + 0.63\sqrt{3}\,x_2 + 0.96\sqrt{3}\,x_3 - 1.5x_1^2 - 0.6x_2^2 - \tfrac{17}{15}x_3^2$$
$$- 0.9\sqrt{3}\,x_1x_2 - 1.4\sqrt{3}\,x_1x_3 - 1.4x_2x_3 \tag{11.30}$$

The value of η is stationary when $\partial\eta/\partial x_i = 0$, $i = 1, 2, 3$. This gives the following simultaneous linear equations for the determination of the stationary point:

$$\begin{aligned} 3.0x_1 + 0.9\sqrt{3}\,x_2 + 1.4\sqrt{3}\,x_3 &= 1.83 \\ 0.9\sqrt{3}\,x_1 + 1.2x_2 + 1.4x_3 &= 0.63\sqrt{3} \\ 1.4\sqrt{3}\,x_1 + 1.4x_2 + \tfrac{34}{15}x_3 &= 0.96\sqrt{3} \end{aligned} \tag{11.31}$$

The solution of these equations is given by $x_1 = 0.1$, $x_2 = 0.1\sqrt{3}$, $x_3 = 0.3\sqrt{3}$. We therefore make the following change of variables:

$$z_1 = x_1 - 0.1, \qquad z_2 = x_2 - 0.1\sqrt{3}, \qquad z_3 = x_3 - 0.3\sqrt{3}. \tag{11.32}$$

In terms of these variables, the expression for η becomes

$$\eta = 58.44 - 1.5z_1^2 - 0.6z_2^2 - \tfrac{17}{15}z_3^2 - 0.9\sqrt{3}\,z_1z_2 - 1.4\sqrt{3}\,z_1z_3 - 1.4z_2z_3. \tag{11.33}$$

In symbols, the analysis, so far, is a particular case of the following. If

$$\eta = c + 2(\mathbf{x}, \mathbf{b}) - (\mathbf{x}, \mathbf{A}\mathbf{x}),$$

the stationary value is given by $\mathbf{x} = \mathbf{x}_0$, where $\mathbf{A}\mathbf{x}_0 = \mathbf{b}$. We make the change of variable $\mathbf{z} = \mathbf{x} - \mathbf{x}_0$, obtaining

$$\eta = [c + (\mathbf{x}_0, \mathbf{b})] - (\mathbf{z}, \mathbf{A}\mathbf{z}).$$

The point of the transformation is to eliminate the linear terms in η.

The next step is to diagonalize the quadratic form in (11.33) which, for convenience, we write as $\eta = 58.44 - (\mathbf{z}, \mathbf{A}\mathbf{z})$, where

$$\mathbf{A} = \begin{bmatrix} 1.5 & 0.45\sqrt{3} & 0.7\sqrt{3} \\ 0.45\sqrt{3} & 0.6 & 0.7 \\ 0.7\sqrt{3} & 0.7 & \tfrac{17}{15} \end{bmatrix}. \tag{11.34}$$

The eigenvalues of this matrix are found to be $\lambda_1 = 3$, $\lambda_2 = \frac{1}{12}$, $\lambda_3 = 0.15$, with corresponding normalized eigenvectors

$$\mathbf{z}_1 = \begin{bmatrix} 0.4\sqrt{3} \\ 0.4 \\ 0.6 \end{bmatrix}, \quad \mathbf{z}_2 = \begin{bmatrix} 0.3\sqrt{3} \\ 0.3 \\ -0.8 \end{bmatrix}, \quad \mathbf{z}_3 = \begin{bmatrix} 0.5 \\ -0.5\sqrt{3} \\ 0 \end{bmatrix} \tag{11.35}$$

As described in Theorem 11.1 above, we now set $\mathbf{z} = \mathbf{P}\mathbf{y}$, where $\mathbf{P} = [\mathbf{z}_1, \mathbf{z}_2, \mathbf{z}_3]$.

We then find that (11.33) gives

$$\eta = 58.44 - 3y_1^2 - \tfrac{1}{12}y_2^2 - 0.15y_3^2.$$

Since the coefficient for y_2^2 and for y_3^2 above are so much smaller than for y_1^2, y_1 is the important variable and the yield η is given approximately by

$$\eta = 58.44 - 3y_1^2.$$

The maximum response is then given by $y_1 = 0$. Since $\mathbf{y} = \mathbf{P}^T\mathbf{z}$, we see, on using (11.32), (11.35), that

$$y_1 = 0.4\sqrt{3}(x_1 - 0.1) + 0.4(x_2 - 0.1\sqrt{3}) + 0.6(x_3 - 0.3\sqrt{3}).$$

Expressing this in terms of the original variables by means of (11.29), we see that the relation $y_1 = 0$ is

$$\sqrt{3}(T - 167) + 2(c - 27.5) + 5(t - 6.5) = 3.25\sqrt{3} \qquad (11.36)$$

The statement that the yield is a maximum when $y_1 = 0$ is equivalent to the statement that any combination of T, c, t on the plane (11.36) will maximize the yield.

The conclusion of this analysis is that the level surfaces are approximately planes in the region in which we are interested, as illustrated graphically in Figure 11.3, which shows the level surfaces corresponding to the maximum value of η (58.44) and also $\eta = 55$. Over the range considered we can obtain maximum yield by working with any combination of the variables T, c, t that satisfies (11.36). This is important in practice, since it means that the operating conditions can be chosen to satisfy criteria in addition to maximum yield.

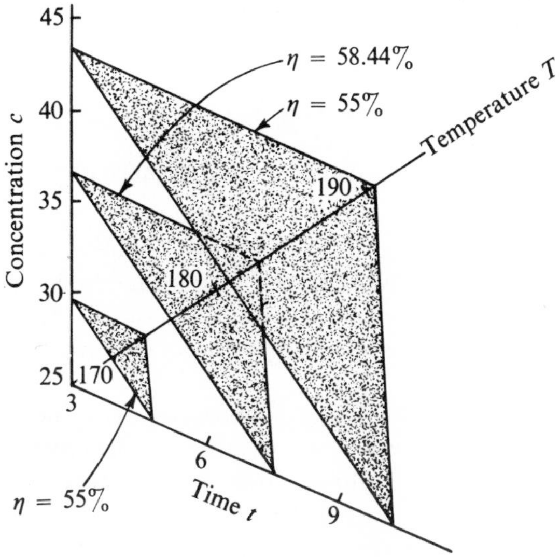

Figure 11.3. Planes of constant yield.

One of the important consequences of Theorem 11.1 is that it enables us to make a fundamental classification of quadratic forms. We consider only the case where **A** is real symmetric and **x** is real. Then the orthogonal transformation used in Theorem 11.1 corresponds to a rigid motion of the coordinate system, distances remaining unchanged.

DEFINITION 11.2. The surface in real n-dimensional space defined by the equation

$$(\mathbf{x}, \mathbf{A}\mathbf{x}) = 1,$$

where **A** is a real symmetric matrix of order n, is known as a *central quadric*. In the case $n = 2$ the word "surface" must be replaced by "curve."

Theorem 11.1 [equation (11.27)] states that, relative to the transformed axes, the equation of the quadric becomes

$$\lambda_1 y_1^2 + \lambda_2 y_2^2 + \cdots + \lambda_n y_n^2 = 1.$$

We have already examined the two-dimensional case in Section 11.2 (Figure 11.2),

$$\lambda_1 y_1^2 + \lambda_2 y_2^2 = 1,$$

and found that there were then three possibilities:

1. Ellipse: λ_1, λ_2 have the same sign.
2. Hyperbola: λ_1, λ_2 have opposite signs.
3. Two parallel straight lines: one zero eigenvalue.

In the three-variable case there are five possibilities. Consider the surface represented by

$$\lambda_1 y_1^2 + \lambda_2 y_2^2 + \lambda_3 y_3^2 = 1.$$

The following cases arise (the reader should visualize these surfaces in three-dimensional space):

1. Ellipsoid: λ_1, λ_2, λ_3 positive. When two λ's are equal, the surface is a rotational ellipsoid. When three λ's are equal, we have a sphere.
2. Hyperboloid with one sheet: λ_1, λ_2 positive, λ_3 negative. When the two positive λ's are equal, we have a rotational hyperboloid or "cooling tower."
3. Hyperboloid with two sheets: λ_1 positive, λ_2, λ_3 negative. When the two negative λ's are equal, we have a rotational hyperboloid.
4. Elliptic or hyperbolic cylinder: one of the λ's is zero.
5. Two parallel planes: two of the λ's are zero.

We can carry out a similar classification when more than three variables are involved, though then, of course, the results no longer have an obvious direct geometrical significance.

EXERCISE 11.7. Diagonalize the quadratic form

$$F = 11x_1^2 + 2x_1x_2 + 3x_2^2.$$

EXERCISE 11.8. Diagonalize the following quadratic form and sketch and describe the surface $F = 1$, where F is defined by

$$F = -x_1^2 - x_2^2 + 6x_1x_2 - 4x_1x_3.$$

11.4 DEFINITE QUADRATIC FORMS

Some quadratic forms have the property that $(\mathbf{x}, \mathbf{Ax}) > 0$ for all $\mathbf{x} \neq \mathbf{0}$. A simple example is

$$F = x_1^2 + x_2^2 + 2x_3^2.$$

Another important type of quadratic form has the property that $(\mathbf{x}, \mathbf{Ax}) \geq 0$ for all $\mathbf{x}$, although for this class of form it is not necessarily true that $(\mathbf{x}, \mathbf{Ax}) = 0$ implies $\mathbf{x} = \mathbf{0}$. A simple example is

$$F = (x_1 - x_2)^2 + 2x_3^2,$$

which is zero for the nonzero vector given by $x_1 = k$, $x_2 = k$, $x_3 = 0$, for any k. In general, of course, a quadratic form can assume positive, zero, and negative values. The special cases that are one-signed are important, and we introduce the following terminology.

DEFINITION 11.3. (The matrix $\mathbf{A}$ in this definition is always hermitian; this means symmetric in the real case.) The quadratic form $(\mathbf{x}, \mathbf{Ax})$ is said to be *positive definite* if $(\mathbf{x}, \mathbf{Ax}) > 0$ for all $\mathbf{x} \neq \mathbf{0}$, and *positive semidefinite* if $(\mathbf{x}, \mathbf{Ax}) \geq 0$ for all $\mathbf{x}$. The matrix $\mathbf{A}$ is said to be *positive definite* or *semidefinite* if the corresponding quadratic form is positive definite or semidefinite. Negative forms and matrices are defined similarly. A quadratic form is said to be *indefinite* if it assumes positive values for some $\mathbf{x}$ and negative values for others.

According to this definition, positive definite matrices are also positive semidefinite. If it is necessary to say that $(\mathbf{x}, \mathbf{Ax}) \geq 0$ and there exists a nonzero $\mathbf{x}$ such that $(\mathbf{x}, \mathbf{Ax}) = 0$, we shall say that $\mathbf{A}$ is positive semidefinite and singular (see Ex. 11.37).

In the last section we confined our attention to unitary transformations because we were interested in keeping lengths unchanged. When studying whether a quadratic form is positive definite or not it is convenient to relax this restriction and consider any transformation of the type $\mathbf{B} = \mathbf{S}^H\mathbf{AS}$, where $\mathbf{S}$ is nonsingular, since $\mathbf{B}$ then describes our quadratic form with respect to a new basis.

DEFINITION 11.4. If there exists a nonsingular matrix $\mathbf{S}$ such that $\mathbf{S}^H\mathbf{AS} = \mathbf{B}$ we say that $\mathbf{B}$ is *hermitian-congruent* to $\mathbf{A}$, and that $\mathbf{B}$ is obtained from $\mathbf{A}$ by a *hermitian-congruence* transformation. When $\mathbf{A}$ and $\mathbf{S}$ are real we can omit the word hermitian. In this case, if $\mathbf{A}$, $\mathbf{B}$ are *congruent*, a real nonsingular matrix $\mathbf{S}$ exists such that $\mathbf{S}^T\mathbf{AS} = \mathbf{B}$.

It is important to note that the property of being positive definite is not affected by a congruence transformation; this is intuitively obvious since $\mathbf{A}$ and $\mathbf{B}$ represent the "same" quadratic form only with respect to two different bases.

THEOREM 11.2. If $(\mathbf{x}, \mathbf{Ax})$ *is a positive definite quadratic form and* $\mathbf{B} = \mathbf{S}^H\mathbf{AS}$, *where* $\mathbf{S}$ *is a nonsingular square matrix, then* $(\mathbf{y}, \mathbf{By})$ *is also a positive definite quadratic form. Similarly for positive semidefinite forms, etc.*

Proof: We have

$$(\mathbf{y}, \mathbf{By}) = (\mathbf{y}, \mathbf{S}^H\mathbf{ASy}) = (\mathbf{Sy}, \mathbf{A}(\mathbf{Sy})).$$

Since $\mathbf{S}$ is nonsingular, $\mathbf{x} = \mathbf{Sy} \neq \mathbf{0}$ if $\mathbf{y} \neq \mathbf{0}$. But if $\mathbf{x} \neq \mathbf{0}$, we know from the definition of a positive definite form that $(\mathbf{x}, \mathbf{Ax}) > 0$. Hence, if $\mathbf{y} \neq \mathbf{0}$, we have $(\mathbf{y}, \mathbf{By}) > 0$, and this form is positive definite. The other cases can be proved similarly.

If we know the eigenvalues of $\mathbf{A}$ we can immediately say whether $(\mathbf{x}, \mathbf{Ax})$ is definite or not.

● *KEY THEOREM 11.3. Let* $\mathbf{A}$ *be a hermitian matrix.*

(i) $(\mathbf{x}, \mathbf{Ax})$ *is positive definite if and only if all the eigenvalues of* $\mathbf{A}$ *are positive.*
(ii) $(\mathbf{x}, \mathbf{Ax})$ *is positive semidefinite if and only if all the eigenvalues of* $\mathbf{A}$ *are non-negative.*
(iii) $(\mathbf{x}, \mathbf{Ax})$ *is indefinite if and only if* $\mathbf{A}$ *has both positive and negative eigenvalues.*

Proof: From Theorem 11.1 we can introduce new variables $\mathbf{y}$ defined by $\mathbf{x} = \mathbf{Sy}$, where $\mathbf{S}$ is nonsingular, such that

$$(\mathbf{x}, \mathbf{Ax}) = \lambda_1\bar{y}_1y_1 + \lambda_2\bar{y}_2y_2 + \cdots + \lambda_n\bar{y}_ny_n. \tag{11.37}$$

If every eigenvalue λ_i is positive, then the form on the right of this equation is obviously positive definite so that, by Theorem 11.2, the form on the left, namely

$(\mathbf{x}, \mathbf{Ax})$, is also positive definite. This proves (i). Parts (ii) and (iii) can be proved similarly.

The criterion for definiteness in **Key Theorem 11.3,** which depends on knowing the eigenvalues of $\mathbf{A}$, is not very convenient in practice, so we state a simpler set of conditions for determining whether a matrix is positive definite; a proof of this result can be based on the LU decomposition of $\mathbf{A}$, but we leave this for the reader (see Exs. 11.38–11.40). First, however, as motivation for the forthcoming conditions, we state some simpler necessary conditions whose validity is easily proved.

THEOREM 11.4. Necessary conditions for a hermitian matrix $\mathbf{A}$ *to be positive definite are:*

(i) *The diagonal elements of* $\mathbf{A}$ *must be positive.*
(ii) $a_{ii}a_{jj} > |a_{ij}|^2 \qquad (i \neq j)$.
(iii) *The element of* $\mathbf{A}$ *of largest absolute value must lie on the diagonal.*
(iv) *det* $\mathbf{A} > 0$. *(In particular,* $\mathbf{A}$ *is nonsingular.)*

Proof: In the quadratic form $(\mathbf{x}, \mathbf{Ax})$, choose all the x_k to be zero except x_i. Then $(\mathbf{x}, \mathbf{Ax}) = a_{ii}|x_i|^2$, and, since $x_i \neq 0$, we must have $a_{ii} > 0$, proving (i). To prove (ii), choose all the x_k to be zero except x_i and x_j. Then

$$(\mathbf{x}, \mathbf{Ax}) = a_{ii}|x_i|^2 + a_{ij}\bar{x}_i x_j + \bar{a}_{ij}x_i\bar{x}_j + a_{jj}|x_j|^2$$
$$= a_{ii}\left|x_i + \frac{a_{ij}x_j}{a_{ii}}\right|^2 + \frac{\{a_{ii}a_{jj} - |a_{ij}|^2\}|x_j|^2}{a_{ii}}$$

By choosing $x_i = -a_{ij}x_j/a_{ii}$ in this expression, we see that, since $a_{ii} > 0$, a necessary condition for $(\mathbf{x}, \mathbf{Ax})$ to be positive is that $a_{ii}a_{jj} - |a_{ij}|^2 > 0$. To prove (iii), suppose that for some i, j we have $|a_{ij}| \geq a_{ii}$, $|a_{ij}| \geq a_{jj}$, where, from (i), $a_{ii} > 0$, $a_{jj} > 0$. In this case, $|a_{ij}|^2 \geq a_{ii}a_{jj}$, which contradicts (ii). To prove (iv) we remind the reader that det $\mathbf{A}$ is precisely equal to the product of the eigenvalues. From **Key Theorem 11.3,** the eigenvalues of a positive definite matrix are all positive so that det $\mathbf{A} > 0$, which proves (iv).

We now state without proof (see Exs. 11.38–11.40) our fundamental characterization of positive definite matrices, greatly strengthening Theorem 11.4.

THEOREM 11.5. Either of the following sets of conditions is necessary and sufficient for $(\mathbf{x}, \mathbf{Ax})$ *to be positive definite (recall that* $\mathbf{A}$ *is hermitian):*

(i) *When* $\mathbf{A}$ *is reduced to row-echelon form working systematically along the main diagonal without interchanging rows, all the pivots are positive.*
(ii) *The principal minors consisting of the determinants of the* $k \times k$ *matrices in the top left-hand corner of* $\mathbf{A}$ *(*$k = 1$ *to* n*) are all positive:*

$$a_{11} > 0, \qquad \begin{vmatrix} a_{11} & a_{12} \\ a_{21} & a_{22} \end{vmatrix} > 0, \qquad \begin{vmatrix} a_{11} & a_{12} & a_{13} \\ a_{21} & a_{22} & a_{23} \\ a_{31} & a_{32} & a_{33} \end{vmatrix} \geq 0, \ldots .$$

The point of this theorem is that we can find out whether a quadratic form is positive definite merely by reducing the corresponding matrix to row-echelon form, checking whether all the pivots are positive, as illustrated in the following example. We do not need to know anything about the eigenvalues or eigenvectors.

EXERCISE 11.9. Is the following quadratic form positive definite?

$$F = 2x_1^2 + x_2^2 + 6x_3^2 + 2x_1x_2 + x_1x_3 + 4x_2x_3.$$

SOLUTION: We write down the matrix of the quadratic form, and reduce it to upper triangular form, *not* dividing the resulting rows by the pivots, so that the pivots appear on the diagonal of the final matrix,

$$\mathbf{A} = \begin{bmatrix} 2 & 1 & \frac{1}{2} \\ 1 & 1 & 2 \\ \frac{1}{2} & 2 & 6 \end{bmatrix} \longrightarrow \begin{bmatrix} 2 & 1 & \frac{1}{2} \\ 0 & \frac{1}{2} & 1\frac{3}{4} \\ 0 & 0 & -\frac{1}{4} \end{bmatrix}.$$

The pivots are 2, $\frac{1}{2}$, $-\frac{1}{4}$ and, since one is negative, the quadratic form is *not* positive definite.

EXERCISE 11.10. Use Theorem 11.5 to determine whether or not the quadratic forms of Exs. 11.5 and 11.7 are positive definite.

EXERCISE 11.11. Use Theorem 11.5 to derive and prove necessary and sufficient conditions for a hermitian matrix $\mathbf{A}$ to define a negative definite quadratic form.

EXERCISE 11.12. Sometimes it is important to use *one* nonsingular matrix $\mathbf{S}$ so as to diagonalize simultaneously two different hermitian matrices $\mathbf{A}$ and $\mathbf{B}$, where $\mathbf{B}$ is positive definite. To do this, first show that

$$\mathbf{B} = \mathbf{P}\mathbf{D}^2\mathbf{P}^H = (\mathbf{P}\mathbf{D})(\mathbf{P}\mathbf{D})^H,$$

where $\mathbf{P}$ is an appropriate unitary matrix and $\mathbf{D}$ is an appropriate real diagonal matrix with positive diagonal elements $d_{ii} > 0$. If $\mathbf{Q}$ is defined to equal $\mathbf{P}\mathbf{D}^{-1}$, this gives $\mathbf{Q}^H\mathbf{B}\mathbf{Q} = \mathbf{I}$. Next show that there exists a unitary matrix $\mathbf{R}$ such that

$$\mathbf{R}^H\mathbf{Q}^H\mathbf{A}\mathbf{Q}\mathbf{R} = \mathbf{\Lambda}$$

where $\mathbf{\Lambda}$ is diagonal. Let $\mathbf{S} = \mathbf{Q}\mathbf{R}$ and show that

$$\mathbf{S}^H\mathbf{A}\mathbf{S} = \mathbf{\Lambda}, \qquad \mathbf{S}^H\mathbf{B}\mathbf{S} = \mathbf{I}.$$

EXERCISE 11.13. In Ex. 11.12, let the ith diagonal element of $\mathbf{\Lambda}$ be λ_i and let the ith column of $\mathbf{S}$ be $\mathbf{s}_i$. Show that λ_i and $\mathbf{s}_i$ solve the generalized eigenvalue problem $\mathbf{A}\mathbf{s}_i = \lambda_i\mathbf{B}\mathbf{s}_i$.

EXERCISE 11.14. Reduce the matrices **A** and **B** simultaneously to diagonal form, where

$$\mathbf{A} = \begin{bmatrix} 75 & 35 \\ 35 & -117 \end{bmatrix}, \qquad \mathbf{B} = \begin{bmatrix} 5 & -3 \\ -3 & 5 \end{bmatrix}.$$

11.5 EXTREMIZING QUADRATIC FORMS: RAYLEIGH'S PRINCIPLE

One of the ways in which we initially motivated our study of quadratic forms in Section 11.1 was by considering a statistical problem in data analysis which required the extremization (actually maximization) of a quadratic form subject to a normalization of its variables. Now that we have developed the necessary understanding of quadratic forms and their simplification, we can return to our earlier concern of extremizing such a form

$$(\mathbf{x}, \mathbf{A}\mathbf{x}) \tag{11.38}$$

subject to the restriction

$$(\mathbf{x}, \mathbf{x}) = 1. \tag{11.39}$$

As we saw in Section 11.2, the equation $(\mathbf{x}, \mathbf{A}\mathbf{x}) = c$ describes a conic section when $\mathbf{x}$ is a 2×1 column vector; maximizing $c = (\mathbf{x}, \mathbf{A}\mathbf{x})$ subject to $(\mathbf{x}, \mathbf{x}) = 1$ is thus equivalent to maximizing c over all those values c for which the curves $(\mathbf{x}, \mathbf{x}) = 1$ and $(\mathbf{x}, \mathbf{A}\mathbf{x}) = c$ intersect. From Figure 11.2 it is geometrically obvious that this corresponds to "expanding" the curve $(\mathbf{x}, \mathbf{A}\mathbf{x}) = 1$ until it tangentially circumscribes the circle $(\mathbf{x}, \mathbf{x}) = 1$; for the elliptic case in Figure 11.2(a) this will occur when the circle is tangent to the ellipse at the point R so that the length OR of the semi-minor axis of the ellipse $(\mathbf{x}, \mathbf{A}\mathbf{x}) = c$ equals the radius of the circle $(\mathbf{x}, \mathbf{x}) = 1$, that is, $\sqrt{c/\lambda_1} = 1$ so that $\lambda_1 = c$. Thus we deduce geometrically that the maximum of (11.38) subject to (11.39) is the largest eigenvalue λ_1 of $\mathbf{A}$; this maximum occurs at the point R which is in the direction $\overrightarrow{OR}$ of the eigenvector $\mathbf{x}_1$ associated with λ_1. It is interesting that this eigenvector solves the problem since we noted earlier that it also gave the direction in which $(\mathbf{x}, \mathbf{x})$ was minimized subject to $(\mathbf{x}, \mathbf{A}\mathbf{x}) = 1$; that is, as is geometrically obvious in Figure 11.2(a), R is a closest point on the ellipse $(\mathbf{x}, \mathbf{A}\mathbf{x}) = 1$ to the origin, and its squared distance is $1/\lambda_1$. These problems are, in fact, closely related in general for the "elliptic" case where $\mathbf{A}$ is positive definite even when we do not have available the intuitively helpful setting of two-dimensional geometry.

THEOREM 11.6. Let $\mathbf{A}$ *be hermitian. Then* $\mathbf{x}_0$ *maximizes* $(\mathbf{x}, \mathbf{A}\mathbf{x})/(\mathbf{x}, \mathbf{x})$ *subject to* $\mathbf{x} \neq \mathbf{0}$ *and yields a maximum value equal to M if and only if* $\mathbf{x}_1 = \mathbf{x}_0/\sqrt{(\mathbf{x}_0, \mathbf{x}_0)}$ *maximizes* $(\mathbf{x}, \mathbf{A}\mathbf{x})$ *subject to* $(\mathbf{x}, \mathbf{x}) = 1$ *and yields a maximum value equal to M.*

Moreover, if $\mathbf{A}$ *is positive definite, the latter condition holds if and only if* $\mathbf{x}_2 = \alpha\mathbf{x}_1$ *[with* $\alpha = 1/\sqrt{(\mathbf{x}_1, \mathbf{A}\mathbf{x}_1)}$*] minimizes* $(\mathbf{x}, \mathbf{x})$ *subject to* $(\mathbf{x}, \mathbf{A}\mathbf{x}) = 1$ *and yields a minimum value of* $m = 1/M$.

Proof: If $\mathbf{x}_0$ maximizes $(\mathbf{x}, \mathbf{A}\mathbf{x})/(\mathbf{x}, \mathbf{x})$ subject to $\mathbf{x} \neq \mathbf{0}$ then $\mathbf{x}_1$ has $(\mathbf{x}_1, \mathbf{x}_1) = 1$ and for any $\mathbf{x}$ with $(\mathbf{x}, \mathbf{x}) = 1$ we have

$$(\mathbf{x}, \mathbf{A}\mathbf{x}) = \frac{(\mathbf{x}, \mathbf{A}\mathbf{x})}{(\mathbf{x}, \mathbf{x})} \leq \frac{(\mathbf{x}_0, \mathbf{A}\mathbf{x}_0)}{(\mathbf{x}_0, \mathbf{x}_0)} = (\mathbf{x}_1, \mathbf{A}\mathbf{x}_1) = M$$

as desired. Conversely, if $\mathbf{x}_1$ with $(\mathbf{x}_1, \mathbf{x}_1) = 1$ maximizes $(\mathbf{x}, \mathbf{A}\mathbf{x})$ subject to $(\mathbf{x}, \mathbf{x}) = 1$, then for any nonzero scalar α and any nonzero vector $\mathbf{x}$ the vector $\mathbf{y} = \mathbf{x}/\sqrt{(\mathbf{x}, \mathbf{x})}$ has $(\mathbf{y}, \mathbf{y}) = 1$ and hence with $\mathbf{x}_0 = \alpha\mathbf{x}_1$ we have

$$\frac{(\mathbf{x}_0, \mathbf{A}\mathbf{x}_0)}{(\mathbf{x}_0, \mathbf{x}_0)} = (\mathbf{x}_1, \mathbf{A}\mathbf{x}_1) \geq (\mathbf{y}, \mathbf{A}\mathbf{y}) = \frac{(\mathbf{x}, \mathbf{A}\mathbf{x})}{(\mathbf{x}, \mathbf{x})}$$

as desired. If in addition $\mathbf{A}$ is positive definite then M is positive and $\mathbf{x}_2 = \mathbf{x}_1/\sqrt{M}$ satisfies $(\mathbf{x}_2, \mathbf{A}\mathbf{x}_2) = 1$ and $(\mathbf{x}_2, \mathbf{x}_2) = 1/M$; for any other $\mathbf{x}$ with $(\mathbf{x}, \mathbf{A}\mathbf{x}) = 1$ the vector $\mathbf{y} = \mathbf{x}/\sqrt{(\mathbf{x}, \mathbf{x})}$ has $(\mathbf{y}, \mathbf{y}) = 1$ so that $M \geq (\mathbf{y}, \mathbf{A}\mathbf{y}) = 1/(\mathbf{x}, \mathbf{x})$ and hence $(\mathbf{x}, \mathbf{x}) \geq 1/M = (\mathbf{x}_2, \mathbf{x}_2)$ as desired. Conversely, if $\mathbf{A}$ is positive definite and $\mathbf{x}_2$ minimizes $(\mathbf{x}, \mathbf{x})$ subject to $(\mathbf{x}, \mathbf{A}\mathbf{x}) = 1$ and yields the minimum value $m = (\mathbf{x}_2, \mathbf{x}_2)$, then $\mathbf{x}_1 = \mathbf{x}_2/\sqrt{m}$ has $(\mathbf{x}_1, \mathbf{x}_1) = 1$ and $(\mathbf{x}_1, \mathbf{A}\mathbf{x}_1) = 1/m$; for any other $\mathbf{x}$ with $(\mathbf{x}, \mathbf{x}) = 1$ we know that $(\mathbf{x}, \mathbf{A}\mathbf{x}) > 0$ so that $\mathbf{y} = \mathbf{x}/\sqrt{(\mathbf{x}, \mathbf{A}\mathbf{x})}$ satisfies $(\mathbf{y}, \mathbf{A}\mathbf{y}) = 1$, and therefore

$$\frac{(\mathbf{x}_1, \mathbf{x}_1)}{(\mathbf{x}_1, \mathbf{A}\mathbf{x}_1)} = m \leq (\mathbf{y}, \mathbf{y}) = \frac{(\mathbf{x}, \mathbf{x})}{(\mathbf{x}, \mathbf{A}\mathbf{x})}$$

so that

$$\frac{(\mathbf{x}, \mathbf{A}\mathbf{x})}{(\mathbf{x}, \mathbf{x})} \leq \frac{(\mathbf{x}_1, \mathbf{A}\mathbf{x}_1)}{(\mathbf{x}_1, \mathbf{x}_1)} = \frac{1}{m}$$

as desired, completing the proof of the theorem.

EXERCISE 11.15. Describe how the above proof of the equivalence of the maximization problem and the minimization problem breaks down if $\mathbf{A}$ is not positive definite.

EXERCISE 11.16. Let

$$\mathbf{A} = \begin{bmatrix} -2 & 0 \\ 0 & -1 \end{bmatrix}$$

and show that the maximum of $(\mathbf{x}, \mathbf{A}\mathbf{x})$ subject to $(\mathbf{x}, \mathbf{x}) = 1$ equals -1 while there are no vectors $\mathbf{x}$ for which $(\mathbf{x}, \mathbf{A}\mathbf{x}) = 1$. Explain why this does not contradict Theorem 11.6.

EXERCISE 11.17. State and prove a modification of Theorem 11.6 for the case in which $\mathbf{A}$ is negative definite.

EXERCISE 11.18. State and prove a modification of Theorem 11.6 for the case in which **A** is indefinite.

While the preceding theorem extended to n dimensions what we observed geometrically in two dimensions on the equivalence of the two extremization problems, we have not yet extended our geometrical argument to show that the extremizing vector is $\mathbf{x}_1$, an eigenvector associated with the largest eigenvalue of **A**; we proceed with this **key** extension.

DEFINITION 11.5. The *Rayleigh quotient* corresponding to a hermitian matrix **A** is the expression

$$\rho = \rho(x) = \frac{(\mathbf{x}, \mathbf{A}\mathbf{x})}{(\mathbf{x}, \mathbf{x})}. \tag{11.40}$$

• *KEY THEOREM 11.7. If* **A** *is hermitian with eigenvalues* $\lambda_1 \leq \lambda_2 \leq \cdots \leq \lambda_n$ *and associated orthonormalized eigenvectors* $\mathbf{x}_1, \ldots, \mathbf{x}_n$, *then*

$$\lambda_1 \leq \rho(\mathbf{x}) \leq \lambda_n,$$

where $\rho(\mathbf{x})$ *is the Rayleigh quotient for any* $\mathbf{x} \neq \mathbf{0}$, *and*

$$\lambda_1 = \min_{\mathbf{x}\neq\mathbf{0}} \rho(\mathbf{x}) = \rho(\mathbf{x}_1), \qquad \lambda_n = \max_{\mathbf{x}\neq\mathbf{0}} \rho(\mathbf{x}) = \rho(\mathbf{x}_n). \tag{11.41}$$

Proof: Suppose that the expansion of an arbitrary vector **x** in terms of the $\mathbf{x}_i$ is

$$\mathbf{x} = \sum_{i=1}^{n} \alpha_i \mathbf{x}_i. \tag{11.42}$$

Then

$$\mathbf{A}\mathbf{x} = \sum_{i=1}^{n} \alpha_i \lambda_i \mathbf{x}_i,$$

$$\rho = \frac{(\mathbf{x}, \mathbf{A}\mathbf{x})}{(\mathbf{x}, \mathbf{x})} = \frac{\lambda_1\bar{\alpha}_1\alpha_1 + \lambda_2\bar{\alpha}_2\alpha_2 + \cdots + \lambda_n\bar{\alpha}_n\alpha_n}{\bar{\alpha}_1\alpha_1 + \bar{\alpha}_2\alpha_2 + \cdots + \bar{\alpha}_n\alpha_n}. \tag{11.43}$$

We have

$$\rho - \lambda_1 = \frac{(\lambda_2 - \lambda_1)\bar{\alpha}_2\alpha_2 + \cdots + (\lambda_n - \lambda_1)\bar{\alpha}_n\alpha_n}{\bar{\alpha}_1\alpha_1 + \bar{\alpha}_2\alpha_2 + \cdots + \bar{\alpha}_n\alpha_n}. \tag{11.44}$$

Since $\lambda_i - \lambda_1 \geq 0$ for all i, we have $\rho \geq \lambda_1$. Also, if we choose $\mathbf{x} = \mathbf{x}_1$ this gives $\rho = \lambda_1$, which proves the first statement in (11.41). The remainder of the theorem is proved similarly by considering $\rho - \lambda_n$.

The Rayleigh quotient can also be used in another way, slightly different from that of **Key Theorem 11.7**, in order to compute approximate eigenvalues. If $\mathbf{x}_i$ is an eigenvector associated with the (real) eigenvalue λ_i of a hermitian matrix **A**, and if we define the vector **x** to be

$$\mathbf{x} = \mathbf{x}_i + \epsilon\mathbf{z}, \tag{11.45}$$

then a direct calculation shows that

$$\rho(\mathbf{x}) = \lambda_i + [\rho(\mathbf{z}) - \lambda_i]\frac{(\mathbf{z}, \mathbf{z})}{(\mathbf{x}, \mathbf{x})}|\epsilon|^2. \tag{11.46}$$

This tells us that if we know an approximation $\mathbf{x}$ to the eigenvector $\mathbf{x}_i$ that is accurate to first order in a small parameter ϵ, then the Rayleigh quotient $\rho(\mathbf{x})$ will give an approximation to λ_i that is accurate to *second* order in ϵ [since the error term in (11.46) is of order $\bar{\epsilon}\epsilon$]. This is not very useful for the intermediate eigenvalues $\lambda_2, \ldots, \lambda_{n-1}$, since for these eigenvalues the sign of the error term in (11.46) is unknown ($\lambda_1 \leq \rho \leq \lambda_n$). However, if we are trying to estimate λ_1, we know that $\rho(\mathbf{z}) - \lambda_1 \geq 0$ for any $\mathbf{z}$, so that the estimate $\rho(\mathbf{x})$ will always be an *upper bound* for λ_1. Similarly, $\rho(\mathbf{x})$ will always be a *lower bound* for λ_n. In vibration problems we can often guess the shape of the mode of vibration corresponding to the lowest frequency of vibration, and use the Rayleigh quotient to give an accurate upper bound for this lowest frequency of vibration. It is usually much more difficult to use the Rayleigh quotient to estimate the largest eigenvalue since it is difficult in general to make a good guess at the shape of the largest eigenvector.

Note that when we refer to the *largest* and *smallest* eigenvalues in the context of the Rayleigh quotient we mean that if the eigenvalues are ordered $\lambda_1 \leq \lambda_2 \leq \cdots \leq \lambda_n$ then λ_1 is the smallest eigenvalue and λ_n the largest. There is no implication concerning the *absolute values* of λ_1 and λ_n.

EXERCISE 11.19. Estimate the smallest eigenvalue of the following matrix by means of the Rayleigh quotient, given that it arises in a vibration problem similar to that discussed in Section 8.1.

$$\mathbf{A} = \begin{bmatrix} 1.7 & -1 & 0 \\ -1 & 2 & -1 \\ 0 & -1 & 2 \end{bmatrix}$$

SOLUTION: The Rayleigh quotient is

$$\rho = \frac{1.7x_1^2 + 2x_2^2 + 2x_3^2 - 2x_1x_2 - 2x_2x_3}{x_1^2 + x_2^2 + x_3^2}$$

From the physics of the problem we know that the eigenvector corresponding to the lowest eigenvalue will have elements that are all of the same sign. If we try $\mathbf{x}_1 \approx [1, 1, 1]^T$ and $[1, 2, 1]^T$, we find $\rho = 0.57$ and 0.62, respectively, to two decimals. We know that these estimates are upper bounds, so the best estimate we have obtained is 0.57 corresponding to a trial eigenvector $[1, 1, 1]^T$. The exact eigenvalue is 0.5, corresponding to an eigenvector $[1, 1.2, 0.8]^T$.

EXERCISE 11.20. Use the Rayleigh quotient to obtain information concerning the eigenvalues of

$$\begin{bmatrix} 0 & -1 & 0 \\ -1 & -1 & 1 \\ 0 & 1 & 0 \end{bmatrix}.$$

SOLUTION: The Rayleigh quotient is

$$\rho = \frac{-x_2^2 - 2x_1x_2 + 2x_2x_3}{x_1^2 + x_2^2 + x_3^2}.$$

We have no additional information in this case, and all we can do is try various forms for $\mathbf{x}$. For example, if we try $[1, 1, 1]^T$, $[1, 0, 0]^T$, $[0, 1, 0]^T$, $[1, 1, 0]^T$, $[1, 1, -1]^T$, we obtain, respectively, $\rho = -\frac{1}{3}, 0, -1, -\frac{3}{2}, -\frac{5}{3}$, so that the smallest eigenvalue is not greater than $-\frac{5}{3}$ and the largest one is not less than 0. (The smallest and largest eigenvalues are -2 and $+1$ with corresponding eigenvectors $[1, 2, -1]^T$, $[1, -1, -1]^T$.) Note that the results are much less satisfactory than in the previous example, and the knowledge that the Rayleigh quotient gives an upper bound for λ_1 and a lower bound for λ_3 is essential in deciding which of our trial functions gives the best estimates.

EXERCISE 11.21. Given that the eigenvector corresponding to the lowest eigenvalue of

$$\begin{bmatrix} 3 & -1 & 0 \\ -1 & 2 & -1 \\ 0 & -1 & 3 \end{bmatrix}$$

is of the form $[1, k, 1]^T$ find the exact value of k by forming the Rayleigh quotient ρ and setting $d\rho/dk = 0$.

EXERCISE 11.22. Given that a good approximation to the eigenvector corresponding to the lowest eigenvalue of a real symmetric matrix $\mathbf{A}$ is of the form $\alpha_1\mathbf{z}_1 + \alpha_2\mathbf{z}_2$, where $\mathbf{z}_1$ and $\mathbf{z}_2$ are given real vectors, but α_1, α_2 are unknown, show how to find optimum values for α_1, α_2 by forming the Rayleigh quotient ρ with $\mathbf{x} = \alpha_1\mathbf{z}_1 + \alpha_2\mathbf{z}_2$, setting $\partial\rho/\partial\alpha_1 = \partial\rho/\partial\alpha_2 = 0$, then solving a 2×2 eigenvalue problem. Generalize to the situation where the trial vector is $\alpha_1\mathbf{z}_1 + \cdots + \alpha_s\mathbf{z}_s$, $s < n$.

As another application of **Key Theorem 11.7**, we recall again the statistical problem of Section 11.1 wherein we wished to find that linear combination

$$y_1 = (\mathbf{a}, \mathbf{x}) = a_1x_1 + \cdots + a_nx_n$$

of the variables $x_1, \ldots, x_n$ so as to maximize the variance $(\mathbf{a}, \mathbf{Sa})$ of y_1 subject to $(\mathbf{a}, \mathbf{a}) = 1$; we now see by **Key Theorem 11.7** that we merely need to choose $\mathbf{a}$ to be the eigenvector of the covariance matrix $\mathbf{S}$ associated with the largest eigenvalue of $\mathbf{S}$. The new variable y_1 is called the first *principal component*.

If the single component y_1 does not adequately represent the total variation $s_{11} + \cdots + s_{nn}$, then we might reasonably introduce a second principal component

$$y_2 = b_1x_1 + \cdots + b_nx_n = (\mathbf{b}, \mathbf{x})$$

chosen so that $(\mathbf{b}, \mathbf{b}) = 1$ and so that y_2 accounts for as much variation as possible independent of that already accounted for by y_1; it is known from statistics that we should require the orthogonality of **b** and **a**, that is, $(\mathbf{b}, \mathbf{a}) = 0$. Thus we should seek **b** to maximize $(\mathbf{b}, \mathbf{Sb})$ subject to $(\mathbf{b}, \mathbf{b}) = 1$ and $(\mathbf{b}, \mathbf{a}) = 0$. It will follow from **Key Theorem 11.8**, to be stated shortly, that **b** should be chosen to be an eigenvector associated with the second largest eigenvalue of **S**; more generally, successive principal components should be chosen as eigenvectors associated with successively smaller eigenvalues of S.

For the reader unfamiliar with statistics, we use the language of linear algebra to restate our new problem of trying to find the second principal component **b**. Given that **A** is a hermitian matrix with eigenvalues

$$\lambda_1 \leq \lambda_2 \leq \cdots \leq \lambda_n$$

and associated orthonormalized eigenvectors $\mathbf{x}_1, \ldots, \mathbf{x}_n$, maximize $(\mathbf{x}, \mathbf{Ax})$ subject to

$$(\mathbf{x}, \mathbf{x}) = 1, \qquad (\mathbf{x}, \mathbf{x}_n) = 0. \tag{11.47}$$

Geometrically we observed earlier that the solving of our original extremum problem by $\mathbf{x}_n$ was essentially equivalent to the fact that $\mathbf{x}_n$ was the direction of the shortest semi-axis of the surface $(\mathbf{x}, \mathbf{Ax}) = 1$; if we now add the restriction (11.47) of only considering vectors in the subspace perpendicular to $\mathbf{x}_n$, it would seem that our new solution would be in the direction of the shortest semi-axis of the surface formed by the intersection of this subspace with the surface $(\mathbf{x}, \mathbf{Ax}) = 1$ and that this therefore should somehow involve again the eigenvectors of **A**. In fact, *every* eigenvalue of **A** can be described by extremization problems for the quadratic form $(\mathbf{x}, \mathbf{Ax})$.

- *KEY THEOREM 11.8 (Rayleigh's Principle). Let* **A** *be an* $n \times n$ *hermitian matrix with eigenvalues* $\lambda_1 \leq \lambda_2 \leq \cdots \leq \lambda_n$ *and associated orthonormalized eigenvectors* $\mathbf{x}_1, \ldots, \mathbf{x}_n$. *For each* j *with* $1 \leq j \leq n$ *let*

$$S_j = \text{the set of all } \mathbf{x} \neq \mathbf{0} \text{ which satisfy } (\mathbf{x}, \mathbf{x}_n) = (\mathbf{x}, \mathbf{x}_{n-1}) = \cdots = (\mathbf{x}, \mathbf{x}_{n-j+1}) = 0,$$

$$T_j = \text{the set of all } \mathbf{x} \neq \mathbf{0} \text{ which satisfy } (\mathbf{x}, \mathbf{x}_1) = (\mathbf{x}, \mathbf{x}_2) = \cdots = (\mathbf{x}, \mathbf{x}_{j-1}) = 0,$$

$$\rho(\mathbf{x}) = \frac{(\mathbf{x}, \mathbf{Ax})}{(\mathbf{x}, \mathbf{x})} \quad \text{for} \quad \mathbf{x} \neq \mathbf{0}.$$

Then

$$\max_{S_j} \rho(\mathbf{x}) = \lambda_{n-j} = \rho(\mathbf{x}_{n-j}) \tag{11.48}$$

$$\min_{T_j} \rho(\mathbf{x}) = \lambda_j = \rho(\mathbf{x}_j). \tag{11.49}$$

Proof: The proof is essentially as in **Key Theorem 11.7.** Any $\mathbf{x}$ in S_j must be of the form

$$\mathbf{x} = \sum_{i=1}^{n-j} \alpha_i \mathbf{x}_i$$

so that

$$\rho(\mathbf{x}) - \lambda_{n-j} = \frac{(\lambda_1 - \lambda_{n-j})\alpha_1\bar{\alpha} + \cdots + (\lambda_{n-j} - \lambda_{n-j})\alpha_{n-j}\bar{\alpha}_{n-j}}{\alpha_1\bar{\alpha}_1 + \cdots + \alpha_{n-j}\bar{\alpha}_{n-j}}$$

and hence $\rho(\mathbf{x}) - \lambda_{n-j} \leq 0$ with $\rho(\mathbf{x}_{n-j}) - \lambda_{n-j} = 0$ as desired. A similar argument succeeds for $\mathbf{x}$ in T_j, completing the proof.

To clarify the above theorem, we restate two special cases:

1. $\mathbf{x}_{n-1}$ maximizes $\rho(\mathbf{x})$ subject to $(\mathbf{x}, \mathbf{x}) = 1$ and $(\mathbf{x}, \mathbf{x}_n) = 0$, yielding the maximum value $\rho(\mathbf{x}_{n-1}) = \lambda_{n-1}$;
2. $\mathbf{x}_2$ minimizes $\rho(\mathbf{x})$ subject to $(\mathbf{x}, \mathbf{x}) = 1$ and $(\mathbf{x}, \mathbf{x}_1) = 0$, yielding the minimum value $\rho(\mathbf{x}_2) = \lambda_2$.

EXERCISE 11.23. Use (11.48) to obtain a lower bound on the second eigenvalue λ_2 in Ex. 11.20.

SOLUTION: In Ex. 11.20 it was stated that $\mathbf{x}_3 = [1 \quad -1 \quad -1]^T$, so that our vectors $[x_1 \quad x_2 \quad x_3]^T$ must satisfy $x_1 - x_2 - x_3 = 0$. The vector $[2 \quad 1 \quad 1]^T$ yields $-\frac{1}{2} \leq \lambda_2$, the vector $[3 \quad 1 \quad 2]^T$ yields $-0.22 \leq \lambda_2$, and $[5 \quad 1 \quad 4]^T$ yields $-0.072 \leq \alpha_2$. Actually $\lambda_2 = 0$.

EXERCISE 11.24. Use (11.49) to obtain an upper bound on the second eigenvalue λ_2 in Exs. 11.20 and 11.23.

EXERCISE 11.25. By considering the Rayleigh quotient for

$$\begin{bmatrix} a & b \\ 0 & d \end{bmatrix},$$

show that it is essential in **Key Theorems 11.7** and **11.8** that **A** be hermitian.

EXERCISE 11.26. Extend **Key Theorems 11.7** and **11.8** to the equation $\mathbf{Ax} = \lambda\mathbf{Bx}$, where **A**, **B** are hermitian and **B** is positive definite, by considering

$$\rho(\mathbf{x}) = \frac{(\mathbf{x}, \mathbf{Ax})}{(\mathbf{x}, \mathbf{Bx})}.$$

See Ex. 11.12.

EXERCISE 11.27. Suppose that a given experiment produced the following covariance matrix **S** for three variables x_1, x_2, x_3;

$$\mathbf{S} = \begin{bmatrix} 0.4 & 0.1 & 0.1 \\ 0.1 & 0.3 & 0.2 \\ 0.1 & 0.2 & 0.3 \end{bmatrix}.$$

Find principal components to account for at least 55% of the total variance $0.4 + 0.3 + 0.3 = 1.0$. Find principal components to account for a least 85% of the total variance.

SOLUTION: The largest eigenvalue λ_3 of **S** can be found to be $\lambda_3 = 0.6$ with eigenvector

$$\mathbf{x}_3 = \begin{bmatrix} \frac{\sqrt{3}}{3} \\ \frac{\sqrt{3}}{3} \\ \frac{\sqrt{3}}{3} \end{bmatrix}.$$

Since $0.6 \geq 55\% \times 1.0$, we only require the first principal component $y_1 = (\sqrt{3}/3)(x_1 + x_2 + x_3)$ to account for at least 55% of the variance. To account for 85% we need the second principal component as well, and the second largest eigenvalue λ_2 of **S** can be found to be $\lambda_2 = 0.3$ with eigenvector

$$\mathbf{x}_2 = \begin{bmatrix} \frac{\sqrt{6}}{3} \\ -\frac{\sqrt{6}}{6} \\ -\frac{\sqrt{6}}{6} \end{bmatrix}.$$

Since $\lambda_2 + \lambda_3 = 0.3 + 0.6 \geq 85\% \times 1.0$, we only require the additional second principal component $y_2 = (\sqrt{6}/6)(2x_1 - x_2 - x_3)$ to account for at least 85% of the total variance by means of y_1 and y_2.

EXERCISE 11.28. Repeat Ex. 11.27 with 55%, 85%, and **S** replaced, respectively, by 80%, 95%, and

$$\begin{bmatrix} 0.6 & 0.3 & 0.3 \\ 0.3 & 0.5 & 0.4 \\ 0.3 & 0.4 & 0.5 \end{bmatrix}.$$

11.6 EXTREMIZING QUADRATIC FORMS: THE MIN-MAX PRINCIPLE

As illustrated in Exs. 11.23 and 11.24, in order to use **Key Theorem 11.8** to obtain information on all eigenvalues λ_j, we need to have already found the eigenvectors $\mathbf{x}_{j+1}, \ldots, \mathbf{x}_n$ or $\mathbf{x}_1, \ldots, \mathbf{x}_{j-1}$; if we know $\mathbf{x}_{j+1}, \ldots, \mathbf{x}_n$, then we can use (11.48) to obtain lower bounds on λ_j, while from $\mathbf{x}_1, \ldots, \mathbf{x}_{j-1}$ we can obtain upper bounds on λ_j via (11.49). We cannot, however, obtain any information on λ_j from **Key Theorem 11.8** without first finding one or the other of these sets of eigenvectors. We now present another characterization of eigenvalues by means of extremization problems, but in this case the characterization of λ_j will be independent of the other eigenvectors so as to allow us to analyze λ_j directly.

We know from Rayleigh's principle that the Rayleigh quotient (11.40) satisfies

$$\lambda_1 \leq \rho(\mathbf{x}) \leq \lambda_n$$

for all $\mathbf{x}$, and that ρ is maximized by $\mathbf{x}_n$ with $\rho(\mathbf{x}_n) = \lambda_n$. Geometrically, for the case in which $\mathbf{A}$ is positive definite (see Theorem 11.6), this is equivalent to observing that the closest point to the origin on the ellipsoid E defined by $(\mathbf{x}, \mathbf{Ax}) = 1$ occurs in the direction $\mathbf{x}_n$ at a distance of $1/\sqrt{\lambda_n}$, the length of the shortest semi-axis of E. Now suppose that we impose the constraint

$$(\mathbf{p}, \mathbf{x}) = 0 \tag{11.50}$$

for some arbitrary but fixed vector $\mathbf{p}$. The constraint (11.50) restricts $\mathbf{x}$ to lie on the plane through the origin perpendicular to $\mathbf{p}$; this plane intersects the ellipsoid E in some ellipsoid $E_\mathbf{p}$ of one lower dimension. The point $\mathbf{x}(\mathbf{p})$ on $E_\mathbf{p}$ closest to the origin is at a distance $d(\mathbf{p})$ and of course, generally, $d(\mathbf{p}) \geq 1\sqrt{\lambda_n}$, the distance from the origin to the closest point on *all* of E; the *largest* that $d(\mathbf{p})$ could become is the length of the second smallest semi-axis of E, and this would occur when $\mathbf{p}$ is chosen in the direction of the shortest semi-axis of E, namely in the direction $\mathbf{x}_n$. The second shortest semi-axis of E, however, has length $1/\sqrt{\lambda_{n-1}}$. Therefore we see geometrically that

$$\begin{aligned}\frac{1}{\sqrt{\lambda_{n-1}}} &= \max_{\text{all } \mathbf{p}} d(\mathbf{p}) = \max_{\mathbf{p}} [\min \|\mathbf{x}\|_2 \mid \mathbf{x} \text{ is in } E_\mathbf{p}] \\ &= \max_{\text{all } \mathbf{p}} [\min \|\mathbf{x}\|_2 \mid (\mathbf{x}, \mathbf{Ax}) = 1, (\mathbf{p}, \mathbf{x}) = 0].\end{aligned}$$

Recalling from Theorem 11.6 that *minimizing* length on an ellipsoid corresponds to *maximizing* $\rho(\mathbf{x})$ on the unit sphere, the above expression for $1/\sqrt{\lambda_{n-1}}$ translates into

$$\lambda_{n-1} = \min_{\text{all } \mathbf{p}} [\max \{\rho(\mathbf{x}) \mid (\mathbf{p}, \mathbf{x}) = 0\}]. \tag{11.51}$$

The important fact is that (11.51) gives a description of the second largest eigenvalue λ_{n-1} that is independent of all the other eigenvalues and eigenvectors, as opposed to the description of λ_{n-1} in **Key Theorem 11.8** and its special case (1) following that theorem.

To help the reader visualize the meanings of these max-min and min-max problems, we consider four different $\mathbf{p}$ values, namely $\mathbf{p}'$, $\mathbf{p}''$, $\mathbf{p}'''$, and $\mathbf{p}^*$ (the optimal value). For each $\mathbf{p}$ value, $\rho(\cdot)$ is a function whose graph is (symbolically) indicated in Figure 11.4; the points $\mathbf{x}(\mathbf{p}')$, $\mathbf{x}(\mathbf{p}'')$, $\mathbf{x}(\mathbf{p}''')$, and $\mathbf{x}(\mathbf{p}^*)$ denote those values of $\mathbf{x}$ which maximize ρ for that associated value of $\mathbf{p}$. The solution of the min-max problem is then at $\mathbf{x}(\mathbf{p}^*)$ since this makes the maximum $\rho(\mathbf{x}(\mathbf{p}))$ as small as possible. Make a similar sketch describing the max-min problem.

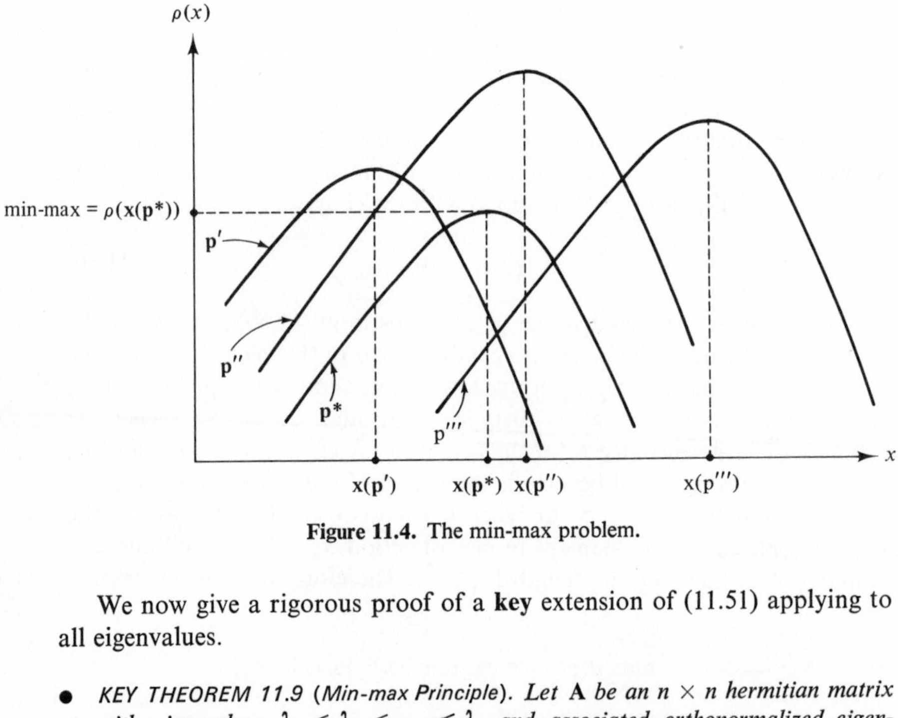

Figure 11.4. The min-max problem.

We now give a rigorous proof of a **key** extension of (11.51) applying to all eigenvalues.

- *KEY THEOREM 11.9 (Min-max Principle). Let* $\mathbf{A}$ *be an* $n \times n$ *hermitian matrix with eigenvalues* $\lambda_1 \leq \lambda_2 \leq \ldots \leq \lambda_n$ *and associated orthonormalized eigenvectors* $\mathbf{x}_1, \ldots, \mathbf{x}_n$. *Then if* ρ *is the Rayleigh quotient (11.40), we have*

$$\lambda_j = \min_{\mathbf{p}} \{\max_{\mathbf{x}} \rho(\mathbf{x})\} \tag{11.52}$$

where the minimum is with respect to all sets containing precisely $n - j$ *vectors* $\mathbf{p}_{j+1}, \ldots, \mathbf{p}_n$ *and the maximum is with respect to all those vectors* $\mathbf{x}$ *satisfying* $(\mathbf{p}_{j+1}, \mathbf{x}) = (\mathbf{p}_{j+2}, \mathbf{x}) = \cdots = (\mathbf{p}_n, \mathbf{x}) = 0$. *Moreover,*

$$\lambda_j = \max_{\mathbf{p}} \{\min_{\mathbf{x}} \rho(\mathbf{x})\} \tag{11.53}$$

where, in this second characterization, the maximum is with respect to all sets *containing precisely $j - 1$ vectors $\mathbf{p}_1, \ldots, \mathbf{p}_{j-1}$ and the minimum is with respect to all those vectors $\mathbf{x}$ satisfying* $(\mathbf{p}_1, \mathbf{x}) = (\mathbf{p}_2, \mathbf{x}) = \cdots = (\mathbf{p}_{j-1}, \mathbf{x}) = 0$.

Proof: We prove only (11.52), since the proof of (11.53) is so similar. Let $\mathbf{X}$ denote the unitary matrix whose columns are $\mathbf{x}_1, \ldots, \mathbf{x}_n$ and let $\mathbf{\Lambda}$ denote the diagonal matrix whose (i, i) element is λ_i for $i = 1, \ldots, n$; then $\mathbf{X}^H\mathbf{A}\mathbf{X} = \mathbf{\Lambda}$. For any vector $\mathbf{x}$ and any vectors $\mathbf{p}_{j+1}, \ldots, \mathbf{p}_n$ we define $\mathbf{y} = \mathbf{X}^H\mathbf{x}$ and $\mathbf{q}_i = \mathbf{X}^H\mathbf{p}_i$ for $i = j + 1, \ldots, n$, so that

$$(\mathbf{x}, \mathbf{A}\mathbf{x}) = (\mathbf{y}, \mathbf{\Lambda}\mathbf{y}), \qquad (\mathbf{x}, \mathbf{x}) = (\mathbf{y}, \mathbf{y}), \qquad (\mathbf{p}_i, \mathbf{x}) = (\mathbf{q}_i, \mathbf{y}).$$

Therefore the expression on the right-hand side of (11.52), whose value we denote by L_j (so that we need to show that $L_j = \lambda_j$), can be replaced by

$$\min_{\mathbf{q}} \left\{ \max_{\mathbf{y}} \left[\frac{\lambda_1 |y_1|^2 + \cdots + \lambda_n |y_n|^2}{|y_1|^2 + \cdots + |y_n|^2} \right] \right\} \tag{11.54}$$

where the minimum is with respect to all sets containing precisely $n - j$ vectors $\mathbf{q}_{j+1}, \ldots, \mathbf{q}_n$ and the maximum is with respect to all those vectors $\mathbf{y}$ satisfying $(\mathbf{q}_{j+1}, \mathbf{y}) = \cdots = (\mathbf{q}_n, \mathbf{y}) = 0$. For any given fixed set of vectors $\mathbf{q}_{j+1}, \ldots, \mathbf{q}_n$ we can regard the conditions $(\mathbf{q}_i, \mathbf{y}) = 0$ for $i = j + 1, \ldots, n$ as a set of $n - j$ equations in n unknowns $y_1, \ldots, y_n$. If we set $y_1 = y_2 = \cdots = y_{j-1} = 0$ then we still have $n - (j - 1) = n - j + 1$ unknowns $y_j, \ldots, y_n$ involved in the $n - j$ homogeneous equations $(\mathbf{q}_i, \mathbf{y}) = 0$ for $i = j + 1, \ldots, n$; since k homogeneous equations in $k + 1$ unknowns $(k = n - j)$ always have at least one solution not identically zero, there is a solution $y_1 = \cdots = y_{j-1} = 0$, $y_j = z_j$, $y_{j+1} = z_{j+1}, \ldots, y_n = z_n$ to $(\mathbf{q}_i, \mathbf{y}) = 0$ for $i = j + 1, \ldots, n$, with $|z_j|^2 + \cdots + |z_n|^2 \neq 0$. Therefore the term in the braces { } in (11.54) satisfies

$$\begin{aligned}
\max_{\mathbf{y}} \left[\frac{\lambda_1 |y_1|^2 + \cdots + \lambda_n |y_n|^2}{|y_1|^2 + \cdots + |y_n|^2} \right] &\geq \frac{\lambda_j |z_j|^2 + \lambda_{j+1} |z_{j+1}|^2 + \cdots + \lambda_n |z_n|^2}{|z_j|^2 + |z_{j+1}|^2 + \cdots + |z_n|^2} \\
&\geq \frac{\lambda_j |z_j|^2 + \lambda_j |z_{j+1}|^2 + \cdots + \lambda_j |z_n|^2}{|z_j|^2 + |z_{j+1}|^2 + \cdots + |z_n|^2} \\
&= \lambda_j.
\end{aligned}$$

Thus for each set of vectors $\mathbf{q}$ the term in braces { } in (11.54) is greater than or equal to λ_j, and hence the minimum of the term in braces as the vectors $\mathbf{q}$ vary must also exceed λ_j; in other words, $L_j \geq \lambda_j$. On the other hand, if we let $\mathbf{p}_i = \mathbf{x}_i$ so that $\mathbf{q}_i = \mathbf{e}_i$, the unit column vector, for $i = j + 1, \ldots, n$, then by Rayleigh's principle (**Key Theorem 11.8**) the term in braces is precisely equal to λ_j, so that the minimum L_j of the term in braces as the vectors $\mathbf{q}$ vary must be less than or equal to λ_j, that is, $L_j \leq \lambda_j$. The two inequalities $L_j \geq \lambda_j$ and $L_j \leq \lambda_j$ of course give us $L_j = \lambda_j$, as we wanted to prove.

One of the reasons that **Key Theorem 11.9**, the min-max principle, is important is because it allows the approximate computation of λ_j independently of the other eigenvalues and eigenvectors.

EXERCISE 11.29. Find a lower bound for λ_2 for the matrix given in Ex. 11.19.

SOLUTION: We use (11.53). The first step is to minimize the Rayleigh quotient with a single constraint. It is clear from the proof above that to obtain the exact result, the constraint we should apply is $(\mathbf{p}, \mathbf{x}) = 0$ where $\mathbf{p}$ is the eigenvector corresponding to the smallest eigenvalue. We do not know $\mathbf{p}$ exactly, but we do know an approximation to $\mathbf{p}$, namely $[1 \quad 1 \quad 1]^T$, found in Ex. 11.19, so we impose the constraint $x_1 + x_2 + x_3 = 0$. If we set $x_1 = -x_2 - x_3$ we find that we now have to minimize

$$R = \frac{5.7x_2^2 + 3.4x_2x_3 + 3.7x_3^2}{2(x_2^2 + x_2x_3 + x_3^2)}.$$

This involves only two variables and minimization involves the solution of a quadratic equation. The minimum is found to be 1.81, and since according to (11.53) λ_2 is the *maximum* of all such minima, $\lambda_2 \geq 1.81$. We could try other vectors for $\mathbf{p}$. The largest resulting lower bound for λ_2 will be the best.

EXERCISE 11.30. As in Ex. 11.29, let $\mathbf{p} = [1 \quad 2 \quad 3]^T$ and find a lower bound on the eigenvalue λ_2 for the matrix given in Ex. 11.19.

EXERCISE 11.31. Use the min-max principle to obtain upper and lower bounds on the eigenvalue λ_2 of Ex. 11.20. Note that the vectors $[1 \quad 1 \quad -1]^T$ and $[1 \quad 0 \quad 0]^T$ furnished the best approximations found in Ex. 11.20 for $\mathbf{x}_1$ and $\mathbf{x}_3$, respectively.

Equally important, however, is the use of the min-max principle to deduce how the eigenvalues (and hence frequencies) of some vibrating system change when the parameters of the system vary; this allows us to conclude from a mathematical model that, for example, the pitch of a guitar string is higher when the string is more tightly stretched. The basis of this result is the following.

THEOREM 11.10. Let the two $n \times n$ hermitian matrices $\mathbf{A}$ and $\mathbf{B}$ satisfy

$$(\mathbf{x}, \mathbf{A}\mathbf{x}) \leq (\mathbf{x}, \mathbf{B}\mathbf{x}) \tag{11.55}$$

for all $\mathbf{x}$, and let $\lambda_1 \leq \lambda_2 \leq \cdots \leq \lambda_n$ and $\mu_1 \leq \mu_2 \leq \cdots \leq \mu_n$ be the eigenvalues of $\mathbf{A}$ and of $\mathbf{B}$, respectively. Then

$$\lambda_i \leq \mu_i \qquad (\text{for } i = 1, \ldots, n).$$

Proof: Let $\rho_{\mathbf{A}}$ and $\rho_{\mathbf{B}}$ denote the Rayleigh quotient for $\mathbf{A}$ and for $\mathbf{B}$, respectively; by (11.55) we have

$$\rho_{\mathbf{A}}(\mathbf{x}) \leq \rho_{\mathbf{B}}(\mathbf{x})$$

for all $\mathbf{x}$. Therefore, in the notation of (11.52), for each set of vectors $\mathbf{p}$ we have

$$\max_{\mathbf{x}} \rho_{\mathbf{A}}(\mathbf{x}) \leq \max_{\mathbf{x}} \rho_{\mathbf{B}}(\mathbf{x})$$

and hence the minimum of the left-hand side (as the vectors **p** vary) is less than or equal to the minimum of the right-hand side (as the vectors **p** vary), which says that

$$\min_{\mathbf{p}} \{\max_{\mathbf{x}} \rho_{\mathbf{A}}(\mathbf{x})\} \leq \min_{\mathbf{p}} \{\max_{\mathbf{x}} \rho_{\mathbf{B}}(\mathbf{x})\}.$$

By (11.52) this gives us $\lambda_j \leq \mu_j$ as required.

EXERCISE 11.32. For the vibration problem depicted in Figure 8.1, discuss what happens if the length between the second and third beads is decreased from $2l$ to $2\alpha l$ for $\alpha < 1$ while the other three lengths stay as depicted.

SOLUTION: The differential equations, of course, must be modified; the equation involving d^2X_1/dt^2 becomes, for example,

$$\frac{d^2X_1}{dt^2} = \frac{-TX_1}{l} + \frac{T(X_2 - X_1)}{2\alpha l}.$$

Eventually we are led to the eigenvalue problem for

$$\mathbf{A}_\alpha = \begin{bmatrix} \dfrac{2\alpha + 1}{\alpha} & -\dfrac{1}{\alpha} & 0 \\ -\dfrac{1}{\alpha} & \dfrac{\alpha + 1}{\alpha} & -1 \\ 0 & -1 & 3 \end{bmatrix}$$

analogous to (8.5) for $\mathbf{A}_1$. We note that

$$\mathbf{A}_\alpha - \mathbf{A}_1 = \begin{bmatrix} \dfrac{1}{\alpha} - 1 & 1 - \dfrac{1}{\alpha} & 0 \\ 1 - \dfrac{1}{\alpha} & \dfrac{1}{\alpha} - 1 & 0 \\ 0 & 0 & 0 \end{bmatrix}$$

has as the values of the principal 1×1, 2×2, and 3×3 minors the determinants

$$\frac{1}{\alpha} - 1, \left(\frac{1}{\alpha} - 1\right)^2 - \left(\frac{1}{\alpha} - 1\right)^2 = 0, 0; \quad \text{for } \alpha < 1$$

these are all nonnegative and hence $\mathbf{A}_\alpha - \mathbf{A}_1$ is nonnegative semi-definite. [Perhaps it is easier to see this by observing that the eigenvalues of $\mathbf{A}_\alpha - \mathbf{A}_1$ are simply 0, 0, and $2[(1/\alpha) - 1] \geq 0$ for $\alpha \leq 1$ so that $\mathbf{A}_\alpha - \mathbf{A}_1$ is semi-definite as claimed.] Therefore,

$$(\mathbf{x}, \mathbf{A}_1\mathbf{x}) = (\mathbf{x}, \mathbf{A}_\alpha\mathbf{x}) + (\mathbf{x}, [\mathbf{A}_1 - \mathbf{A}_\alpha]\mathbf{x}) \leq (\mathbf{x}, \mathbf{A}_\alpha\mathbf{x})$$

and hence by Theorem 11.10 each eigenvalue of $\mathbf{A}_\alpha$ is no less than the corresponding eigenvalue of $\mathbf{A}_1$. Since eigenvalues λ are related to frequencies ω of vibration via $\lambda = 2\omega^2 ml/T$, we conclude that the fundamental frequencies of vibration of the system with X_1 and X_2 separated by the shorter distance $2\alpha l$ are greater than the corresponding frequencies when $\alpha = 1$.

EXERCISE 11.33. For the vibration problem depicted in Figure 8.1, discuss as in Ex. 11.32 what happens if the lengths l from the first and third beads to the left and right ends of the string, respectively, are decreased to αl and βl, respectively, with $\alpha < 1$ and $\beta < 1$.

EXERCISE 11.34. Use Rayleigh's principle to show, under the hypotheses of Theorem 11.10, that $\lambda_1 \leq \mu_1$ and $\lambda_n \leq \mu_n$. Explain why Rayleigh's principle cannot also be used to show that $\lambda_i \leq \mu_i$ for $i = 2, \ldots, n - 1$ while the min-max principle can be so used.

EXERCISE 11.35. Let $\mathbf{A}$ be an $m \times n$ matrix. Prove that $\|\mathbf{A}\|_2$ is just the largest singular value of $\mathbf{A}$.

SOLUTION: We know that

$$\|\mathbf{A}\|_2^2 = \max_{\mathbf{x}\neq\mathbf{0}} \frac{\|\mathbf{Ax}\|_2^2}{\|\mathbf{x}\|_2^2} = \max_{\mathbf{x}\neq\mathbf{0}} \frac{(\mathbf{Ax}, \mathbf{Ax})}{(\mathbf{x}, \mathbf{x})} = \max_{\mathbf{x}\neq\mathbf{0}} \frac{(\mathbf{x}, \mathbf{A}^H\mathbf{Ax})}{(\mathbf{x}, \mathbf{x})}.$$

Since $\mathbf{A}^H\mathbf{A}$ is hermitian, this last maximum equals the largest eigenvalue of the positive semidefinite matrix $\mathbf{A}^H\mathbf{A}$, and the largest singular value (see Definition 9.2) is simply the square root of that eigenvalue.

EXERCISE 11.36. Let $\mathbf{A}$ be an $n \times n$ hermitian matrix with spectral radius r (see Definition 10.5). Prove that $\|\mathbf{A}\|_2 = r$.

MISCELLANEOUS EXERCISES 11

EXERCISE 11.37. Suppose that the hermitian matrix $\mathbf{A}$ is semidefinite. Prove that a nonzero $\mathbf{x}$ exists for which $(\mathbf{x}, \mathbf{Ax}) = 0$ if and only if $\mathbf{A}$ is singular.

EXERCISE 11.38. In Section 6.4 we discussed the LU-decomposition. Prove that an $n \times n$ matrix $\mathbf{A}$ can be written $\mathbf{A} = \mathbf{LU}$ with $\mathbf{L}$ a lower triangular matrix with ones on the diagonal and $\mathbf{U}$ an upper triangular matrix with no zeros on the diagonal if and only if it holds that the principal minors consisting of the determinants of the $k \times k$ submatrices of $\mathbf{A}$ in the top-left corner of $\mathbf{A}$ are nonzero for $k = 1, \ldots, n$.

EXERCISE 11.39. Let the $n \times n$ hermitian matrix $\mathbf{A}$ satisfy the condition of Ex. 11.38 and let $\mathbf{A} = \mathbf{LU}$ as in Ex. 11.38. Let $\mathbf{U} = \mathbf{DU}_0$, where $\mathbf{D}$ is a diagonal matrix and $\mathbf{U}_0$ is upper triangular with ones on its diagonal, so that $\mathbf{A} = \mathbf{LDU}_0$. Prove that $\mathbf{L} = \mathbf{U}_0^H$ so that $\mathbf{A} = \mathbf{LDL}^H$.

EXERCISE 11.40. Prove Theorem 11.5.

EXERCISE 11.41. Prove that (11.49) is valid.

EXERCISE 11.42. Prove that (11.53) is valid.

EXERCISE 11.43. Prove that if **A** is positive definite, then $\mathbf{A}^{-1}$ is also positive definite.

EXERCISE 11.44. Prove that if **A** is a positive definite matrix of order m and **B** is $m \times n$, then $\mathbf{B}^H\mathbf{AB}$ is positive semidefinite. Prove that if the rank of **B** is n, then $\mathbf{B}^H\mathbf{AB}$ is positive definite; prove that if the rank of **B** is less than n, then $\mathbf{B}^H\mathbf{AB}$ is positive semidefinite and singular.

EXERCISE 11.45. If **A** is positive definite, prove that so also is any principal submatrix. (The diagonal elements of a *principal* submatrix of **A** are diagonal elements of **A**.)

EXERCISE 11.46. If $\det(\mathbf{A} + \lambda\mathbf{B} + \lambda^2\mathbf{C}) = 0$, where **A**, **B**, **C** are positive definite, prove that the real part of λ is less than zero. If **C** is only hermitian, prove that the real part of any complex λ is less than zero. [The reader might be interested to compare the brevity of the treatment using matrix notation with the longhand version of this problem, and other matters, in A. G. Webster, *The Dynamics of Particles*, etc., Hafner (1949), pp. 157–166, and appendices.]

EXERCISE 11.47. If **A** is positive semidefinite, prove that

(a) $\lambda_i \geq 0$.
(b) $\mathbf{B}^H\mathbf{AB}$ is also positive semidefinite, where **A** is $m \times m$, and **B** is $m \times n$.
(c) The determinant of any principal submatrix of **A** is greater than or equal to zero. In particular, $a_{ii} \geq 0$, $a_{ii}a_{jj} - |a_{ij}|^2 \geq 0$.

EXERCISE 11.48. If $\mathbf{A} = \mathbf{B} + \mathbf{C}$, where **B** is positive definite and **C** is positive semidefinite, show that

(a) **A** is positive definite.
(b) $\det \mathbf{B} \leq \det \mathbf{A}$.
(c) $\mathbf{B}^{-1} - \mathbf{A}^{-1}$ is positive semidefinite.

EXERCISE 11.49. Prove that if **B** is positive definite and **A** is hermitian, then the eigenvalues of **BA** are real.

EXERCISE 11.50. Reduce the following quadratic forms to a sum of multiples of squares by an orthogonal transformation.

(a) $5x_1^2 + 6x_2^2 + 7x_3^2 - 4x_1x_2 - 4x_2x_3$.
(b) $3x_1^2 + 2x_2^2 + 2x_3^2 + 2x_1x_2 + 4x_2x_3 + 2x_3x_1$.
(c) $x_1^2 + x_2^2 + x_3^2 + 4x_1x_2 + 4x_2x_3 + 4x_3x_1$.

Sketch, relative to their principal axes, the surfaces given by setting these quadratic forms equal to a positive constant.

EXERCISE 11.51. If **A** is positive definite, prove that

$$\det \mathbf{A} \leq a_{11}a_{22} \cdots a_{nn}.$$

Deduce that, if **B** is an arbitrary square matrix,

$$|\det \mathbf{B}|^2 \leq \prod_{i=1}^{n} \left(\sum_{j=1}^{n} |b_{ij}|^2 \right).$$

This is known as *Hadamard's inequality*. The absolute value of the determinant of a matrix can be interpreted as the volume in n-dimensional space of the solid whose edges are described by the row vectors forming the rows of the matrix. Hadamard's inequality then says that this volume is less than or equal to that of the n-dimensional *rectangular* solid whose sides have the same lengths. Verify this for $n = 2, 3$ from your knowledge of the geometry of parallelograms and parallelepipeds in two and three dimensions. Under what conditions is equality attained in Hadamard's inequality? Give an independent proof of the equality in this case. Show that if $a = \max|a_{ij}|$ then

$$|\det \mathbf{A}| \leq a^n n^{n/2}.$$

EXERCISE 11.52. If $d\mathbf{x}/dt = -\mathbf{A}\mathbf{x}$, where **A** is positive definite, prove that $\mathbf{x} \longrightarrow \mathbf{0}$ as $t \longrightarrow \infty$. (See Section 10.7.)

EXERCISE 11.53. A matrix is said to be *idempotent* if $\mathbf{A}^2 = \mathbf{A}$. Prove:

(a) The eigenvalues of an idempotent matrix are either 0 or 1.
(b) The only nonsingular idempotent matrix is the identity matrix.
(c) A necessary and sufficient condition that a *hermitian* matrix of order n be idempotent is that k of its eigenvalues are equal to 1, and the remaining $n - k$ are equal to zero, where k is the rank of the matrix.
(d) The trace of a hermitian idempotent matrix is equal to its rank.
(e) A singular 2×2 matrix with $a_{11} + a_{22} = 1$ is idempotent.
(f) If $\mathbf{AGA} = \mathbf{A}$, then $\mathbf{GA}$ is idempotent.
(g) If **A** is idempotent and **P** is unitary, then $\mathbf{P}^H\mathbf{AP}$ is idempotent.
(h) $\mathbf{A} = \mathbf{U}(\mathbf{VU})^{-1}\mathbf{V}$ is idempotent.

EXERCISE 11.54. Suppose that n measurements are represented by a vector $\mathbf{x} = [x_1, \ldots, x_n]^T$. The mean and variance of the measurements are given by

$$\bar{x} = \frac{1}{n}\sum_{i=1}^{n} x_i, \qquad s^2 = \frac{1}{n}\sum_{i=1}^{n}(x_i - \bar{x})^2.$$

Show that

$$s^2 = \left(\mathbf{x}, \left(\mathbf{I} - \frac{1}{n}\mathbf{J}\right)\mathbf{x}\right),$$

where **J** is a square matrix, all of whose elements are unity. Prove that $\mathbf{I} - n^{-1}\mathbf{J}$ is idempotent. (See Ex. 11.53.) (Applications of idempotent matrices occur in statistics and regression theory. See, for instance, Graybill [85].)

EXERCISE 11.55. Show that, for any hermitian matrix **A**, a hermitian-congruence transformation **S** exists such that

$$\mathbf{S}^H\mathbf{A}\mathbf{S} = \begin{bmatrix} \mathbf{I}_r & \mathbf{0} & \mathbf{0} \\ \mathbf{0} & -\mathbf{I}_s & \mathbf{0} \\ \mathbf{0} & \mathbf{0} & \mathbf{0} \end{bmatrix},$$

where $\mathbf{I}_r$, $\mathbf{I}_s$ are unit matrices of orders r and s, and the integers r, s are always the same for a given **A**, independent of **S**, which is not unique. If the quadratic form $(\mathbf{x}, \mathbf{A}\mathbf{x})$ is reduced to a sum of squares $(\mathbf{y}, \mathbf{D}\mathbf{y})$, where **D** is diagonal, by a nonsingular transformation $\mathbf{x} = \mathbf{P}\mathbf{y}$, prove that the number of positive, negative, and zero diagonal elements of **D** is always the same, independent of the transformation. (The diagonal elements of **D** must, of course, be real.) This is known as *Sylvester's law of inertia*.

EXERCISE 11.56. Show that, when reducing a real symmetric square matrix **A** to row-echelon form as in Theorem 11.5, the numbers of positive, negative, and zero pivots are equal to the numbers of positive, negative, and zero eigenvalues.

EXERCISE 11.57. If **A** is a positive definite hermitian matrix, prove that there exists a unique positive definite hermitian matrix **B** such that $\mathbf{A} = \mathbf{B}^2$. If **A** is semidefinite, prove that **B** is semidefinite. **B** is called the *square root* of *A*.

EXERCISE 11.58. Prove:

(a) If **A** is hermitian, then $\mathbf{U} = e^{i\mathbf{A}}$ is unitary. Conversely, any unitary matrix **U** can be expressed in the form $\mathbf{U} = e^{i\mathbf{A}}$, where **A** is hermitian (see Section 10.7).

(b) If **A** is any square matrix, then prove that there exist positive definite or semidefinite hermitian matrices **M** and **N** and unitary matrices **U** and **V** such that $\mathbf{A} = \mathbf{U}\mathbf{M} = \mathbf{N}\mathbf{V}$. This is known as the *polar* representation of **A**, by analogy with the representation $a + ib = re^{i\theta}$ for any complex number. The matrices **M**, **N** are unique. If **A** is nonsingular, $\mathbf{U} = \mathbf{V}$ and this matrix is unique.

EXERCISE 11.59. An orthogonal matrix **A** is called *proper* if $\det \mathbf{A} = 1$ and *improper* if $\det \mathbf{A} = -1$. Prove that all proper and improper 2×2 orthogonal matrices can be written in the following forms, respectively, where $0 \leq \alpha < 2\pi$:

$$\mathbf{L} = \begin{bmatrix} \cos\alpha & -\sin\alpha \\ \sin\alpha & \cos\alpha \end{bmatrix} \quad \text{and} \quad \begin{bmatrix} \cos\alpha & \sin\alpha \\ \sin\alpha & -\cos\alpha \end{bmatrix}$$

Prove that any proper 2×2 orthogonal matrix represents a unique rotation; also that any improper 2×2 orthogonal matrix represents rotation followed (or preceded) by a reflection in a line through the origin (see Section 8.5).

APPENDIX ONE

HINTS AND ANSWERS TO SELECTED EXERCISES

CHAPTER 1

EXERCISE 1.1. When parentheses are used, operations inside the parentheses are performed first.

EXERCISE 1.6. $a_{3j} = 4a_{1j}$. Third row of **AB** is $\sum_j a_{3j}b_{jk} = 4\sum_j a_{1j}b_{jk}$.

EXERCISE 1.7. Since equation is true for *all* b_1, b_2, choose (a) $b_1 = 1$, $b_2 = 0$, (b) $b_1 = 0$, $b_2 = 1$.

EXERCISE 1.13. See Ex. 1.11. Write $\mathbf{A} = \mathbf{B} + i\mathbf{C}$, $i = \sqrt{-1}$.

EXERCISE 1.22.

$$\frac{1}{12}\begin{bmatrix} -7 & -6 & 5 \\ 2 & 0 & 2 \\ 1 & -6 & 1 \end{bmatrix}.$$

EXERCISE 1.23.

$$\frac{1}{2}\begin{bmatrix} 1+\alpha & 1-5\alpha & 2\alpha \\ -1+\beta & 1-5\beta & 2\beta \end{bmatrix}, \qquad \text{any} \quad \alpha, \beta.$$

EXERCISE 1.37. Use induction.

EXERCISE 1.38. Only null matrices. (Consider the diagonal elements of $\mathbf{A}^T\mathbf{A}$.)

EXERCISE 1.41. Use the fact that **A** must commute with

$$\begin{bmatrix}1 & 0\\0 & 0\end{bmatrix},\quad \begin{bmatrix}0 & 1\\0 & 0\end{bmatrix},\quad \begin{bmatrix}0 & 0\\1 & 0\end{bmatrix},\quad \begin{bmatrix}0 & 0\\0 & 1\end{bmatrix}.$$

EXERCISE 1.43. "Proof" assumes that **HA** has a left-inverse.

EXERCISE 1.44. (a) Let $\mathbf{A}^{-1} = \mathbf{B}$. Then $\mathbf{AB} = \mathbf{BA} = \mathbf{I}$, so that $\mathbf{B}^T\mathbf{A}^T = \mathbf{A}^T\mathbf{B}^T = \mathbf{I}$. Hence $\mathbf{B}^T = (\mathbf{A}^T)^{-1}$, which is the required result.

EXERCISE 1.46. $(\mathbf{I} + \mathbf{K})^T = \mathbf{I} - \mathbf{K}$; $(\mathbf{I} - \mathbf{K})(\mathbf{I} + \mathbf{K}) = (\mathbf{I} + \mathbf{K})(\mathbf{I} - \mathbf{K})$.

EXERCISE 1.47. Equations $\mathbf{A}\mathbf{x}_i = \mathbf{e}_i$ or $\mathbf{A}^T\mathbf{y}_i = \mathbf{e}_i$ will be contradictory.

EXERCISE 1.50. (b) $\begin{bmatrix}\cos\theta & \sin\theta\\ -\sin\theta & \cos\theta\end{bmatrix}$ or $\begin{bmatrix}\cos\theta & \sin\theta\\ \sin\theta & -\cos\theta\end{bmatrix}$.

EXERCISE 1.58. Proof of (d) follows directly from (c).

CHAPTER 2

EXERCISE 2.1.

$$\begin{bmatrix}0.39\\0.19\\0.42\end{bmatrix},\quad \begin{bmatrix}0.381\\0.183\\0.436\end{bmatrix},\quad \frac{1}{6}\begin{bmatrix}2\\1\\3\end{bmatrix}.$$

EXERCISE 2.7. A "war-like" society.

EXERCISE 2.14. Maximize $M = 40x_1 + 60x_2$ subject to $2x_1 + x_2 \leq 55$, $x_1 + x_2 \leq 40$, $x_1 + 3x_2 \leq 100$ for the first question.

CHAPTER 3

EXERCISE 3.4. When Gauss-Jordan is applied to $[\mathbf{A}, \mathbf{e}_i]$ we obtain $[\mathbf{I}, \mathbf{x}_i]$.

EXERCISE 3.8. The row-echelon form of the augmented matrix is

$$\begin{bmatrix}1 & -1 & 0 & 0 & 4\\0 & 0 & 1 & 1 & -3\\0 & 0 & 0 & 0 & 0\end{bmatrix}.$$

EXERCISE 3.9. x_1; x_2 or x_3; any two of x_4, x_5, x_6, x_7 except the pair x_5 and x_6.

EXERCISE 3.10. (a) 0. (b) ∞. (c) 1.

EXERCISE 3.11. (a) ∞. (b) ∞. (c) 0. (d) ∞. (e) 0. (f) 1.

EXERCISE 3.13. The row-echelon form is $\begin{bmatrix} 1 & 0 \\ 0 & 1 \\ 0 & 0 \end{bmatrix}$.

EXERCISE 3.15.

(a) $\begin{bmatrix} 1 & 0 & -5 & -8 \\ 0 & 1 & -3 & -6 \\ 0 & 0 & 0 & 0 \end{bmatrix}$ and rank = 2.

(d) $\begin{bmatrix} 1 & 0 & \frac{9}{19} & \frac{27}{19} \\ 0 & 1 & \frac{-2}{19} & \frac{-6}{19} \\ 0 & 0 & 0 & 0 \end{bmatrix}$ and rank = 2.

EXERCISE 3.18. $[\mathbf{I}, \mathbf{A}]$ already is in row-echelon form.

EXERCISE 3.25. Use Theorems 3.1, 3.2, and **Key Theorem 3.11**, and show that $\mathbf{I}$ is the only $m \times m$ matrix of rank m already in row-echelon form.

EXERCISE 3.28. $\alpha = 5$.

EXERCISE 3.31. Examine the question of whether the equations $\mathbf{A}\mathbf{x}_i = \mathbf{e}_i$ possess solutions by reducing the matrix $[\mathbf{A}, \mathbf{I}]$ to row-echelon form.

EXERCISE 3.32. See hint for Ex. 3.31.

EXERCISE 3.33. Reduce $\mathbf{A}$ to row-echelon form to show that $\mathbf{B} - \mathbf{C} = \mathbf{x}\boldsymbol{\alpha}^T$ where $\mathbf{x} = [-3 \quad 0 \quad 1]^T$ and $\boldsymbol{\alpha}$ is an arbitrary 4×1 vector.

EXERCISE 3.34. $\mathbf{b} = [b_1, b_2, b_1 + b_2]^T$.

EXERCISE 3.35. Contradictory for $k = -4$. For $k = 0$, solution is $x_1 = 3 - 3\alpha$, $x_2 = 1$, $x_3 = \alpha$, any α.

EXERCISE 3.40. $\mathbf{x}^T(\mathbf{K}\mathbf{x}) = (\mathbf{K}\mathbf{x})^T\mathbf{x}$ and $(\mathbf{x}^T\mathbf{K})\mathbf{x} = (\mathbf{K}^T\mathbf{x})^T\mathbf{x}$. If $(\mathbf{I} + \mathbf{K})\mathbf{x} = \mathbf{0}$, consider $\mathbf{x}^T(\mathbf{I} + \mathbf{K})\mathbf{x}$.

EXERCISE 3.45. $\mathbf{A} = \mathbf{u}\mathbf{v}^T$, $c = \mathbf{v}^T\mathbf{u}$.

EXERCISE 3.46. Use the fact that $\mathbf{B}$ has rank less than m to find an $\mathbf{x} \neq \mathbf{0}$ for which $\mathbf{A}\mathbf{B}\mathbf{x} = \mathbf{0}$. See Ex. 3.23.

EXERCISE 3.50. See hint for Ex. 1.41.

EXERCISE 3.51. For the first part, use Theorems 3.1, 3.2, and 1.5(ii). For the second part, consider the sth column of $\mathbf{A}^{-1}\mathbf{A}$.

CHAPTER 4

EXERCISE 4.1. (a) No. (b) No. (c) Yes. (d) No. (e) No. (f) Yes.

EXERCISE 4.5. See that (c) of Definition 4.1 fails because $\mathbf{x} + \mathbf{y}$ is not in V.

EXERCISE 4.12. $1 - \beta + \alpha\beta \neq 0$.

EXERCISE 4.15.

$$[x \quad y \quad z] = \tfrac{1}{4}(x + y + z)[1 \quad 2 \quad 1] + \tfrac{1}{2}(x - z)[1 \quad 0 \quad -1] + \tfrac{1}{4}(x - y + z)[1 \quad -2 \quad 1].$$

EXERCISE 4.16. The second one.

EXERCISE 4.19. The set of vectors $[1 \quad 1 \quad 1 \quad 0]^T$, $[0 \quad 0 \quad 0 \quad 1]^T$ is such a basis.

EXERCISE 4.32. (a) Yes. (b) No.

EXERCISE 4.33. $k = 1$, 3, or 4.

EXERCISE 4.35. Use **Key Theorems 3.11** and **4.8.**

EXERCISE 4.37. Since $\mathbf{x}^T\mathbf{AB}$ is a linear combination of the rows of $\mathbf{AB}$, and $\mathbf{ABx}$ is a linear combination of the columns of $\mathbf{AB}$, deduce that the row space (column space) of $\mathbf{AB}$ is a subspace of that of $\mathbf{B}(\mathbf{A})$. Use **Key Theorem 4.8** next.

EXERCISE 4.43. $\begin{bmatrix} 1 \\ 0 \end{bmatrix}$; $\begin{bmatrix} 1 \\ 0 \end{bmatrix}$; $\begin{bmatrix} 1 \\ t \end{bmatrix}$ for $-1 \leq t \leq 1$.

EXERCISE 4.53. $\{\mathbf{e}_1, \ldots, \mathbf{e}_n\}$ works fine.

EXERICSE 4.75. Use the first three vectors.

EXERCISE 4.76. Add any unit vector.

EXERCISE 4.77. Add $[0 \quad 0 \quad 0 \quad 1]^T$.

EXERCISE 4.78. Use method in **Key Theorem 4.8**(ii).

EXERCISE 4.79. dim = 3; columns 1, 2, 4.

EXERCISE 4.80. (b) or (c).

EXERCISE 4.82. Basis $[2 \quad -7 \quad -1]^T$.

EXERCISE 4.83. Let the columns of the $n \times p$ matrix $\mathbf{U}$ be a basis for the null space of $\mathbf{Ax} = \mathbf{0}$, so that the rank of $\mathbf{U}$ is p. The equations $\mathbf{U}^T\mathbf{x} = \mathbf{0}$ have an independent set of $n - p$ solutions. These, transposed, are the rows of one possible $\mathbf{A}$ (why?). For the given example, one answer is $x_1 = x_3 = x_5, x_2 = x_4$.

EXERCISE 4.84. The first two are, the third is not. (Reduce to row-echelon form the 4×6 matrix whose first three columns are the spanning vectors. Alternatively, use the method in **Key Theorem 4.8** to show that $\{[1 \quad 0 \quad 0 \quad \frac{1}{2}], [0 \quad 1 \quad 0 \quad 0], [0 \quad 0 \quad 1 \quad \frac{1}{2}]\}$ is a basis in standard form, and check whether the vectors in question are multiples of these.)

EXERCISE 4.85. The row-echelon form of $[\mathbf{u}_1\ \mathbf{u}_2\ \mathbf{u}_3\ \mathbf{v}_1\ \mathbf{v}_2\ \mathbf{v}_3\ \mathbf{x}] = [\mathbf{U}, \mathbf{V}, \mathbf{x}]$ is

$$\begin{bmatrix} 1 & 0 & 0 & 1 & \frac{1}{2} & 1 & \frac{1}{2}x_2 \\ 0 & 1 & 0 & -1 & \frac{1}{2} & 2 & \frac{1}{6}(6x_1 - 5x_2 + 4x_4) \\ 0 & 0 & 1 & 0 & \frac{1}{2} & -1 & \frac{1}{6}(x_2 - 2x_4) \\ 0 & 0 & 0 & 0 & 0 & 0 & x_3 - 2x_1 \end{bmatrix} = \begin{bmatrix} \mathbf{I} & \mathbf{K} & \boldsymbol{\alpha} \\ \mathbf{0}^T & \mathbf{0}^T & \alpha_4 \end{bmatrix}$$

EXERCISE 4.86. Show that the last three vectors in each of the expressions for $\mathbf{x}$ span the same space and the difference of the first vectors belongs to this space.

EXERCISE 4.87. $\|\mathbf{x} + \mathbf{y}\|^2 = \|\mathbf{x}\|^2 + \|\mathbf{y}\|^2 + 2(\mathbf{x}, \mathbf{y})$.

EXERCISE 4.88. $\alpha_i = (\mathbf{v}, \mathbf{v}_i)$

CHAPTER 5

EXERCISE 5.2.

$$\mathbf{A} = \begin{bmatrix} 1 & 0 & 0 & 0 \\ 0 & 2 & 0 & 0 \\ 0 & 0 & 3 & 0 \\ 0 & 0 & 0 & 4 \end{bmatrix}.$$

EXERCISE 5.12. Show that $\mathbf{M}(A^{-1})\mathbf{M}(A) = \mathbf{I}$.

EXERCISE 5.15. $\|\mathbf{A}\|_1 = 1.7; \|\mathbf{A}\|_\infty = 1.38$.

EXERCISE 5.18. (a) 3;6. (b) 6;3. (c) 10;11. (d) 4;6.

EXERCISE 5.22. Use **Key Lemma 5.1** and $\|\cdot\|_\infty$.

EXERCISE 5.33. The second row nearly equals (1.1 × the first row) − (the third row).

CHAPTER 6

EXERCISE 6.1. Consider whether or not this problem is ill-conditioned.

EXERCISE 6.2. Yes.

EXERCISE 6.6. Use (6.9), (6.10), and a count of the number of operations needed to multiply an $n \times n$ matrix by k different $n \times 1$ vectors.

EXERCISE 6.10. Note that $\mathbf{x} = \mathbf{A}^{-1}\mathbf{b}$ and $\mathbf{x}'$, respectively, solve $\mathbf{Ax} = \mathbf{b}$ and $\mathbf{Ax}' = \mathbf{b} + \mathbf{r}$, and use **Key Theorem 5.9.**

EXERCISE 6.13. The numerical solution is $x_1 = 0.051$, $x_2 = 2.0$.

EXERCISE 6.14. If n is large enough, $2^{n-1} + \delta$ will round to 2^{n-1}, so that the computer will produce the same answer to both sets of equations, although the true answers can differ significantly.

EXERCISE 6.15. The computer should end up with $\epsilon x_i = \beta$ for inconsistent equations as opposed to $\epsilon x_i = \delta$ for consistent equations, for some i, where ϵ and δ are of comparable size, but are *much* smaller than is β.

EXERCISE 6.23. Note that $\mathbf{K}$ is simply 1×1.

EXERCISE 6.30. Expand by the last row or column.

EXERCISE 6.31. Multiply each column by x_1 and subtract the result from the next column to the right, starting at the right-hand end. Deduce $V_n = (x_n - x_1) \cdots (x_2 - x_1)V_{n-1}$ and use induction.

EXERCISES 6.45, 6.46. Note that $\det(\mathbf{A}) = \det(\mathbf{A}^T)$.

EXERCISE 6.49. Write the matrix as

$$\begin{bmatrix} \mathbf{A} & \mathbf{0} \\ \mathbf{C} & \mathbf{D} \end{bmatrix} = \begin{bmatrix} \mathbf{A} & \mathbf{0} \\ \mathbf{0} & \mathbf{I} \end{bmatrix} \begin{bmatrix} \mathbf{I} & \mathbf{0} \\ \mathbf{C} & \mathbf{D} \end{bmatrix}.$$

EXERCISE 6.50. Note that

$$\begin{bmatrix} \mathbf{A} & \mathbf{0} \\ \mathbf{C} & \mathbf{D} \end{bmatrix} \begin{bmatrix} \mathbf{A}^{-1} & \mathbf{0} \\ \mathbf{0} & \mathbf{D}^{-1} \end{bmatrix} \begin{bmatrix} \mathbf{A} & \mathbf{B} \\ -\mathbf{C} & \mathbf{D} \end{bmatrix} = \begin{bmatrix} \mathbf{A} & \mathbf{B} \\ \mathbf{0} & \mathbf{D} + \mathbf{C}\mathbf{A}^{-1}\mathbf{B} \end{bmatrix},$$

and that

$$\begin{bmatrix} \mathbf{A} & \mathbf{B} \\ -\mathbf{C} & \mathbf{D} \end{bmatrix}\begin{bmatrix} \mathbf{A}^{-1} & \mathbf{0} \\ \mathbf{0} & \mathbf{D}^{-1} \end{bmatrix}\begin{bmatrix} \mathbf{A} & \mathbf{0} \\ \mathbf{C} & \mathbf{D} \end{bmatrix} = \begin{bmatrix} \mathbf{A} + \mathbf{BD}^{-1}\mathbf{C} & \mathbf{B} \\ \mathbf{0} & \mathbf{D} \end{bmatrix}.$$

EXERCISE 6.51. Choose $\mathbf{A} = \mathbf{I}_m$, $\mathbf{D} = \mathbf{I}_1$, $\mathbf{B} = \mathbf{u}$, and $\mathbf{C} = \mathbf{v}^T$ in Ex. 6.50.

CHAPTER 7

EXERCISE 7.1. $x_1 = 10$, $x_2 = 20$, $M = 50$.

EXERCISE 7.5. $x_1 = 40$, $x_2 = 20$, $M = 80$.

EXERCISE 7.11. They are not degenerate.

EXERCISE 7.14. (7.15) describes the problem of maximizing

$$-20x_1 + 20x_5 + 1800$$

subject to

$$\tfrac{5}{3}x_1 + x_3 - \tfrac{1}{3}x_5 = 40,\ \tfrac{2}{3}x_1 + x_4 - \tfrac{1}{3}x_5 = 10,\ \tfrac{1}{3}x_1 + x_2 + \tfrac{1}{3}x_5 = 30 \qquad \text{all } x_i \geq 0,$$

which in turn is equivalent to

$$\tfrac{5}{3}x_1 - \tfrac{1}{3}x_5 \leq 40,\ \tfrac{2}{3}x_1 - \tfrac{1}{3}x_5 \leq 10,\ \tfrac{1}{3}x_1 + \tfrac{1}{3}x_5 \leq 30,\ x_1 \geq 0,\ x_5 \geq 0.$$

EXERCISE 7.22. The program is unbounded above.

EXERCISE 7.27. $x_1 = 10$, $x_2 = 20$, $M = 30$.

EXERCISE 7.54. There are no feasible vectors.

EXERCISE 7.56. The dual is to minimize $\mathbf{d}^T\mathbf{z}$ subject to $\mathbf{B}^T\mathbf{z} \geq \mathbf{c}$ but without sign restrictions on $\mathbf{z}$.

CHAPTER 8

EXERCISE 8.5. Eigenvalues are 1, 4 and -1, 0, 2, respectively.

EXERCISE 8.8. $[1 \quad \sqrt{2} \quad 1]^T$, $[1 \quad 0 \quad -1]^T$, $[1 \quad -\sqrt{2} \quad 1]^T$, $\lambda_i = 2 - \sqrt{2}, 2, 2 + \sqrt{2}$.

EXERCISE 8.9. $[1 \quad -2 \quad 0]^T$, $[0 \quad 3 \quad 1]^T$; $[2 \quad 1 \quad -1]^T$; $\lambda_3 = 6$.

EXERCISE 8.12. If $\mathbf{x}$ is an eigenvector corresponding to λ, $f(\mathbf{A})\mathbf{x} = f(\lambda)\mathbf{x}$.

EXERCISE 8.14. Complex zeros of real polynomials occur in conjugate pairs. Real polynomials of odd degree always have a real zero.

EXERCISE 8.15. The eigenvalues are 6, 6, 8.

EXERCISE 8.19. The representation is

$$\begin{bmatrix} 1 & 0 \\ 3 & 3 \end{bmatrix}.$$

EXERCISE 8.23. See Ex. 8.7 and **Key Theorem 8.3**.

EXERCISE 8.27. The statement is false.

EXERCISE 8.30. Recall Ex. 8.29.

EXERCISE 8.34. Use Theorem 8.7.

EXERCISE 8.38. For the first part, $\bar{\mathbf{P}}\bar{\mathbf{P}}^H = \overline{[\mathbf{P}\mathbf{P}^H]} = \bar{\mathbf{I}} = \mathbf{I}$.

EXERCISE 8.39. Use an identity analogous to $ab = \frac{1}{4}[(a+b)^2 - (a-b)^2]$.

EXERCISE 8.46. Note that the first column of **HA** is just **H** times the first column of **A**.

EXERCISE 8.52. Show that the Gerschgorin circles do not contain the origin.

EXERCISE 8.65. $\lambda_i = k + 2\cos\{i\pi/(n+1)\}$.

EXERCISE 8.67. det **A** is a product of eigenvalues and a product of complex conjugate numbers is positive.

EXERCISE 8.69. Since $\mathbf{ABx} = \lambda\mathbf{x}$, $\lambda \neq 0$, $\mathbf{x} \neq 0$ then $\mathbf{Bx} \neq \mathbf{0}$ and $\mathbf{BA}(\mathbf{Bx}) = \lambda(\mathbf{Bx})$. Let **AB** and **BA** have r, s independent eigenvectors corresponding to $\lambda \neq 0$. Since $\mathbf{A}\Sigma\alpha_i\mathbf{Bx}_i = \lambda\Sigma\alpha_i\mathbf{x}_i$ the $\mathbf{Bx}_i$ are linearly independent if the $\mathbf{x}_j$ are, and $s \geq r$. Interchange roles of **A** and **B** and deduce $r = s$. **AB** and **BA** have orders m, n and the same numbers of eigenvectors corresponding to nonzero eigenvalues.

EXERCISE 8.74. $\omega_1 = 0$ corresponds to a translation.

CHAPTER 9

EXERCISE 9.7. $\mathbf{A}^H\mathbf{A} = \mathbf{I} = \mathbf{A}\mathbf{A}^H$.

EXERCISE 9.10. The eigenvectors, for example, can be taken to be

$$\begin{bmatrix} \frac{1}{\sqrt{2}} \\ \frac{1}{\sqrt{2}} \\ 0 \end{bmatrix}, \quad \begin{bmatrix} \frac{1}{\sqrt{6}} \\ -\frac{1}{\sqrt{6}} \\ -\frac{2}{\sqrt{6}} \end{bmatrix}, \quad \begin{bmatrix} \frac{1}{\sqrt{3}} \\ -\frac{1}{\sqrt{3}} \\ \frac{1}{\sqrt{3}} \end{bmatrix}.$$

EXERCISE 9.14. Use **Key Theorem 9.2.**

EXERCISE 9.15. Use Ex. 9.10.

EXERCISE 9.18. The conditions are $b_1 - b_2 = 0$ and $b_1 + b_2 = 0$, respectively.

EXERCISE 9.21. Use **Key Theorem 9.2** to write $\mathbf{A} = \mathbf{PDP}^H$ for diagonal $\mathbf{D}$, and then apply the hypotheses of the problem. For example, if $\mathbf{D}$ is real, then $\mathbf{D}^H = \mathbf{D}$ and therefore $\mathbf{A}^H = \mathbf{PD}^H\mathbf{P}^H = \mathbf{A}$.

EXERCISE 9.28.

$$\mathbf{A} = \begin{bmatrix} \frac{1}{\sqrt{2}} & \frac{-1}{\sqrt{3}} & \frac{1}{\sqrt{6}} \\ 0 & \frac{1}{\sqrt{3}} & \frac{2}{\sqrt{6}} \\ \frac{1}{\sqrt{2}} & \frac{1}{\sqrt{3}} & \frac{-1}{\sqrt{6}} \end{bmatrix} \begin{bmatrix} \sqrt{2} & 3\sqrt{2} & \frac{9}{2}\sqrt{2} \\ 0 & \sqrt{3} & \frac{4}{3}\sqrt{3} \\ 0 & 0 & -\frac{1}{6}\sqrt{6} \end{bmatrix}.$$

EXERCISE 9.30. $\mathbf{A}$ and $\mathbf{R}$ have the same null space, since the independence of the columns of $\mathbf{Q}$ implies that $\mathbf{Ax} = \mathbf{0}$ if and only if $\mathbf{Q}(\mathbf{Rx}) = \mathbf{0}$ if and only if $\mathbf{Rx} = \mathbf{0}$. Then use Theorem 5.2.

EXERCISE 9.37. Try **I**, for example.

EXERCISE 9.41. For the second case,

$$\mathbf{Q}_1 = \begin{bmatrix} \frac{2}{\sqrt{5}} & \frac{1}{\sqrt{5}} \\ \frac{1}{\sqrt{5}} & \frac{-2}{\sqrt{5}} \end{bmatrix} \begin{bmatrix} \frac{1}{\sqrt{2}} & \frac{1}{\sqrt{2}} \\ \frac{1}{\sqrt{2}} & \frac{-1}{\sqrt{2}} \end{bmatrix}.$$

EXERCISE 9.48. We find, for example, that $x_1 = 0.58$ and $x_2 = 0.50$, to two figures.

EXERCISE 9.54. We have $m = n = k$, so that in (9.27) certainly $\mathbf{\Sigma}^+ = \mathbf{\Sigma}^{-1}$.

EXERCISE 9.57. The columns of $\mathbf{U\Sigma}$ are $\sigma_1\mathbf{u}_1, \sigma_2\mathbf{u}_2, \ldots, \sigma_k\mathbf{u}_k, \mathbf{0}, \ldots, \mathbf{0}$, and $\mathbf{y} = \mathbf{Ax}$ if and only if $\mathbf{y} = (\mathbf{U\Sigma})\mathbf{x}'$ where $\mathbf{x}' = \mathbf{V}^H\mathbf{x}$.

EXERCISE 9.63. Note that $\mathbf{P}^2 = \mathbf{Q}^2 = \mathbf{I}$.

EXERCISE 9.74. $\mathbf{Q} = \mathbf{P}'^H\mathbf{P}$.

EXERCISE 9.81. $(\mathbf{I} - 2\mathbf{w}\mathbf{w}^H)^H = \mathbf{I}^H - 2(\mathbf{w}^H)^H\mathbf{w}^H = \mathbf{I} - 2\mathbf{w}\mathbf{w}^H$.

EXERCISE 9.86. Describe $\mathbf{R}_1 = \mathbf{Q}\mathbf{R}_2$ one column at a time.

EXERCISE 9.87. Find the QR decomposition of $\mathbf{A}^H$, for example.

CHAPTER 10

EXERCISE 10.9. The eigenvectors for $\lambda = 2$ are $\mathbf{p}_1$ and $\mathbf{p}_2$.

EXERCISE 10.16. The eigenvalues are 1, -2, -3.

EXERCISE 10.19. The form is

$$\begin{bmatrix} 1 & -2 & 3 & 0 \\ 0 & 1 & -1 & 0 \\ 0 & 0 & 1 & 0 \\ 0 & 0 & 0 & -3 \end{bmatrix}.$$

EXERCISE 10.24. In the notation of **Key Theorem 10.2**, the characteristic polynomial is $f(\lambda) = (\lambda_1 - \lambda)^{n_1}(\lambda_2 - \lambda)^{n_2} \cdots (\lambda_k - \lambda)^{n_k}$, which reduces the problem essentially to that in Ex. 10.23.

EXERCISE 10.29. The eigenvalues are 2, 2, -1. The only eigenvectors for $\lambda = 2$ are of the form $[z \quad 0 \quad 0]^T$, while those for $\lambda = 1$ have the form $[0 \quad z \quad -z]^T$.

EXERCISE 10.33. $\mathbf{y}^H\mathbf{x}_{i+1} = \mathbf{y}^H(\mathbf{A}\mathbf{x}_i) = (\mathbf{y}^H\mathbf{A})\mathbf{x}_i = (\lambda\mathbf{y}^H)\mathbf{x}_i = \lambda(\mathbf{y}^H\mathbf{x}_i)$.

EXERCISE 10.37. The left-eigenvectors may be taken to be $\mathbf{e}_1, \mathbf{e}_2, \mathbf{e}_3, \mathbf{e}_4$.

EXERCISE 10.42. $5b_1 + b_2 - 7b_4 = 0$ and $9b_1 + b_3 - 12b_4 = 0$.

EXERCISE 10.48. $||\mathbf{A}_i||_\infty = 2^i + (\frac{1}{2})^i$ for $i \geq 2$.

EXERCISE 10.55. Choose k so that no eigenvalue exceeds unity in magnitude while one of them equals unity.

EXERCISE 10.59. If λ is an eigenvalue of $\mathbf{A}$ with $|\lambda| = 1$ and eigenvector $\mathbf{v}'$, then $\mu = \lambda^p$ is an eigenvalue of $\mathbf{A}^p$ with eigenvector $\mathbf{v}'$, and $|\mu| = 1$ also. Build from this.

EXERCISE 10.76. By Theorem 10.12, $p = \frac{1}{2}\{1 + (1 - \frac{1}{4})^{1/2}\} \approx 0.93$ and the relevant eigenvalue is

$$\frac{1 - (\frac{3}{4})^{1/2}}{1 + (\frac{3}{4})^{1/2}} \approx \frac{0.133}{1.866} \approx 0.0712.$$

EXERCISE 10.77. For large N, the spectral radii are approximately

$$1 - \frac{\pi^2}{2(N+1)^2},\ 1 - \frac{\pi^2}{(N+1)^2},\ \text{and } 1 - \frac{\pi\sqrt{2}}{N+1},$$

the last of which is by far the smallest for large N.

EXERCISE 10.82.

$$x(t) = \begin{bmatrix} 5 & 50 \\ 3 & 40 \end{bmatrix} e^{-0.1t} \begin{bmatrix} 1 & t \\ 0 & 1 \end{bmatrix} \begin{bmatrix} 0.8 & -1 \\ -0.06 & 0.1 \end{bmatrix} \mathbf{x}_0.$$

EXERCISE 10.85. $x_1(t) = (1 + t)e^{2t}$, $x_2(t) = e^{2t}$.

EXERCISE 10.90.

$$\exp(\mathbf{A}) = \begin{bmatrix} 2e^4 & \frac{3}{2}e^4 - \frac{1}{2}e^{-2} & \frac{1}{2}e^4 - \frac{1}{2}e^{-2} \\ -e^4 & \frac{1}{2}e^{-2} - \frac{1}{2}e^4 & \frac{1}{2}e^{-2} - \frac{1}{2}e^4 \\ e^4 & \frac{1}{2}e^{-2} + \frac{1}{2}e^4 & \frac{1}{2}e^{-2} + \frac{1}{2}e^4 \end{bmatrix}.$$

EXERCISE 10.92. Choose k so that no eigenvalue exceeds unity in magnitude but one equals unity in magnitude.

EXERCISE 10.94. Use Ex. 10.93 and show that $[\exp(\mathbf{A})][\exp(-\mathbf{A})] = \mathbf{I}$.

EXERCISE 10.104. $\mathbf{x}_0$ must be orthogonal to the left-eigenvector of $\mathbf{A}$ associated with $\lambda = 1$.

CHAPTER 11

EXERCISE 11.4. (a) Hyperbola. (b) Two straight lines. (c) Ellipse. (d) Ellipse centered at $x = 1$, $y = -1$. (e) Parabola (see Ex. 11.3).

EXERCISE 11.11. Obviously $-\mathbf{A}$ must be positive definite, allowing us to apply Theorem 11.5 directly to $-\mathbf{A}$.

EXERCISE 11.22. Solve $(\mathbf{A} - \rho\mathbf{B})\boldsymbol{\alpha} = \mathbf{0}$, where $\mathbf{A} = [(\mathbf{z}_i, \mathbf{A}\mathbf{z}_j)]$, $\mathbf{B} = [(\mathbf{z}_i, \mathbf{z}_j)]$.

EXERCISE 11.43. $(\mathbf{x}, \mathbf{A}\mathbf{x}) = (\mathbf{y}, \mathbf{A}^{-1}\mathbf{y})$, $\mathbf{y} = \mathbf{A}\mathbf{x}$.

EXERCISE 11.44. $(\mathbf{x}, \mathbf{B}^H\mathbf{ABx}) = (\mathbf{z}, \mathbf{Az})$, $\mathbf{z} = \mathbf{Bx}$. If rank $\mathbf{B} < n$, there exists a nonzero $\mathbf{x}$ with $\mathbf{Bx} = \mathbf{0}$.

EXERCISE 11.46. $\alpha + \beta\lambda + \gamma\lambda^2 = 0$, where $\alpha = (\mathbf{x}, \mathbf{Ax})$, $\beta = (\mathbf{x}, \mathbf{Bx})$, $\gamma = (\mathbf{x}, \mathbf{Cx})$, and $(\mathbf{A} + \mathbf{B}\lambda + \mathbf{C}\lambda^2)\mathbf{x} = \mathbf{0}$, $\mathbf{x} \neq \mathbf{0}$.

EXERCISE 11.48. Let $\mathbf{A}^{-1} = \mathbf{M}^H\mathbf{M}$, $\mathbf{A} = \mathbf{M}^{-1}(\mathbf{M}^H)^{-1}$. Then $\mathbf{I} - \mathbf{MBM}^H = \mathbf{MCM}^H$ and the eigenvalues of $\mathbf{MBM}^H$ are such that $0 < \lambda_i \leq 1$. Hence det $\mathbf{MBM}^H \leq 1$ from which (b) follows. Eigenvalues of $(\mathbf{MBM}^H)^{-1}$ are ≥ 1 so that $(\mathbf{MBM}^H)^{-1} - \mathbf{I}$, and hence $\mathbf{B}^{-1} - \mathbf{A}^{-1}$ is positive semidefinite.

EXERCISE 11.49. $\mathbf{B}$ is nonsingular.

EXERCISE 11.50. $3y_1^2 + 6y_2^2 + 9y_3^2$; $2x_1^2 + 5x_2^2$; $5x_1^2 - x_2^2 - x_3^2$.

EXERCISE 11.51.

$$\begin{bmatrix} \mathbf{I} & \mathbf{0} \\ -\boldsymbol{\alpha}^H\mathbf{A}_1^{-1} & 1 \end{bmatrix}\begin{bmatrix} \mathbf{A}_1 & \boldsymbol{\alpha} \\ \boldsymbol{\alpha}^H & a_{nn} \end{bmatrix} = \begin{bmatrix} \mathbf{A}_1 & \boldsymbol{\alpha} \\ \mathbf{0} & a_{nn} - \boldsymbol{\alpha}^H\mathbf{A}_1^{-1}\boldsymbol{\alpha} \end{bmatrix}.$$

Hence det $\mathbf{A} = (a_{nn} - \boldsymbol{\alpha}^H\mathbf{A}_1^{-1}\boldsymbol{\alpha})$ det $\mathbf{A}_1$. Since $\mathbf{A}_1^{-1}$ is positive definite, the first formula follows by induction. Apply to $\mathbf{BB}^H$. Equality is attained if $\mathbf{B}$ is unitary.

EXERCISE 11.53. (a) $(\mathbf{A} - \lambda\mathbf{I})\mathbf{x} = \mathbf{0}$ implies $(\mathbf{A}^2 - \lambda\mathbf{A})\mathbf{x} = \mathbf{0}$ or $(1 - \lambda)\mathbf{Ax} = \mathbf{0}$. Hence $\lambda = 1$ or $\mathbf{Ax} = \mathbf{0}$ which implies $\lambda = 0$. (b) If $\mathbf{A}$ is nonsingular, $\mathbf{A}^2 = \mathbf{A}$ implies $\mathbf{A} = \mathbf{I}$. (d). The trace equals the sum of the eigenvalues.

EXERCISE 11.57. $\mathbf{A} = \mathbf{P}^H\mathbf{\Lambda P}$, $\mathbf{B} = \mathbf{P}^H\mathbf{\Lambda}^{1/2}\mathbf{P}$. To prove uniqueness, if $\mathbf{B}$ is hermitian and $\mathbf{B}^2 = \mathbf{A}$ this means $\mathbf{B} = \mathbf{Q}^H\mathbf{DQ}$, $\mathbf{B}^2 = \mathbf{Q}^H\mathbf{D}^2\mathbf{Q}$ and diagonal elements must be the eigenvalues of $\mathbf{A}$ in some (unimportant) order.

EXERCISE 11.58. (b) Prove $\mathbf{A}^H\mathbf{A} = \mathbf{M}^2$, $\mathbf{AA}^H = \mathbf{N}^2$, so that $\mathbf{M}$, $\mathbf{N}$ are the unique square roots (Ex. 11.57). Use theory of singular values, $\mathbf{A} = \mathbf{PDQ}^H$, unitary $\mathbf{P}$, $\mathbf{Q}$. Find $\mathbf{M} = \mathbf{QDQ}^H$, $\mathbf{N} = \mathbf{PDP}^H$. If $\mathbf{A}$ nonsingular, prove $\mathbf{U} = \mathbf{V} = \mathbf{PQ}^H$. If $\mathbf{A}$ singular, $\mathbf{A} = \mathbf{UM}$ gives $\mathbf{PD} = \mathbf{UQD}$ and can still construct (nonunique) $\mathbf{U}$.

APPENDIX TWO

BIBLIOGRAPHY

REFERENCES ORIENTED TOWARD PURE MATHEMATICS

[1] A. C. Aitken, *Determinants and Matrices*, Oliver and Boyd, Interscience (1956).

[2] F. Ayres, *Matrices*, Schaum outlines (1962).

[3] G. Birkhoff and S. MacLane, *A Survey of Modern Algebra*, 3rd Ed., Macmillan (1965)

[4] C. G. Cullen, *Matrices and Linear Transformations*, Addison-Wesley (1966).

[5] C. W. Curtis, *Linear Algebra*, Allyn and Bacon (1963).

[6] P. J. Davis, *The Mathematics of Matrices*, Blaisdell (1965).

[7] D. T. Finkbeiner, *Introduction to Matrices and Linear Transformations*, W. H. Freeman (1960).

[8] L. E. Fuller, *Basic Matrix Theory*, Prentice-Hall (1961).

[9] F. R. Gantmacher, *Theory of Matrices*, Vols. I, II, Chelsea (1959).

[10] I. M. Gel'fand, *Lectures on Linear Algebra*, Interscience (1961).

[11] P. R. Halmos, *Finite-dimensional Vector Spaces*, Van Nostrand (1958).

[12] K. Hoffman and R. Kunze, *Linear Algebra*, Prentice-Hall (1961).

[13] F. E. Hohn, *Elementary Matrix Algebra*, Macmillan (1957).

[14] N. Jacobson, *Lectures in Abstract Algebra*, Vol. II, Van Nostrand (1953).

[15] S. Lang, *Linear Algebra*, Addison-Wesley (1966).

[16] C. C. MacDuffee, *Vectors and Matrices* (Carus Monograph No. 7), Mathematical Association of America, Open Court (1943).

[17] M. Marcus and H. Minc, *Introduction to Linear Algebra*, Macmillan (1965).

[18] ______, *A Survey of Matrix Theory and Matrix Inequalities*, Allyn and Bacon (1964).

[19] L. Mirsky, *Introduction to Linear Algebra*, Oxford (1955).

[20] T. Muir, *Determinants*, Dover (1960).

[21] J. R. Munkres, *Elementary Linear Algebra*, Addison-Wesley (1964).

[22] D. C. Murdoch, *Linear Algebra for Undergraduates*, Wiley (1957).

[23] E. D. Nering, *Linear Algebra and Matrix Theory*, Wiley (1963).

[24] L. J. Paige and J. D. Swift, *Elements of Linear Algebra*, Blaisdell (1965).

[25] S. Perlis, *Theory of Matrices*, Addison-Wesley (1952).

[26] H. Schneider and G. P. Barker, *Matrices and Linear Algebra*, Holt, Rinehart & Winston (1968).

[27] V. I. Smirnov, *Linear Algebra and Group Theory*, McGraw-Hill (1961).

[28] R. R. Stoll, *Linear Algebra and Matrix Theory*, McGraw-Hill (1952).

[29] F. M. Stewart, *Introduction to Linear Algebra*, Van Nostrand (1963).

[30] R. M. Thrall and L. Tornheim, *Vector Spaces and Matrices*, Wiley (1957).

[31] H. W. Turnbull and A. C. Aitken, *An Introduction to the Theory of Canonical Matrices*, Dover (1961).

[32] N. V. Yefimov, *Quadratic Forms and Matrices*, Academic (1964).

REFERENCES ORIENTED MORE TOWARD APPLIED MATHEMATICS

[33] N. R. Amundsen, *Mathematical Methods in Chemical Engineering, Matrices and Their Applications*, Prentice-Hall (1966).

[34] R. E. Bellman, *Introduction to Matrix Algebra*, McGraw-Hill (1960).

[35] W. G. Bickley and R. S. H. G. Thompson, *Matrices, Their Meaning and Manipulation*, Van Nostrand (1964).

[36] R. Braae, *Matrix Algebra for Electrical Engineers*, Addison-Wesley (1963).

[37] J. N. Franklin, *Matrix Theory*, Prentice-Hall (1968).

[38] R. A. Frazer, W. J. Duncan, and A. R. Collar, *Elementary Matrices and Some Applications to Dynamics and Differential Equations*, Cambridge Univ. Press (1938).

[39] A. Gewirtz, H. Sitomer, A. W. Tucker, *Constructive Linear Algebra*, Prentice-Hall (1974).

[40] G. Hadley, *Linear Algebra*, Addison-Wesley (1961).

[41] G. G. Hall, *Matrices and Tensors*, Macmillan and Pergamon (1963).

[42] J. Heading, *Matrix Theory for Physicists*, Wiley (1960).

[43] J. B. Johnston, G. B. Price, and F. S. Van Vleck, *Linear Equations and Matrices*, Addison-Wesley (1966).

[44] S. Karlin, *Mathematical Methods and Theory in Games, Programming, and Economics*, Vols. I, II, Addison-Wesley (1959).

[45] J. G. Kemeny, J. L. Snell, and G. L. Thompson, *Introduction to Finite Mathematics*, Prentice-Hall (1957).

[46] M. C. Pease, *Methods of Matrix Algebra*, Academic Press (1965).

[47] L. A. Pipes, *Matrix Methods in Engineering*, Prentice-Hall (1963).

[48] S. R. Searle, *Matrix Algebra for the Biological Sciences (Including Applications in Statistics)*, Wiley (1960).

[49] D. I. Steinberg, *Computational Matrix Algebra*, McGraw-Hill (1974).

[50] G. Strang, *Linear Algebra with Applications*, Academic Press (1976).

[51] A. V. Weiss, *Matrix Analysis for Electrical Engineers*, Van Nostrand (1964).

[52] R. Zurmuhl, *Matrizen und ihre technischen anwendungen*, 3rd Ed., Springer (1961).

REFERENCES ON NUMERICAL METHODS

[53] E. Bodewig, *Matrix Calculus*, 2nd Ed., North-Holland (1959).

[54] B. L. Buzbee, F. W. Dorr, J. A. George, G. H. Golub, "The Direct Solution of the Discrete Poisson Equation on Irregular Regions," SIAM J. Num. Anal., vol. 8 (1971), 722–736.

[55] V. N. Faddeeva, *Computational Methods of Linear Algebra*, Dover (1959).

[56] D. K. Faddeev and V. N. Faddeeva, *Computational Methods of Linear Algebra*, W. H. Freeman and Co. (1963).

[57] G. E. Forsythe and C. B. Moler, *Computer Solution of Linear Algebraic Systems*, Prentice-Hall (1967).

[58] L. Fox, *An Introduction to Numerical Linear Algebra*, Oxford (1957).

[59] R. W. Hockney, "The Potential Calculation and Some Applications," Meths. in Comput. Physics, vol. 9 (1970), 135–211.

[60] A. S. Householder, *Principles of Numerical Analysis*, McGraw-Hill (1953).

[61] ______, *The Theory of Matrices in Numerical Analysis*, Blaisdell (1964).

[62] E. Isaacson and H. B. Keller, *Analysis of Numerical Methods*, Wiley (1966).

[63] *Modern Computing Methods*, 2nd Ed., H. M. Stationery Office, London (1961).

[64] A. Ralston and H. S. Wilf, Eds., *Mathematical Methods for Digital Computers*, Vol. II, Wiley (1967).

[65] G. W. Stewart, *Introduction to Matrix Computations*, Academic Press (1973).

[66] R. S. Varga, *Matrix Iterative Analysis*, Prentice-Hall (1962).

[67] J. H. Wilkinson, *Rounding Errors in Algebraic Processes*, Prentice-Hall (1963).

[68] _____, *The Algebraic Eigenvalue Problem*, Oxford (1965).

REFERENCES CONTAINING APPLICATIONS OF LINEAR ALGEBRA AND MATRICES. (See also references quoted in exercises at the end of Chapter 2.)

[69] R. G. D. Allen, *Mathematical Economics*, St. Martin's Press (1957).

[70] H. H. Argyris, *Energy Theorems and Structural Analysis*, Butterworth (1960).

[71] R. Bellman, K. L. Cooke, *Modern Elementary Differential Equations, 2nd edition*, Addison-Wesley (1971).

[72] S. F. Borg, *Matrix-Tensor Methods in Continuum Mechanics*, Van Nostrand (1963).

[73] W. Brouwer, *Matrix Methods in Optical Instrument Design*, W. A. Benjamin (1964).

[74] H. B. Chenery and P. B. Clark, *Interindustry Economics*, Wiley (1959).

[75] G. B. Dantzig, *Linear Programming and Extensions*, Princeton Univ. Press (1963).

[76] R. Dorfman, P. A. Samuelson, and R. M. Solow, *Linear Programming and Economic Analysis*, McGraw-Hill (1958).

[77] D. Gale, *The Theory of Linear Economic Models*, McGraw-Hill (1960).

[78] F. R. Gantmacher and M. G. Krein, *Oszillationsmatrizen, Oszillationskerne und kleine Schwingungen mechanischer Systeme*, Akademic-Verlag, Berlin (1960).

[79] S. I. Gass, *Linear Programming*, McGraw-Hill (1964).

[80] J. J. Gennaro, *Computer Methods in Solid Mechanics*, Macmillan (1965).

[81] W. J. Gibbs, *Tensors in Electrical Machine Theory*, Chapman and Hall (1952).

[82] A. M. Glicksman, *An Introduction to Linear Programming and the Theory of Games*, Wiley (1963).

[83] A. Goldberger, *Econometric Theory*, Wiley (1964).

[84] H. Goldstein, *Classical Mechanics*, Addison-Wesley (1950).

[85] F. A. Graybill, *An Introduction to Linear Statistical Models*, Vol. I, McGraw-Hill (1961).

[86] E. A. Guillemin, *The Mathematics of Circuit Analysis*, Wiley (1949).

[87] G. Hadley, *Linear Programming*, Addison-Wesley (1962).

[88] B. Higman, *Applied Group-Theoretic and Matrix Methods*, Oxford (1955).

[89] W. C. Hurty and M. F. Rubinstein, *Dynamics of Structures*, Prentice-Hall (1964).

[90] J. Johnston, *Econometric Methods*, McGraw-Hill (1963).

[91] J. G. Kemeny and J. L. Snell, *Finite Markov Chains*, Van Nostrand (1960).

[92] J. G. Kemeny and J. L. Snell, *Mathematical Models in the Social Sciences*, MIT Press (1972).

[93] ______, *Mathematical Models in the Social Sciences*, Ginn (1962).

[94] G. Kron, *Tensor Analysis of Networks*, Wiley (1939).

[95] ______, *Tensors for Circuits*, Dover (1959).

[96] ______, *Diakoptics*, Macdonald (1963).

[97] P. Lancaster, *Lambda-Matrices and Vibrating Systems*, Pergamon (1966).

[98] P. LeCorbeiller, *Matrix Analysis of Electrical Networks*, Harvard Univ. Press (1950).

[99] D. P. Maki, M. Thompson, *Mathematical Models and Applications*, Prentice-Hall (1973).

[100] W. Murray, *Unconstrained Optimization*, Academic Press (1972).

[101] B. Noble, *Applications of Undergraduate Mathematics in Engineering*, Mathematical Association of America, Macmillan (1967).

[102] H. M. Nodelman and F. W. Smith, *Mathematics for Electronics*, McGraw-Hill (1956).

[103] J. F. Nye, *Physical Properties of Crystals: Their Representation by Tensors and Matrices*, Oxford (1957).

[104] E. C. Pestel and F. A. Leckie, *Matrix Methods in Elastomechanics*, McGraw-Hill (1963).

[105] C. R. Rao, *Advanced Statistical Methods in Biometric Research*, Wiley (1952).

[106] J. Robinson, *Structural Matrix Analysis for the Engineer*, Wiley (1966).

[107] J. T. Schwartz, *Lectures on the Mathematical Method in Analytical Economics*, Gordon and Breach (1961).

[108] M. Simonnard, *Linear Programming*, translated by W. S. Jewell, Prentice-Hall (1966).

[109] S. Vajda, *An Introduction to Linear Programming and the Theory of Games*, Methuen and Wiley (1960).

[110] E. B. Wilson, J. C. Decius, and P. Gross, *Molecular Vibrations*, McGraw-Hill (1955).

[111] L. A. Zadeh and C. A. DeSoer, *Linear System Theory*, McGraw-Hill (1963).

INDEX TO NUMBERED STATEMENTS

INDEX